ASYMPTOTIC ANALYSIS FOR PERIODIC STRUCTURES

Asymptotic Analysis for Periodic Structures

A. Bensoussan
J.-L. Lions
G. Papanicolaou

AMS CHELSEA PUBLISHING
American Mathematical Society • Providence, Rhode Island

2010 *Mathematics Subject Classification.* Primary 80M40, 35B27, 74Q05, 74Q10, 60H10, 60F05.

For additional information and updates on this book, visit
www.ams.org/bookpages/chel-374

Library of Congress Cataloging-in-Publication Data

Bensoussan, Alain.
  Asymptotic analysis for periodic structures / A. Bensoussan, J.-L. Lions, G. Papanicolaou.
    p. cm.
  Originally published: Amsterdam ; New York : North-Holland Pub. Co., 1978.
  Includes bibliographical references.
  ISBN 978-0-8218-5324-5 (alk. paper)
  1. Boundary value problems—Numerical solutions. 2. Differential equations, Partial—Asymptotic theory. 3. Probabilities. I. Lions, J.-L. (Jacques-Louis), 1928–2001. II. Papanicolaou, George. III. Title.

QA379.B45 2011
515′.353—dc23

                                                                        2011029403

Contents

Preface

In the thirty three years since this book appeared, homogenization, or the theory of partial differential equations with rapidly oscillating coefficients, has flourished. The book has been out of print for many years and many other book-level expositions of various aspects of homogenization have appeared in the meantime. We decided to re-print the book, with minor corrections and bibliographical additions, for two reasons. First, we are very fond of the book since it contains work in our favorite subject, which was done at an early part of our career and has cemented our life-long friendship. Second, we want to pay homage to our senior co-author and mentor Jacques-Louis Lions, who is no longer with us. He introduced us to this field and he was the driving force behind this book, with his own contributions and his enthusiasm for carrying out this endeavor.

We hope that the book will still be useful to those interested in homogenization. We would like to thank all our colleagues with whom we have worked on problems in the area of homogenization in the past. The book was typed in LaTeX by Simon Rubinstein-Salzedo at Stanford. We thank him and appreciate very much his help. We also thank the American Mathematical Society for including the book in their publications.

Alain Bensoussan
George Papanicolaou

June 2011

Introduction

In Mechanics, Physics, Chemistry and Engineering, in the study of composite materials, macroscopic properties of crystalline or polymer structures, nuclear reactor design, etc., one is led to the study of boundary value problems in media with periodic structure.

If the period of the structure is small compared to the size of the region in which the system is to be studied, then an asymptotic analysis is called for: to obtain an asymptotic expansion of the solution in terms of a small parameter ε which is the ratio of the period of the structure to a typical length in the region. In other words, to obtain by systematic expansion procedures the passage from a **microscopic** description to a **macroscopic** description of the behavior of the system.

In mathematical terms, the above problems can be formulated, typically, as follows. A family of partial differential operators A^ε, depending on the small parameter ε, is given. The partial differential operators may be time independent or time dependent, steady or of evolution typic, linear or nonlinear, etc. These operators have coefficients which are periodic (or sometimes almost periodic) functions in all or in some variables with periods proportional to ε. Since ε is assumed to be small, we have a family of operators with **rapidly oscillating coefficients**.

In a domain Ω, we have a boundary value problem

$$(0.0.1) \qquad\qquad A^\varepsilon u_\varepsilon = f \text{ in } \Omega,$$

$$(0.0.2) \qquad\qquad u_\varepsilon \text{ subject to appropriate boundary conditions,}$$

which we assume is well set when $\varepsilon > 0$ is fixed. The problem is now to obtain, if possible, an expansion[1] of u_ε

$$(0.0.3) \qquad\qquad u_\varepsilon = u_0 + \varepsilon u_1 + \cdots,$$

which would be asymptotic in general, or at least obtain the first term of this expansion along with a convergence theorem as $\varepsilon \to 0$.

The type of results that one obtains in many cases is that with a suitable definition of convergence (necessarily of a **weak** type as we shall see i.e., convergence of suitable averages), u_ε converges as $\varepsilon \to 0$ to u_0 where u_0 is the solution of

$$(0.0.4) \qquad\qquad \mathcal{A}u_0 = f \text{ in } \Omega,$$

$$(0.0.5) \qquad\qquad u_0 \text{ subject to appropriate boundary conditions.}[2]$$

---

[1]The form of the expansion may be more complicated than (0.0.3).

In $(0.0.4)$ $\mathcal{A}$ is, in general,[3] a partial differential operator with simple[4] coefficients; it is called the **homogenized** operator of the family A^ε because, in a well defined sense, we approximate u_ε by u_0 which satisfies an equation with simple coefficients. The coefficients of $\mathcal{A}$ are called, by definition, **the effective coefficients** or **effective parameters** that describe the macroscopic properties of the underlying medium.

The most important aspect of the passage from $(0.0.1)$, $(0.0.2)$ to $(0.0.4)$, $(0.0.5)$ is the **explicit** analytical construction of $\mathcal{A}$ (i.e., its coefficients) and not merely the assertion that such an operator exists. Throughout the book we give analytical formulas for the construction of the coefficients of $\mathcal{A}$. This construction requires, typically, the solution of a boundary value problem within a single period cell. We call this the **cell problem**.

Thus, the solutions of $(0.0.1)$, $(0.0.2)$ when ε is small is replaced by the solution of a cell problem and then of $(0.0.4)$, $(0.0.5)$. It is hardly surprising that both the solution of a cell problem and of $(0.0.4)$, $(0.0.5)$ must be treated numerically since, except in trivial cases, exact solutions in closed form are not available. However, whereas the direct numerical solution of $(0.0.1)$, $(0.0.2)$, when ε is small, is an ill-conditioned and complicated computational problem, the solution of the cell problem plus $(0.0.4)$, $(0.0.5)$ is, usually, a standard problem in numerical analysis.

In order to obtain $\mathcal{A}$ from A^ε, in this book we use four methods which we now describe. The distinctions between these methods are not, however, sharp or clear-cut and the description that follows should be considered as a rough one.

The **first method** is based on the construction of asymptotic expansions using **multiple scales**. The use of multiple scales is well known in many specialized contexts (but may not be clearly articulated) as well as in modern perturbation theory. For many problems in ordinary differential equations, multiple scale methods give the same results as the well known methods of averaging.[5] In the present context there are at least **two natural spatial length scales**. One measuring variations within one period cell (the **fast** scale) and the one measuring variations within the region of interest (the **slow** scale).

The use of multiple space scales (along with multiple time scales) to treat systematically boundary value problems with rapidly varying periodic structure was introduced and exploited by us. Its effectiveness in this context was anticipated by J. B. Keller.[6] The method was also used by E. Larsen[7] in transport theory (we do not treat transport problems here).

The **second method** is based on energy estimates. Since the coefficients of A^ε are rapidly oscillating, derivatives of the coefficients are multiplied by powers of ε^{-1}. This makes it difficult to obtain estimates independent of ε. One must then pass to the limit in a weak sense and for this one uses integration by parts and suitable test functions. The prototype of this argument is due to L. Tartar.

---

[2]Which, of course, depend on the conditions imposed on u_ε.

[3]We shall see examples where the A^ε are partial differential operators but $\mathcal{A}$ is an integrodifferential operator. Also $\mathcal{A}$ itself may depend on ε, so we write sometimes $\mathcal{A}^\varepsilon$.

[4]In many cases they are constants.

[5]N. N. Bogoliubov and Yu. Mitropolsky, *Asymptotic methods in non-linear mechanics*, Gordon and Breach, New York, 1961.

[6]Private communication; see also, S. Kogelman and J. B. Keller, SIAM J. Appl. Math., **24** (1973), pp. 352–361, for a general expansion procedure using multiple space scales.

[7]J. Math. Phys., **16** (1975), pp. 1421–1427.

Very often we use the two methods together, especially in Chapters 1 and 2. The multiple scales method is used to obtain the operator $\mathcal{A}$ and expansions under liberal regularity conditions on the data and the coefficients. It is also used to construct special test functions in order to pass to the limit by the energy methods yielding convergence results (without expansions) under minimal regularity hypotheses.

The **third method** is based on probabilistic arguments and works whenever the problem admits a probabilistic formulation or has a probabilistic origin. Again, one can construct expansions using multiple scales when a lot of regularity is assumed. With the probabilistically natural notions of weak solution and weak convergence one can also obtain convergence results, without expansions, using test functions constructed by multiple scales under minimal regularity conditions.

The **fourth method** is based on the spectral decomposition of operators with periodic coefficients, the so-called expansion in Bloch waves. This method is not intended as an alternative to the above methods since its applicability is more restricted. It is, however, indispensable in the study of high frequency wave propagation in rapidly varying periodic media. By high frequency we mean here that another length scale becomes relevant in the problem, namely the typical wavelength, which is now assumed to be small and comparable to the period of the structure of the medium.

The material in this book is organized in three units that have been written so that they could be read independently by readers with more specialized interests.

Chapters 1 and 2 form the first unit which deals with elliptic, parabolic and hyperbolic (but not high frequency) problems with emphasis on the **energy methods** (methods 1 and 2 mostly). Sections 1, 2 and 3 of Chapter 1 are basic to the whole book, however, and should be at least looked at by readers more interested in Chapters 3 or 4.

Chapter 3 is the second unit which treats problems probabilistically i.e., by method 3. A separate introduction to the contents is provided at the beginning of this chapter.

Chapter 4 is the third unit which deals with high frequency problems, i.e., method 4 mostly. Here again a separate introduction to the contents is provided at the beginning of the chapter.

In Chapter 1 we consider, among other things, elliptic systems, operators with coefficients that have multiple periodic structure (with coefficients that depend on several variables and are periodic with periods $\varepsilon, \varepsilon^2, \ldots, \varepsilon^N$, in each variable, respectively), some nonlinear operators and variational inequalities. Chapter 2 follows essentially the lines of Chapter 1, first for parabolic equations and next for hyperbolic operators. In the parabolic case we study operators with period of order ε in the space variables and of order ε^k in the time variable. For second order parabolic operators one has essentially the 3 cases $k < 2$, $k = 2$, $k > 2$, at least as far as the **first** term in the expansion (0.0.3) is concerned. In both Chapters 1 and 2 we give examples of partial differential operators for which the corresponding homogenized operator $\mathcal{A}$ is an integrodifferential operator.

A number of questions which can be analyzed by methods similar to the ones in this book are not studied here. In particular, for the analysis of **transport problems** we refer to a paper by us[8] and the references cited therein. For problems

---

[8] J. Publ. RIMS, Kyoto Univ., Japan, 1978.

that deal with periodic distribution of holes, we refer to D. Cioranescu[9] and to our own papers.[10]

A systemtic treatment of the **numerical** and **computational** aspects of the problems considered here would go beyond the scope of this book. Some references are cited in the bibliography of Chapter 1.

The problems considered in this book have a rather long history. The first attempt to construct "effective parameters" for complicated media seems to go back to Poisson. We refer to I. Babuška[11] where a brief account of the historical development is given (along with several references). In Babuška's paper one sees clearly that the notion of "effective parameters" or "effective coefficients" depends very much on how one chooses to **model** a physical problem.

This means that a given physical problem may be modelled by imbedding it into a family of problems (parametrized by ε) of the form (0.0.1), (0.0.2) in many different ways. For example, modelling a problem multiple periodic structure (cf. Chapter 1, Section 8) time variations proportional to ε^k with $k < 2$, $k = 2$ or $k > 2$, static or high frequency excitation, etc., constitutes a specific choice. The homogenized problems, and hence the effective parameters, are different in each case.

The modeling question is not, in this form,[12] a mathematical one and it is important to keep in mind that the definition of effective parameters is a relative one. The formulas change by changing the scaling of a problem, i.e., by adopting another family of problems (0.0.1), (0.0.2) in which to imbed a given physical problem.

For interesting partly heuristic arguments leading to effective parameters, we refer to E. Sanchez Palencia.[13]

There are connections between the asymptotic problems considered here and the very general viewpoint of E. de Giorgi which is called Γ and G convergence. We refer to the work of E. de Giorgi and S. Spagnolo, S. Spagnolo,[14] C. Sbordone,[15] and to the references cited in a recent paper[16] of de Giorgi.

Some questions studied in Chapters 1 and 2 have also been studied by I. Babuška and N. S. Bakbalov (cf. the Bibliography at the end of Chapters 1 and 2).

A particular case of the probabilistic problems analyzed in Chapter 3 has been studied previously by M. I. Freidlin (cf. the Bibliography of Chapter 3). The method of multiple scales that is followed frequently here was not used by Freidlin. His motivation seems to have been the generalization of the averaging method to stochastic equations with **spatially** rapidly oscillating periodic coefficients.

Some of the material of Chapters 1, 2 and 3 has been announced in the notes at the C.R.A.S., Paris, and has been presented in various lectures by the authors, since 1975.

---

[9]Thesis, Paris, October 1977.

[10]To appear.

[11]I. Babuška, Technical Note BN-821, July 1975, Institute of Fliud Dynamics and Appl. Math., Univ. of Maryland, College Park, Maryland 20742.

[12]The **inverse** problem: given a particular homogenization algorithm, find a family of problems (scaling) in some class, whose asymptotic limit is the given homogenized problem, is frequently interesting, difficult and need not have a "solution" in general.

[13]Int. J. Engng. Sci., **12** (1974), pp. 331–351.

[14]Boll. U.M.I., **8** (1973), pp. 391–411, and in *Numerical Methods of Partial Differential Equations, III*, B. Hubbard, editor, Academic Press, New York, 1976, pp. 469–498.

[15]Annali Scuola Norm. Sup. Pisa, IV (1975), pp. 617–638.

[16]Boll. U.M.I., **5** (1977).

Some results in Chapter 4 are well known in solid-state physics (for example, the notion of effective mass). We give a systematic treatment here using the W.K.B or geometrical optics methods[17] combined with multiple scale methods. A preliminary version of this chapter circulated among colleagues as a preprint during 1977.

A large number of questions remain open. Some of them are indicated in the text and follow the present lines of development. Of particular importance is the analysis of the behavior of solutions near boundaries and, possibly, any associated boundary layers. Relatively little seems to be known[18] about this problem.

Note added in 2011

The literature on homogenization has grown enormously since this book was published in 1978. This includes not only partial differential equations with periodic coefficients but also with random coefficients. We list here just a few references, in book form, to guide the reader:

1. Bakhvalov, N. S., Panasenko, G. (1989), Homogenisation: Averaging Processes in Periodic Media, Berlin-Heidelberg-New York: Springer-Verlag.

2. Kozlov, S. M., Oleinik, O. A., Zhikov, V. V. (1994), Homogenization of differential operators and integral functionals, Berlin-Heidelberg-New York: Springer-Verlag.

3. Cioranescu, D., Donato, P. (1999), An introduction to homogenization, Oxford University Press.

4. Allaire, G. (2002), Shape optimization by the homogenization methods, Berlin-Heidelberg-New York: Springer-Verlag.

---

[17] A brief, self-contained introduction to these methods, along with references to the voluminous literature, is given in Chapter 4, Section 2. W.K.B. stands for Wentzel, Kramers and Brillouin who used these methods in the 1920's for the solution of some quantum mechanical problems. The methods were widely used much earlier by Liouville, Rayleigh and others but the terminology WKB persists.

[18] A. Bensoussan, J.L. Lions, G. Papanicolaou, Boundary Layer Analysis of the Dirichlet problem for elliptic equations with rapidly varying coefficients. Proceedings of the Kyoto Conference on Stochastic differential equations, Kyoto July 1976, to be published.

CHAPTER 1

Elliptic Operators

Orientation

Sections 1, 2, 3 are basic for the reading of all other parts of the book. Section 1 presents the simplest "model" problem; the multiple scale method is applied to this problem in Section 2 and an "energy" proof of convergence is given in Section 3.

The reader mainly interested in Probabilistic Methods (Chapter 3) should read section 5 (after Sections 1, 2, 3), some parts of Chapter 2 (indicated in the Orientation of Chapter 2) and the proceed with Chapter 3. The same can be said for those mostly interested in high frequency wave propagation, before reading Chapter 4.

Section 4 gives L^p estimates, $p > 2$ (not too large); these estimates are not indispensable; we use them in some parts of Section 8, but — as we show in that Section — one can avoid these estimates by making some stronger regularity assumptions on the coefficients.

Section 5 gives "correctors" which are improving the approximation. These correctors (or variant of them) are indispensable in numerical computations (not reported in this book).

Sections 6, 7, 9, 10, 11 give extensions and variants to elliptic equations of higher order or to some elliptic systems of interest in the applications.

Section 8 treats the case when there are more than 2 different scales; this Section is technically difficult but the final result is quite simple. Some even more general situations (but for second order operators only) are considered in Chapter 3.

Sections 12, 13, 14 give variants; Section 14 shows (among other things) that one has to be quite careful when working with problems which involve at the same time "homogenization" and "singular perturbations." Section 15 gives examples where the homogenized operator or Partial Differential Operators (**local** operators) is a **non local** (pseudo-differential) operator. Sections 16 and 17 study the homogenization of some non linear problems, in particular of some Variational Inequalities.

1. Setting of the "model" problem

1.1. Setting of the problem (I). Let $\mathcal{O}$ be a bounded open set of $\mathbb{R}^n$; $\mathcal{O}$ is assumed to be bounded to simplify the exposition but this hypothesis is by no means indispensable; we shall return to this point on several occasions.

In $\mathcal{O}$ we are going to consider various **boundary value problems** associated to operators A^ε which are uniformly elliptic when $\varepsilon \to 0$ and the coefficients of A^ε

are rapidly oscillating (with "period" ε). Moe precisely, we define

$$Y = \prod_{j=0}^{n}]0, y_j^{\circ}[\subset \mathbb{R}^n;$$

a function $f : \mathbb{R}^n \to \mathbb{R}^m$ is said to be Y-periodic if it admits period y_j° in the direction y_j, $j = 1, \ldots, n$' we consider functions a_{ij}, $i, j = 1, \ldots, n$, such that

(1.1.1) $a_{ij}(y) \in \mathbb{R}$, a_{ij} is Y-periodic, $a_{ij} \in L^{\infty}(\mathbb{R}^n)$,

$$a_{ij}(y)\xi_i\xi_j \geq \alpha\xi_i\xi_i, \qquad \alpha > 0, \qquad \text{a.e. in } y.^1$$

We **do not assume**, for the time being, that $a_{ij} = a_{ji}$.

We also consider a_0 such that

(1.1.2) $a_0 \in L^{\infty}(\mathbb{R}^n)$, a_0 is Y-periodic,

$$a_0(y) \geq a_0 > 0 \text{ a.e.}$$

To the functions a_{ij} and a_0 we associate the family of operators

(1.1.3) $A^{\varepsilon} = -\dfrac{\partial}{\partial x_i}\left(a_{ij}\left(\dfrac{x}{\varepsilon}\right)\dfrac{\partial}{\partial x_j}\right) + a_0\left(\dfrac{x}{\varepsilon}\right),$

where ε is a **"small" positive parameter**.

REMARK 1.1. As it was said in the Introduction, operators (1.1.3) model the simplest possible situations of **composite materials**, when the period of the structure can be chosen to be ε.

REMARK 1.2. It is immediately seen from (1.1.1) and (1.1.2) that the family A^{ε} consists of second order elliptic operators, which are **uniformly elliptic** in ε.

REMARK 1.3. In many cases, it will be possible to take

(1.1.4) $a_0 = 0$

(one could even consider negative functions a_0, provided they are not too large).

The first problem we want to study in this book is, roughly speaking, as follows: We consider the equation

(1.1.5) $A^{\varepsilon}u_{\varepsilon} = f$ in $\mathcal{O}$,

u_{ε} subject to boundary conditions on $\Gamma = \partial\mathcal{O}^2$,

and we want to study the behavior of u_{ε} as $\varepsilon \to 0$.

A typical "result" we shall obtain (in a more precise form!) is: one can **construct** (with constructive formulas) **a second order elliptic operator** $\mathcal{A}$ **such that** $u_{\varepsilon} \to u$ (in an appropriate topology) **where** u **is the solution of**

(1.1.6) $\mathcal{A}_u = f$ in $\mathcal{O}$,

u is subject to boundary conditions (which of course

will depend on those boundary conditions imposed on u_{ε}).

The operator $\mathcal{A}$ is the so-called **homogenized operator** of the family A^{ε}.

We now proceed with a more precise formulation of the boundary condition in (1.1.5).

---

[1]We adopt here and in what follows the summation convention.

[2]The boundary conditions are made precise in Section 2 below.

1.2. Setting of the problem (II): boundary conditions. We use a variational formulation. The presentation is partially imposed by the structure of (1.1.3) since, even assuming the a_{ij}'s regular (which will not be the case in many results obtained below), the functions $a_{ij}\left(\frac{x}{\varepsilon}\right)$ would have derivatives of order $\frac{1}{\varepsilon}$; therefore the **a priori estimates** on u_ε **which are independent of** ε will not be based on the regularity (if any) of the coefficients. **Therefore a variational formulation (of a weak type) is indispensable here.**

1.2.1. *Sobolev spaces.* (cf. S. L. Sobolev [109], J. Nečas[85], J. L. Lions and E. Magenes [71] and [72], R. Adams [1]).

We shall denote by $H^1(\mathcal{O})$ the space

$$(1.2.1) \qquad H^1(\mathcal{O}) = \left\{ v \,\middle|\, v, \frac{\partial v}{\partial x_1}, \ldots, \frac{\partial v}{\partial x_n} \in L^2(\mathcal{O}) \right\};$$

provided with the norm given by

$$(1.2.2) \qquad \|v\|^2_{H^1(\mathcal{O})} = |v|^2 + \sum_{i=1}^n \left| \frac{\partial v}{\partial x_i} \right|^2,$$
$$|v|^2 = \int v^2 \, dx, \text{[3]}$$

$H^1(\mathcal{O})$ is a Hilbert space.

We define

$$(1.2.3) \qquad H^1_0(\mathcal{O}) = \text{ closure of } C^\infty_0(\mathcal{O}) \text{ in } H^1(\mathcal{O}),$$

$$(1.2.4) \qquad C^\infty_0(\mathcal{O}) = C^\infty \text{ functions with compact support in } \mathcal{O}.$$

One has (cf. Bibliography)

$$(1.2.5) \qquad H^1_0(\mathcal{O}) = \{v \mid v \in H^1(\mathcal{O}), \ v = 0 \text{ on } \Gamma\},$$

and it is known (Poincaré's inequality) that on $H^1_0(\mathcal{O})$, the norm $\|v\|_{H^1(\mathcal{O})}$ and $\left(\sum_{i=1}^n \left| \frac{\partial v}{\partial x_i} \right|^2 \right)^{1/2}$ are **equivalent**.

We shall introduce

$$(1.2.6) \qquad V = \text{ closed subspace of } H^1(\mathcal{O}) \text{ such that } H^1_0(\mathcal{O}) \subseteq V \subseteq H^1(\mathcal{O}).$$

1.2.2. *Bilinear form associated to* A^ε. For $u, v \in H^1(\mathcal{O})$, we define

$$(1.2.7) \qquad a^\varepsilon(u, v) = \int_{\mathcal{O}} a^\varepsilon_{ij}(x) \frac{\partial u}{\partial x_j} \frac{\partial v}{\partial x_i} \, dx + \int_{\mathcal{O}} a^\varepsilon_0 uv \, dx,$$

where we have set

$$(1.2.8) \qquad a^\varepsilon_{ij}(x) = a_{ij}(x/\varepsilon), \qquad a^\varepsilon_0(x) = a_0(x/\varepsilon).$$

We observe that, by virtue of (1.1.1), (1.1.2)

$$(1.2.9) \qquad a^\varepsilon(v, v) \geq \min(\alpha, a_0) \|v\|^2_{H^1(\mathcal{O})} \ \forall v \in H^1(\mathcal{O}).$$

---

[3]For the time being, all functions are supposed to be **real valued**.

REMARK 1.4. Let $\Gamma_0 \subset \Gamma$ be a subset of Γ of positive surface measure (actually one could take Γ_0 of positive capacity), and let us assume that

$$(1.2.10) \qquad V = \{v \mid v \in H^1(\mathcal{O}),\ v = 0 \text{ on } \Gamma_0\}$$

The precise meaning of (1.2.10) is clear if Γ_0 is "smooth"; if not, V consists of the closure in $H^1(\mathcal{O})$ of functions which are zero in a (variable) neighborhood of Γ_0. Then if $a_0^\varepsilon = 0$, i.e. if

$$(1.2.11) \qquad a^\varepsilon(u, v) = \int_{\mathcal{O}} a_{ij}^\varepsilon(x) \frac{\partial u}{\partial x_j} \frac{\partial v}{\partial x_i}\ dx,$$

one has

$$(1.2.12) \qquad a^\varepsilon(v, v) \geq c\|v\|_{H^1(\mathcal{O})}^2,\ c > 0,\ \forall v \in V,$$

where c does not depend on ε.

1.2.3. *Boundary value problem.* The function u_ε is defined as the solution of[4]

$$(1.2.13) \qquad \begin{aligned} &u_\varepsilon \in V, \\ &a^\varepsilon(u_\varepsilon, v) = (f, v),\ \forall v \in V, \end{aligned}$$

where

$$(1.2.14) \qquad (f, v) = \int_{\mathcal{O}} fv\ dx, \qquad f \in L^2(\mathcal{O}).[5]$$

1.2.4. *Examples.*

EXAMPLE 1.5. $V = H_0^1(\mathcal{O})$. Then (1.2.13) is the classical **Dirichlet's problem**.

EXAMPLE 1.6. $V = H^1(\mathcal{O})$. Then (1.2.14) is the Neumann's problem; assuming the boundary Γ and the coefficients a_{ij}'s smooth enough, the problem is

$$(1.2.15) \qquad A^\varepsilon u_\varepsilon = f \text{ in } \mathcal{O},$$

$$\frac{\partial u_\varepsilon}{\partial \nu_{A^\varepsilon}} = 0 \text{ on } \Gamma,$$

where $\frac{\partial v}{\partial \nu_{A^\varepsilon}} = a_{ij}^\varepsilon \frac{\partial v}{\partial x_i} \cos(\nu, x_j)$, $\nu = $ outer normal to Γ.

EXAMPLE 1.7. V is given by (1.2.10). Then (1.2.13) means:

$$(1.2.16) \qquad A^\varepsilon u_\varepsilon = f \text{ in } \mathcal{O},$$

$$u_\varepsilon = 0 \text{ on } \Gamma_0,$$

$$\frac{\partial u_\varepsilon}{\partial \nu_{A^\varepsilon}} = 0 \text{ on } \Gamma_1 = \Gamma - \Gamma_0.$$

EXAMPLE 1.8. Let us suppose that

$$(1.2.17) \qquad V = \{v \in H^1(\mathcal{O}),\ v = \text{ constant on } \Gamma\}$$

(where the value of the constant depends of course on v).

---

[4]We have existence and uniqueness of a solution by the Lax-Milgram's Lemma, since one has (1.2.9) (or (1.2.12)).

[5]We shall consider more general right hand sides later on.

Then (1.2.13) means

$$(1.2.18) \qquad A^\varepsilon u_\varepsilon = f \text{ in } \mathcal{O},$$

$$u_\varepsilon = c_\varepsilon \text{ on } \Gamma,$$

$$c_\varepsilon = \text{ constant (not given)},$$

$$\int_\Gamma \frac{\partial u_\varepsilon}{\partial \nu_{A^\varepsilon}} \, d\Gamma = 0.$$

The first problem we want to study is now in its precise form **to derive the behavior of u_ε, solution of** (1.2.13), **as $\varepsilon \to 0$.**

Before entering the study of the general case, let us consider a very simple particular case.

1.3. An example: a one-dimensional problem. We consider the case $n = 1$ in the above problem, for the Dirichlet's boundary condition (to fix ideas). Therefore if $\mathcal{O} =]x_0, x_1[$, we have

$$(1.3.1) \qquad -\frac{d}{dx}\left[a^\varepsilon(x)\frac{du_\varepsilon}{dx}\right] = f \text{ in } \mathcal{O},$$

$$(1.3.2) \qquad u_\varepsilon(x_0) = u_\varepsilon(x_1) = 0,$$

where $a(y)$ is Y periodic (i.e. admits a period y_0) and $a(y) \geq \alpha > 0$ a.e.

The variational formulation is

$$(1.3.3) \qquad \int_\mathcal{O} a^\varepsilon \frac{du_\varepsilon}{dx}\frac{dv}{dx} \, dx = \int_\mathcal{O} fv \, dx, \ \forall v \in V = H_0^1(\mathcal{O}),$$

$$u_\varepsilon \in H_0^1(\mathcal{O}).$$

By taking $v = u_\varepsilon$ in (1.3.3) one immediately sees that

$$(1.3.4) \qquad \|u_\varepsilon\|_{H^1(\mathcal{O})} \leq C.[6]$$

Therefore one can extract a subsequence, still denoted by u_ε, such that

$$(1.3.5) \qquad u_\varepsilon \to u \text{ in } H_0^1(\mathcal{O}) \text{ weakly.}$$

We also notice (this is a simple exercise) that

$$(1.3.6) \qquad a^\varepsilon \to \mathcal{M}(a) = \frac{1}{y_0}\int_0^{y_0} a(y) \, dy \text{ in } L^\infty(\mathcal{O}) \text{ weak star.}[7]$$

From (1.3.5), (1.3.6), (1.3.1) it is tempting to believe that in the limit one has

$$(1.3.7) \qquad -\frac{d}{dx}\left(\mathcal{M}(a)\frac{du}{dx}\right) = f,$$

u satisfying the boundary conditions analogous to (1.3.2), i.e.

$$(1.3.8) \qquad u(x_0) = u(x_1) = 0.$$

But this is untrue (in general). The correct answer is as follows: we introduce

$$(1.3.9) \qquad \xi^\varepsilon = a^\varepsilon\frac{du_\varepsilon}{dx};$$

---

[6]Where here and in what follows, the c's denote constants which do not depend on ε.

[7]In general, if $g_\varepsilon, g \in L^\infty(\mathcal{O}), g_\varepsilon \to g$ in $L^\infty(\mathcal{O})$ weak star means $\int_\mathcal{O} g_\varepsilon \phi \, dx \to \int_\mathcal{O} g\phi \, dx$, $\phi \in L^1(\mathcal{O})$.

since a^ε remains in a bounded set of $L^\infty(\mathcal{O})$ and since one has (1.3.4), ξ^ε is bounded in $L^2(\mathcal{O})$ and by (1.3.1) we have

$$(1.3.10) \qquad\qquad -\frac{d\xi^\varepsilon}{dx} = f,$$

so that ξ^ε is bounded in $H^1(\mathcal{O})$. Since the identity mapping from $H^1(\mathcal{O}) \to L^2(\mathcal{O})$ is compact,[8] it follows that one can assume

$$(1.3.11) \qquad\qquad \xi^\varepsilon \to \xi \text{ in } L^2(\mathcal{O}) \text{ strongly,}$$

so that

$$(1.3.12) \qquad\qquad \frac{1}{a^\varepsilon}\xi^\varepsilon \to \mathcal{M}\left(\frac{1}{a}\right)\xi \text{ in } L^2(\mathcal{O}) \text{ \textbf{weakly}}$$

(since $\frac{1}{a^\varepsilon} \to \mathcal{M}\left(\frac{1}{a}\right)$ in $L^\infty(\mathcal{O})$ weak star). But $\frac{1}{a^\varepsilon}\xi^\varepsilon = \frac{du_\varepsilon}{dx}$ so that (1.3.12) and (1.3.5) imply

$$\frac{du}{dx} = \mathcal{M}\left(\frac{1}{a}\right)\xi.$$

On the other hand (1.3.12) gives $-\frac{d\xi}{dx} = f$, so that

$$(1.3.13) \qquad\qquad -\frac{d}{dx}\left(\frac{1}{\mathcal{M}\left(\frac{1}{a}\right)}\frac{du}{dx}\right) = f.$$

The **homogenized operator associated to** A^ε is given by

$$(1.3.14) \qquad\qquad -\frac{1}{\mathcal{M}\left(\frac{1}{a}\right)}\frac{d^2}{dx^2} = \mathcal{A}.$$

Since $\mathcal{A}$ is uniquely defined, $u_\varepsilon \to u$ in $H_0^1(\mathcal{A})$ weakly, without extracting a subsequence.

REMARK 1.9. We notice that

$$(1.3.15) \qquad\qquad \mathcal{M}(a) \geq \frac{1}{\mathcal{M}\left(\frac{1}{a}\right)}, \text{ with strict inequality in general.}$$

REMARK 1.10. The periodicity of $a(y)$ did not play a fundamental role in the above result. We can more generally assume that[9]

$$(1.3.16) \qquad\qquad a^\varepsilon \text{ remains in a bounded set of } L^\infty(\mathcal{O}),$$

$$a^\varepsilon(x) \geq \alpha > 0 \text{ a.e.}$$

Then $\frac{1}{a^\varepsilon}$ also remains in a bounded set of $L^\infty(\mathcal{O})$. Therefore we can extract a subsequence $\frac{1}{a^\varepsilon}$, such that

$$(1.3.17) \qquad\qquad \frac{1}{a^{\varepsilon'}} \to \mu \text{ in } L^\infty(\mathcal{O}) \text{ weak star.}$$

Then $u_{\varepsilon'} \to u$ in $H_0^1(\mathcal{O})$ weakly, where u is the solution of

$$(1.3.18) \qquad\qquad -\frac{d}{dx}\left(\frac{1}{\mu(x)}\frac{du}{dx}\right) = f.$$

But this time, contrary to the preceding case, the limit u is not unique.

---

[8]This is true if $\mathcal{O}$ is **bounded**. But in the unbounded case the **local** compactness is sufficient (i.e. the compactness of the injection mapping $H^1(\mathcal{O}) \to L^2(\mathcal{O}')$, $\mathcal{O}'$ bounded, $\overline{\mathcal{O}'} \subset \mathcal{O}$).

[9]We do not assume that $a^\varepsilon(x) = a(x/\varepsilon)$.

2. Asymptotic expansions

2.1. Orientation. We introduce in this section a method based on asymptotic expansions using multiple scales (i.e. "slow" and "fast" variables). As we shall see all over this book, the method we are going to develop is the most convenient and the most useful to **obtain the right answers**. The **justification** of the formulas obtained by this method can sometimes be made directly, but in general other tools will be needed; these tools will be introduced later on in this book.

2.2. Asymptotic expansions using multiple scales. We introduce functions $\phi(x, y)$, $y \in \mathbb{R}^n$, which are Y-periodic in y, and we associate to $\phi(x, y)$ the function $\phi(x, x/\varepsilon)$.

We shall look for $u_\varepsilon = u_\varepsilon(x)$ in the form of the asymptotic expansion

$$(2.2.1) \qquad u_\varepsilon(x) = u_0(x, x/\varepsilon) + \varepsilon u_1(x, x/\varepsilon) + \varepsilon^2 u_2(x, x/\varepsilon) + \cdots,$$

where the functions $u_j(x, y)$ are Y-periodic in y, $\forall x \in \mathcal{O}$.

REMARK 2.1. It is technically complicated to keep track of **boundary** conditions when seeking u_ε in the form (2.2.1) and this is actually the source of serious technical difficulties in justifying the method. The method will nevertheless give the "right answer" because it will turn out that, in this sort of problem, the boundary conditions are somewhat irrelevant.

The idea of the method is (simply) to insert (2.2.1) in equation (1.1.5) and to identify powers of ε.

In order to present these computations in a simple form, it is useful to consider first x and y as independent variables and to replace next y by x/ε.

Applied to a function $\phi(x, x/\varepsilon)$, the operator $\frac{\partial}{\partial x_j}$ becomes $\frac{\partial}{\partial x_j} + \frac{1}{\varepsilon} \frac{\partial}{\partial y_j}$. With this in mind, one can write

$$(2.2.2) \qquad A^\varepsilon = \varepsilon^{-2} A_1 + \varepsilon^{-1} A_2 + \varepsilon^0 A_3,$$

where

$$(2.2.3) \qquad A_1 = -\frac{\partial}{\partial y_i}\left(a_{ij}(y)\frac{\partial}{\partial y_j}\right),$$

$$A_2 = -\frac{\partial}{\partial y_i}\left(a_{ij}(y)\frac{\partial}{\partial x_j}\right) - \frac{\partial}{\partial x_i}\left(a_{ij}(y)\frac{\partial}{\partial y_j}\right),$$

$$A_3 = -\frac{\partial}{\partial x_i}\left(a_{ij}(y)\frac{\partial}{\partial x_j}\right) + a_0.$$

Using (2.2.1), (2.2.2), the equation (1.1.5) becomes

$$(2.2.4) \qquad A_1 u_0 = 0,$$

$$(2.2.5) \qquad A_1 u_1 + A_2 u_0 = 0,$$

$$(2.2.6) \qquad A_1 u_2 + A_2 u_1 + A_3 u_0 = f;$$

one can of course (formally) proceed

$$(2.2.7) \qquad A_1 u_3 + A_2 u_2 + A_3 u_1 = 0, \text{ etc.}$$

We are now going to see that the homogenized operator $\mathcal{A}$ can be constructed from (2.2.4), (2.2.5), (2.2.6).

Let us remark that the equation

(2.2.8) $$A_1\phi = F \text{ in } Y,$$
$$\phi \text{ periodic in } Y,$$

admits a unique solution (up to an additive constant), iff

(2.2.9) $$\int_Y F(y)\,dy = 0.$$

Indeed let us introduce

(2.2.10) $$W(Y) = \{\phi \mid \phi \in H^1(Y), \ \phi \text{ "periodic"}\},$$

i.e. ϕ takes equal values on opposite faces of Y.

For $\phi, \psi \in W(Y)$, let us set

(2.2.11) $$a_1(\phi, \psi) = \int_Y a_{ij}(y) \frac{\partial \phi}{\partial y_j} \frac{\partial \psi}{\partial y_i}\, dy,$$

and let us also set

(2.2.12) $$(F, \psi)_Y = \int_Y F(y)\psi(y)\, dy.$$

Then (2.2.8) is equivalent to:

(2.2.13) $$\phi \in W(y),$$
$$a_1(\phi, \psi) = (F, \psi)_Y, \ \forall \psi \in W(y).$$

Condition (2.2.9) is clearly **necessary** since $\int_Y (A_1\phi)\,dy = 0$ if ϕ is periodic. If (2.2.9) holds, we can consider

(2.2.14) $$W'(Y) = W(Y)/R$$

and $\psi \to (F, \psi)_Y = (F, \psi+c)_Y, \ \forall c \in \mathbb{R}$, becomes a continuous linear form on $W'(Y)$ (assuming $F \in L^2(Y)$). Since $a_1(\phi, \phi) \geq c\|\phi\|^2_{W'(Y)}$, $c > 0$, it follows **that** (2.2.13) **admits a unique solution in** $W'(Y)$ (i.e. ϕ is defined up to an additive constant).

We apply now this remark to the solution of (2.2.4), (2.2.5), (2.2.6).

2.2.1. *Solution of (2.2.4)*. By virtue of (2.2.8), (2.2.9), the only periodic solution of (2.2.6) is $u_0 = $ constant where x is a parameter, i.e.

(2.2.15) $$u_0(x, y) = u(x).$$

REMARK 2.2. One could **directly** justify the fact that, **in this case**, the **first term of the expansion (2.2.1) should not depend on** y but, as we shall see later on, this need not be always the case.

2.2.2. *Solution of (2.2.5)*. Using (2.2.15), (2.2.5) reduces to

(2.2.16) $$A_1 u_1 = \left(\frac{\partial}{\partial y_i} a_{ij}(y)\right) \frac{\partial u}{\partial x_j}(x).$$

Due to the separation of variables in the right hand side of (2.2.16), we can represent u_1 in a simple form.

We define $\chi^j = \chi^j(y)$ as the solution (up to an additive constant) of

(2.2.17) $$A_1 \chi^j = A_1 y_j = -\frac{\partial}{\partial y_i} a_{ij}(y),$$
$$\chi^j \ Y\text{-periodic.}$$

Since $\int_Y (A_1 y_j)\, dy = 0$, χ^j exists by virtue of (2.2.8) and (2.2.9). Then the general solution of (2.2.16) is

$$(2.2.18) \qquad u_1(x, y) = -\chi^j(y)\frac{\partial u}{\partial x_j}(x) + \tilde{u}_1(x).$$

2.2.3. *Solution of (2.2.6).* We now consider (2.2.6) where we think of u_2 as the unknown, x being a parameter. By virtue of (2.2.8), (2.2.9), u_2 exists iff

$$(2.2.19) \qquad \int_Y (A_2 u_1 + A_3 u_0)\, dy = \int_Y f\, dy = |Y| f,$$

where $|Y| =$ measure of Y. **Condition (2.2.19) is the homogenized equation we are looking for.** We observe that

$$\int_Y A_2 u_1\, dy = -\frac{\partial}{\partial x_i}\int_Y a_{ik}(y)\frac{\partial u_1}{\partial y_k}\, dy$$

and using (2.2.18)

$$\int_Y A_2 u_1\, dy = \frac{\partial}{\partial x_i}\int_Y a_{ik}(y)\frac{\partial \chi^i}{\partial y_k}(y)\, dy\, \frac{\partial u}{\partial x_j}$$

so that (2.2.19) becomes

$$(2.2.20) \qquad -\frac{1}{|Y|}\left(\int_Y \left(a_{ij} - a_{ik}\frac{\partial \chi^j}{\partial y_k}\right)\, dy\right)\frac{\partial^2 u}{\partial x_i \partial x_j} + \frac{1}{|Y|}\left(\int_Y a_0(y)\, dy\right) u = f.$$

2.2.4. *Conclusion.* **The formal rule** (which will be justified below) **to compute the homogenized operator $\mathcal{A}$ is as follows:**

(1) **Solve (2.2.17) on the unit cell Y, for $j = 1, \ldots, n$;**
(2) **$\mathcal{A}$ is given by (2.2.19).**

A few remarks are now in order.

2.3. Remarks on the homogenized operator.

REMARK 2.3. In the **one dimensional case** one has ($\chi^j = \chi$)

$$-\frac{d}{dy}\left(a(y)\frac{d\chi}{dy}\right) = -\frac{da}{dy}$$

hence $a(y)\frac{d\chi}{dy} = a(y) + c$ and this equation admits a periodic solution iff

$$\int_Y \left(1 + \frac{c}{a}\right)\, dy = 0,$$

i.e.

$$(2.3.1) \qquad 1 + c\mathcal{M}\left(\frac{1}{a}\right) = 0, \qquad \frac{d\chi}{dy} = 1 + \frac{c}{z}.$$

Then the coefficient of $-\frac{d^2 u}{dx^2}$ in $\mathcal{A}$ is, according to (2.2.20):

$$\frac{1}{|Y|}\int_Y (a - a - c)\, dy = -c = \frac{1}{\mathcal{M}\left(\frac{1}{a}\right)}.$$

and we find (1.3.14).

REMARK 2.4. If we set

$$(2.3.2) \qquad \mathcal{A} = -q_{ij}\frac{\partial^2}{\partial x_i \partial x_j} + \mathcal{M}(a_0),\,^{10}$$

the structure of q_{ij} is interesting:

$$(2.3.3) \qquad q_{ij} = \mathcal{M}(a_{ij}) - \mathcal{M}\left(a_{ik}\frac{\partial \chi^j}{\partial y_k}\right).$$

As we have already seen in the one dimensional case,

$$a_{ij}^{\varepsilon}\frac{\partial u_{\varepsilon}}{\partial x_j} \not\to \mathcal{M}(a_{ij})\frac{\partial u}{\partial x_j}$$

and $-\mathcal{M}\left(a_{ik}\frac{\partial \chi^j}{\partial y_k}\right)$ **appears as the "corrector."** This remark will be made precise later on in this chapter.

REMARK 2.5. In (2.2.20) the term $\mathcal{M}(a_0)u$ is trivial, since $u_{\varepsilon} + u$ in $L^2(\mathcal{O})$ **strongly** (cf. Section 3).

REMARK 2.6. At this stage, the **ellipticity** of $\mathcal{A}$ is not entirely obvious. In order to verify that, indeed, $\mathcal{A}$ is an elliptic operator, let us first prove that

$$(2.3.4) \qquad q_{ij} = \frac{1}{|Y|}a_1(\chi^j - y_j, \chi^i - y_i).$$

Indeed (2.2.17) is equivalent to

$$(2.3.5) \qquad a_1(\chi^j - y_j, \psi) = 0 \;\; \forall \psi \in W(Y).$$

Taking $\psi = \chi^i$ in (2.2.17) we see that

$$\frac{1}{|Y|}a_1(\chi^j - y_j, \chi^i - y_i) = -\frac{1}{|Y|}a_1(\chi^j - y_j, y_i) = q_{ij}$$

which proves (2.3.4).

The ellipticity of $\mathcal{A}$ follows. Indeed

$$(2.3.6) \qquad q_{ij}\xi_i\xi_j = \frac{1}{|Y|}a_1(w, w),$$

$$w = \xi_i(\chi^i - y_i).$$

But $a_1(w, w) \geq \alpha \int_Y \sum \left(\frac{\partial w}{\partial y_k}\right)^2 dy$ so that $q_{ij}\xi_i\xi_j \geq 0$ and $q_{ij}\xi_i\xi_j = 0$ iff $\frac{\partial w}{\partial y_k} = 0$ $\forall k$, i.e. $w = c$. But then $\xi_i y_i = \xi_i\chi^i - c$ is y-periodic, i.e. $\xi_i = 0$ $\forall i$ so that $q_{ij}\xi_i\xi_j$ **is positive definite**.

REMARK 2.7. In case one has

$$(2.3.7) \qquad a_{ij} = a_{ji} \;\; \forall i, j,$$

(i.e. $\mathcal{A}^{\varepsilon}$ is symmetric) then $a_1(\phi, \psi) = a_1(\psi, \phi)$ $\forall \phi, \psi \in H^1(Y)$ so that (2.3.4) implies

$$(2.3.8) \qquad q_{ij} = q_{ji} \;\; \forall i, j,$$

i.e. $\mathcal{A}$ is **also symmetric**.

---

^{10}We set in general $\mathcal{M}(\phi) = -\frac{1}{|Y|}\int_Y \phi(y)\, dy.$

REMARK 2.8. If A^ε is "**diagonal**" (i.e. $a_{ij} = 0$ if $i \neq j$), **this is not neces-sarily the case for** $\mathcal{A}$; indeed for $i \neq j$, one has

$$q_{ij} = -\frac{1}{|Y|} \int_Y a_{ii} \frac{\partial \chi^j}{\partial y_i} \, dy \text{ (no summation in } i\text{!)}$$

which is not necessarily equal to zero.

REMARK 2.9. The operator $\mathcal{A}$ **does not depend on** $\mathcal{O}$.

2.4. Justification of the asymptotic expansion for Dirichlet's bound-ary conditions. We consider $u_{\varepsilon'}$ solution of the Dirichlet's problem

$$(2.4.1) \qquad A^\varepsilon u_\varepsilon = f, \; u_\varepsilon = 0 \text{ on } \Gamma,$$

and we are going to show that, assuming all data and all functions to be smooth enough,[11] u_ε converges in the uniform topology, toward u, solution of

$$(2.4.2) \qquad \mathcal{A}u = f, \; u = 0 \text{ on } \Gamma.$$

We take $u_1(x, y)$ as defined by $(2.2.18)$ with $\tilde{u}_1 = 0$, and we set

$$(2.4.3) \qquad z_\varepsilon = u_\varepsilon - (u + \varepsilon u_1 + \varepsilon^2 w), \; w = w(x, y),$$

where we choose w so that $A^\varepsilon z_\varepsilon$ is "**as small as possible.**" A natural candidate is

$$(2.4.4) \qquad w = u_2;$$

then

$$(2.4.5) \qquad A^\varepsilon z_\varepsilon = -\varepsilon r_\varepsilon,$$

with

$$(2.4.6) \qquad r_\varepsilon = A_2 w + A_3 u_1 + \varepsilon A_3 w.$$

We can define uniquely w if we add the condition

$$(2.4.7) \qquad \int_Y w(x, y) \, dy = 0.$$

If f is smooth, say $f \in C^k(\overline{\mathcal{O}})$,[12] then $x \to w(x, y)$ will be k times continuously differentiable in $\overline{\mathcal{O}}$ with values in $C^1(\overline{Y})$. For this choice of w, $(2.4.6)$ gives

$$(2.4.8) \qquad \|r_\varepsilon\|_{L^\infty(\mathcal{O})} \leq c.$$

On the other hand, **on** Γ,

$$z_\varepsilon = -(\varepsilon u_1 + \varepsilon^2 w) \, |_\Gamma,$$

hence

$$(2.4.9) \qquad \|z_\varepsilon\|_{L^\infty(\Gamma)} \leq c\varepsilon.$$

It follows from $(2.4.5)$, $(2.4.9)$ and from the maximum principle that

$$\|z_\varepsilon\|_{L^\infty(\mathcal{O})} \leq c\varepsilon$$

and therefore

$$(2.4.10) \qquad \|u_\varepsilon - u\|_{L^\infty(\mathcal{O})} \leq c\varepsilon.$$

---

[11] We do not make efforts to minimize here the regularity hypothesis, for reasons explained at the end of this section.

[12] $C^k(\overline{\mathcal{O}})$ = space of k-time continuously differentiable functions in $\mathcal{O}$.

This estimate proves the desired result and moreover, it gives an error estimate between u_ε and the solution u of the homogenized equation.

REMARK 2.10. The difficulty in extending the above proof to other boundary conditions is obvious; if u_ε (respectively u) is the solution of the **Neumann** problem associated to A^ε respectively $\mathcal{A}$), then u_ε and u satisfy **different conditions on** Γ, and z_ε, defined as in (2.4.3), does not satisfy simple boundary conditions. **The way out of this difficulty is to work with weak (variational) solutions.**

REMARK 2.11. The above proof is slightly misleading, since it uses the maximum principle and since we will see below that u_ε converges to the solution of a homogenized equation also for **higher order elliptic operators** (where the maximum principle no longer applies), but u_ε will converge in a different topology (in general).

2.5. Higher order terms in the expansion. We can of course proceed with the computation of $u_2(x,y), u_3(x,y), \ldots$, in the expansion of (2.2.1). If we set

$$(2.5.1) \qquad b_{ij}(y) = a_{ij}(y) - a_{ik}(y)\frac{\partial \chi^j}{\partial y_k} - \frac{\partial}{\partial y_k}(a_{ki}\chi^j)$$

then we see from (2.2.6) that

$$(2.5.2) \qquad A_1 u_2 = f + b_{ij}\frac{\partial^2 u}{\partial x_i \partial x_j} + \left(\frac{\partial}{\partial y_i}a_{ij}\right)\frac{\partial \tilde{u}_1}{\partial x_j}.$$

But

$$\mathcal{A}u = -\mathcal{M}(b_{ij})\frac{\partial^2 u}{\partial x_i \partial x_j}$$

(since $\mathcal{M}\left(\frac{\partial}{\partial y_k}(a_{ki}\chi^j)\right) = 0$) so that (2.5.2) becomes

$$(2.5.3) \qquad A_1 u_2 = +(b_{ij} - \mathcal{M}(b_{ij}))\frac{\partial^2 u}{\partial x_i \partial x_j} + \left(\frac{\partial}{\partial y_i}a_{ij}\right)\frac{\partial \tilde{u}_1}{\partial x_j}.$$

If we introduce χ^{ij} as a Y-periodic solution (defined up to an additive constant) of

$$(2.5.4) \qquad A_1 \chi^{ij} = b_{ij} - \mathcal{M}(b_{ij})$$

then (if $a_0 = 0$):

$$(2.5.5) \qquad u_2 = \chi^{ij}\frac{\partial^2 u}{\partial x_i \partial x_j} - \chi^j\frac{\partial \tilde{u}_1}{\partial x_j} + \tilde{u}_2(x).$$

We can now compute $\tilde{u}_1$ in such a manner that the computation can proceed. We consider (2.2.7). We can solve for u_2 iff

$$(2.5.6) \qquad \int_Y (A_2 u_2 + A_3 u_1)\, dy = 0.$$

But by an explicit computation, this condition is equivalent to

$$(2.5.7) \qquad \mathcal{A}\tilde{u}_1 = \mathcal{M}\left(a_{k\ell}\frac{\partial \chi^{ij}}{\partial y_\ell} - a_{ij}\chi^k\right)\frac{\partial^3 u}{\partial x_i \partial x_j \partial x_k}.$$

If the boundary conditions are of Dirichlet's type, a natural choice for $\tilde{u}_1$ on Γ would be

$$(2.5.8) \qquad \tilde{u}_1 = \chi^j\left(\frac{x}{\varepsilon}\right)\frac{\partial u}{\partial x_j}(x),$$

but it gives a $\tilde{u}_1$ which depends on ε and has "bad" derivatives.

2.6. Extensions. It is clear by now that the method presented above can be extended in many ways and in several directions such as for the solution of higher order partial differential operators, to systems, to nonlinear problems, etc. (and also to evolution equations). We shall therefore return many times to this method.

3. Energy proof of the homogenization formula

3.1. Orientation: Statement of the main result. We now return to (1.2.13). For $u, v, \in H^1(\mathcal{O})$ we set

$$(3.1.1) \qquad \mathcal{A}(u, v) = \int_{\mathcal{O}} q_{ij} \frac{\partial u}{\partial x_j} \frac{\partial v}{\partial x_i} \, dx,$$

q_{ij} defined by (2.3.4), χ^i being defined by (2.2.17).

Let u be the solution of the **homogenized problem**:

$$(3.1.2) \qquad u \in V,$$
$$\mathcal{A}(u, v) = (f, v) \ \forall v \in V.$$

We shall prove:

THEOREM 3.1. *We assume that a_{ij}, a_0 satisfy (1.1.1), (1.1.2). Let u_ε (respectively u) be the solution of (1.2.13) (respectively (3.1.2)). Then as $\varepsilon \to 0$ one has*

$$(3.1.3) \qquad u_\varepsilon \to u \text{ in } V \text{ weakly.}$$

REMARK 3.2. Since the coefficients a_{ij}'s belong to $L^\infty(Y)$ they can be discontinuous on each cell Y i.e. we have a transmission problem on each cell $\varepsilon(Y + \zeta)$, $\zeta \in \mathbb{Z}^n$.

REMARK 3.3. The example of 1.3 shows that in general $\frac{\partial u_\varepsilon}{\partial x_i}$ **does not converge strongly toward** $\frac{\partial u}{\partial x_i}$. Therefore (3.1.3) cannot be improved without adding extra terms (of the "corrector" type).

REMARK 3.4. Let us consider more generally a family $f_\varepsilon \in V'$ (dual of V); let u_ε be the solution of

$$(3.1.4) \qquad a^\varepsilon(u_\varepsilon, v) = (f_\varepsilon, v) \ \forall v \in V, \ u_\varepsilon \in V;$$

let us assume that

$$(3.1.5) \qquad f_\varepsilon \to f \text{ in } V' \text{ strongly}$$

and let u be the solution of

$$(3.1.6) \qquad \mathcal{A}(u, v) = (f, v) \ \forall v \in V, \ u \in V.$$

Then one still has (3.1.3).

3.2. Proof of the convergence theorem. We consider the more general case of Remark 3.4. Taking $v = u_\varepsilon$ in (3.1.4) and using (1.2.12) we have

$$(3.2.1) \qquad \|u_\varepsilon\| \le C$$

where $\| \cdot \|$ denotes the norm in V (or $H^1(\mathcal{O})$).

We set

$$(3.2.2) \qquad \xi_i^\varepsilon = a_{ij}^\varepsilon \frac{\partial u_\varepsilon}{\partial x_j};$$

it follows from (3.2.1) that

$$(3.2.3) \qquad\qquad |\xi_i^\varepsilon| \le C$$

(where $|\cdot|$ denotes the norm in $L^2(\mathcal{O})$). Therefore we can extract a subsequence, still denoted by u_ε, ξ_i^ε such that

$$(3.2.4) \qquad\qquad u_\varepsilon \to u \text{ in } V \text{ weakly,}$$

$$\xi_i^\varepsilon \to \xi_i \text{ in } L^2(\mathcal{O}) \text{ weakly.}$$

Equation (3.1.4) can be written[13]

$$\left(\xi_i^\varepsilon, \frac{\partial v}{\partial x_i}\right) = (f_\varepsilon, v) \ \forall v \in V$$

which gives in the limit

$$(3.2.5) \qquad\qquad \left(\xi_i, \frac{\partial v}{\partial x_i}\right) = (f, v) \ \forall v \in V$$

(here the **weak** convergence in V' of f_ε to f would suffice).

We now compute ξ_i using adjoint equations. Let $P(y)$ be a homogeneous polynomial of degree 1 and let A_1^* be the adjoint of A_1:

$$(3.2.6) \qquad\qquad A_1^* = -\frac{\partial}{\partial y_i}\left(a_{ij}^* \frac{\partial}{\partial y_j}\right), \qquad a_{ij}^* = a_{ji}.$$

We define w as "the" solution of

$$(3.2.7) \qquad\qquad A_1^* w = 0$$

such that $w - P$ is Y periodic. If

$$(3.2.8) \qquad\qquad w - P = -\widehat{\chi}$$

then

$$(3.2.9) \qquad\qquad A_1^*\widehat{\chi} = A_1^* P, \ \widehat{\chi} \text{ is } Y\text{-periodic}$$

($\widehat{\chi}$ is defined up to an additive constant).

We define next

$$(3.2.10) \qquad\qquad w_\varepsilon = \varepsilon w\left(\frac{x}{\varepsilon}\right) = P(x) - \varepsilon\widehat{\chi}\left(\frac{x}{\varepsilon}\right).$$

We have (we can always assume that $a_0 = 0$):

$$(3.2.11) \qquad\qquad (A^\varepsilon)^* w_\varepsilon = 0 \text{ in } \mathcal{O}.$$

We now take in (3.1.4), $v = \phi w_\varepsilon$, $\phi \in C_0^\infty(\mathcal{O})$ (so that $\phi w_\varepsilon \in V$ since $\phi w_\varepsilon \in H_0^1(\mathcal{O}) \subseteq V$) and we take the scalar product of (3.2.11) with ϕu_ε. Taking the difference, we obtain

$$(3.2.12) \qquad\qquad a^\varepsilon(u_\varepsilon, \phi w_\varepsilon) - a^\varepsilon(\phi u_\varepsilon, w_\varepsilon) = (f_\varepsilon, \phi w_\varepsilon).$$

The left hand side of (3.2.12) equals

$$(3.2.13) \qquad\qquad \left(\xi_i^\varepsilon, \left(\frac{\partial \phi}{\partial x_i}\right) w_\varepsilon\right) - \int a_{ij}^\varepsilon \frac{\partial w_\varepsilon}{\partial x_i} \frac{\partial \phi}{\partial x_j} u_\varepsilon \, dx;$$

---

[13]We do not write the term $(a_0^\varepsilon u_\varepsilon, v)$ whose limit is $\mathcal{M}(a_0)(u, v)$.

but $a_{ij}^\varepsilon \frac{\partial w_\varepsilon}{\partial x_i} = \left(a_{ij} \frac{\partial w}{\partial y_i} \right)^\varepsilon$ where $g^\varepsilon(x) = g(x/\varepsilon)$ so that the expression in (3.2.13) converges, as $\varepsilon \to 0$, to

$$(3.2.14) \qquad \left(\xi_i, \left(\frac{\partial \phi}{\partial x_i} \right) P \right) - \mathcal{M} \left(a_{ij} \frac{\partial w}{\partial y_i} \right) \int \frac{\partial \phi}{\partial x_j} u \, dx.$$

indeed $w_\varepsilon \to P$ in $L^2(\mathcal{O})$ strongly, $u_\varepsilon \to u$ in $H^1(\mathcal{O})$ weakly, therefore in $L^2(\mathcal{O}')$ **strongly** for every $\mathcal{O}' \subset \overline{\mathcal{O}'} \subset \mathcal{O}$, $\mathcal{O}'$ bounded and $\left(a_{ij} \frac{\partial w}{\partial y_i} \right)^\varepsilon$ converges to $\mathcal{M} \left(a_{ij} \frac{\partial w}{\partial y_i} \right)$ in $L^\infty(\mathcal{O})$ weak star.

On the other hand

$$(3.2.15) \qquad \phi w_\varepsilon \to \phi P \text{ in } H_0^1(\mathcal{O}) \text{ weakly;}$$

indeed $\phi w_\varepsilon \to \phi P$ in $L^2(\mathcal{O})$ and

$$\phi \frac{\partial w_\varepsilon}{\partial x_i} = \phi \frac{\partial P}{\partial x_i} - \phi \left(\frac{\partial \widehat{\chi}}{\partial y_i} \right)^\varepsilon$$

and

$$\phi \left(\frac{\partial \widehat{\chi}}{\partial y_i} \right)^\varepsilon \to \phi \mathcal{M} \left(\frac{\partial \widehat{\chi}}{\partial y_i} \right) = 0 \text{ in } L^2(\mathcal{O}) \text{ weakly.}$$

Therefore (3.2.15) and (3.1.5) imply

$$(f_\varepsilon, \phi w_\varepsilon) \to (f, \phi P).$$

By using (3.2.5), $(f, \phi P) = \left(\xi_i, \frac{\partial(\phi P)}{\partial x_i} \right)$ so that

$$\left(\xi_i, \left(\frac{\partial \phi}{\partial x_i} \right) P \right) - \mathcal{M} \left(a_{ij} \frac{\partial w}{\partial y_i} \right) \int_{\mathcal{O}} \frac{\partial \phi}{\partial x_j} u \, dx = \left(\xi_i, \frac{\partial(\phi P)}{\partial x_i} \right).$$

Therefore

$$\left(\xi_i \frac{\partial P}{\partial x_i}, \phi \right) = \mathcal{M} \left(a_{ij} \frac{\partial w}{\partial y_i} \right) \left(\frac{\partial u}{\partial x_j}, \phi \right) \quad \forall \phi \in C_0^\infty(\mathcal{O})$$

i.e.

$$(3.2.16) \qquad \xi_i \frac{\partial P}{\partial x_i} = \mathcal{M} \left(a_{ij} \frac{\partial w}{\partial y_i} \right) \frac{\partial u}{\partial x_j}.$$

If we take $P(y) = y_i$, and if we define

$$(3.2.17) \qquad A_i^* \widehat{\chi}^i = A_1^* y_i$$

we have $w = y_i - \widehat{\chi}^i$ and

$$(3.2.18) \qquad \xi_i = \mathcal{M} \left(a_{kj} \frac{\partial}{\partial y_k} (y_i - \widehat{\chi}^i) \right) \frac{\partial u}{\partial x_j}.$$

Consequently, using (3.2.18) into (3.2.5), we see that u satisfies

$$(3.2.19) \qquad \left(\mathcal{M} \left(a_{ij} - a_{kj} \frac{\partial \widehat{\chi}^i}{\partial y_k} \right) \frac{\partial u}{\partial x_j}, \frac{\partial v}{\partial x_i} \right) = (f, v) \; \forall v \in V.$$

It remains only to verify that (3.2.19) **coincides** with (3.1.6), i.e. that

$$(3.2.20) \qquad \mathcal{M} \left(a_{ik} \frac{\partial \chi^j}{\partial y_k} \right) = \mathcal{M} \left(a_{kj} \frac{\partial \widehat{\chi}^i}{\partial y_k} \right).$$

But taking the scalar product of (3.2.17) with χ^j gives

$$(A_1^* \widehat{\chi}^i, \chi^j)_Y = \int_Y a_{ik} \frac{\partial \chi^j}{\partial y_k} \, dy$$

and taking the scalar product of (2.2.17) with $\widehat{\chi}^i$ gives

$$(a_1\chi^j, \widehat{\chi}^i)_Y = \int a_{kj}\frac{\partial \widehat{\chi}^i}{\partial y_k}\, dy.$$

Therefore (3.2.20) is equivalent to saying that

(3.2.21) $$(A_1^*\widehat{\chi}^i, \chi^j)_Y = (A_1\chi^j, \widehat{\chi}^i)_Y$$

which is true. The proof of Theorem 3.1 (and of Remark 3.4) is completed.

REMARK 3.5. It easily follows from the equivalence of (3.2.9) and (3.1.6) that **"the homogenized of the adjoint equals the adjoint of the homogenized operator."**

3.3. A remark on the use of the "adjoint expansion". We consider again (1.2.13) with

(3.3.1) $$V = H_0^1(\mathcal{O}) \text{ (Dirichlet's problem}^{14}).$$

In this case, the subspace

(3.3.2) $$H_{comp}^1(\mathcal{O}) = \{v \mid v \in H^1(\mathcal{O}), v \text{ with compact support in } \mathcal{O}\}$$

is **dense** in V and (1.2.13) is equivalent with

(3.3.3) $$(u_\varepsilon, A^{\varepsilon^*}v) = (f, v) \,\forall v \in H_{comp}^1(\mathcal{O}).$$

We can of course in (3.3.3) take a variable family v_ε of test functions:

(3.3.4) $$(u_\varepsilon, A^{\varepsilon^*}v_\varepsilon) = (f, v_\varepsilon), \ v_\varepsilon \in H_{comp}^1(\mathcal{O}).$$

We now choose v_ε in a particular manner, using the **"adjoint expansion."**[15] We shall verify below that, given $v \in C_0^\infty(\mathcal{O})$, one can construct (**by using the method of asymptotic expansions**) functions v_ε such that

(3.3.5) $$v_\varepsilon \in H_{comp}^1(\mathcal{O}), \ v_\varepsilon \to v \text{ in } L^2(\mathcal{O}) \text{ weakly (for instance)},$$

$$A^{\varepsilon^*}v_\varepsilon \to \mathcal{A}^*v \text{ in } H^{-1}(\mathcal{O}) \textbf{ strongly}.$$

Using this choice of v_ε in (3.3.4) gives in the limit, since $u_\varepsilon \to u$ in $H_0^1(\mathcal{O})$ weakly,

$$(u, \mathcal{A}^*v) = (f, v) \,\forall v \in C_0^\infty(\mathcal{O})$$

i.e. u satisfies (3.1.2) for V given by (3.2.12).

3.3.1. *Construction of v_ε.* We use the notations of Section 2. We look for v_ε in the form

(3.3.6) $$v_\varepsilon = v + \varepsilon_1, \ v \in C_0^\infty(\mathcal{O}), \ v_1 = v(x, y), \ Y \text{ periodic in } y;$$

then

(3.3.7) $$\begin{aligned} A^{\varepsilon^*}v_\varepsilon &= (\varepsilon^{-2}A_1^* + \varepsilon^{-1}A_2^* + \varepsilon^0 A_3^*)(v + \varepsilon v_1) \\ &= \varepsilon^{-1}(A_1^*v_1 + A_2^*v) + \varepsilon^0(A_2^*v_1 + A_3^*v) + \varepsilon A_3^*v_1. \end{aligned}$$

We choose v_1 such that

$$A_1^*v_1 + A_2^*v = 0$$

---

[14]Cf. Remark 3.6 below for the case of other boundary conditions.

[15]We assume the coefficients a_{ij} to be smooth enough. The weakest hypotheses are given by the method of Section 3.2.

and we can take, using (3.2.17),

$$(3.3.8) \qquad v_1 = -\widehat{\chi}^j \frac{\partial v}{\partial x_j}.$$

This choice gives $v_\varepsilon \in H^1_{comp}(\mathcal{O}) \;\forall \varepsilon$. Then (3.3.7) reduces to

$$(3.3.9) \qquad A^{\varepsilon^*} v_\varepsilon = A_2^* v_1 + A_3^* v + \varepsilon A_3^* v_1$$

$$= -\left(a_{ij}^* - a_{ik}^* \frac{\partial \widehat{\chi}^j}{\partial y_k} \right) \frac{\partial^2 v}{\partial x_i \partial x_j} + \frac{\partial}{\partial y_i} (a_{ik}^* \widehat{\chi}^j) \frac{\partial^2 v}{\partial x_j \partial x_k}$$

$$+ \varepsilon a_{ij} \chi^k \frac{\partial^3 v}{\partial x_i \partial x_j \partial x_k}.$$

One easily checks that one has

$$(3.3.10) \quad A^{\varepsilon^*} v_\varepsilon \to -\frac{1}{|Y|} \int \left(a_{ij}^* - a_{ik}^* \frac{\partial \widehat{\chi}^j}{\partial y_k} \right) dy \, \frac{\partial^2 v}{\partial x_i \partial x_j} = \mathcal{A}^* v \text{ in } L^2(\mathcal{O}) \text{ \textbf{weakly}};$$

since the injection of $H^1_0(\mathcal{O}) \to L^2(\mathcal{O})$ is compact, (3.3.10) implies that $A^{\varepsilon^*} v_\varepsilon \to \mathcal{A}^* v$ in $H^{-1}(\mathcal{O})$ **strongly** and the result follows.

REMARK 3.6. If $V \neq H^1_0(\mathcal{O})$, a similar method leads to difficulties in dealing with boundary conditions of Neumann's type to impose (on some part of the boundary) on v_ε. The method of Section 3.2 is then simpler.

REMARK 3.7. In situations where the identity mapping $V \to H$ is not **compact** (we shall meet such a situation in Section 11) then the weak convergence in $L^2(\mathcal{O})$ does not suffice — **and it is then necessary to take higher order terms in the "expansion" (3.3.6)**.

REMARK 3.8. The idea of taking variable test functions constructed in a special manner is useful in several contexts.

3.4. Comparison results. We are going to show:

THEOREM 3.9. *If we assume (1.1.1) and if q_{ij} are the homogenized coefficients then one has*

$$(3.4.1) \qquad q_{ij} \xi_i \xi_j \geq \alpha |\xi|^2,$$

with the same α in (3.3.6) and in (1.1.1).

This inequality can be directly verified, but we are going to show it is a particular case of a general "comparison theorem," due to L. Tartar.

THEOREM 3.10. *Let us consider a second family of operators*

$$(3.4.2) \qquad B^\varepsilon = -\frac{\partial}{\partial x_i} \left(b_{ij}(x/\varepsilon) \frac{\partial}{\partial x_j} \right).$$

We denote by $[a_{ij}(y)]$, $[b_{ij}(y)]$, ..., the $n \times n$ matrices with coefficients $a_{ij}(y), b_{ij}(y), \ldots,$ we suppose that

$$(3.4.3) \qquad [b_{ij}]^* = [b_{ij}]^{16}$$

$$(3.4.4) \qquad [b_{ij}] \leq [a_{ij}].^{17}$$

---

[16]We do not assume that $[a_{ij}]^* = [a_{ij}]$.

Then if $[\mathcal{A}]$, $[\mathcal{B}]$ *denote the homogenized matrices, one has*

(3.4.5) $$[\mathcal{B}] \leq [\mathcal{A}].$$

REMARK 3.11. If we take $[b_{ij}] = \alpha$ (identity matrix) and if the a_{ij}'s satisfy (1.1.1), we are in the situation of Theorem 3.10, and since of course $[\mathcal{B}] = \alpha$ (identity), (3.4.1) follows.

REMARK 3.12. We are going to give a proof of Theorem 3.10 which is somewhat longer than it could be but which can be extended to the "generalized" convergence in the sense of de Giorgi and Spagnolo [**30**], cf. L. Tartar [**116**]. It can also be applied to situations met in the subsequent sections.

PROOF OF THEOREM 3.10. Given $\lambda \in \mathbb{R}^n$, one can find a sequence of functions u_ε such that

(3.4.6) $u_\varepsilon \to u$ in $H^1(\mathcal{O})$ weakly,

$\operatorname{grad} u_\varepsilon \to \lambda$ in $(L^2(\mathcal{O}))^n$ weakly,

$[a_{ij}^\varepsilon] \operatorname{grad} u_\varepsilon \to [\mathcal{A}]\lambda$ in $(L^2(\mathcal{O}))^n$ weakly,

$\operatorname{div}[a_{ij}^\varepsilon] \operatorname{grad} u_\varepsilon$ converges in H^{-1} strongly,

and such that

(3.4.7) $[a_{ij}^\varepsilon] \operatorname{grad} u_\varepsilon \cdot \operatorname{grad} u_\varepsilon \to [\mathcal{A}]\lambda \cdot \lambda$ **"vaguely"**

i.e.,

(3.4.8) $$\int [a_{ij}^\varepsilon] \operatorname{grad} u_\varepsilon \cdot \operatorname{grad} u_\varepsilon \phi \, dx \to \int [\mathcal{A}]\lambda \cdot \lambda \phi \, dx, \ \forall \phi \in C_0^\infty(\mathcal{O}).$$

This will be verified below.

We can also find a sequence v_ε with similar properties but with a_{ij} (respectively $\mathcal{A}$) replaced by b_{ij} (respectively $\mathcal{B}$). We have also (this is verified below)

(3.4.9) $[b_{ij}^\varepsilon] \operatorname{grad} v_\varepsilon \cdot \operatorname{grad} u_\varepsilon \to [\mathcal{B}]\lambda \cdot \lambda$ "vaguely."

Assuming these properties to hold true, we observe that

$$X_\varepsilon = [b_{ij}^\varepsilon](\operatorname{grad} v_\varepsilon - \operatorname{grad} u_\varepsilon) \cdot (\operatorname{grad} v_\varepsilon - \operatorname{grad} u_\varepsilon) \geq 0.$$

But

$$X_\varepsilon = [b_{ij}^\varepsilon] \operatorname{grad} v_\varepsilon \cdot \operatorname{grad} v_\varepsilon - [b_{ij}^\varepsilon] \operatorname{grad} v_\varepsilon \cdot \operatorname{grad} u_\varepsilon$$
$$- [b_{ij}^\varepsilon] \operatorname{grad} u_\varepsilon \cdot \operatorname{grad} v_\varepsilon + [b_{ij}^\varepsilon] \operatorname{grad} u_\varepsilon \cdot \operatorname{grad} u_\varepsilon.$$

By virtue of the symmetry of $[b_{ij}]$, one can write

$$X_\varepsilon = [b_{ij}^\varepsilon] \operatorname{grad} v_\varepsilon \cdot \operatorname{grad} v_\varepsilon - 2[b_{ij}^\varepsilon] \operatorname{grad} v_\varepsilon \cdot \operatorname{grad} u_\varepsilon$$
$$+ [b_{ij}^\varepsilon] \operatorname{grad} u_\varepsilon \cdot \operatorname{grad} u_\varepsilon$$

and by virtue of (3.4.4), we have

(3.4.10) $$0 \leq X_\varepsilon \leq [b_{ij}^\varepsilon] \operatorname{grad} v_\varepsilon \cdot \operatorname{grad} v_\varepsilon - 2[b_{ij}^\varepsilon] \operatorname{grad} v_\varepsilon \cdot \operatorname{grad} u_\varepsilon$$
$$+ [a_{ij}^\varepsilon] \operatorname{grad} u_\varepsilon \cdot \operatorname{grad} u_\varepsilon.$$

---

[17] i.e., the matrix $M = [a_{ij}] - [b_{ij}]$ is ≥ 0, i.e. $M\lambda \cdot \lambda \geq 0$, $\lambda \in \mathbb{R}^n$.

Since positivity is conserved in the vague topology, it follows from (3.4.7), (3.4.9) and (3.4.10) that

$$0 \leq -[\mathcal{B}]\lambda \cdot \lambda + [\mathcal{A}]\lambda \cdot \lambda \,\, \forall \lambda, \text{ i.e. (3.3.10).}$$

It remains to verify (3.4.6), (3.4.7), (3.4.9).

Given $\lambda \in \mathbb{R}^n$, we define

$$(3.4.11) \qquad u_\varepsilon(x) = \varepsilon u(x/\varepsilon),$$

$$u(y) = \lambda_j(y_j - \chi^j), \,\, \chi^j \text{ defined as in Section 2.}$$

We observe that $A^\varepsilon u_\varepsilon = 0$ and (3.4.7) follows easily.

Let us prove more than (3.4.7). Let w_ε be given in $H^1(\mathcal{O})$ such that $w_\varepsilon \to w$ in $H^1(\mathcal{O})$ weakly. Then

$$\int_{\mathcal{O}} ([a_{ij}^\varepsilon] \operatorname{grad} u_\varepsilon \cdot \operatorname{grad} w_\varepsilon)\phi \, dx = -\int_{\mathcal{O}} \frac{\partial \phi}{\partial x_i} \left(a_{ij}^\varepsilon \frac{\partial u_\varepsilon}{\partial x_j} \right) w_\varepsilon \, dx - \int_{\mathcal{O}} \phi (A^\varepsilon u_\varepsilon) w_\varepsilon \, dx$$

$$= -\int_{\mathcal{O}} \frac{\partial \phi}{\partial x_i} \left(a_{ij}^\varepsilon \frac{\partial u_\varepsilon}{\partial x_j} \right) w_\varepsilon \, dx$$

and by virtue of (3.4.6), the integral converges to

$$-\int_{\mathcal{O}} \frac{\partial \phi}{\partial x_i} ([\mathcal{A}]\lambda)_i w \, dx = \int_{\mathcal{O}} ([\mathcal{A}]\lambda \cdot \operatorname{grad} w)\phi \, dx$$

so that

$$(3.4.12) \qquad [a_{ij}^\varepsilon] \operatorname{grad} u_\varepsilon \cdot \operatorname{grad} w_\varepsilon \to [\mathcal{A}]\lambda \cdot \operatorname{grad} w \text{ "}\textbf{vaguely.}\text{"}$$

Taking $w_\varepsilon = u_\varepsilon$ gives (3.4.7).

In the same manner

$$[b_{ij}^\varepsilon] \operatorname{grad} v_\varepsilon \cdot \operatorname{grad} w_\varepsilon \to [\mathcal{B}]\lambda \cdot \operatorname{grad} w \text{ "vaguely"}$$

and taking $w_\varepsilon = u_\varepsilon$ gives (3.4.7). The proof is completed. ∎

4. L^p estimates

4.1. Estimates for the Dirichlet problem. We are going to show in this section that one can slightly improve the estimate (3.2.1) for the Dirichlet problem (the case of other boundary conditions is considered in Section 4.4 below). More precisely we shall prove the following.

THEOREM 4.1. *We assume that $\mathcal{O}$ is bounded with a smooth boundary. Let $u_\varepsilon \in H_0^1(\mathcal{O})$ be the solution of*

$$(4.1.1) \qquad a^\varepsilon(u_\varepsilon, v) = (f, v), \,\, \forall v \in H_0^1(\mathcal{O}),$$

under the hypothesis of Theorem 3.1. Then there exists a number $p > 2$ (and independent of ε) such that if

$$(4.1.2) \qquad f = f_0 + \frac{\partial f_i}{\partial x_i}, \,\, f_0, f_1, \ldots, f_n \in L^p(\mathcal{O})$$

then $\frac{\partial u_\varepsilon}{\partial x_i} \in L^p(\mathcal{O})$ and

$$(4.1.3) \qquad \sum \left\| \frac{\partial u_\varepsilon}{\partial x_i} \right\|_{L^p(\mathcal{O})} \leq c \left[\|f_0\|_{L^p(\mathcal{O})} + \sum \|f_i\|_{L^p(\mathcal{O})} \right].$$

The number p can be (roughly) estimated in terms of the a_{ij}'s and the dimension (cf. Section 4.3).

In (4.1.3) the constant c does not depend on ε.

REMARK 4.2. The estimate (4.1.3) is indeed an improvement with respect to (3.2.1); more importantly this estimate will be used as a **tool** to prove other results.

Actually Theorem 4.1 is a particular case of a more general result due to Meyers [**81**], which is relative to general second order elliptic operators **with L^∞ coefficients**.

In order to present this result, it is necessary to introduce classical Sobolev spaces $W^{1,p}(\mathcal{O})$ by taking some care in the precise definition of the norm of these spaces.

4.1.1. *Spaces* $W^{1,p}(\mathcal{O}), W_0^{1,p}(\mathcal{O})$.

$$(4.1.4) \qquad W^{1,p}(\mathcal{O}) = \left\{ v \ \middle| \ v, \frac{\partial v}{\partial x_i} \in L^p(\mathcal{O}) \right\};$$

$W^{1,p}(\mathcal{O})$ is a Banach space with the norm

$$(4.1.5) \qquad \left(\|v\|_{L^p(\mathcal{O})}^p + \sum_{i=1}^n \left\| \frac{\partial v}{\partial x_i} \right\|_{L^p(\mathcal{O})}^p \right)^{1/p}.$$

We introduce next

$$(4.1.6) \qquad W_0^{1,p}(\mathcal{O}) = \text{ closure of } C_0^\infty(\mathcal{O}) \text{ in } W^{1,p}(\mathcal{O}).$$

The norm we put on $W_0^{1,p}(\mathcal{O})$ (which is equivalent to (4.1.5)) is as follows: we set

(4.1.7)

$$\nabla \phi = \left\{ \frac{\partial \phi}{\partial x_1}, \dots, \frac{\partial \phi}{\partial x_n} \right\},$$

$$\|f\|_{(L^p(\mathcal{O}))^n} = \left(\int_\mathcal{O} |f(x)|^p \, dx \right)^{1/p} \quad \text{where } |f(x)| = \text{Euclidean norm of } f(x) \text{ in } \mathbb{R}^n,$$

and we **define**

$$(4.1.8) \qquad \|\phi\|_{W_0^{1,p}(\mathcal{O})} = \|\nabla \phi\|_{(L^p(\mathcal{O}))^n}.$$

By virtue of Poincaré's inequality, (4.1.8) is equivalent to (4.1.5) on $W_0^{1,p}(\mathcal{O})$.

4.1.2. *Space* $W^{-1,p}(\mathcal{O})$. We take $1 < p < \infty$ and we define p' by $1/p + 1/p' = 1$. We define

$$(4.1.9) \qquad W^{-1,p}(\mathcal{O}) = (W_0^{1,p'}(\mathcal{O}))' = \text{ dual space of } W_0^{1,p'} \text{ where}$$

$$L^2(\mathcal{O}) \text{ is identified with its dual.}$$

We observe that the mapping

$$(4.1.10) \qquad \phi \to \text{div } \phi \text{ from } (L^p(\mathcal{O}))^n \to W^{-1,p}(\mathcal{O})$$

is **onto**. We provide then $W^{-1,p}(\mathcal{O})$ with the "quotient" norm associated to (4.1.10), namely

$$(4.1.11) \qquad \|f\|_{W^{-1,p}}(\mathcal{O}) = \inf \|g\|_{(L^p(\mathcal{O}))^n}, \qquad \text{div } g = f,$$

which is of course one way of defining the norm on $W^{-1,p}(\mathcal{O})$.

4.1.3. *The operator A.* We consider now a family of functions a_{ij} such that

(4.1.12) $a_{ij} \in L^\infty(\mathcal{O})$,

$$a_{ij}(x)\xi_i\xi_j \geq \alpha\xi_i\xi_i \text{ a.e. in } \mathcal{O}, \ \alpha > 0,$$

and we define

(4.1.13) $$A = -\frac{\partial}{\partial x_i}\left(a_{ij}(x)\frac{\partial}{\partial x_j}\right).$$

We consider the Dirichlet's problem

(4.1.14) $Au = f, \quad u \in H_0^1(\mathcal{O}), \quad f \in H^{-1}(\mathcal{O}).$

We shall prove

THEOREM 4.3. *We assume that (4.1.12) holds true and that $\mathcal{O}$ is bounded with a smooth boundary. There exists a number $p > 2$ (which depends on α, on the L^∞ norm of the a_{ij}'s and on $\mathcal{O}$ and on the dimension) such taht, if $f \in W^{-1,p}(\mathcal{O})$ then the solution u of (4.1.14) belongs to $W_0^{1,p}(\mathcal{O})$ and satisfies*

(4.1.15) $\|u\|_{W_0^{1,p}(\mathcal{O})} \leq C\|f\|_{W^{-1,p}(\mathcal{O})}$

(where C depends on the same quantities as p does).

REMARK 4.4. Theorem 4.3 contains Theorem 4.1.1 as a particular case.

The proof of Theorem 4.3 is presented in two steps: in the first step (Section 4.2) we make an "algebraic" reduction of (4.1.14) to a special form and the theorem itself is proved in Section 4.3.

4.2. Reduction of the equation. Let us introduce the following notations: we denote by $[a_{ij}(x)]$ the matrix wtih entries the a_{ij}'s and we set

(4.2.1) $M(x) = [a_{ij}(x)].$

We denote by $\|M(x)\|$ the norm defined by

$$\|M(x)\| = \sup_{|\xi|\leq 1} |M(x)\xi|,$$

and we set

(4.2.2) $\beta = \sup_{x \in \mathcal{O}} \operatorname{ess} \|M(x)\|.$

We define $M_1(x) = \frac{1}{2}[M(x) + M^*(x)]$, $M_2(x) = \frac{1}{2}[M(x) - M^*(x)]$ and we write the identity

(4.2.3) $M = M_1 + cI + M_2 - cI,$

                     c **to be chosen later**, I = identity matrix.

We repalce (4.1.14) by

(4.2.4) $$\frac{1}{\beta+c}Au = \frac{1}{\beta+c}f.$$

We set

(4.2.5) $P_1 = \frac{1}{\beta+c}(M_1 + cI), \qquad P_2 = \frac{1}{\beta+c}(M_2 - cI)$

and we denote by A_1 and A_2 the operators corresponding to P_1 and P_2; if

(4.2.6) $P_1(x) = [a_{ij}^1(x)], \qquad P_2(x) = [a_{ij}^2(x)];$

then

$$(4.2.7) \qquad A_1 = -\frac{\partial}{\partial x_i}\left(a_{ij}^1(x)\frac{\partial}{\partial x_j}\right), \qquad A_2 = -\frac{\partial}{\partial x_i}\left(a_{ij}^2(x)\frac{\partial}{\partial x_j}\right).$$

We are going to verify that we can choose c in such a fashion that

$$(4.2.8) \qquad a_{ij}^1(x) = a_{ji}^1(x),\ a_{ij}^1(x)\xi_i\xi_j \geq \mu\xi_i\xi_i,\ \mu > 0,\ \mu < 1,$$

$$(4.2.9) \qquad \|[a_{ij}^2(x)]\| \leq \nu \text{ a.e. in } \mathcal{O},$$

$$(4.2.10) \qquad 1 - \mu + \nu < 1 \text{ (i.e. } \nu < \mu).$$

Indeed $P_1(x)^* = P_1(x)$ and

$$(P_1(x)\xi, \xi) = \frac{1}{\beta + c}[(M(x)\xi, \xi) + c|\xi|^2] \geq \frac{\alpha + c}{\beta + c}|\xi|^2$$

so that (since $\alpha < \beta$) we shall have (4.2.8) with

$$(4.2.11) \qquad \mu = \frac{\alpha + c}{\beta + c}, \qquad c \geq 0.$$

We next have

$$(\beta + c)^2|P_2(x)\xi|^2 = |M_2(x)\xi|^2 + c^2|\xi|^2 + c^2|\xi|^2 \leq (\beta^2 + c^2)|\xi|^2$$

hence (4.2.9) with

$$\nu^2 = \frac{\beta^2 + c^2}{(\beta + c)^2}.$$

It remains to choose c such that $\frac{\beta^2 + c^2}{(\beta+c)^2} < \frac{(\alpha+c)^2}{(\beta+c)^2}$, i.e.

$$(4.2.12) \qquad c > \frac{\beta^2 - \alpha^2}{2\alpha}.$$

Summarizing: **the equation (4.1.14) is equivalent to**

$$(4.2.13) \qquad (A_1 + A_2)u = \tilde{f}, \qquad \tilde{f} = \frac{1}{\beta + c}f,$$

where c is chosen to satisfy (4.2.12) and where A_1, A_2 satisfy (4.2.8), (4.2.9), (4.2.10).

4.3. Proof of Theorem 4.3. We replace (4.2.13) by

$$(4.3.1) \qquad -\Delta u + (\Delta + A_1 + A_2)u = \tilde{f}$$

and we use the fact that (cf. Agmon-Douglis-Nirenberg [2], J. Peetre [92], J. L. Lions-E. Magenes [70])
(4.3.2)

$\forall p$ such that $1 < p < \infty$, $-\Delta$ is an isomorphism from $W_0^{1,p}(\mathcal{O}) = W^{-1,p}(\mathcal{O})$.

We set

$$(4.3.3) \qquad (-\Delta)^{-1} = G.$$

Then (assuming that $\tilde{f} \in W^{-1,p}(\mathcal{O})$) (4.3.1) is equivalent to

$$(4.3.4) \qquad u + G(\Delta + A_1 + A_2)u = G\tilde{f}.$$

We are going to show that there exists $p > 2$ such that

$$(4.3.5) \qquad \|G(\Delta + A_1 + A_2)\|_{\mathcal{L}(W_0^{1,p}(\mathcal{O});W_0^{1,p}(\mathcal{O}))} = k(p) < 1.$$

For such a p, $I + G(\Delta + A_1 + A_2)$ is invertible in $W_0^{1,p}(\mathcal{O})$ and the result follows. Therefore everything rests on (4.3.5). We have

(4.3.6)
$$k(p) \leq \|G\|_{\mathcal{L}(W^{-1,p}(\mathcal{O}); W_0^{1,p}(\mathcal{O}))}$$
$$\cdot \|\Delta + A_1 + A_2\|_{\mathcal{L}(W_0^{1,p}(\mathcal{O}); W^{-1,p}(\mathcal{O}))}.$$

We are going to verify that

(4.3.7)
$$\|(\Delta + A_1)\|_{\mathcal{L}(W_0^{1,p}(\mathcal{O}); W^{-1,p}(\mathcal{O}))} \leq 1 - \mu$$

and

(4.3.8)
$$\|A_2\|_{\mathcal{L}(W_0^{1,p}(\mathcal{O}); W^{-1,p}(\mathcal{O}))} \leq \nu.$$

If we admit these estimates for a moment, and if we set

(4.3.9)
$$g(p) = \|G\|_{\mathcal{L}(W^{-1,p}(\mathcal{O}); W_0^{1,p}(\mathcal{O}))}$$

then (4.3.6), (4.3.7), (4.3.8) imply

(4.3.10)
$$k(p) \leq g(p)(1 - \mu + \nu).$$

It is straightforward to verify that

(4.3.11)
$$g(2) = 1.$$

Since $1 - \mu + \nu < 1$ (cf. (4.2.10)), we shall have proved (4.3.5) if we can prove that[18]

(4.3.12)   **there exists a function $\hat{g}(p)$, continuous of $p \geq 2$,**

such that $g(p) \geq \hat{g}(p)$ and $\hat{g}(2) = g(2) = 1$.

PROOF OF (4.3.12). We take $p_0 > 2$ arbitrarily large and we take p such that $2 \leq p \leq p_0$. For $h \in (L^p(\mathcal{O}))^n$, we define $u \in W_0^{1,p}$ by

(4.3.13)
$$-\Delta u = \operatorname{div} h$$

and we consider the mapping

(4.3.14)
$$h \to \pi h = \nabla u$$

for $(L^p(\mathcal{O}))^n$ into itself. Let $\overline{\omega}(p)$ be the norm of this mapping. By virtue of the Riesz-Thorin interpolation theorem, we have

(4.3.15)
$$\overline{\omega}(p) \leq \overline{\omega}(p_0)^{1-\theta(p)} \overline{\omega}(2)^{\theta(p)} = \overline{\omega}(p_0)^{1-\theta(p)},$$
$$\frac{1}{p} = \frac{1-\theta(p)}{p_0} + \frac{\theta(p)}{2},$$

(since $\overline{\omega}(2) = 1$).

For $f \in W^{-1,p}(\mathcal{O})$, we write $f = \operatorname{div} h$; then

$$Gf = \pi h \text{ so that } \|Gf\|_{W_0^{1,p}(\mathcal{O})} = \|\nabla u\|_{W_0^{1,p}(\mathcal{O})} \leq \overline{\omega}(p_0)^{1-\theta(p)} \|h\|_{(L^p(\mathcal{O}))^n}$$

hence

$$\|Gf\|_{W_0^{1,p}(\mathcal{O})} \leq \overline{\omega}(p_0)^{1-\theta(p)} \inf_{\substack{h \\ \operatorname{div} h = f}} \|h\|_{(L^p(\mathcal{O}))^n}$$
$$= \overline{\omega}(p_0)^{1-\theta(p)} \|f\|_{W^{-1,p}(\mathcal{O})}$$

---

[18]It seems likely — but it does not appear to be proven somewhere in the literature — that the function $p \to g(p)$ itself is continuous. But (4.3.12) is sufficient for our purpose.

so that we can take in (4.3.12)

(4.3.16) $\widehat{g}(p) = \overline{\omega}(p_0)^{1-\theta(p)}.$

(For $p = 2$, $\theta(p) = 1$ and $\widehat{g}(2) = 1$).

■

PROOF OF (4.3.7). Let u be in $W_0^{1,p}(\mathcal{O})$, we have

(4.3.17) $(\Delta + A_1)u = \mathrm{div}\, g, \qquad g = \{g_i\},$

$$g_i = (\delta_i^j - a_{ij}^1(x))\frac{\partial u}{\partial x_j}(x), \qquad \delta_i^j = \text{Kronecker index}.$$

Therefore

(4.3.18) $\|(\Delta + A_1)u\|_{W^{-1,p}(\mathcal{O})} \leq \|g\|_{(L^p(\mathcal{O}))^n}.$

But $|g(x)| \leq (1 - \mu)|\nabla u(x)|$ (here $|\cdot|$ denotes the Euclidean norm in $\mathbb{R}^n$) so that

$$\|g\|_{(L^p(\mathcal{O}))^n} \leq (1 - \mu)\|\nabla u\|_{(L^p(\mathcal{O}))^n}$$
$$= (1 - \mu)\|u\|_{W_0^{1,p}(\mathcal{O})}$$

hence (4.3.7) follows.

■

PROOF OF (4.3.8). We have, for $u \in W_0^{1,p}(\mathcal{O})$,

(4.3.19) $A_2 u = -\mathrm{div}\, g,$

$$g = \{g_i\}, \qquad g_i = a_{ij}^2(x)\frac{\partial u}{\partial x_j}(x)$$

so that

$$\|A_2 u\|_{W^{-1,p}(\mathcal{O})} \leq \|g\|_{(L^p(\mathcal{O}))^n}$$

hence (4.3.8) follows since $|g(x)| \leq \nu|\nabla u(x)|$.

■

4.4. Local estimates. We consider

(4.4.1) $a_0 \in L^\infty(\mathcal{O}), \qquad a_0(x) \geq \alpha_0 > 0$

and we set, $\forall u, v \in H^1(\mathcal{O})$,

(4.4.2) $a(u,v) = \int_{\mathcal{O}} a_{ij}(x)\frac{\partial u}{\partial x_i}\frac{\partial v}{\partial x_i}\,dx + \int_{\mathcal{O}} a_0 uv\,dx.$

For $f \in V'$, we denote by u the solution of (notations of (1.2.6)):

(4.4.3) $a(u,v) = (f,v), \; \forall v \in V, \; u \in V.$

We observe that if $\phi \in C_0^\infty(\mathcal{O})$ we define $\phi f \in H^{-1}(\mathcal{O})$ by

$$(\phi f, v) = (f, \phi v), \; \forall v \in H_0^1(\mathcal{O}).$$

We have

THEOREM 4.5. *We assume that the hypothesis of Theorem 4.3 hold true[19] and that (4.4.1) takes place. Then there exists p such that*

$$(4.4.4) \qquad 2 < p \le \frac{2n}{n-2},\text{[20]}$$

so that, if $\phi \in C_0^\infty(\mathcal{O})$ is given and $\phi f \in W^{-1,p}(\mathcal{O})$, then $\phi u \in W_0^{1,p}(\mathcal{O})$ and

$$(4.4.5) \qquad \|\phi u\|_{W_0^{1,p}(\mathcal{O})} \le c \left[\|\phi f\|_{W^{-1,p}(\mathcal{O})} + \|f\|_{V'} \right],$$

where c depends only on $\mathcal{O}$, ϕ and on the a_{ij}'s.

REMARK 4.6. *If we take $\phi = 1$ on a set $\mathcal{O}'$, $\overline{\mathcal{O}}' \subseteq \mathcal{O}$, (4.4.5) gives a local estimate.*

PROOF OF THEOREM 4.5. By a simple computation, we have

$$(4.4.6) \qquad A(\phi u) = \phi f - a_{ij}\frac{\partial u}{\partial x_j}\frac{\partial \phi}{\partial x_i} - \frac{\partial}{\partial x_i}\left(a_{ij} u \frac{\partial \phi}{\partial x_j} \right);$$

we observe that $a_{ij}\frac{\partial u}{\partial x_j}\frac{\partial \phi}{\partial x_i} \in L^2(\mathcal{O}) \subset W^{-1,p}(\mathcal{O})$[21] and that $a_{ij} u \frac{\partial \phi}{\partial x_j} \in L^q(\mathcal{O})$, $\frac{1}{q} = \frac{1}{2} - \frac{1}{n}$ (Sobolev's theorem), so that, by virtue of (4.4.4),

$$\frac{\partial}{\partial x_i}\left(a_{ij} u \frac{\partial \phi}{\partial x_j} \right) \in W^{-1,p}(\mathcal{O}).$$

The result then follows from Theorem 4.3. ∎

4.5. Extensions. The above results can be extended to higher order elliptic equations or systems. Let us give one example. We consider

$$(4.5.1) \qquad A = (-1)^m D^\alpha(a_{\alpha\beta}(x)D^\beta),$$

where $|\alpha| = |\beta| = m$, and where

$$(4.5.2) \qquad a_{\alpha\beta} \in L^\infty(\mathcal{O}),$$

$$a_{\alpha\beta}\xi^\alpha\xi^\beta \ge c|\xi|^{2m} \ \forall \xi \in \mathbb{R}^n, \ c > 0.$$

We set

$$(4.5.3) \qquad D^m v = \{D^\alpha v \mid |\alpha| = m\};$$

let N = number of indices α such that $|\alpha| = m$. We define

$$(4.5.4) \qquad W^{m,p}(\mathcal{O}) = \{v \mid D^\gamma v \in L^p(\mathcal{O}), \ \forall \gamma \text{ such that } |\gamma| \le m\},$$

which is a Banach space for the norm

$$(4.5.5) \qquad \left(\sum_{|\gamma| \le m} \|D^\gamma v\|^p_{L^p(\mathcal{O})} \right)^{1/p}.$$

We then introduce $W_0^{m,p}(\mathcal{O})$ by

$$(4.5.6) \qquad W_0^{m,p}(\mathcal{O}) = \text{ closure of } C_0^\infty(\mathcal{O}) \text{ into } W^{m,p}(\mathcal{O})$$

---

[19]But no smoothness on $\Gamma = \partial\mathcal{O}$ is needed here.

[20]This upper bound becomes irrelevant if $n = 2$.

[21]This inclusion follows from the inclusion $W_0^{1,p'}(\mathcal{O}) \subset L^2(\mathcal{O})$ which, according to the Sobolev's theorem is true if $\frac{1}{p'} - \frac{1}{n} \le \frac{1}{2}$, which is equivalent to the upper bound in (4.4.4).

and we put on $W_0^{m,p}(\mathcal{O})$ the norm (equivalent to (4.5.5))

(4.5.7) $$\|v\|_{W_0^{m,p}(\mathcal{O})} = \|D^m v\|_{(L^p(\mathcal{O}))^N},$$

where

(4.5.8) $$\|f\|_{(L^p(\mathcal{O}))^N} = \left(\int_{\mathcal{O}} |f(x)|_{\mathbb{R}^n}^p \, dx \right)^{1/p}.$$

The operator $(D^m)^*$, defined by

(4.5.9) $$(D^m)^* f = (-1)^m D^\alpha f_\alpha$$

maps $(L^p(\mathcal{O}))^N$ **onto** $W^{-m,p}(\mathcal{O})$, where

(4.5.10) $$W^{-m,p}(\mathcal{O}) = \text{ dual space of } W_0^{m,p'}(\mathcal{O});$$

therefore we can provide $W^{-m,p}(\mathcal{O})$ with the norm

(4.5.11) $$\|f\|_{W^{-m,p}(\mathcal{O})} = \inf \|g\|_{(L^p(\mathcal{O}))^N}, \ (D^m)^* g = f.$$

(All the remarks are entirely parallel to those presented in Section 4.1.)

We consider now the Dirichlet's problem

(4.5.12) $$Au = f, \qquad f \in H^{-m}(\mathcal{O})^{22}, \qquad u \in H_0^m(\mathcal{O}).$$

We have

THEOREM 4.7. *We assume that (4.5.2) holds true and that $\mathcal{O}$ is bounded with a smooth boundary. Then there exists a number $p > 2$ such that if $f \in W^{-m,p}(\mathcal{O})$, then the solution u of (4.5.12) belongs to $W_0^{m,p}(\mathcal{O})$ and satisfies*

(4.5.13) $$\|u\|_{W_0^{m,p}(\mathcal{O})} \le c\|f\|_{W^{-m,p}(\mathcal{O})}.$$

PROOF. The proof is along the lines of the proof of Theorem 4.3. One replaces first (4.5.12) by

(4.5.14) $$(A_1 + A_2)u = \widetilde{f},$$

where

$$A_1 = (-1)^m D^\alpha (a_{\alpha\beta}^1(x) D^\beta), \qquad |\alpha| = |\beta| = m,$$
$$A_2 = (-1)^m D^\alpha (a_{\alpha\beta}^2(x) D^\beta), \qquad |\alpha| = |\beta| = m,$$

(4.5.15) $$a_{\alpha\beta}^1(x) = a_{\beta\alpha}^1 x) \ \forall \alpha, \beta,$$
$$a_{\alpha\beta}^1(x)\xi^\alpha \xi^\beta \ge \mu |\xi|^{2m}, \qquad \mu < 1,$$

(4.5.16) $$\|[a_{\alpha\beta}^2(x)]\|_{\mathcal{L}(\mathbb{R}^N)} \le \nu,$$

(4.5.17) $$1 - \mu + \nu < 1 \ (\text{i.e. } \nu < \mu).$$

We introduce next

(4.5.18) $$E = (-1)^m D^\alpha D^\alpha, \qquad |\alpha| = m,$$

and we use the fact (cf. Agmon, Douglis, Nirenberg [2]; the regularity hypothesis on Γ is used here) that E is an isomorphism from $W_0^{m,p}(\mathcal{O})$ onto $W^{-m,p}(\mathcal{O})$, $\forall p$ finite.

---

[22] $H^{-m}(\mathcal{O}) = W^{-m,2}(\mathcal{O})$, $H_0^m(\mathcal{O}) = W_0^{m,2}(\mathcal{O})$.

We replace (4.5.14) by

(4.5.19) $$(E + (A_1 - E) + A_2)u = \tilde{f}$$

or, equivalently, if

(4.5.20) $$E^{-1} = G,$$

(4.5.21) $$[I + G(A_1 - E + A_2)]u = G\tilde{f}.$$

We then show, by an argument similar to that of Section 4.3, that one can find $p > 2$ such that

(4.5.22) $$\|G(A_1 - E + A_2)\|_{\mathcal{L}(W_0^{m,p}(\mathcal{O}); W^{-m,p}(\mathcal{O}))} < 1,$$

and the Theorem follows. ∎

5. Correctors

5.1. Orientation. We now return to the notations of Section 3.

As we already said, $u_\varepsilon - u$ does not converge to 0 in V **strongly** (but only **weakly**); the results just obtained show that $\frac{\partial}{\partial x_i}(u_\varepsilon - u) \to 0$ in $L^p(\mathcal{O})$ (or in $L^p_{\text{loc}}(\mathcal{O})$[23]) **weakly** for some $p > 2$, but this does not imply strong convergence in V of $u_\varepsilon - u$ (and this is actually **false**). But we can introduce correctors, say θ_ε, **such that**

(5.1.1) $$u_\varepsilon - u - \theta_\varepsilon \to 0 \text{ in } V \text{ strongly}.$$

Of course a remark like (5.1.1) can be entirely trivial, if one takes $\theta_\varepsilon = u_\varepsilon - u$!! But we are going to show that **one can achieve (5.1.1) with a corrector θ_ε which is of order ε in the $L^2(\mathcal{O})$ norm.**

5.2. Structure of the first corrector — Statement of theorem. The first idea which comes to mind is to take

(5.2.1) $$\theta_\varepsilon = \varepsilon u_1, \qquad u_1 = u_1(x, x/\varepsilon),$$

$$u_1 \text{ defined in } (2.2.18), u_1 = -\chi^j \frac{\partial u}{\partial x_j}.$$

Assuming that $\chi^j \in W^{1,\infty}(Y)$ and that $u \in H^2(\mathcal{O})$, θ_ε belongs to $H^1(\mathcal{O})$ but in general does not belong to V.

Let us first remark that one can always — by a small transformation — take a corrector in V. We introduce to this effect cut-off functions m_ε; they have the following properties

$$M_\varepsilon \in \mathcal{D}(\mathcal{O}),[24]$$

(5.2.2)

$m_\varepsilon(x) = 0$ if $d(x, \Gamma) = $ (distance from x to Γ) $\leq \varepsilon$,

$m_\varepsilon(x) = 1$ if $d(x, \Gamma) \geq 2\varepsilon$,

$\varepsilon^{|\gamma|} |D^\gamma m_\varepsilon(x)| \leq c_{\gamma'} \; \forall \gamma$, where c_γ depends on γ but does not depend on ε.

Such functions exist, provided Γ is smooth enough.

---

[23] $L^p_{\text{loc}}(\mathcal{O}) = $ space of functions which are **locally** L^p in $\mathcal{O}$.

[24] Actually it would be sufficient here to take $m_\varepsilon \in C^1(\mathcal{O})$ and to take the γ's with $|\gamma| = 1$.

We then define

$$(5.2.3) \qquad \theta_\varepsilon = \varepsilon m_\varepsilon u_1 \; (= -\varepsilon m_\varepsilon \chi^j \left(\frac{x}{\varepsilon}\right) \frac{\partial u}{\partial x_j}).$$

We shall prove

THEOREM 5.1. *We assume that the hypothesis of Theorem 3.1 hold true. Moreover we assume that*

$$(5.2.4) \qquad f \in L^2(\mathcal{O}),$$

$$(5.2.5) \qquad \chi^j \text{ defined by (2.2.17) belongs to } W^{1,\infty}(Y) \; \forall j,$$

and the boundary Γ of $\mathcal{O}$ is smooth enough. Then if θ_ε is defined by (5.2.3) we have

$$(5.2.6) \qquad z_\varepsilon = u_\varepsilon - u - \theta_\varepsilon \to 0 \text{ in } V \text{ \textbf{strongly}}.$$

The function θ_ε is called a **first order corrector**.
We shall prove this result in Section 5.3 below.

REMARK 5.2. It follows from (5.2.4) and from the smoothness of Γ that the solution u satisfies

$$(5.2.7) \qquad u \in H^2(\mathcal{O}).$$

Therefore $\theta_\varepsilon \in V$ and so does Z_ε.

REMARK 5.3. If we do not introduce the cut-off functions m_ε, and if we consider

$$(5.2.8) \qquad \zeta_\varepsilon = u_\varepsilon - u - \varepsilon u_1$$

then we have

$$(5.2.9) \qquad \zeta_\varepsilon \to 0 \text{ in } H^1(\mathcal{O}) \text{ strongly}.$$

(Therefore if $V = H^1(\mathcal{O})$ it is clear that we should take $m_\varepsilon = 1$, although we can **also** introduce m_ε even in that case.)
Indeed

$$z_\varepsilon - \zeta_\varepsilon = \varepsilon(1 - m_\varepsilon)u_1,$$

we have to show that

$$\frac{\partial}{\partial x_i}(\varepsilon(1 - m_\varepsilon)u_1) \to 0 \text{ in } L^2(\mathcal{O}) \text{ strongly}.$$

But this equals

$$-\varepsilon \frac{\partial m_\varepsilon}{\partial x_i} u_1 + \varepsilon(1 - m_\varepsilon)\frac{\partial(u_1)}{\partial x_i} = -\varepsilon\left(\frac{\partial m_\varepsilon}{\partial x_i}\right) u_1 + (1 - m_\varepsilon)\frac{\partial u_1}{\partial y_i} + \varepsilon(1 - m_\varepsilon)\frac{\partial u_1}{\partial x_i}.$$

By virtue of the hypothesis (5.2.5), (5.2.2) it follows from the Lebesgue theorem that $\varepsilon\left(\frac{\partial m_\varepsilon}{\partial x_i}\right) u_1$ and $(1 - m_\varepsilon)\frac{\partial u_1}{\partial y_i} \to 0$ in $L^2(\mathcal{O})$ strongly and since $u \in H^2(\mathcal{O})$, $\left| \varepsilon(1 - m_\varepsilon)\frac{\partial u_1}{\partial x_i} \right|_{L^2(\mathcal{O})} = O(\varepsilon).$

REMARK 5.4. The corrector θ_ε (which is of course not unique) is a **smoothing operator**: it corrects for the rapid oscillations in the derivatives of $u_\varepsilon - u$, and the boundary does not play any role; θ_ε is completely different from a boundary layer corrector.

REMARK 5.5. We can "change the scaling" in m_ε (there are really **two** small parameters, that we took equal in Theorem 5.1.1). Let λ be a positive number. We can choose m_ε such that

$$\varepsilon^{\lambda|\gamma|}|D^\gamma m_\varepsilon(x)| \leq c_\gamma \ \forall \gamma,$$

(5.2.10)
$$m_\varepsilon(x) = 0 \text{ if } d(x, \Gamma) \leq \varepsilon^\lambda,$$

$$m_\varepsilon(x) = 1 \text{ if } d(x, \Gamma) \geq 2\varepsilon^\lambda$$

without changing Theorem 5.1. A convenient choice of λ can be useful in other situations.

REMARK 5.6. As we said in Section 5.1, one has $|\theta_\varepsilon|_{L^2(\mathcal{O})} \leq C\varepsilon$.

REMARK 5.7. This remark completes Remark 2.6. With the notations of Section 3, $\xi_i^\varepsilon = a_{ij}^\varepsilon \frac{\partial u_\varepsilon}{\partial x_j} \rightarrow q_{ij}\frac{\partial u}{\partial x_j}$ in $L^2(\mathcal{O})$ weakly. By virtue of Theorem 5.1, ξ_i^ε will hav the same limit that $a_{ij}^\varepsilon \frac{\partial}{\partial x_j}(u + \theta_\varepsilon)$, and this converges to $q_{ij}\frac{\partial u}{\partial x_j}$.

5.3. Proof of Theorem 5.1. We consider two cases; we begin with the **symmetric case**

(5.3.1) $a_{ij} = a_{ji}, \ \forall i, j$

where the proof is slightly simpler, and we consider next the **nonsymmetric case**.

We write $a^\varepsilon(\phi)$ for $a^\varepsilon(\phi, \phi)$. By virtue of (4.1.15), it will be sufficient to prove that

(5.3.2) $a^\varepsilon(z_\varepsilon) \rightarrow 0$

where we do not consider zero order terms.

But, **using the symmetry** (5.3.1), we can write

$$a^\varepsilon(z_\varepsilon) = a^\varepsilon(u_\varepsilon, z_\varepsilon) - a^\varepsilon(u + \theta_\varepsilon, u_\varepsilon) + a^\varepsilon(u + \theta_\varepsilon)$$

$$= a^\varepsilon(u_\varepsilon, z_\varepsilon) - a^\varepsilon(u_\varepsilon, u + \theta_\varepsilon) + a^\varepsilon(u + \theta_\varepsilon)$$

and since $z_\varepsilon, u + \theta_\varepsilon \in V$, we have

(5.3.3) $a^\varepsilon(z_\varepsilon) = (f, u_\varepsilon - 2(u + \theta_\varepsilon)) + a^\varepsilon(u + \theta_\varepsilon).$

Let us admit for a moment that

(5.3.4) $a^\varepsilon(u + \theta_\varepsilon) \rightarrow \mathcal{A}(u, u).$

Then (5.3.3) gives: $a^\varepsilon(z_\varepsilon) \rightarrow -(f, u) + \mathcal{A}(u, u) = 0$, and (5.3.2) is proven.

PROOF OF (5.3.4). We have

$$a^\varepsilon(u + \theta_\varepsilon) = \int_\mathcal{O} a_{k\ell}^\varepsilon \frac{\partial(u + \theta_\varepsilon)}{\partial x_k} \frac{\partial(u + \theta_\varepsilon)}{\partial x_\ell} \, dx.$$

But

$$\frac{\partial\theta_\varepsilon}{\partial x_k} = -\lambda_{\varepsilon k} + r_{\varepsilon k},$$

$$\lambda_{\varepsilon k} = -\frac{\partial\chi^i}{\partial y_k}\frac{\partial u}{\partial x_i}m_\varepsilon,$$

$$r_{\varepsilon k} = -\varepsilon\chi^i\frac{\partial^2 u}{\partial x_i\partial x_k}m_\varepsilon = -\varepsilon\frac{\partial m_\varepsilon}{\partial x_k}\chi^i\frac{\partial u}{\partial x_i}.$$

Using the properties of m_ε, u, χ^i and Lebesgue's theorem, one sees that $r_{\varepsilon k} \to 0$ in $L^2(\mathcal{O})$ strongly, so that $a^\varepsilon(u + \theta_\varepsilon)$ **has the same limit as**

$$X_\varepsilon = \int_{\mathcal{O}} a_{k\ell}^\varepsilon \left(\frac{\partial u}{\partial x_k} - \lambda_{\varepsilon k} \right) \left(\frac{\partial u}{\partial x_\ell} - \lambda_{\varepsilon \ell} \right) \, dx.$$

But

$$X_\varepsilon = \int_{\mathcal{O}} a_{k\ell}^\varepsilon \frac{\partial u}{\partial x_k} \frac{\partial u}{\partial x_\ell} \, dx - \int_{\mathcal{O}} \left(a_{k\ell} \frac{\partial \chi^i}{\partial y_k} \right)^\varepsilon \frac{\partial u}{\partial x_i} \frac{\partial u}{\partial x_\ell} m_\varepsilon \, dx$$
$$- \int_{\mathcal{O}} \left(a_{k\ell} \frac{\partial \chi^j}{\partial y_\ell} \right)^\varepsilon \frac{\partial u}{\partial x_k} \frac{\partial u}{\partial x_j} m_\varepsilon \, dx$$
$$+ \int_{\mathcal{O}} \left(a_{k\ell} \frac{\partial \chi^i}{\partial y_k} \frac{\partial \chi^j}{\partial y_\ell} \right)^\varepsilon \frac{\partial u}{\partial x_i} \frac{\partial u}{\partial x_j} m_\varepsilon^2 \, dx,$$

where we have set in general $\psi^\varepsilon(x) = \psi(x/\varepsilon)$. Therefore

$$(5.3.5) \qquad \lim_{\varepsilon \to 0} X_\varepsilon = \mathcal{M} \left(a_{ij} - a_{i\ell} \frac{\partial \chi^i}{\partial y_\ell} \right) \int_{\mathcal{O}} \frac{\partial u}{\partial x_i} \frac{\partial u}{\partial x_j} \, dx$$
$$+ \mathcal{M} \left(a_{k\ell} \frac{\partial \chi^j}{\partial y_\ell} \frac{\partial \chi^i}{\partial y_k} - a_{kj} \frac{\partial \chi^i}{\partial y_k} \right) \int_{\mathcal{O}} \frac{\partial u}{\partial x_i} \frac{\partial u}{\partial x_j} \, dx.$$

By virtue of the definition of χ^j, we have

$$\mathcal{M} \left(a_{k\ell} \frac{\partial \chi^j}{\partial y_\ell} \frac{\partial \chi^i}{\partial y_k} - a_{kj} \frac{\partial \chi^i}{\partial y_k} \right) = 0$$

and $(5.3.5)$ gives the desired result. $\qquad\qquad\qquad\qquad\qquad\blacksquare$

REMARK 5.8. **Let us emphasize that the symmetry was not used in (5.3.4).**

We now pass to the nonsymmetric case. According to Remark 5.3, it amounts to the same thing to work with ζ_ε. We therefore introduce

$$a^\varepsilon(\zeta_\varepsilon) = a^\varepsilon(u_\varepsilon) - a^\varepsilon(u + \varepsilon u_1, u_\varepsilon) - a^\varepsilon(u_\varepsilon, u + \varepsilon u_1)$$
$$+ a^\varepsilon(u + \varepsilon u_1);$$

we have $a^\varepsilon(u_\varepsilon) = (f, u_\varepsilon)$ so that we already know that

$$a^\varepsilon(u_\varepsilon) + a^\varepsilon(u + \varepsilon u_1) \to (f, u) + \mathcal{A}(u, u) = 2\mathcal{A}(u, u).$$

Therefore we shall have the desired result if we show that

$$(5.3.6) \qquad\qquad\qquad a^\varepsilon(u + \varepsilon u_1, u_\varepsilon) \to \mathcal{A}(u, u),$$

$$(5.3.7) \qquad\qquad\qquad a^\varepsilon(u_\varepsilon, u + \varepsilon u_1) \to \mathcal{A}(u, u).$$

Let ϕ be given in $C_0^\infty(\mathcal{O})$, with $\phi = 1$ except on a set E of measure $|E|$ which tends to zero. We set

$$a_\phi^\varepsilon(u, v) = \int_{\mathcal{O}} a_{ij}^\varepsilon \frac{\partial u}{\partial x_j} \frac{\partial}{\partial x_i} \phi \, dx,$$

and we use the analogous notation for $\mathcal{A}_\phi(u, v)$. We are going to verify that

$$(5.3.8) \qquad a^\varepsilon(u + \varepsilon u_1, u_\varepsilon) - a_\phi^\varepsilon(u + \varepsilon u_1, u_\varepsilon) \to 0 \text{ as } |E| \to 0, \text{ uniformly in } \varepsilon.$$

Let us notice that there exists q finite such that

$$(5.3.9) \qquad |a^\varepsilon(u + \varepsilon u_1, u_\varepsilon) - a_\phi^\varepsilon(u + \varepsilon u_1, u_\varepsilon)| \leq C(\varepsilon + |E|^{1/q}),$$

and that we have a similar estimate for $a^\varepsilon(u_\varepsilon, u + \varepsilon u_1) - a_\phi^\varepsilon(u_\varepsilon, u + \varepsilon u_1)$.

The difference in (5.3.8) equals

$$(5.3.10) \quad \int_{\mathcal{O}} a_{ij}^\varepsilon \left(\frac{\partial u}{\partial x_i} - \frac{\partial \chi^k}{\partial y_i} \frac{\partial u}{\partial x_k} \right) \frac{\partial u_\varepsilon}{\partial x_j}(1-\phi)\, dx - \int_{\mathcal{O}} a_{ij}^\varepsilon \varepsilon \chi^k \frac{\partial^2 u}{\partial x_i \partial x_k} \frac{\partial u_\varepsilon}{\partial x_j}(1-\phi)\, dx.$$

The second term in (5.3.10) is bounded by $C\varepsilon$. Since $\chi^k \in W^{1,\infty}(Y)$, the first term in (5.3.10) is bounded by

$$C \left(\int_E |\nabla u|^2\, dx \right)^{1/2};$$

since $u \in H^2(\mathcal{O})$ we have $\frac{\partial u}{\partial x_i} \in L^p(\mathcal{O})$ for $p > 2$ and (5.3.9) follows.

Therefore, in order to prove (5.3.6), (5.3.7), it suffices to prove the similar results for a_p^ε, $\mathcal{A}_\phi$, with a fixed $\phi \in \mathcal{D}(\mathcal{O})$, i.e. that

$$(5.3.11) \qquad\qquad a_\phi^\varepsilon(u + \varepsilon u_1, u) \to \mathcal{A}_\phi(u, u),$$

$$(5.3.12) \qquad\qquad a_\phi^\varepsilon(u_\varepsilon, u + \varepsilon u_1) \to \mathcal{A}_\phi(u, u).$$

We have

$$(5.3.13) \qquad a_\phi^\varepsilon(u + \varepsilon u_1, u_\varepsilon) = \int_{\mathcal{O}} a_{ij}^\varepsilon \frac{\partial}{\partial x_j}(u + \varepsilon u_1)$$
$$\cdot \left[\frac{\partial}{\partial x_i}(u_\varepsilon \phi) - u_\varepsilon \frac{\partial \phi}{\partial x_i} \right]\, dx.$$

The second term in (5.3.13) converges to

$$(5.3.14) \qquad\qquad -\int q_{ij} \frac{\partial u}{\partial x_j} u \frac{\partial \phi}{\partial x_i}\, dx.$$

The first term in (5.3.13) equals $\int A^\varepsilon(u + \varepsilon u_1)(u_\varepsilon \phi)\, dx$. By virtue of the construction of u_1, we have

$$A^\varepsilon(u + \varepsilon u_1) = A_3 u + A_2 u_1 \to \mathcal{A}u \text{ in } L^2(\mathcal{O}) \text{ weakly}$$

so that, with (5.3.14), we obtain

$$a_\phi^\varepsilon(u + \varepsilon u_1, u_\varepsilon) \to (\mathcal{A}u, u\phi) - \int_{\mathcal{O}} q_{ij} \frac{\partial u}{\partial x_j} u \frac{\partial \phi}{\partial x_i}\, dx = \mathcal{A}_\phi(u, u)$$

which proves (5.3.11).

For proving (5.3.12) we observe that

$$a_\phi^\varepsilon(u_\varepsilon, u + \varepsilon u_1) = \int_{\mathcal{O}} a_{ij}^\varepsilon \frac{\partial u_\varepsilon}{\partial x_j} \left(\frac{\partial((u + \varepsilon u_1)\phi)}{\partial x_i} - (u + \varepsilon u_1) \frac{\partial \phi}{\partial x_i} \right)\, dx$$
$$= \int_{\mathcal{O}} (A^\varepsilon u_\varepsilon)(u + \varepsilon u_1)\phi\, dx - \int_{\mathcal{O}} \xi_i^\varepsilon (u + \varepsilon u_1) \frac{\partial \phi}{\partial x_i}\, dx$$
$$\to (f, u\phi) - \int_{\mathcal{O}} q_{ij} \frac{\partial u}{\partial x_j} u \frac{\partial \phi}{\partial x_i}\, dx = \mathcal{A}(u, u).$$

COROLLARY 5.9. *We suppose that the hypothesis of Theorem 5.1 hold true. Let F be any measurable set $\subseteq \mathcal{O}$. Then*

$$(5.3.15) \qquad\qquad \int_F a_{ij}^\varepsilon \frac{\partial u_\varepsilon}{\partial x_j} \frac{\partial u_\varepsilon}{\partial x_i}\, dx \to \int_F q_{ij} \frac{\partial u}{\partial x_j} \frac{\partial u}{\partial x_i}\, dx.$$

PROOF. According to Theorem 5.1, the expression in (5.3.15) has the same limit as

$$\int_F a_{ij}^\varepsilon \frac{\partial(u + \theta_\varepsilon)}{\partial x_j} \frac{\partial(u + \theta_\varepsilon)}{\partial x_i} \, dx$$

and one completes the proof as in the proof of (5.3.4).   ■

Orientation. A natural goal is now to obtain **error estimates** for z_ε or for ζ_ε in $H^1(\mathcal{O})$. We shall give in Section 5.5 below **some** results for the Dirichlet's boundary conditions (and we shall return to this question in Section 18). We found it convenient for getting these error estimates to write the equation $A^\varepsilon u_\varepsilon = f$ as **a first order system**; this is made in Section 5.4, as we shall see it leads to two results of independent interest:

(1) the so-called **dual formulas** (cf. Remarks 5.10 and 5.11 below);
(2) a strong approximation in $(L^2(\mathcal{O}))^n$ **for** $a^\varepsilon \operatorname{grad} u_\varepsilon$ (cf. Theorem 5.13, Section 5.5).

5.4. First order system and asymptotic expansion. We consider equation (1.1.5), where for simplicity (the zero order terms do not cause here any difficulty) we assume that

$$(5.4.1) \qquad\qquad a_0(y) = \lambda > 0.$$

We suppose that we are in $\mathbb{R}^3$, but this is just for the convenience of writing. We write (1.1.5) as a **first order system with the idea of getting a better approximation of the first derivatives.**[25] Since (1.1.5) is equivalent to[26]

$$(5.4.2) \qquad\qquad -\operatorname{div}(a^\varepsilon \operatorname{grad} u_\varepsilon) + \lambda u_\varepsilon = f$$

we can write the first order system

$$(5.4.3) \qquad\qquad a^\varepsilon \operatorname{grad} u_\varepsilon - v_\varepsilon = 0,$$
$$-\operatorname{div} v_\varepsilon + \lambda u_\varepsilon = f.$$

We look for an expansion of the form (as in Section 2)

$$(5.4.4) \qquad\qquad u_\varepsilon = u_0 + \varepsilon u_1 + \cdots,$$
$$v_\varepsilon = v_0 + \varepsilon v_1 + \cdots,$$

where $u_j(x, y)$, $v_j(x, y)$ are Y-periodic.

By identification, we obtain

$$(5.4.5) \qquad\qquad a \operatorname{grad}_y u_0 = 0,$$

$$(5.4.6) \qquad\qquad -\operatorname{div}_y v_0 = 0,$$

$$(5.4.7) \qquad\qquad a \operatorname{grad}_y u_1 + a \operatorname{grad}_x u_0 - v_0 = 0,$$

$$(5.4.8) \qquad\qquad \operatorname{div}_y v_1 - \operatorname{div}_x v_0 + \lambda u_0 = f.$$

Since a is invertible, (5.4.5) is equivalent to $\operatorname{grad}_y u_0 = 0$, i.e. (as it should!)

$$(5.4.9) \qquad\qquad u_0 = u(x).$$

Using (5.4.9) in (5.4.7) gives:

$$(5.4.10) \qquad\qquad \operatorname{grad}_y u_1 + \operatorname{grad}_x u = a^{-1} v_0$$

---

[25] We write here a^ε instead of $[a^\varepsilon]$ as in Section 3.4.
[26] It is this idea which leads to the **mixed** approximation techniques in Finite Elements.

which can be solved in u_1 iff

$$\text{rot}_y \, \text{grad}_x \, u = \text{rot}_y(a^{-1}v_0)$$

i.e.

(5.4.11) $\text{rot}_y(a^{-1}v_0) = 0.$

We have now two ways of proceeding, either by exploiting (5.4.11) first, or by exploiting (5.4.6) first.

Let us start with (5.4.11). We set

(5.4.12) $a^{-1}v_0 = w_0,$

and we define $\mathcal{M}(w_0)$ by taking the average of each component. Then

(5.4.13) $\text{rot}_y(w_0 - \mathcal{M}(w_0)) = 0,$
$$\mathcal{M}(w_0 - \mathcal{M}(w_0)) = 0,$$

hence it follows that there exists p such that

(5.4.14) $w_0 - \mathcal{M}(w_0) = \text{grad}_y \, p,$ $p(x,y)$ **being Y-periodic.**

We now use (5.4.12) and (5.4.14) in (5.4.6) to get

(5.4.15) $- \text{div}_y \, a \, \text{grad}_y \, p = - \text{div} \, a\mathcal{M}(w_0).$

We can solve (5.4.7) for u_1 iff

(5.4.16) $\text{grad} \, u = \mathcal{M}(a^{-1}v_0) = \mathcal{M}(w_0).$

Therefore (5.4.15), (5.4.16) give, with the usual notations

(5.4.17) $p = -\chi^j \dfrac{\partial u}{\partial x_j}$ (defined up to the addition of a function of x).

We can solve (5.4.8) for v_1 iff

(5.4.18) $- \text{div} \, \mathcal{M}(v_0) + \lambda u = f.$

Using $v_0 = aw_0$ and (5.4.14), (5.4.16), (5.4.17), (5.4.18) gives the usual equation $\mathcal{A}u + \lambda u = f$.

Let us start now with (5.4.6). We set

(5.4.19) $\mathcal{M}(v_0) = \rho,$

(5.4.20) $\tilde{v}_0 = v_0 - \rho.$

Then

(5.4.21) $\text{div}_y \, \tilde{v}_0 = 0,$ $\mathcal{M}(\tilde{v}_0) = 0.$

We claim that (5.4.21) is **equivalent** to

(5.4.22) $tildev_0 = \text{rot}_y \, \phi,$ $\text{div}_y \, \phi = 0,$ ϕ Y-periodic in y.

It is obvious that (5.4.22) implies (5.4.21). Conversely, if $\tilde{v}_0$ satisfies (5.4.21), we can find ϕ such that (5.4.22) holds true. We begin by observing that — by the standard projection theorem — there exists ϕ, such that ϕ is Y-periodic and $\text{div}_y \, \phi = 0$ and

(5.4.23) $(\text{rot}_y \, \phi, \text{rot}_y \, \psi) = (\text{rot}_y \, \tilde{v}_0, \psi)$

$\forall \psi$ such that ψ is Y-periodic, $\text{div}_y \, \psi = 0$, $\text{rot}_y \, \psi \in (L^2(Y))^3$.

Then

$$\text{rot}_y(\text{rot}_y \, \phi - \tilde{v}_0) = \text{grad}_y \, \pi,$$

π being Y periodic, and div $\text{grad}_y \, \pi = \Delta \pi = 0$ so that $\pi = $ constant in y. Therefore

$$(5.4.24) \qquad \qquad \text{rot}_y(\text{rot}_y \, \phi - \tilde{v}_0) = 0.$$

By virtue of (5.4.21)

$$(5.4.25) \qquad \qquad \text{div}_y(\text{rot}_y \, \phi - \tilde{v}_0) = 0$$

and

$$(5.4.26) \qquad \qquad \mathcal{M}(\text{rot}_y \, \phi - \tilde{v}_0) = 0.$$

But (5.4.24), (5.4.25), (5.4.26) are equivalent to $\text{rot}_y \, \phi - \tilde{v}_0 = 0$ and (5.4.22) follows.

We now use (5.4.22) in (5.4.11):

$$(5.4.27) \qquad \qquad \text{rot}_y \, a^{-1}(\text{rot}_y \, \phi + \rho) = 0,$$
$$\text{div}_y \, \phi = 0, \qquad \phi \; Y\text{-periodic}.$$

We use now (5.4.27) to express ϕ in terms of ρ. We define $\mathbf{e}_1 = (1,0,0)$, $\mathbf{e}_2 = (0,1,0)$, $\mathbf{e}_3 = (0,0,1)$ **and we define** $\tilde{\chi}^p(y)$ **as the solution of**

$$(5.4.28) \qquad \begin{aligned} & \text{rot} \, a^{-1} \, \text{rot} \, \tilde{\chi}^p(y) = \text{rot}(a^{-1}\mathbf{e}_p), \\ & \text{div} \, \tilde{\chi}^p = 0, \\ & \tilde{\chi}^p \text{ is } Y\text{-periodic}, \; \mathcal{M}(\tilde{\chi}^p) = 0. \end{aligned}$$

Then

$$(5.4.29) \qquad \qquad \phi = -\tilde{\chi}^p(y)\rho_p(x) + \widehat{\phi}(x).$$

Therefore

$$(5.4.30) \qquad \qquad v_0 = \rho - \text{rot}_y \, \tilde{\chi}^p(y)\rho_p(x).$$

If we **define** the **matrix** $\text{rot}_y \, \tilde{\chi}$ by

$$(5.4.31) \qquad \qquad \text{rot}_y \, \tilde{\chi}(y) \cdot \xi = \text{rot}_y \, \tilde{\chi}^p(y)\xi_p,$$

we can write (5.4.30) in the equivalent form

$$(5.4.32) \qquad \qquad v_0 = (I - \text{rot}_y \, \tilde{\chi}(y))\rho(x).$$

We take the y averages in (5.4.7) and (5.4.8) after multiplying by a^{-1}; we obtain (5.4.16) which can now be written

$$(5.4.33) \qquad \qquad \text{grad} \, u = \mathcal{M}(a^{-1}(I - \text{rot}_y \, \tilde{\chi})\rho(x)).$$

Also (5.4.18) with use of (5.4.19) becomes

$$(5.4.34) \qquad \qquad) - \text{div} \, \rho + \lambda u = f.$$

If we define the average of a matrix by

$$\mathcal{M}([b_{ij}]) = [\mathcal{M}(b_{ij})],$$

(5.4.33) gives

$$\text{grad} \, u = \mathcal{M}(a^{-1}(I - \text{rot}_y \, \tilde{\chi}))\rho$$

i.e.,

$$\rho = \mathcal{M}(a^{-1}(I - \text{rot}_y \, \tilde{\chi}))^{-1} \, \text{grad} \, u.$$

Then (5.4.34) gives

$$(5.4.35) \qquad \qquad -\text{div} \, \mathcal{M}(a^{-1}(I - \text{rot}_y \, \tilde{\chi}))^{-1} \, \text{grad} \, u + \lambda u = f.$$

REMARK 5.10. Since we necessarily obtain the same homogenized operator, it follows — **if (5.4.35) is justified** — that

$$(5.4.36) \qquad \mathcal{M}(a^{-1}(I - \mathrm{rot}_y\,\tilde{\chi}))^{-1} = [\mathcal{A}].$$

But, according to usual formulae

$$(5.4.37) \qquad [\mathcal{A}] = \mathcal{M}(a(I - \mathrm{grad}_y\,\chi))$$

$$\mathrm{grad}_y\,\chi = \mathrm{matrix}(\mathrm{grad}_y\,\chi^j).$$

This is indeed true. We can obtain a little bit more, namely that

$$(5.4.38) \qquad a^{-1}(I - \mathrm{rot}_y\,\tilde{\chi}) = (I - \mathrm{grad}_y\,\chi)[\mathcal{A}]^{-1}$$

which gives (5.4.36) after taking averages.

Since both sides of (5.4.38) consist of periodic matrices, it is enough to verify that

$$(5.4.39) \qquad \mathrm{rot}_y(a^{-1}(I - \mathrm{rot}_y\,\tilde{\chi})) = \mathrm{rot}_y(I - \mathrm{grad}_y\,\chi)[\mathcal{A}]^{-1},$$

$$(5.4.40) \qquad \mathrm{div}_y(a^{-1}(I - \mathrm{rot}_y\,\tilde{\chi})) = \mathrm{div}_y\,a(I - \mathrm{grad}_y\,\chi)[\mathcal{A}]^{-1};$$

but using $\mathrm{div}_y\,\mathrm{rot}_y = 0$, $\mathrm{rot}_y\,\mathrm{grad}_y = 0$ and the definitions of χ and $\tilde{\chi}$, **one sees that all expressions which appear in (5.4.39) and in (5.4.40) are zero,** hence the result follows.

REMARK 5.11. We have therefore two formulas for computing $[\mathcal{A}]$; we shall call these **dual formulas**.

REMARK 5.12. We can choose a solution v_1 of (5.4.8)

$$(5.4.41) \qquad v_1 = -\,\mathrm{rot}_x(\tilde{\chi}\rho).$$

Indeed, we note that $\mathrm{div}_y\,\mathrm{rot}_x = -\,\mathrm{div}_x\,\mathrm{rot}_y$ so that if v_1 is **defined** by (5.4.41) then

$$-\,\mathrm{div}_y\,v_1 = \mathrm{div}_x\,\mathrm{rot}_y(\tilde{\chi}\rho)$$

$$= (\text{according to } (5.4.30))\ \ \mathrm{div}_\rho - \mathrm{div}_x\,v_0$$

$$= (\text{according to } (5.4.34))\ \ \lambda u - f - \mathrm{div}_x\,v_0$$

i.e., v_1 satisfies (5.4.8).

5.5. Correctors: Error estimates for the Dirichlet's problem. We now return to the question of correctors, in the sense of Section 5.2. We confine ourselves to the Dirichlet's problem, but we do not assume symmetry. The notations are those of Theorem 5.1.

THEOREM 5.13. *The hypotheses are those of Theorem* 5.1.
We consider the Dirichlet's problem $(V = H_0^1(\mathcal{O}))$. Then, m_ε being chosen as in (5.2.2), one has

$$(5.5.1) \qquad z_\varepsilon = u_\varepsilon - u - \varepsilon m_\varepsilon u_1 \to 0 \ \ in \ H_0^1(\mathcal{O}) \ strongly,$$

$$(5.5.2) \qquad n_\varepsilon = a^\varepsilon\,\mathrm{grad}\,u_\varepsilon - v_0 - \varepsilon v_1 \to 0 \ \ in \ (L^2(\mathcal{O}))^3 \ strongly,$$

where

$$(5.5.3) \qquad u_1 = -\chi^j\,\frac{\partial u}{\partial x_j},$$

$$v_0 = (I - \mathrm{rot}_y\, \tilde{\chi})\rho,$$

(5.5.4)
$$\rho = \mathcal{M}(a^{-1}(I - \mathrm{rot}_y\, \tilde{\chi}))^{-1}\, \mathrm{grad}\, u,$$

$$v_1 = -\mathrm{rot}_x(\tilde{\chi}\rho).$$

REMARK 5.14. For general boundary conditions, we have a similar result (without m_ε) on every $\mathcal{O}'$ such that $\overline{\mathcal{O}'} \subset \mathcal{O}$.

REMARK 5.15. The result (5.5.2) does not follow from (5.5.1). We may also suppress the term εu_1 in (5.5.2).

REMARK 5.16. We obtain **error estimates** as follows.
Since we know that $u \in H^2(\mathcal{O})$, it follows that

$$\frac{\partial u}{\partial x_i} \in L^q(\mathcal{O}), \qquad \frac{1}{q} = \frac{1}{2} - \frac{1}{n} \text{ if } n \geq 3,$$

q finite if $n = 2$. Therefore we can assume that there exists $p > 2$ such that

(5.5.5)
$$\frac{\partial u}{\partial x_i} \in L^p(\mathcal{O}).$$

Then

(5.5.6)
$$\|z_\varepsilon\|_{H_0^1(\mathcal{O})} + \|n_\varepsilon\|_{(L^2(\mathcal{O}))^3} \leq C\varepsilon^{1/2p'}, \quad \frac{1}{p} + \frac{1}{p'} = 1.$$

Therefore the best possible result (by the techniques used here) is obtained when we have (5.5.5) with $p = +\infty$, in which case the error estimate is $C\varepsilon^{1/2}$.

PROOF OF THEOREM 5.13. We set

(5.5.7)
$$a^\varepsilon\, \mathrm{grad}\, z_\varepsilon - n_\varepsilon = g_{1\varepsilon},$$

$$-\mathrm{div}\, n_\varepsilon + \lambda z_\varepsilon = g_{2\varepsilon},$$

and we shall show below that

(5.5.8)
$$\|g_{1\varepsilon}\|_{(L^2(\mathcal{O}))^3} \leq C\varepsilon^{1/2p'},$$

(5.5.9)
$$\|g_{2\varepsilon}\|_{L^2(\mathcal{O})} \leq C\varepsilon.$$

Let us assume these estimates for a moment. We multiply the first (respectively, second) equation (5.5.7) by $\mathrm{grad}\, z_\varepsilon$ (respectively, z_ε). Since $z_\varepsilon \in H_0^1(\mathcal{O})$ we have $(n, \mathrm{grad}\, z_\varepsilon) + (\mathrm{div}\, n_\varepsilon, z_\varepsilon) = 0$, so that

$$(a^\varepsilon\, \mathrm{grad}\, z_\varepsilon, \mathrm{grad}\, z_\varepsilon) + \lambda(z_\varepsilon, z_\varepsilon) = (g_{1\varepsilon}, \mathrm{grad}\, z_\varepsilon) + (g_{2\varepsilon}, z_\varepsilon),$$

hence it follows that

$$\|z_\varepsilon\|^2 \leq C(\varepsilon + \varepsilon^{1/2p'})\|z_\varepsilon\|$$

hence (5.5.6) follows (the estimate in n_ε follows from $n_\varepsilon = a^\varepsilon\, \mathrm{grad}\, z_\varepsilon - g_{1\varepsilon}$).

PROOF OF (5.5.9). We have

$$g_{2\varepsilon} = -\mathrm{div}(a^\varepsilon\, \mathrm{grad}\, u_\varepsilon) + \lambda u_\varepsilon + \varepsilon^{-1}\, \mathrm{div}_y\, v_0$$

$$+ (\mathrm{div}_x\, v_0 + \mathrm{div}_y\, v_1 - \lambda u) + \varepsilon(\mathrm{div}_x\, v_1 - \lambda m_\varepsilon u_1).$$

But $-\mathrm{div}(a^\varepsilon\, \mathrm{grad}\, u_\varepsilon) + \lambda u_\varepsilon = f$ and using (5.4.6), (5.4.8) we obtain

$$g_{2\varepsilon} = \varepsilon(\mathrm{div}_x\, v_1 - \lambda m_\varepsilon u_1),$$

hence (5.5.9) follows. ∎

PROOF OF (5.5.8). We have

$$g_{1\varepsilon} = a^\varepsilon \operatorname{grad} u_\varepsilon - v_\varepsilon - a^\varepsilon \operatorname{grad} u$$
$$- \varepsilon a^\varepsilon \operatorname{grad}(m_\varepsilon u_1) + v_0 + \varepsilon v_1$$
$$= -a^\varepsilon \operatorname{grad} u - \varepsilon a^\varepsilon \operatorname{grad} u_1 + \varepsilon a^\varepsilon \operatorname{grad}(1 - m_\varepsilon)u_1 + v_0 + \varepsilon v_1.$$

Using (5.4.7), we obtain

$$g_{1\varepsilon} = b_\varepsilon + c_\varepsilon + d_\varepsilon, \text{ where}$$
$$b_\varepsilon = \varepsilon(v_1 - m_\varepsilon a^\varepsilon \operatorname{grad}_x u_1),$$
$$c_\varepsilon = -\varepsilon a^\varepsilon (\operatorname{grad} m_\varepsilon)u_1,$$
$$d_\varepsilon = a^\varepsilon(1 - m_\varepsilon) \operatorname{grad}_y u_1.$$

We have $\|b_\varepsilon\|_{(L^2(\mathcal{O}))^3} \leq C\varepsilon$. Since $|\varepsilon \operatorname{grad} m_\varepsilon(x)| \leq C$, and $\varepsilon \operatorname{grad} m_\varepsilon = 0$ if $m_\varepsilon = 1$, the estimation of c_ε and of d_ε reduces (assuming $\chi^j \in W^{1,\infty}(Y)$) to the estimation of

$$Y^\varepsilon = \left(\int_{d(x,\Gamma) \leq 2\varepsilon} \left\| \frac{\partial u}{\partial x_j} \right\|^2 dx \right)^{1/2}.$$

Using (5.5.5), $Y_\varepsilon \leq C\varepsilon^{1/2p'}$.   ■

■

We consider now the situation of Theorem 5.13 but **without introducing** m_ε. We have

THEOREM 5.17. *The hypotheses are those of Theorem 5.2.1. We define*

(5.5.10) $$\zeta_\varepsilon = u_\varepsilon - u - \varepsilon u_1,$$

and n_ε as in (5.5.2), (5.5.4), u_1 being defined as in (5.5.3). We have

(5.5.11) $$\|\zeta_\varepsilon\|_{H^1(\mathcal{O})} + \|n_\varepsilon\|_{L^2(\mathcal{O})} \leq C\varepsilon^{1/4}.$$

REMARK 5.18. Since we did not introduce m_ε, $\zeta_\varepsilon \notin V = H_0^1(\mathcal{O})$; but on the other hand we obtain a better $H^1(\mathcal{O})$ approximation.

PROOF. We use here spaces $H^s(\Gamma)$ for $a = \pm\frac{1}{2}$; we refer, for instance to Lions-Magenes [70] for the main properties of these spaces.

We will need

(5.5.12) $$\|u_1\|_{H^{1/2}(\Gamma)} \leq C\varepsilon^{-1/2}.$$

(5.5.13) $$\|n_\varepsilon \cdot \nu\|_{H^{-1/2}(\Gamma)} \leq C.$$

We assume these estimates for a moment and we show that they imply (5.5.11). We observe that

(5.5.14) $$a^\varepsilon \operatorname{grad} \zeta_\varepsilon - n_\varepsilon = h_{1\varepsilon},$$
$$- \operatorname{div} n_\varepsilon + \lambda \zeta_\varepsilon = h_{2\varepsilon},$$

with

$$h_{1\varepsilon} = \varepsilon(v_1 - a^\varepsilon \operatorname{grad}_x u_1),$$
$$h_{2\varepsilon} = \varepsilon(\operatorname{div}_x v_1 - \lambda u_1).$$

Therefore $|h_{1\varepsilon}| + |h_{2\varepsilon}| < C\varepsilon$ and (5.5.14) gives
(5.5.15)
$$(a^\varepsilon \operatorname{grad} \zeta_\varepsilon, \operatorname{grad} \zeta_\varepsilon) + \lambda|\zeta_\varepsilon|^2 - [(n_\varepsilon, \operatorname{grad} \zeta_\varepsilon) + (\operatorname{div} n_\varepsilon, \zeta_\varepsilon)] = (h_{1\varepsilon}, \operatorname{grad} \zeta_\varepsilon) + (h_{2\varepsilon}, \zeta_\varepsilon)$$

with a right hand side bounded by $C\varepsilon\|\zeta_\varepsilon\|$.

But
$$(n_\varepsilon, \operatorname{grad} \zeta_\varepsilon) + (\operatorname{div} n_\varepsilon, \zeta_\varepsilon) = \int_\Gamma (n_\varepsilon \cdot \nu)\zeta_\varepsilon \, d\Gamma$$

o that
$$|(n_\varepsilon, \operatorname{grad} \zeta_\varepsilon) + (\operatorname{div} n_\varepsilon, \zeta_\varepsilon)| \leq \|n_\varepsilon \cdot \nu\|_{H^{-1/2}(\Gamma)} \cdot \|\zeta_\varepsilon\|_{H^{1/2}(\Gamma)}.$$

By virtue of (5.5.13), and since $\zeta_\varepsilon = -u_1$ on Γ, we have

(5.5.16) $$|(n_\varepsilon, \operatorname{grad} \zeta_\varepsilon) + (\operatorname{div} n_\varepsilon, \zeta_\varepsilon)| \leq C\varepsilon\|u_1\|_{H^{1/2}(\Gamma)} \leq C\varepsilon^{1/2}.$$

Using (5.5.16) into (5.5.15) gives
$$\|\zeta_\varepsilon\|^2_{H^1(\mathcal{O})} \leq C\varepsilon\|\zeta_\varepsilon\|_{H^1(\mathcal{O})} + C\varepsilon^{1/2}$$

hence the result follows.   ■

PROOF OF (5.5.12). It follows from (5.5.3) (where $\chi^j = \chi^j(x/\varepsilon)$) that
$$\|u_1\|_{L^2(\Gamma)} \leq C, \qquad \|u_1\|_{H^1(\Gamma)} \leq C/\varepsilon,$$

so that
$$\|u_1\|_{H^{1/2}(\Gamma)} \leq C\|u_1\|^{1/2}_{L^2(\Gamma)}\|u_1\|^{1/2}_{H^1(\Gamma)} \leq C\varepsilon^{-1/2}.$$
    ■

PROOF OF (5.5.13). Since $\|v_0 + \varepsilon v_1\|_{L^2(\Gamma)} \leq C$, we have only to verify that

(5.5.17) $$\|a^\varepsilon \operatorname{grad} u_\varepsilon \cdot v\|_{H^{-1/2}(\Gamma)} \leq C.$$

But $a^\varepsilon \operatorname{grad} u_\varepsilon = v_\varepsilon$ is bounded in $(L^2(\mathcal{O}))^3$ and $\operatorname{div} v_\varepsilon = \lambda u_\varepsilon - f$ is bounded in $L^2(\mathcal{O})$. It is known (cf. Lions-Magenes, loc. cit.) that
$$\|v_\varepsilon \cdot \nu\|_{H^{-1/2}(\Gamma)} \leq C\left[\|v_\varepsilon\|_{(L^2(\mathcal{O}))^3} + \|\operatorname{div} v_\varepsilon\|_{L^2(\mathcal{O})}\right]$$

hence (5.5.17) follows.   ■

6. Second order elliptic operators with non-uniformly oscillating coefficients

6.1. Setting of the problem and general families. We consider now a family of operators of the type

(6.1.1) $$A^\varepsilon = -\frac{\partial}{\partial x_i}\left(a_{ij}\left(x, \frac{x}{\varepsilon}\right)\frac{\partial}{\partial x_j}\right) + a_0\left(x, \frac{x}{\varepsilon}\right)$$

with the following hypotheses on the coefficients a_{ij}, a_0. We denote by $C^0(\overline{\mathcal{O}}; L^\infty_p(\mathbb{R}^n))$ the space of functions $\phi(x, y)$ defined in $\overline{\mathcal{O}} \times \mathbb{R}^n$, real valued, such that $\phi(x, \cdot) \in L^\infty(\mathbb{R}^n)$, $x \in \overline{\mathcal{O}}$, $x \to \phi(x, \cdot)$ is continuous from $\overline{\mathcal{O}} \to L^\infty(\mathbb{R}^n)$, and ϕ is Y-periodic in y, $\forall x$ (we recall this fact by the index "p" in $L^\infty_p(\mathbb{R}^n)$). We provide $C^0(\overline{\mathcal{O}}; L^\infty_p(\mathbb{R}^n))$ with the norm

(6.1.2) $$\sup_x \|\phi(x, \cdot)\|_{L^\infty(\mathbb{R}^n)} = \sup_x \operatorname{ess\,sup}_y |\phi(x, y)|;$$

we now assume that

$$a_{ij} \in C^0(\overline{\mathcal{O}}; L_p^\infty(\mathbb{R}^n)), \; i, j,$$

(6.1.3)
$$a_{ij}(x, y)\xi_i\xi_j \geq \alpha\xi_i\xi_i, \; \forall\xi \in \mathbb{R}^n, \; \alpha > 0,$$

$$x \in \overline{\mathcal{O}}, \text{ a.e. in } y,$$

and that

(6.1.4)
$$a_0 \in C^0(\overline{\mathcal{O}}; L_p^\infty(\mathbb{R}^n)),$$

$$a_0(x, y) \geq \alpha_0 > 0.$$

(In what follows, we can take $\alpha_0 = 0$ if V consists of functions in $H^1(\mathcal{O})$ which are zero on a part Γ_0 of Γ, measure $\Gamma_0 > 0$.)

For $u, v \in H^1(\mathcal{O})$, we set

(6.1.5)
$$a^\varepsilon(u, v) = \int a_{ij}\left(x, \frac{x}{\varepsilon}\right) \frac{\partial u}{\partial x_j} \frac{\partial v}{\partial x_i} \, dx + \int a_0\left(x, \frac{x}{\varepsilon}\right) uv \, dx,$$

and we denote by u_ε the solution of

$$u_\varepsilon \in V,$$

(6.1.6)
$$a^\varepsilon(u_\varepsilon, v) = (f, v), \; \forall v \in V,$$

where f is given in V'.

We want to study the behavior of u_ε when $\varepsilon \to 0$.

REMARK 6.1. In case the a_{ij} (and a_0) are **constant** in x, the problem reduces to the one solved in Sections 1 to 3.

We shall prove the existence of an **homogenized operator** constructed as follows:

(6.1.7)
$$\mathcal{A} = -\frac{\partial}{\partial x_i}\left(q_{ij}(x)\frac{\partial}{\partial x_j}\right) + q_0(x),$$

(6.1.8)
$$q_0(x) = \mathcal{M}_y a_0(x, y) = \frac{1}{|Y|}\int_Y a_0(x, y) \, dy;$$

$q_{ij}(x)$ is constructed as follows; we define

(6.1.9)
$$A_1(x) = -\frac{\partial}{\partial y_i}\left(a_{ij}(x, y)\frac{\partial}{\partial y_j}\right),$$

and we define $\chi^j(x, y)$ as the unique solution (up to an additive constant) of

$$A_1(x)(\chi^j(x, y) - y_j) = 0.$$

(6.1.10)
$$\chi^j(x, y) \; Y\text{-periodic in } y,$$

$$\chi^j \text{ depends continuously on } x \text{ with values in } W(Y).[27]$$

For $\phi, \psi \in H^1(Y)$ we set

(6.1.11)
$$a_1(x; \phi, \psi) = \int_Y a_{ij}(x, y)\frac{\partial\phi}{\partial y_j}\frac{\partial\psi}{\partial y_i} \, dy,$$

---

[27]This will be the case if we define for instance $\chi^j(x, y)$ by $\int_Y \chi^j(x, y) \, dy = 0$.

then $q_{ij}(x)$ is given by

(6.1.12) $$q_{ij}(x) = \frac{1}{|Y|} a_1 (\chi^j(x,y) - y_j, \chi^i(x,y) - y_i).$$

REMARK 6.2. The functions $x \to q_{ij}(x)$ are continuous in $\overline{\mathcal{O}}$. The polynomial $q_{ij}(x)\xi_i\xi_j$ being positive definite for every x (cf. Remark 2.6) it follows that there exists $\gamma > 0$ such that

(6.1.13) $$q_{ij}(x)\xi_i\xi_j \geq \gamma\xi_i\xi_i, \ \forall x \in \mathcal{O}.$$

We set, $\forall u, v \in H^1(\mathcal{O})$

(6.1.14) $$\mathcal{A}(u,v) = \int_{\mathcal{O}} q_{ij}(x) \frac{\partial u}{\partial x_j} \frac{\partial v}{\partial x_i} \, dx + \int_{\mathcal{O}} q_0(x) uv \, dx.$$

By virtue of (6.1.13) and since (6.1.4) implies $q_0(x) \geq \alpha_0$, it follows that there exists a unique u such that

$$u \in V,$$
(6.1.15) $$\mathcal{A}(u,v) = (f,v), \ \forall v \in V.$$

The main result of this Section is as follows:

THEOREM 6.3. *Under the hypotheses (6.1.3), (6.1.4), the solution u_ε of (6.1.6) converges in V weakly toward u, solution of (6.1.15), where $\mathcal{A}$ (and A) are constructed by (6.1.7), ..., (6.1.12). The operator $\mathcal{A}$ is still called the homogenized form of A^ε.*

REMARK 6.4. With the notations of Section 2, one can construct an asymptotic expansion. We have

$$A^\varepsilon = \varepsilon^{-2} A_1(x) + \varepsilon^{-1} A_2 + \varepsilon^0 A_3,$$
(6.1.16) $$A_1(x) = -\frac{\partial}{\partial y_i}\left(a_{ij}(x,y) \frac{\partial}{\partial y_j} \right),$$
$$A_2 = -\frac{\partial}{\partial y_i}\left(a_{ij}(x,y) \frac{\partial}{\partial x_j} \right) - \frac{\partial}{\partial x_i}\left(a_{ij}(x,y) \frac{\partial}{\partial y_j} \right),$$
$$A_3 = -\frac{\partial}{\partial x_i}\left(a_{ij}(x,y) \frac{\partial}{\partial x_j} \right).$$

If we look for u_ε in the form

(6.1.17) $$u_\varepsilon = u_0(x,y) + \varepsilon u_1(x,y) + \varepsilon^2 u_2(x,y) + \cdots,$$

we find (as in Section 2) that

(6.1.18) $$A_1(x) u_0(x,y) = 0,$$

(6.1.19) $$A_1(x) u_1 + A_2 u_0 = 0,$$

(6.1.20) $$A_1(x) u_2 + A_2 u_1 + A_3 u_0 = f,$$

etc. Equation (6.1.18) gives

$$u_0(x,y) = u(x)$$

so that (6.1.20) reduces to

$$A_1(x) u_1 = \left(\frac{\partial}{\partial y_i} a_{ij}(x,y) \right) \frac{\partial u_1}{\partial x_j}(x).$$

Since x plays here the role of a parameter, it follows that

$$(6.1.21) \qquad u_1(x,y) = -\chi^j(x,y)\frac{\partial u}{\partial x_j}(x) + \tilde{u}_1(x).$$

Equation (6.1.20) admits a Y-periodic solution in u_2, iff

$$\int_Y (A_2 u_1 + A_3 u_0)\, dy = |Y| f(x)$$

which leads to

$$\mathcal{A}u = f.$$

REMARK 6.5. Assuming all data to be smooth, one can prove, as in Section 2.4 that, **for the Dirichlet's problem**

$$\|u_\varepsilon - u\|_{L^\infty(\mathcal{O})} \le C\varepsilon.$$

REMARK 6.6 (Correctors). Let us assume that A^ε is **symmetric** ($a_{ij} = a_{ji}$, $\forall i,j$) and let us assume that

$$(6.1.22) \qquad a_{ij} \in C^1(\overline{\mathcal{O}}; L_p^\infty(\mathbb{R}^n)).$$

We use the same notations as in Section 5 and we introduce

$$(6.1.23) \qquad \theta_\varepsilon = -\varepsilon\chi^j\left(x, \frac{x}{\varepsilon}\right)\frac{\partial u}{\partial x_j}(x) m_\varepsilon(x).$$

Then we have

THEOREM 6.7. *We assume that the hypotheses of Theorem 6.3 hold true and that $a_{ij} = a_{ji}$, $\forall i,j$ and that we have (6.1.22). Then if $f \in V' \cap L^2_{\mathrm{loc}(\mathcal{O})}$,*

$$(6.1.24) \qquad z_\varepsilon = u_\varepsilon - (u + \theta_\varepsilon) \to 0 \text{ in } V \text{ strongly.}$$

PROOF. (Assuming Theorem 6.3 proven.) We argue as in Section 5. We have

$$(6.1.25) \qquad a^\varepsilon(z_\varepsilon) = (f, u_\varepsilon - 2(u + \theta_\varepsilon)) + X_\varepsilon,$$
$$X_\varepsilon = a^\varepsilon(u + \theta_\varepsilon).$$

We observe that by virtue of (6.1.22) one can check that

$$(6.1.26) \qquad q_{ij} \in C^1(\overline{\mathcal{O}})$$

so that the solution u of (6.1.15) satisfies (if Γ is smooth and for suitable V)

$$(6.1.27) \qquad u \in H^2(\mathcal{O}).$$

It remains, as in Section 5, to prove that $X_\varepsilon \to \mathcal{A}(u)$. But

$$(6.1.28) \qquad \frac{\partial \theta_\varepsilon}{\partial x_k} = -\frac{\partial \chi^j}{\partial y_k}\left(x, \frac{x}{\varepsilon}\right)\frac{\partial u}{\partial x_j}(x) m_\varepsilon(x) + r_{\varepsilon k}$$

where $r_{\varepsilon k}$ contains the same terms as in the proof of (5.3.4) and the term

$$-\varepsilon\frac{\partial \chi^j}{\partial x_k}\left(x, \frac{x}{\varepsilon}\right)\frac{\partial u}{\partial x_j}(x) m_\varepsilon(x),$$

which tends to 0 in $L^2(\mathcal{O})$. Hence the result follows. ∎

Before proceeding to the proof of Theorem 6.3, we prove a variant of this theorem, which is of independent interest.

6.2. Homogenization of transmission problems. Let us consider a partition of $\mathcal{O}$[28]

$$(6.2.1) \qquad\qquad \mathcal{O} = \bigcup_{k=1}^{N} \mathcal{O}^k,$$

and let us give a family of functions $a_{ij}^k(y)$ such that

$$(6.2.2) \qquad a_{ij}^k(y) \in L^\infty(\mathbb{R}^n), \qquad a_{ij}^k(y) \text{ is } Y\text{-periodic } \forall i,j,k$$

$$a_{ij}^k(y)\xi_i\xi_j \geq \alpha\xi_i\xi_i, \ \alpha > 0, \text{ a.e. in } y, \ k = 1,\dots,N$$

and also a family of functions $a_0^k(y)$:

$$(6.2.3) \qquad a_0^k \in L^\infty(\mathbb{R}^n), \quad a_0^k \text{ is } Y \text{ periodic},$$

$$a_0^k(y) \geq \alpha_0 > 0, \text{ a.e. in } y.$$

For $u, v \in H^1(\mathcal{O}^k)$, we define

$$(6.2.4) \qquad a^\varepsilon(u,v) = \sum_{k=1}^{N}\left[\int_{\mathcal{O}^k} a_{ij}^k\left(\frac{x}{\varepsilon}\right)\frac{\partial u}{\partial x_j}\frac{\partial v}{\partial x_i}\,dx + \int_{\mathcal{O}^k} a_0^k\left(\frac{x}{\varepsilon}\right)uv\,dx\right].$$

We take V as in the preceding sections. By virtue of (6.2.2), (6.2.3), there exists a unique $u_\varepsilon \in V$ such that

$$(6.2.5) \qquad\qquad a^\varepsilon(u_\varepsilon,v) = (f,v), \ \forall v \in V, \ f \in V'.$$

We want to "homogenize" this problem, which is indeed possible.

REMARK 6.8. If we define

$$(6.2.6) \qquad a_{ij}(x,y) = a_{ij}^k(y) \text{ in } \mathcal{O}^k, \qquad k = 1,\dots,N,$$

$$(6.2.7) \qquad a_0(x,y) = a_0^k(y) \text{ in } \mathcal{O}^k,$$

problem (6.2.5) enters into the framework of (6.1.6), but with functions a_{ij} **which are not in** $C^0(\overline{\mathcal{O}}; L_p^\infty(\mathbb{R}^n))$, but step-functions with values in $L_p^\infty(\mathbb{R}^n)$. Actually Theorem 6.9 below will show that (as one could easily surmise) the "regularity" condition $a_{ij} \in C^0(\overline{\mathcal{O}}; L_p^\infty(\mathbb{R}^n))$ is **not** necessary in order that Theorem 6.3 be true. (The condition $a_{ij} \in L^\infty(\mathcal{O}; L_p^\infty(\mathbb{R}^n))$ is too general, since then $a_{ij}\left(x,\frac{x}{\varepsilon}\right)$ need not even be measurable in x.)

The homogenization of (6.2.5) proceeds as follows. We define
$(6.2.8)$

$$\mathcal{A}^k = -q_{ij}^k\frac{\partial^2}{\partial x_i\partial x_j} = \text{ homogenized operator associated to } \frac{\partial}{\partial x_i}\left(a_{ij}^k\left(\frac{x}{\varepsilon}\right)\frac{\partial}{\partial x_j}\right)$$

and we define

$$(6.2.9) \qquad \mathcal{A}(u,v) = \sum_{k=1}^{N}\left[\int_{\mathcal{O}^k} q_{ij}^k\frac{\partial u}{\partial x_j}\frac{\partial v}{\partial x_i}\,dx + \int_{\mathcal{O}^k} q_0^k uv\,dx\right],$$

where $q_0^k = \mathcal{M}(a_0^k)$.

Let u be the solution of

$$(6.2.10) \qquad\qquad u \in V, \ \mathcal{A}(u,v) = (f,v), \ \forall v \in V.$$

---

[28]Equality (6.2.1) below does not take into account the interfaces between the sets $\mathcal{O}^k$.

One has

THEOREM 6.9. *We assume that (6.2.1), (6.2.2), (6.2.3) hold true. Let u_ε (respectively, u) be the solution of (6.2.5) (respectively, (6.2.10)). Then one has*

$$u_\varepsilon \to u \text{ in } V \text{ weakly.}$$

REMARK 6.10. Problem (6.2.5) is a **transmission problem**; there are **transmission boundary conditions** on the interfaces between adjacent sets $\mathcal{O}^k$ and $\mathcal{O}^p$. Problem (6.2.10) is the **homogenized problem**.

REMARK 6.11. Applying (formally) the rule of Theorem 6.3 to the present situation leads to (6.2.10).

REMARK 6.12. One can also introduce (at least in the symmetric case, i.e., $a_{ij}^k = a_{ji}^k \ \forall i, j, \forall k$) **correctors**. One has then to introduce

(6.2.11) $m_\varepsilon^k(x) = $ "truncation" function in $\mathcal{O}^k$ (cf. (5.2.2)).

We define

$$A_1^k = -\frac{\partial}{\partial y_i}\left(a_{ij}^k(y)\frac{\partial}{\partial y_j}\right)$$

and $\chi^{j,k}$ by

$$A_1^k(\chi^{j,k} - y_j) = 0.$$

Then we introduce the corrector

(6.2.12) $$\theta_\varepsilon(x) = -\varepsilon \sum_{k=1}^N \chi^{j,k}\left(\frac{x}{\varepsilon}\right)\frac{\partial u}{\partial x_j}(x)m_\varepsilon^k(x)$$

and we can prove that

(6.2.13) $$u_\varepsilon - (u + \theta_\varepsilon) \to 0 \text{ in } V \text{ strongly.}$$

PROOF OF THEOREM 6.9. We have

(6.2.14) $$\|u_\varepsilon\|_V \leq C.$$

If we set

(6.2.15) $$\xi_i^{\varepsilon,k} = a_{ij}^k\left(\frac{x}{\varepsilon}\right)\frac{\partial u_\varepsilon}{\partial x_j},$$

we can extract a subsequence, still denoted by u_ε, such that

(6.2.16) $$u_\varepsilon \to u \text{ in } V \text{ weakly,}$$

$$\xi_i^{\varepsilon,k} \to \xi_1^k \text{ in } L^2(\mathcal{O}^k) \text{ weakly, } k = 1, 2, \ldots, N.$$

Equation (6.2.5) gives in the limit

(6.2.17) $$\sum_k (\xi_i^k, v)_{L^2(\mathcal{O}^k)} = (f, v), \ \forall v \in V,$$

and it suffices to prove that

(6.2.18) $$\xi_i^k = q_{ij}^k \frac{\partial u}{\partial x_j} \text{ in } \mathcal{O}^k.$$

But this is proven exactly like we proved (3.2.18). We introduce w such that

$$(A_1^k)^* w = 0, \ w - P(y) \text{ is } Y \text{ periodic;}$$

We define next

$$w_\varepsilon(x) = \varepsilon w(x/\varepsilon).$$

We have then

(6.2.19) $(A^{\varepsilon,k})^* w_\varepsilon = 0$ in $\mathcal{O}^k$ (or in $\mathbb{R}^n$!),

where $A^{\varepsilon,k} = -\frac{\partial}{\partial x_i}\left(a_{ij}^k(x/\varepsilon)\frac{\partial}{\partial x_j}\right)$. We then take $v = \phi w_\varepsilon \in \mathcal{D}(\mathcal{O}^k)$, in (6.2.5). We multiply (6.2.19) by ϕm_ε and we integrate over $\mathcal{O}^k$; after subtracting, we can pass to the limit in ε and we obtain (6.2.18) as in Section 3.   ∎

6.3. Proof of Theorem 6.3. [29] We are going to "approximate" the a_{ij}'s by step functions and use Theorem 6.9. We introduce a "small" parameter h[30] and a "triangulation" of $\mathcal{O}$:

(6.3.1) $\mathcal{O} = \bigcup_{k=1}^{N(h)} \mathcal{O}_h^k$ (equality up to interfaces),

$$\mathcal{O}_h^k \cap \mathcal{O}_h^{k'} = \varnothing, \text{ if } k \neq k',$$

$$(\text{diameter of } \mathcal{O}_h^k) \leq Ch.$$

Of course $N(h) \to \infty$ as $h \to 0$.

For every h we define step functions $a_{ijh}(x, y)$ by

$$a_{ijh}(x, y) = a_{ijh}^k(y) \text{ in } \mathcal{O}_h^k,$$

(6.3.2) $a_{ijh}^k(y) = \dfrac{1}{|\mathcal{O}_h^k|} \displaystyle\int_{\mathcal{O}_h^k} a_{ij}(x, y)\, dx,$

$$|\mathcal{O}_h^k| = \text{ measure of } \mathcal{O}_h^k,$$

and we define

(6.3.3) $a_h^\varepsilon(u, v) = \displaystyle\int_\mathcal{O} a_{ijh}\left(x, \dfrac{x}{\varepsilon}\right) \dfrac{\partial u}{\partial x_j} \dfrac{\partial v}{\partial x_i}\, dx + \displaystyle\int_\mathcal{O} a_{0h}\left(x, \dfrac{x}{\varepsilon}\right) uv\, dx$

where a_{0h} is defined in a similar manner.[31]

Let $u_{\varepsilon h}$ be the solution of

(6.3.4) $a_h^\varepsilon(u_{\varepsilon h}, v) = (f, v), \ \forall v \in V, \qquad u_{\varepsilon h} \in V.$

Then since the a_{ij}'s are uniformly continuous from $\overline{\mathcal{O}}$ $L_p^\infty(\mathbb{R}^n)$, one has

(6.3.5) $u_\varepsilon - u_{\varepsilon h} \to 0$ in V as $h \to 0$, uniformly in ε.

For $h > 0$, **fixed**, we can homogenize (6.3.4), using Theorem 6.9. We find that

(6.3.6) $u_{\varepsilon h} \to u_h$ in V weakly,

where u_h is the solution of

(6.3.7) $\mathcal{A}_h(u_h, v) = (f, v), \ \forall v \in V, \ u_h \in V,$

and where $\mathcal{A}_h$ is computed as follows. We define

(6.3.8) $A_{1h}^k(y) = -\dfrac{\partial}{\partial y_i}\left(a_{ijh}^k(y)\dfrac{\partial}{\partial y_j}\right),$

---

[29] Another approach is presented in Section 6.4 below.
[30] Which plays the role of the parameter h in the **finite element** methods.
[31] For the symmetry of the procedure. The zero order term is straightforward anyway.

and we introduce $\chi_h^{jk}(y)$ as the solution (up to an additive constant) of

(6.3.9) $\qquad A_{ih}^k(y)(\chi_h^{jk}(y) - y_j) = 0, \qquad \chi_h^{jk}(y)$ Y-periodic.

Then if we set

(6.3.10) $\qquad q_{ijh}^k = \dfrac{1}{|Y|} a_{1h}^k(\chi_h^{jk}(y) - y_j, \chi_h^{ik}(y) - y_i),$

$$a_{1h}^k(\phi, \psi) = \int_Y a_{ijh}^k(y) \frac{\partial \phi}{\partial y_j} \frac{\partial \psi}{\partial y_i} \, dy,$$

then

(6.3.11) $\qquad \mathcal{A}_h(u, v) = \displaystyle\sum_{k=1}^{N(h)} \left[\int_{\mathcal{O}_h^k} q_{ijh}^k \frac{\partial u}{\partial x_j} \frac{\partial v}{\partial x_i} \, dx + \int_{\mathcal{O}_h^k} q_{0h}^k uv \, dx \right].$

One then verifies that

(6.3.12) $\qquad u_h - u \to 0$ in V as $h \to 0$.

and the theorem follows. ∎

6.4. Another approach to Theorem 6.3. We can give a proof of Theorem 6.3 along the lines of Section 3. The solution u_ε of (6.1.6) satisfies

(6.4.1) $\qquad \|u_\varepsilon\|_V \leq C.$

If we set

$$\xi_i^\varepsilon = a_{ij}\left(x, \frac{x}{\varepsilon}\right) \frac{\partial u_\varepsilon}{\partial x_j}(x),$$

we have

(6.4.2) $\qquad |\xi_i^\varepsilon|(L^2(\mathcal{O}) \text{ norm}) \leq C$

and we can extract a subsequence, still denoted by u_ε, such that

(6.4.3) $\qquad u_\varepsilon \to u$ in V weakly,

$$\xi_i^\varepsilon \to \xi_i \text{ in } L^2(\mathcal{O}) \text{ weakly.}$$

The equation (6.1.6), which can be written $\left(\xi_i^\varepsilon, \frac{\partial v}{\partial x_i}\right) = (f, v)$, $v \in V$ gives in the limit

(6.4.4) $\qquad \left(\xi_i, \dfrac{\partial v}{\partial x_i}\right) = (f, v), \quad \forall v \in V.$

As in Section 3.2, the main point is now to use "adjoint" functions. In Section 3.2 we introduced w_ε such that[32]

(6.4.5) $\qquad (A^\varepsilon)^* w_\varepsilon = 0,$

and w_ε has a known behavior as $\varepsilon \to 0$.

Due to the fact that the coefficients depend on x **and on** y, the construction of w_ε as in Section 3.2 is no longer possible. But we are going to construct w_ε with a known behavior as $\varepsilon \to 0$ and satisfying

(6.4.6) $\qquad (A^\varepsilon)^* w_\varepsilon = g_\varepsilon$

---

[32] Assuming that $a_0 = 0$, which does not restrict the generality.

where g_ε has a known behavior.[33] It suffices to look for w_ε in the form

(6.4.7) $$w_\varepsilon = \varepsilon(y_p - \beta), \qquad p = \text{ given index.}$$

Using the "adjoint identity" of (6.1.16), an identification computation leads to

(6.4.8) $$(A_1(x))^*(\beta - y_p) = 0,$$

and then

(6.4.9) $$g_\varepsilon = A_2^*(y_p - \beta) + \varepsilon A_3^*(y_p - \beta).$$

If we define (compare to (6.1.10)), $\widehat{\chi}^p(x, y)$ by

(6.4.10) $$A_1(x)^*(\chi^p - y_p) = 0, \qquad \widehat{\chi}^p \ Y\text{-periodic in } y,$$

$$\text{and } \widehat{\chi}^p \text{ satisfies for instance } \int_Y \widehat{\chi}^p(x, y)\, dy = 0,$$

we can take $\beta = \widehat{\chi}^p$. We have

(6.4.11) $$w_\varepsilon \to x_p \text{ in } L^2(\mathcal{O}),$$

(6.4.12) $$g_\varepsilon \to -\frac{\partial}{\partial x_i} \mathcal{M}_y \left(a_{ij}^*(x, y) \frac{\partial}{\partial y_j}(y_p - \widehat{\chi}^p(x, y)) \right) \text{ in } L^2(\mathcal{O}) \text{ weakly,}$$

where $a_{ij}^* = a_{ji}$ and where in (6.4.12) we assume all functions to be smooth enough[34] so that

(6.4.13) $$\varepsilon A_3^*(y_p - \beta) = -A_3^*\widehat{\chi}^p \to 0 \text{ in } L^2(\mathcal{O}).$$

We now take $v = \phi w_\varepsilon$ in (6.1.6), we multiply (6.1.5) by ϕu_ε and we integrate over $\mathcal{O}$; after subtracting, we obtain

(6.4.14) $$\left(\xi_i^\varepsilon, \left(\frac{\partial \phi}{\partial x_i} \right) w_\varepsilon \right) - \int a_{ij}\left(x, \frac{x}{\varepsilon} \right) \frac{\partial w_\varepsilon}{\partial x_i} \frac{\partial \phi}{\partial x_j} u_\varepsilon = (f, \phi w_\varepsilon) - (g_\varepsilon, \phi u_\varepsilon).$$

Passing to the limit, we obtain

(6.4.15) $$\left(\xi_i, \frac{\partial \phi}{\partial x_i} x_p \right) - \int_\mathcal{O} \mathcal{M}_y \left(a_{ij}(x, y) \frac{\partial}{\partial y_i}(y_p - \widehat{\chi}^p) \right) \frac{\partial \phi}{\partial x_j} u\, dx$$

$$= (f, \phi w_p) + \int_\mathcal{O} \left(\frac{\partial}{\partial x_i} \mathcal{M}_y \left(a_{ij}^*(x, y) \frac{\partial}{\partial y_j}(y_p - \widehat{\chi}^p) \right) \right) \phi u\, dx.$$

---

[33] We will use several times in the following of this book this important method.

[34] We have

$$g_\varepsilon = A_2^*(y_p - \widehat{\chi}^p) - \varepsilon \frac{\partial}{\partial x_i}\left(a_{ij}^*(x, y) \frac{\partial \widehat{\chi}^p}{\partial x_j}(x, y) \right).$$

The last term in $g_\varepsilon \to 0$, in, say, $L^2(\mathcal{O})$ (this can be improved) if $\widehat{\chi}^p$ is C^2 in $x \in \overline{\mathcal{O}}$ with values in $W(Y)$. In obtaining (6.4.12) we also use the fact that if $\phi \in C^0(\overline{\mathcal{O}}; L_p^\infty(\mathbb{R}^n))$, then

$$\phi\left(x, \frac{x}{\varepsilon} \right) \to \mathcal{M}_y(\phi(x, y)) \text{ in } L^\infty(\mathcal{O}) \text{ weak star}$$

where

$$\mathcal{M}_y(\phi(x, y)) = \frac{1}{|Y|} \int_Y \phi(x, y)\, dy.$$

The regularity hypotheses needed here on the a_{ij}'s can then be weakened by a density argument as in Section 8 below.

But using (6.4.14), (6.4.15) reduces to

$$\int_{\mathcal{O}} \mathcal{M}_y \left(a_{ij} \frac{\partial}{\partial y_i} (y_p - \widehat{\chi}^p) \right) \phi \frac{\partial u}{\partial x_j} \, dx + \int_{\mathcal{O}} \left(\frac{\partial}{\partial x_j} \mathcal{M}_y (a_{ij} \frac{\partial}{\partial y_i} (y_p - \widehat{\chi}^p)) \right) \phi u \, dx$$

$$= (\xi_p, \phi) + \int_{\mathcal{O}} \frac{\partial}{\partial x_i} \mathcal{M}_y (a_{ij}^* \frac{\partial}{\partial y_j} (y_p - \widehat{\chi}^p)) \phi u \, dx$$

and therefore

$$(6.4.16) \qquad \xi_p = \mathcal{M}_y \left(a_{ij}(x,y) \frac{\partial}{\partial y_i} (y_p - \widehat{\chi}^p) \right) \frac{\partial u}{\partial x_j}.$$

It is now a simple exercise (similar to Section 3.2) to verify that we obtain the same formulas as in (6.1.12), (6.1.14).

7. Complements on boundary conditions

7.1. A remark on the nonhomogeneous Neumann's problem. Let us consider problem (3.1.4), Remark 3.4, Section 3.1, with $V = H^1(\mathcal{O})$ and with

$$(7.1.1) \qquad (f_\varepsilon, v) = \int_{\mathcal{O}} f_0 \left(x, \frac{x}{\varepsilon} \right) v(x) \, dx + \int_\Gamma f_1 \left(x, \frac{\varepsilon}{x} \right) v(x) \, d\Gamma(x)$$

where

$$(7.1.2) \qquad f_0(x,y) \in C^0(\overline{\mathcal{O}}; L^\infty(\mathbb{R}^n_y)), \qquad f_1 \in C^0(\overline{\mathcal{O}} \times \mathbb{R}^n_y),$$
$$f_1 \text{ is } Y\text{-periodic in } y,$$

and where we assume that the boundary Γ is smooth enough.

The interpretation of (3.1.4) is formally[35]

$$(7.1.3) \qquad A^\varepsilon u_\varepsilon = f_0 \left(x, \frac{x}{\varepsilon} \right) \text{ in } \mathcal{O},$$

$$(7.1.4) \qquad \frac{\partial u_\varepsilon}{\partial \nu_{A^\varepsilon}} = f_1 \left(x, \frac{x}{\varepsilon} \right) \text{ on } \Gamma.$$

Then one can apply Theorem 3.1 **if (3.1.5) takes place**.

As we already noticed

$$(7.1.5) \qquad f_0 \left(x, \frac{x}{\varepsilon} \right) \to \mathcal{M}_y(f_0(x,y)) \text{ in } L^\infty \text{ weak star}$$

where $\mathcal{M}_y(f_0(x,y)) = \frac{1}{|Y|} \int_Y f_0(x,y) \, dy$.

The limit of the **surface integral** in (7.1.1) is now completed. We have to distinguish a "generic" case from an "exceptional" one:

$(7.1.6)$ (**Generic behavior**): We assume that Γ does not contain flat pieces or that it contains finitely many flat pieces with conormal not proportional to any $m \in \mathbb{Z}^n$; then

$$f_1 \left(x, \frac{x}{\varepsilon} \right) \to \mathcal{M}_y(f_1(x,y)) \text{ in } L^\infty(\Gamma) \text{ weak star.}[36]$$

---

[35]The interpretation if formal under the hypothesis $a_{ij} \in L^\infty$.

(7.1.7)   (**Exceptional behavior**): We assume that Γ contains one[37]
                flat piece, say Γ_0, with conormal $m_0 \in \mathbb{Z}^n$, the remaining
                part of Γ, say Γ_1, being generic. Then one has of course
                the generic behavior on Γ_1; on Γ_0, one can extract
                subsequences $\varepsilon', \varepsilon'', \ldots$, such that if $\varepsilon = \varepsilon'$
                respectively, $\varepsilon'', \ldots$,

$$f_1\left(x, \frac{x}{\varepsilon}\right) \to f'(x) \text{ (respectively, } f''(x), \ldots) \text{ in } L^\infty(\Gamma_0)$$
weak star (where the $f', f'', \ldots$, are different).

The proof of (7.1.6), (7.1.7) is given in Section 7.3 below. The application to
the problem (7.1.3), (7.1.4) is straightforward.

(7.1.8)   In the generic case (7.1.6), one has $u_\varepsilon \to u$ in $H^1(\mathcal{O})$
                weakly where u is the solution of

$$\mathcal{A}u = \mathcal{M}_y(f_0(x, y)) \text{ in } \mathcal{O},$$
$$\frac{\partial u}{\partial \nu_{\mathcal{A}}} = \mathcal{M}_y(f_1(x, y)) \text{ on } \Gamma;$$

(7.1.9)   In the exceptional case (7.1.7), one can extract subsequences
                $u_{\varepsilon'}, u_{\varepsilon''}, \ldots$, which converge weakly in $H^1(\mathcal{O})$ to $u', u'', \ldots$, where

$$\mathcal{A}u' = \mathcal{A}u'' = \cdots = \mathcal{M}_y(f_0(x, y)) \text{ in } \mathcal{O},$$
$$\frac{\partial u'}{\partial \nu_{\mathcal{A}}} = \frac{\partial u''}{\partial \nu_{\mathcal{A}}} = \cdots = \mathcal{M}_y(f_1(x, y)) \text{ on } \Gamma_1,$$
$$\frac{\partial u'}{\partial \nu_{\mathcal{A}}} = f', \qquad \frac{\partial u''}{\partial \nu_{\mathcal{A}}} = f'', \qquad \text{on } \Gamma_0.$$

7.2. Higher order boundary conditions. We consider a boundary value
problem in the **half space**

(7.2.1) $\mathcal{O} = \{x \mid x_n > 0\}, \qquad \Gamma = \{x \mid x_n = 0\}.$

We shall set

$$x' = \{x_1, \ldots, x_{n-1}\}, \ y' = x'/\varepsilon;$$
we consider functions

(7.2.2) a_{ij}, a_0 satisfying (1.1.1), (1.1.2),

---

[36] Here $f_1\left(x, \frac{x}{\varepsilon}\right)$ denotes in fact its **restriction** to Γ.
[37] Extension to the case of a finite number of pieces is obvious.

$$b_{ij}(y') \in L^\infty(\mathbb{R}^{n-1}), \ b_{ij} \text{ is } Y'\text{-periodic},$$

(7.2.3)
$$Y' = \prod_{j=1}^{n-1}]0, y_j'^0[,$$

$$b_{ij}(y')\xi_i\xi_j \geq \beta\xi_i\xi_i, \ \beta > 0, \text{ a.e. in } y',$$

(7.2.4)
$$b_0 \text{ is } Y' \text{ periodic}, \ b_0 \in L^\infty(\mathbb{R}^{n-1}),$$

$$b_0(y') \geq \beta_0 > 0 \text{ a.e. in } y'.$$

REMARK 7.1. The period $y_j'^0$ $(1 \leq j \leq n-1)$ can be different from the periods y_j^0.

NOTATIONS. We introduce the space

(7.2.5)
$$V = \{v \mid v \in H^1(\mathcal{O}), \ v(x', 0) \in H^1(\mathbb{R}^{n-1})\},$$

which is a Hilbert space for the norm given by

(7.2.6)
$$\|v\|_V^2 = \|v\|_{H^1(\mathcal{O})}^2 + \|v(x', 0)\|_{H^1(\mathbb{R}^{n-1})}^2.$$

For $u, v \in V$, we set

(7.2.7)
$$a^\varepsilon(u, v) = \int_{mc\mathcal{O}} a_{ij}\left(\frac{x}{\varepsilon}\right) \frac{\partial u}{\partial x_j} \frac{\partial v}{\partial x_i} \, dx + \int_{\mathcal{O}} a_0 uv \, dx,$$

(7.2.8)
$$b^\varepsilon(u, v) = \int_{\mathbb{R}^{n-1}} b_{ij}\left(\frac{x'}{\varepsilon}\right) \frac{\partial u}{\partial x_j}(x', 0) \frac{\partial v}{\partial x_i}(x', 0) \, dx'$$
$$+ \int_{\mathbb{R}^{n-1}} b_0\left(\frac{x'}{\varepsilon}\right) uv(x', 0) \, dx',$$

and we consider the boundary value problem

(7.2.9)
$$a^\varepsilon(u_\varepsilon, v) + b^\varepsilon(u_\varepsilon, v) = (f, v), \ \forall v \in V,$$

$$u_\varepsilon \in V.$$

By virtue of (7.2.2), (7.2.3), (7.2.4), problem (7.2.9) admits a unique solution.
The interpretation of (7.2.9) is as follows:

$$A^\varepsilon u_\varepsilon = f \text{ in } \mathcal{O},$$

(7.2.10)
$$\frac{\partial u_\varepsilon}{\partial \nu_{A^\varepsilon}} + B^\varepsilon u_\varepsilon = 0 \text{ on } \Gamma = \mathbb{R}^{n-1},$$

where

(7.2.11)
$$B^\varepsilon u_\varepsilon = -\frac{\partial}{\partial x_i} b_{ij}\left(\frac{x'}{\varepsilon}\right) \frac{\partial}{\partial x_j} u_\varepsilon(x', 0) + b_0\left(\frac{x'}{\varepsilon}\right) u_\varepsilon(x', 0)$$

where the summation in i, j extends here from 1 to $n-1$.

Let us denote by

(7.2.12)
$$\mathcal{A} = \text{ homogenized operator (in } \mathbb{R}^n) \text{ of } A^\varepsilon,$$

$$\mathcal{B} = \text{ homogenized operator (in } \mathbb{R}^{n-1}) \text{ of } B^\varepsilon.$$

Let $\mathcal{A}(u, v)$, $\mathcal{B}(u, v)$ be the bilinear forms associated to $\mathcal{A}$ and $\mathcal{B}$. Then the **homogenized problem** is

(7.2.13)
$$\mathcal{A}(u, v) + \mathcal{B}(u(x', 0), v(x', 0)) = (f, v), \ \forall v \in V, \ u \in V,$$

and

(7.2.14) $u_\varepsilon \to u$ in V weakly when $\varepsilon \to 0$.

The proof is as follows. One has first $\|u_\varepsilon\|_V \leq C$; we can therefore extract a subsequence, still denoted by u_ε, such that

$$u_\varepsilon \to u \text{ in } V \text{ weakly,}$$

(7.2.15) $\xi_i^\varepsilon = a_{ij}\left(\dfrac{x}{\varepsilon}\right) \dfrac{\partial u_\varepsilon}{\partial x_j} \to x_i$ in $L^2(\mathcal{O})$ weakly;

if we set

(7.2.16) $w_\varepsilon(x') = u_\varepsilon(x', 0)$

we have

(7.2.17) $w_\varepsilon \to w = u(x', 0)$ in $H^1(\mathbb{R}^{n-1})$ weakly.

We define $\dfrac{\partial u_\varepsilon}{\partial \nu_{A^\varepsilon}} \in H^{-1/2}(\mathbb{R}^{n-1})^{38}$ by

(7.2.18)

$$\int_{\mathbb{R}^{n-1}} \frac{\partial u_\varepsilon}{\partial \nu_{A^\varepsilon}} \phi(x') \, dx' = a^\varepsilon(u_\varepsilon, \Phi) - (f, \Phi) = \left(\xi_i^\varepsilon, \frac{\partial \Phi}{\partial x_i}\right) + (a_0^\varepsilon u_\varepsilon, \Phi) - (f, \Phi),$$

$\phi \in H^{1/2}(\mathbb{R}^{n-1})$, $\Phi \in H^1(\mathcal{O})$, $\Phi(x', 0) = \phi$,

$\phi \to \Phi$ being a continuous linear mapping from $H^{1/2}(\mathbb{R}^{n-1}) \to H^1(\mathcal{O})$.

This gives a precise meaning to the boundary condition in (7.2.10), and it implies that

(7.2.19) $\dfrac{\partial u_\varepsilon}{\partial \nu_{A^\varepsilon}} \to \zeta$ in $H^{-1/2}(\mathbb{R}^{n-1})$ weakly,

where ζ is given by

(7.2.20) $\displaystyle\int_{\mathbb{R}^{n-1}} \zeta \phi \, dx' = \left(\xi_i, \frac{\partial \Phi}{\partial x_i}\right) - (f, \Phi) + (\mathcal{M}(a_0)u, \Phi).$

The boundary condition in (7.2.10) is written

(7.2.21) $B^\varepsilon w_\varepsilon = -\dfrac{\partial u_\varepsilon}{\partial \nu_{A^\varepsilon}}$

and we think of (7.2.21) as **an equation** in $\mathbb{R}^{n-1}$.

We can apply the theory of Sections 1 to 3. Indeed by virtue of

(7.2.22) $\dfrac{\partial u_\varepsilon}{\partial \nu_{A^\varepsilon}} \to -\zeta$ in $H_{local}^{-1}(\mathbb{R}^{n-1})$ **strongly**

which is sufficient to prove the desired results (in Section 3), we have

(7.2.23) $w_\varepsilon \to w$ in $H^1(\mathbb{R}^{n-1})$ weakly, where w is the solution of $\mathcal{B}w = -\zeta$.

On the other hand the reasoning of Section 3 gives

(7.2.24) $\xi_i = q_{ij}\dfrac{\partial u}{\partial x_j}$, $q_{ij} =$ coefficient of $(-\mathcal{A})$;

---

[38] For every real s, we define

$$H^s(\mathbb{R}^{n-1}) = \{\phi \mid (1 + |\xi|)^s \hat{\phi}(\xi) \in L^2(\mathbb{R}_\xi^{n-1})\},$$

where $\hat{\phi} =$ Fourier transform of ϕ. (cf. Lions–Magenes [**71**] and [**72**].)

using (7.2.23) and (7.2.24) in (7.2.20) gives

$$-\int_{\mathbb{R}^{n-1}} (\mathcal{B}w)\Phi(x',0)\,dx' = \left(q_{ij}\frac{\partial u}{\partial x_j}, \frac{\partial \Phi}{\partial x_i}\right) - (\mathcal{M}(a_0)u, \Phi) - (f, \Phi)$$

$\forall \Phi \in V$, i.e., since $w = u(x',0)$, the equation (7.2.13).

7.3. Proof of (7.1.6), (7.1.7). Let us prove first (7.1.6) (generic behavior). Since $f_1(x, x/\varepsilon)$ belongs (under assumption (7.1.2)) to a bounded set of $L^\infty(\Gamma)$, we have only to show that

$$(7.3.1) \qquad \int_\Gamma f_1\left(x, \frac{x}{\varepsilon}\right) v(x)\,d\Gamma(x) \to \int_\Gamma \mathcal{M}_y(f_1)v(x)\,d\Gamma(x),$$

$\forall v$ smooth and with compact support on Γ.

Since $C^k(\overline{\mathcal{O}}) \otimes C^k(\mathbb{R}^n_y)$ (with periodic functions in y) is dense in $C^0(\overline{\mathcal{O}} \times \mathbb{R}^n_y)$ (with periodic functions in y), and since formulas in (7.3.1) are continuous for the $C^0(\overline{\mathcal{O}} \times \mathbb{R}^n_y)$ topology, it suffices to prove (7.3.1) for

$$(7.3.2) \qquad f_1(x, y) = \text{ finite sum of functions } \alpha(x)\beta(y), \ \alpha \text{ and } \beta \text{ being } C^k;$$

here we can take k as large as we want and even $k = \infty$. Replacing αv by v, we are led to proving that

$$(7.3.3) \qquad \int_\Gamma \beta\left(\frac{x}{\varepsilon}\right) v(x)\,d\Gamma(x) = \int_\Gamma \mathcal{M}(\beta)v(x)\,d\Gamma(x),$$

where β is C^∞ and Y-periodic with all its derivatives.

We do not restrict the generality by assuming $Y =]0, 1[^n$ and we can use Fourier series:

$$(7.3.4) \qquad \beta(y) = \sum_p \beta_p \exp(2\pi i p y), \ p \in \mathbb{Z}^n.$$

In (7.3.4) the series is absolutely and uniformly convergent (together with all its derivatives) and one is reduced to proving (7.3.3) for

$$(7.3.5) \qquad \beta(y) = \textbf{finite} \text{ sum of } \beta_p \exp(2\pi i p y),$$

i.e., that

$$(7.3.6) \qquad \int_\Gamma \exp\left(2\pi i p \frac{x}{\varepsilon}\right) v(x)\,d\Gamma(x) \to 0, \text{ if } p \neq 0.$$

This is a classical result (stationary phase approximation; cf. Bibliography of Chapter 4) under the hypothesis made on Γ in (7.1.6).

PROOF OF (7.1.7). Let us consider a single particular case, where

$$(7.3.7) \qquad f_1(x, y) = \exp(2\pi i p y), \ p \text{ fixed in } \mathbb{Z}^n,$$

and let us assume that Γ is contained in the hyperplane

$$(7.3.8) \qquad m_0 x = \gamma, \ m_0 \in \mathbb{Z}^n.$$

If $p \neq k m_0$, $k \in \mathbb{Z}$, then $\int_\Gamma f_1\left(x, \frac{x}{\varepsilon}\right) v(x)\,d\Gamma(x) \to 0$. But if $p = k m_0$, then

$$(7.3.9) \qquad \int_\Gamma f_1\left(x, \frac{x}{\varepsilon}\right) v(x)\,d\Gamma(x) = \exp\left(\frac{2\pi i \gamma}{\varepsilon}\right) \int_\Gamma v(x)\,d\Gamma(x),$$

which admits indeed infinitely many different limits depending on the way ε tends to 0. ∎

8. Reiterated homogenization

8.1. Setting of the problem: Statement of the main result. We define

$$Y = \prod_{j=1}^{n}]0, y_j^0[,$$

$$Z = \prod_{j=1}^{n}]0, z_j^0[,$$

and we consider functions $a_{ij}(y, z)$ such that

$$a_{ij} \in L^\infty(\mathbb{R}^n \times \mathbb{R}^n),$$

(8.1.1) a_{ij} is $Y - Z$ periodic,

$$a_{ij}(y, z)\xi_i\xi_j \geq \alpha\xi_i\xi_i, \quad \alpha > 0, \text{ a.e. in } y, z.$$

We also consider $a_0(y, z)$ such that

$$a_0 \in L^\infty(\mathbb{R}^n \times \mathbb{R}^n),$$

(8.1.2) a_0 is $Y - Z$ periodic,

$$a_0(y, z) \geq \alpha_0 > 0, {}^{39}$$

and we define now[40] A^ε by

$$(8.1.3) \qquad A^\varepsilon = -\frac{\partial}{\partial x_i}\left(a_{ij}\left(\frac{x}{\varepsilon}, \frac{x}{\varepsilon^2}\right)\frac{\partial}{\partial x_j}\right) + a_0\left(\frac{x}{\varepsilon}, \frac{x}{\varepsilon^2}\right).$$

We want to study the behavior of the solution u_ε of boundary value problems associated to A^ε as $\varepsilon \to 0$.

REMARK 8.1. Under the mere hypotheses (8.1.1), (8.1.2), the functions $a_{ij}\left(\frac{x}{\varepsilon}, \frac{x}{\varepsilon^2}\right)$, $a_0\left(\frac{x}{\varepsilon}, \frac{x}{\varepsilon^2}\right)$ need not even be measurable. We shall make from now on the hypothesis that

$$(8.1.4) \qquad a_{ij}(y, z) \in C^0(\mathbb{R}^n_z; L^\infty(\mathbb{R}^n_y)),$$

$$a_0(y, z) \in C^0(\mathbb{R}^n_z; L^\infty(\mathbb{R}^n_y)).$$

REMARK 8.2. If functions a_{ij}, a_0 are constant in z, the problem reduces to the one considered in Sections 1 to 3.

REMARK 8.3. All that we are going to say readily extends to the case of operators with coefficients

$$a_{ij}\left(\frac{x}{\varepsilon}, \frac{x}{\varepsilon^2}, \ldots, \frac{x}{\varepsilon^N}\right)$$

which are periodic — with different periods — in $y = x/\varepsilon$, $z^{(1)} = x/\varepsilon^2, \ldots, z^{(N-1)} = x/\varepsilon^N$. We would assume that

$$(8.1.5) \qquad a_{ij} \in C^0(\mathbb{R}^n_{z^{(1)}} \times \cdots \times \mathbb{R}^n_{z^{(N-1)}}; L^\infty(\mathbb{R}^n_y)).$$

REMARK 8.4. We can also extend the results to follow to the case of coefficients $a_{ij}\left(x, \frac{x}{\varepsilon}, \frac{x}{\varepsilon^2}\right)$, assuming that

$$(8.1.6) \qquad a_{ij}(x, y, z) \in C^0(\overline{\mathcal{O}}_x \times \mathbb{R}^n_z; L^\infty(\mathbb{R}^n_y)).$$

---

[39]As usual this hypothesis is not necessary if, for instance, V consists of functions which are zero on a set Γ_0 of positive measure on Γ.

[40]See Remark 8.1 below.

8.1.1. *Statement of main result: Homogenization with parameter.* Let us consider first a family of functions $a_{ij}(\lambda, y)$, λ = parameter; $\lambda \in \Lambda$, Λ = topological set. We assume that

(8.1.7)
$$a_{ij}(\lambda, y) \text{ is } Y \text{ periodic, } \forall \lambda \in \Lambda,$$
$$a_{ij}(\lambda, y)\xi_i\xi_j \geq \alpha\xi_i\xi_j, \ \alpha > 0,$$

and we also consider

(8.1.8)
$$a_0(\lambda, y) \ Y \text{ periodic, } \forall \lambda \in \Lambda, \ a_0(\lambda, y) \geq \alpha_0 > 0,$$

Then the operator

(8.1.9)
$$A^\varepsilon_{(\lambda)} = -\frac{\partial}{\partial x_i}\left(a_{ij}\left(\lambda, \frac{x}{\varepsilon}\right)\frac{\partial}{\partial x_j}\right) + a_0\left(\lambda, \frac{x}{\varepsilon}\right)$$

is **homogenized by the standard formulas, λ being a parameter.** This (entirely obvious) operation is called "**homogenization with parameter.**"

We homogenize A^ε — as given by (8.1.3) — in two steps, by "**reiteration of the homogenization.**" We consider first the operator

(8.1.10)
$$-\frac{\partial}{\partial x_i}a_{ij}\left(\lambda, \frac{x}{\varepsilon}\right)\frac{\partial}{\partial x_j} + a_0\left(\lambda, \frac{x}{\varepsilon}\right)$$

which is homogenized with λ as a parameter. More precisely we introduce

(8.1.11)
$$A_{1,\lambda} = -\frac{\partial}{\partial z_i}a_{ij}(\lambda, z)\frac{\partial}{\partial z_j} + a_0(\lambda, z)$$

and we define $\chi^j(z)$ as the Z-periodic solution of

(8.1.12)
$$A_{1,\lambda}(\chi^j_\lambda(z) - z_j) = 0 \text{ such that } \int_Z \chi^j(z)\, dz = 0;\,[41]$$

for $\phi, \psi \in H^1(Z)$, we set

(8.1.13)
$$a_{1,\lambda}(\phi, \psi) = \int_Z a_{ij}(\lambda, z)\frac{\partial \phi}{\partial z_j}\frac{\partial \psi}{\partial z_i}\, dz + \int_Z a_0(\lambda, z)\phi\psi\, dz,$$

then the homogenized operator of (8.1.10) is given by

(8.1.14)
$$\mathcal{A}_{(\lambda)} = -\frac{\partial}{\partial x_i}q^1_{ij}(\lambda)\frac{\partial}{\partial x_j} + a_0(\lambda),$$

where

(8.1.15)
$$q^1_{ij}(\lambda) = -\frac{1}{|Z|}a_{1,\lambda}(\chi^j_\lambda - z_j, \chi^i_\lambda - z_i)$$

and

(8.1.16)
$$q^1_0(\lambda) = \frac{1}{|Z|}\int_Z a_0(\lambda, z)\, dz.$$

We consider then the operator

(8.1.17)
$$\mathcal{A}^\varepsilon = -\frac{\partial}{\partial x_i}\left(q^1_{ij}\left(\frac{x}{\varepsilon}\right)\frac{\partial}{\partial x_j}\right) + q^1_0\left(\frac{x}{\varepsilon}\right)$$

which is homogenized by

(8.1.18)
$$\mathcal{A} = -\frac{\partial}{\partial x_i}q_{ij}\frac{\partial}{\partial x_j} + q_0$$

---

[41]This implies that χ^j_λ depends continuously on λ.

where q_{ij} is obtained as follows. We introduce

(8.1.19) $$\mathcal{A}_1 = -\frac{\partial}{\partial y_i} q_{ij}(y) \frac{\partial}{\partial y_j} + q_0^1(y);$$

for $\phi, \psi \in H^1(Y)$, we set

(8.1.20) $$\mathcal{A}_1(\phi, \psi) = \int_Y q_{ij}^1(y) \frac{\partial \phi}{\partial y_j} \frac{\partial \psi}{\partial y_i} \, dy + \int_Y q_0^1(y) \phi \psi \, dx,$$

and we define χ^j as a solution of

(8.1.21) $$\mathcal{A}_1(\chi^j - y_j) = 0, \ \chi^j \text{ is } Y \text{ periodic.}$$

Then

(8.1.22) $$q_{ij} = \frac{1}{|Y|} \mathcal{A}_1(\chi^j - y_j, \chi^i - y_i)$$

and

(8.1.23) $$q_0 = \frac{1}{|Y|} \int_Y q_0^1(y) \, dy.$$

It can be verified that $\mathcal{A}$ is **elliptic**. For $u, v \in H^1(\mathcal{O})$ we set

(8.1.24) $$a^\varepsilon(u, v) = \int a_{ij}\left(\frac{x}{\varepsilon}, \frac{x}{\varepsilon^2}\right) \frac{\partial u}{\partial x_j} \frac{\partial v}{\partial x_i} \, dx + \int a_0\left(\frac{x}{\varepsilon}, \frac{x}{\varepsilon^2}\right) uv \, dx,$$

(8.1.25) $$\mathcal{A}(u, v) = \int q_{ij} \frac{\partial u}{\partial x_j} \frac{\partial v}{\partial x_i} \, dx + \int q_0 uv \, dx.$$

The boundary value problems we consider are:

(8.1.26) $$u_\varepsilon \in V, a^\varepsilon(u_\varepsilon, v) = (f, v), \ \forall v \in V, \ f \in V',$$

(8.1.27) $$u \in V, \ \mathcal{A}(u, v) = (f, v), \ \forall v \in V;$$

the problem (8.1.27) is called the **homogenized problem** of (8.1.26).

According to Section 4, **there exists $p > 2$ such that**

(8.1.28) $$\text{if } V = H_0^1(\mathcal{O}), \text{ and if } f \in W^{-1,p}(\mathcal{O}),$$

$$\text{then } \left\|\frac{\partial u_\varepsilon}{\partial x_i}\right\|_{L^p(\mathcal{O})} \leq c, \ \forall i;$$

(8.1.29) $$\text{if } \phi f \in W^{-1,p}(\mathcal{O}),[42]$$

$$\text{then } \left\|\phi \frac{\partial u_\varepsilon}{\partial x_i}\right\|_{L^p(\mathcal{O})} \leq c, \ \forall i, \text{ where } \phi \in C_0^\infty(\mathcal{O}).$$

The value "p" (just a little larger than 2) is used in Theorem 8.5 below.

THEOREM 8.5. *We assume that (8.1.1), (8.1.2), (8.1.4) hold true. We assume that*

(8.1.30) $$f \in W^{-1,p}(\mathcal{O}), \ if \ V = H_0^1(\mathcal{O})$$

or that

(8.1.31) $$\phi f \in W^{-1,p}(\mathcal{O}), \ \forall \phi \in C_0^\infty(\mathcal{O}), \ V \ arbitrary.$$

---

[42]Here p is restricted to $p \leq \frac{2n}{n-2}$, but one is interested in taking p **as close to 2 as possible**.

Let $\mathcal{A}$ be constructed by (8.1.10), ..., (8.1.25) and let u_ε (respectively, u) be the solution of (8.1.26) (respectively, (8.1.27)). Then

(8.1.32) $$u_\varepsilon \to u \text{ in } V \text{ weakly;}$$

moreover

(8.1.33) $$\frac{\partial u_\varepsilon}{\partial x_i} \to \frac{\partial u}{\partial x_i} \text{ in } L^p(\mathcal{O}) \text{ weakly, in case of (8.1.30);}$$

(8.1.34) $$\phi \frac{\partial u_\varepsilon}{\partial x_i} \to \phi \frac{\partial u}{\partial x_i} \text{ in } L^p(\mathcal{O}) \text{ weakly, } \phi \in C_0^\infty(\mathcal{O}), \text{ in case of (8.1.31).}$$

The hypothesis (8.1.30) (or (8.1.31)) can be suppressed if one reinforces (8.1.4). We have

THEOREM 8.6. *We assume that (8.1.1), (8.1.2) hold true, together with*

(8.1.35) $$a_{ij} \in C^0(\mathbb{R}_y^n \times \mathbb{R}_z^n).$$

Then, for $f \in V'$, one has (8.1.32).

REMARK 8.7. The practical meaning of the construction (8.1.10), ..., (8.1.15) of $\mathcal{A}$ can be summarized in saying that **one homogenizes first with respect to z, considering y as a parameter and that one homogenizes next with respect to y. This is the "reiterated homogenization."**

REMARK 8.8. Theorem 8.6 is much simpler to state than Theorem 8.5 and its proof will not use (contrary to Theorem 8.5) the results of Section 4. Its defect is that it does not contain the "model" problem $a_{ij}(x/\varepsilon)$ with $a_{ij} \in L^\infty(\mathbb{R}^n)$.

REMARK 8.9. One obtains results similar to those of Theorem 8.5 and Theorem 8.6 when considering coefficients of the form, say:

(8.1.36) $$a_{ij}\left(\frac{x}{\varepsilon}, \frac{x}{\varepsilon^2}\right).$$

But one would have then to modify the first corrector.

The proof of Theorems 8.5 and 8.6 is given in two steps. We first show (Section 8.2) that one can reduce the problem to the case when the coefficients a_{ij} are C^∞ in y and in z, and we prove the Theorems in Section 8.4 under this assumption, after using in Section 8.3 asymptotic expansions.

8.2. Approximation by smooth coefficients.

8.2.1. 1st *Case: Case of Theorem 8.6.* By virtue of (8.1.35) we can find a sequence of functions

(8.2.1)
$$a_{ij}^{(\beta)} \in C^\infty(\mathbb{R}_y^n \times \mathbb{R}_z^n)$$
$$Y - Z \text{ periodic } \forall i, j,$$
$$a_{ij}^{(\beta)}(y, z)\xi_i\xi_j \geq \alpha_1\xi_i\xi_i, \ \alpha_1 > 0,$$

such that

(8.2.2) $$\|a_{ij}^{(\beta)} - a_{ij}\|_{C^0(\overline{Y} \times \overline{Z})} \to 0 \text{ as } \beta \to \infty.$$

We can do the same thing with a_0, approximated by smooth functions $a_0^{(\beta)}$, although this is by no means indispensable.

We then define

$$(8.2.3) \qquad a^{\varepsilon(\beta)}(u,v) = \int_{\mathcal{O}} a_{ij}^{(\beta)}\left(\frac{x}{\varepsilon}, \frac{x}{\varepsilon^2}\right) \frac{\partial u}{\partial x_j} \frac{\partial v}{\partial x_i} \, dx + \int_{\mathcal{O}} a_0^{(\beta)}\left(\frac{x}{\varepsilon}, \frac{x}{\varepsilon^2}\right) uv \, dx,$$

and we introduce $u_\varepsilon^{(\beta)}$ as the solution of

$$(8.2.4) \qquad\qquad u_\varepsilon^{(\beta)} \in V, \; a^{\varepsilon(\beta)}(u_\varepsilon^{(\beta)}, v) = (f, v) \; \forall v \in V.$$

Then it is standard to verify that

$$(8.2.5) \qquad \|u_\varepsilon^{(\beta)} - u_\varepsilon\|_V \le C \left[\sup_{i,j} \|a_{ij}^{(\beta)} - a_{ij}\|_{C^0(\overline{Y} \times \overline{Z})} + \|a_0^{(\beta)} - a_0\|_{C^0(\overline{Y} \times \overline{Z})} \right].$$

Let us assume Theorem 8.6 proven when the coefficients are C^∞. Then we know that

$$(8.2.6) \qquad\qquad u_\varepsilon^{(\beta)} \to u^{(\beta)} \text{ in } V \text{ weakly, as } \varepsilon \to 0, \; \beta \text{ fixed,}$$

where $u^{(\beta)}$ is the solution of

$$(8.2.7) \qquad\qquad \mathcal{A}(\beta)(u^{(\beta)}, v) = (f, v) \; \forall v \in V,$$
$$u^{(\beta)} \in V$$

where $\mathcal{A}^{(\beta)}$ is obtained by the reiterated homogenization procedure. It is easily verified that if $q_{ij}^{(\beta)}$, $q_0^{(\beta)}$ **denote the coefficients of $\mathcal{A}^{(\beta)}$, then**

$$(8.2.8) \qquad\qquad q_{ij}^{(\beta)}, q_0^{(\beta)} \to q_{ij}, q_0 \text{ as } \beta \to \infty.$$

Therefore

$$(8.2.9) \qquad\qquad \|u^{(\beta)} - u\|_V \to 0 \text{ if } \beta \to \infty,$$

and we shall have proved (8.1.32).

8.2.2. 2nd *Case: Case of Theorem* 8.5. Smooth functions being not dense in L^∞, (8.2.1), (8.2.2) no longer apply to the case where we assume (8.1.4) and we have then to rely on the results of Section 4.

Let us consider first the case $V = H_0^1(\mathcal{O})$ under hypothesis (8.1.30). We use (8.1.28) and we introduce q such that

$$(8.2.10) \qquad\qquad \frac{1}{q} + \frac{1}{p} = \frac{1}{2}.$$

We can find a sequence $a_{ij}^{(\beta)}$ satisfying (8.2.1) and such that

$$(8.2.11) \qquad\qquad \|a_{ij}^{(\beta)} - a_{ij}\|_{L^q(Y; C^0(\overline{Z}))} \to 0 \text{ as } \beta \to \infty.$$

Condition (8.2.11) means that

$$(8.2.12) \qquad\qquad \int_Y \left(\sup_{z \in Z} |a_{ij}^{(\infty)}(y, z) - a_{ij}(y, z)| \right)^q dy \to 0.$$

We construct (for the symmetry of the procedure) a sequence $a_0^{(\beta)}$ with similar properties.

We define next $a^{\varepsilon(\beta)}(u, v)$ and $u_\varepsilon(\beta)$ by (8.2.3) and (8.2.4). We have then

$$(8.2.13) \quad \|u_\varepsilon^{(\beta)} - u_\varepsilon\|_V \le C \left[\sup_{i,j} \|a_{ij}^{(\beta)} - a_{ij}\|_{L^q(Y; C^0(\overline{Z}))} + \|a_0^{(\beta)} - a_0\|_{L^q(Y; C^0(\overline{Z}))} \right].$$

Indeed we have the identity

$$(8.2.14) \qquad a^\varepsilon(u_\varepsilon - u_\varepsilon^{(\beta)}) = a^{\varepsilon(\beta)}(u_\varepsilon^{(\beta)}, u_\varepsilon - u_\varepsilon^{(\beta)}) - a^\varepsilon(u_\varepsilon^{(\beta)}, u_\varepsilon - u_\varepsilon^{(\beta)})$$

$$= \int_{\mathcal{O}} \left[a_{ij}\left(\frac{x}{\varepsilon}, \frac{x}{\varepsilon^2}\right) - a_{ij}\left(\frac{x}{\varepsilon}, \frac{x}{\varepsilon^2}\right) \right]$$

$$\cdot \frac{\partial u_\varepsilon^{(\beta)}}{\partial x_j} \frac{\partial(u_\varepsilon - u_\varepsilon^{(\beta)})}{\partial x_i} \, dx$$

$$+ \int_{\mathcal{O}} \left[a_0^{(\beta)}\left(\frac{x}{\varepsilon}, \frac{x}{\varepsilon^2}\right) - a_0\left(\frac{x}{\varepsilon}, \frac{x}{\varepsilon^2}\right) \right]$$

$$u_\varepsilon^{(\beta)}(u_\varepsilon - u_\varepsilon^{(\beta)}) \, dx.$$

Let us estimate the general term of the right hand side of (8.2.14) containing first order derivatives. It is bounded by

$$\left\| a_{ij}^{(\beta)}\left(\frac{x}{\varepsilon}, \frac{x}{\varepsilon^2}\right) - a_{ij}\left(\frac{x}{\varepsilon}, \frac{x}{\varepsilon^2}\right) \right\|_{L^q(\mathcal{O})} \cdot \left\| \frac{\partial u_\varepsilon^{(\beta)}}{\partial x_j} \right\|_{L^p(\mathcal{O})} \left\| \frac{\partial}{\partial x_i}(u_\varepsilon - u_\varepsilon^{(\beta)}) \right\|_{L^2(\mathcal{O})} ;$$

since $\left\| \frac{\partial u_\varepsilon^{(\beta)}}{\partial x_j} \right\|_{L^p(\mathcal{O})} \leq C$, it remains to estimate

$$(8.2.15) \qquad \int_{\mathcal{O}} \left| a_{ij}^{(\beta)}\left(\frac{x}{\varepsilon}, \frac{x}{\varepsilon^2}\right) - a_{ij}\left(\frac{x}{\varepsilon}, \frac{x}{\varepsilon^2}\right) \right|^q \, dx \leq \int_{\mathcal{O}} \phi^{(\beta)}\left(\frac{x}{\varepsilon}\right) \, dx$$

where we have set

$$(8.2.16) \qquad \phi^{(\beta)}\left(\frac{x}{\varepsilon}\right) = \left(\sup_z \left| a_{ij}^{(\beta)}\left(\frac{x}{\varepsilon}, z\right) - a_{ij}\left(\frac{x}{\varepsilon}, z\right) \right| \right)^q.$$

But

$$\int_{\mathcal{O}} \phi^{(\beta)}\left(\frac{x}{\varepsilon}\right) \, dx = \varepsilon^n \int_{\frac{1}{\varepsilon}\mathcal{O}} \phi^{(\beta)}(y) \, dy \leq C \int_Y \phi^{(\beta)}(y) \, dy,$$

and $\int_Y \phi^{(\beta)}(y) \, dy$ is the expression (8.2.12), hence (8.2.13) follows.

Assuming Theorem 8.5 to be proven for a_{ij} smooth, we can therefore "homogenize" problem (8.2.4) and we shall have (8.2.6), (8.2.7).

It remains to verify (8.2.8) under (8.2.12). We define (compare to (8.1.11))

$$(8.2.17) \qquad A_{1,y}^{(\beta)} = -\frac{\partial}{\partial z_i}\left(a_{ij}^{(\beta)}(y, z) \frac{\partial}{\partial z_j} \right) + a_0^{(\beta)}(y, z)$$

and define $\chi_y^{j(\beta)}(z)$ by

$$(8.2.18) \qquad A_{1,y}^{(\beta)}(\chi_y^{j(\beta)}(z) - z_j) = 0,$$

$$\chi_y^{j(\beta)}(z) \text{ is } Z \text{ periodic}, \quad \int_Z \chi_y^{j(\beta)}(z) \, dz = 0.$$

We set

$$(8.2.19) \qquad q_{ij}^{1(\beta)}(y) = \frac{1}{|Z|} \left[\int_Z a_{ik}^{(\beta)}(y, z) \frac{\partial \chi_y^{j(\beta)}}{\partial z_k} \, dz - \int_Z a_{ij}^{(\beta)}(y, z) \, dz \right].$$

If we can show that

$$(8.2.20) \qquad q_{ij}^{1(\beta)}(y) \to q_{ij}^1(y) \text{ in } L^q(Y)$$

then, by an argument similar to the one leading to (8.2.13) we shall have (8.2.9) and the desired result.

In order to verify (8.2.20), and by virtue of (8.2.12), it remains to show that

$$(8.2.21) \qquad r_{ij}^{1(\beta)}(y) = \int_Z a_{ik}^{(\beta)}(y,z) \frac{\partial \chi_y^{j(\beta)}}{\partial z_k} \, dz \to r_{ij}^1(y) \text{ in } L^q(Y),$$

where

$$\hat{r}_{ij}(y) = \int_Z a_{ik}(y,z) \frac{\partial \chi_y^j}{\partial z_k} \, dz.$$

But

$$(8.2.22) \qquad r_{ij}^{1(\beta)} - r_{ij}^1 = \int_Z [a_{ik}^{(\beta)}(y,z) - a_{ik}(y,z)] \frac{\partial \chi_y^{j(\beta)}}{\partial z_j} \, dz$$

$$+ \int_Z a_{ik}(y,z) \frac{\partial (\chi_y^{j(\beta)} - \chi_y^j)}{\partial z_k} \, dz.$$

We have

$$\|\chi_y^{j(\beta)} - \chi_y^j\|_{H^1(Z)} \le c \sup_z |a_{k\ell}^{(\beta)}(y,z) - a_{k\ell}(y,z)|$$

so that (8.2.22) gives

$$|r_{ij}^{1(\beta)}(y) - r_{ij}^1(y)|^q \le c \left(\sup_z |a_k^{(\beta)}(y,z) - a_{k\ell}(y,z)| \right)^q$$

and therefore

$$\int_Y |r_{ij}^{1(\beta)}(y) - \hat{r}_{ij}(y)|^q \, dz \to 0.$$

Let us consider now the case when we have (8.1.32). We still approximate a_{ij} by $a_{ij}^{(\beta)}$ as in (8.2.11). The result (8.2.20) (**which does not depend on boundary conditions**) still holds true. We replace (8.2.13) and (8.2.9) by local results:

$$\|\phi(u_\varepsilon^{(\beta)} - u_\varepsilon)\|_V \to 0, \qquad \|\phi(u^{(\beta)} - u)\|_V \to 0.$$

8.3. Asymptotic expansion. We use now the methods of Section 2, with the variables x, y, z. One has

$$(8.3.1) \qquad A^\varepsilon = \varepsilon^{-4} A_1 + \varepsilon^{-3} A_2 + \varepsilon^{-2} A_3 + \varepsilon^{-1} A_4 + \varepsilon^0 A_5,$$

where

$$A_1 = -\frac{\partial}{\partial z_i} \left(a_{ij}(y,z) \frac{\partial}{\partial z_j} \right),$$

$$A_2 = -\frac{\partial}{\partial z_i} \left(a_{ij}(y,z) \frac{\partial}{\partial y_j} \right) - \frac{\partial}{\partial y_i} \left(a_{ij}(y,z) \frac{\partial}{\partial z_j} \right),$$

$$(8.3.2) \qquad A_3 = -\frac{\partial}{\partial y_i} \left(a_{ij}(y,z) \frac{\partial}{\partial y_j} \right) - \frac{\partial}{\partial z_i} \left(a_{ij}(y,z) \frac{\partial}{\partial x_j} \right)$$

$$- \frac{\partial}{\partial x_i} \left(a_{ij}(y,z) \frac{\partial}{\partial z_j} \right),$$

$$A_4 = -\frac{\partial}{\partial y_i} \left(a_{ij}(y,z) \frac{\partial}{\partial x_j} \right) - \frac{\partial}{\partial x_i} \left(a_{ij}(y,z) \frac{\partial}{\partial y_j} \right),$$

$$A_5 = -\frac{\partial}{\partial x_i} \left(a_{ij}(y,z) \frac{\partial}{\partial x_j} \right).$$

We are looking for u_ε in the form

$$(8.3.3) \qquad u = u_0 + \varepsilon u_1 + \varepsilon^2 u_2 + \cdots,$$

where

(8.3.4) $u_j = u_j(x, y, z)$ is Y periodic in y and Z periodic in z.

An identification in powers of ε in the equation $A^\varepsilon u_\varepsilon = f$ leads to

$$(8.3.5) \qquad\qquad A_1 u_0 = 0,$$

$$(8.3.6) \qquad\qquad A_1 u_1 + A_2 u_0 = 0,$$

$$(8.3.7) \qquad\qquad A_1 u_2 + A_2 u_1 + A_3 u_0 = 0,$$

$$(8.3.8) \qquad\qquad A_1 u_3 + A_2 u_2 + A_3 u_1 + A_4 u_0 = 0,$$

$$(8.3.9) \qquad\qquad A_1 u_4 + A_2 u_3 + A_3 u_2 + A_4 u_1 + A_5 u_0 = f,$$

etc.

In the equation (8.3.5), x and y are parameters, and the general solution of (8.3.5) is

$$(8.3.10) \qquad\qquad u_0 = \tilde{u}_0(x, y)$$

Using (8.3.10) into (8.3.6) gives

$$(8.3.11) \qquad\qquad A_1 u_1 = \frac{\partial a_{ij}}{\partial z_i} \frac{\partial \tilde{u}_0}{\partial y_j}.$$

We introduce (notations are consistent with (8.1.12)) $\chi_y^j(z)$ as the Z periodic solution in z of

$$(8.3.12) \qquad\qquad A_1(\chi_y^j(z) - z_j) = 0, \qquad \int_Z \chi_y^j(z)\, dz = 0.$$

Then (8.3.11) gives

$$(8.3.13) \qquad\qquad u_1 = -\chi_y^j(z) \frac{\partial u_0}{\partial y_j} + u_1(x, y).$$

We can solve (8.3.7) for u_2 iff

$$(8.3.14) \qquad\qquad \frac{1}{|Z|} \int (A_2 u_1 + A_3 u_0)\, dz = 0.$$

Using (8.3.13) into (8.3.14) leads to the introduction of

$$(8.3.15) \qquad \mathcal{A}_1 = -\frac{\partial}{\partial y_i} \frac{1}{|Z|} \left[\int_Z a_{ij}(y, z)\, dz - \int_Z a_{ik} \frac{\partial \chi_y^j(z)}{\partial z_k}\, dz \right] \frac{\partial}{\partial y_j}$$

which coincides with (8.1.19) (if $a_0 = 0$ which does not restrict the generality).

With this notation (8.3.14) is equivalent to

$$\mathcal{A}_1 \tilde{u}_0 = 0,$$

i.e.,

$$(8.3.16) \qquad\qquad \tilde{u}_0 = u(x).$$

Then (8.3.13) reduces to

$$(8.3.17) \qquad\qquad u_1 = \tilde{u}_1(x, y)$$

and (8.3.7) reduces to

$$(8.3.18) \qquad\qquad A_1 u_2 = \left(\frac{\partial}{\partial z_i} a_{ij}(y, z) \right) \left(\frac{\partial u}{\partial x_j} + \frac{\partial \tilde{u}_1}{\partial y_j} \right),$$

hence

(8.3.19)
$$u_2 = -\chi_y^j(z)\left(\frac{\partial u}{\partial x_j} + \frac{\partial \tilde{u}_1}{\partial y_j}\right) + \tilde{u}_2(x,y).$$

We can solve (8.3.8) for u_3, iff $\int_Z (A_2 u_2 + A_3 u_1 + A_4 u_0)\, dz = 0$, i.e.,

$$\frac{\partial}{\partial y_i}\left[\left(\int_Z a_{ik}\frac{\partial \chi_y^j(z)}{\partial z_k}\, dz\right)\left(\frac{\partial u}{\partial x_j} + \frac{\partial \tilde{u}_1}{\partial y_j}\right)\right] - \frac{\partial}{\partial y_i}\int_Z a_{ij}(y,z)\, dz\,\frac{\partial \tilde{u}_1}{\partial y_j}$$
$$- \frac{\partial}{\partial y_i}\int_Z a_{ij}(y,z)\, dz\frac{\partial u}{\partial x_j} = 0,$$

i.e.,

(8.3.20)
$$\mathcal{A}_1 \tilde{u}_1 = (\mathcal{A}_1 y_j)\frac{\partial u}{\partial x_j}.$$

We introduce $\chi^j = \chi^j(y)$ by[43]

(8.3.21)
$$\mathcal{A}_1(\chi^j - y_j) = 0, \ \chi^j \ Y\text{-periodic},$$

(χ^j is defined up to an additive constant), and we have

(8.3.22)
$$u_1 = -\chi^j(y)\frac{\partial u}{\partial x_j}(x) + \tilde{u}_1(x).$$

A necessary condition for being able to proceed with the computations in (8.3.9) is:

(8.3.23)
$$\iint_{Y\times Z}(A_1 u_4 + A_2 u_3 + A_3 u_2 + A_4 u_1 + A_5 u_0)\, dy\, dz = |Y||Z|f.$$

Since $\iint_{Y\times Z} A_1 u_4\, dy\, dz = \iint_{Y\times Z} A_2 u_3\, dy\, dz = 0$, (8.3.23) reduces to

(8.3.24)
$$\frac{\partial}{\partial x_i}\iint_{Y\times Z} a_{ik}\frac{\partial \chi_y^j(z)}{\partial z_k}\left(\frac{\partial u}{\partial x_j} + \frac{\partial \tilde{u}_1}{\partial y_j}\right)\, dy\, dz - \frac{\partial}{\partial x_i}\iint_{Y\times Z} a_{ik}\frac{\partial \tilde{u}_1}{\partial y_k}\, dy\, dz$$
$$- \iint_{Y\times Z} a_{ij}\, dy\, dz\frac{\partial^2 u}{\partial x_i \partial x_j} = |Y||Z|f.$$

We use (8.3.22) into (8.3.24) and we obtain in this manner

(8.3.25)
$$-\hat{q}_{ij}\frac{\partial^2 u}{\partial x_i \partial x_j} = f,$$

with

(8.3.26)
$$|Y||Z|\hat{q}_{ij} = \iint_{Y\times Z}\left[a_{ij} - a_{ik}\frac{\partial \chi_y^i}{\partial z_k} - a_{ik}\frac{\partial \chi^j}{\partial y_k} + a_{ik}\frac{\partial \chi_y^\ell}{\partial z_k}\frac{\partial \chi^j}{\partial y_\ell}\right]\, dy\, dz.$$

But (8.1.22) gives

$$|Y||Z|q_{ij} = |Z|\mathcal{A}_1(\chi^j - y_j, -y_i)$$
$$= |Z|\int_Y\left(q_{ij}^1(y) - q_{i\ell}^1(y)\frac{\partial \chi^j}{\partial y_\ell}\right)\, dy$$

and since $|Z|q_{ij}^1(y) = \int_Z\left(a_{ij} - a_{ik}\frac{\partial \chi_y^j}{\partial z_k}\right)\, dz$, we see that (as it should!) $q_{ij} = \hat{q}_{ij}$.

---

[43]This coincides with (8.1.21).

REMARK 8.10. For the **Dirichlet boundary condition** one can prove (as in Section 2.4) that

$$(8.3.27) \qquad \|u_\varepsilon - u\|_{L^\infty(\mathcal{O})} \to 0.$$

8.4. Proof of the reiteration formula for smooth coefficients. We assume now that a_{ij}, a_0 satisfy (8.1.1), (8.1.2) and moreover that

$$(8.4.1) \qquad a_{ij}, a_0 \in C^\infty(\mathbb{R}^n_y \times \mathbb{R}^n_z).$$

We use a technique similar to that of Section 6.4. Let λ be a given index in $[1, 2, \ldots,]$. We look for a function w_ε, solution of

$$(8.4.2) \qquad (A^\varepsilon)^* w_\varepsilon = \varepsilon g_\varepsilon$$

with

$$(8.4.3) \qquad w_\varepsilon = \varepsilon \alpha + \varepsilon^2 (z_\lambda + \beta) + \varepsilon^3 \gamma + \varepsilon^4 \delta,$$

$\alpha, \beta, \gamma, \delta$ to be defined. Using the adjoint expansion of (8.3.1), an identification leads to

$$(8.4.4) \qquad A_1^* \alpha = 0,$$

$$(8.4.5) \qquad A_1^*(\beta + z_\lambda) + A_2^* \alpha = 0,$$

$$(8.4.6) \qquad A_1^* \gamma + A_2^*(\beta + z_\lambda) + A_3^* \alpha = 0,$$

$$(8.4.7) \qquad A_1^* \delta + A_2^* \gamma + A_3^*(\beta + z_\lambda) + A_4^* \alpha = 0,$$

and then we have

$$
\begin{aligned}
(8.4.8) \qquad g_\varepsilon = {} & A_2^* \delta + A_3^* \gamma + A_4^*(\beta + z_\lambda) + A_5^* \alpha \\
& + \varepsilon[A_3^* \delta + A_4^* \gamma + A_5^*(\beta + z_\lambda)] \\
& + \varepsilon^2 [A_4^* \delta + A_5^* \gamma] + \varepsilon^3 A_5^* \delta.
\end{aligned}
$$

Equation (8.4.4) is satisfied iff $\alpha = \alpha(y)$. Equation (8.4.5) in β means

$$A_1^* \beta = \frac{\partial}{\partial z_i} a_{i\lambda}^* + \frac{\partial}{\partial z_i} a_{ij}^* \frac{\partial \alpha}{\partial y_j}$$

(where we write $a_{ij}^* = a_{ji}$); if we introduce (compare to (8.3.12)) $\hat{\chi}_y^j(z)$

$$(8.4.9) \qquad \begin{aligned} & A_1^*(\hat{\chi}_y^j(z) - z_j) = 0, \\ & \hat{\chi}_y^j(z) \text{ is } Z \text{ periodic and} \\ & \int_Z \hat{\chi}_y^j(z) \, dz = 0, \end{aligned}$$

then

$$(8.4.10) \qquad \beta = -\hat{\chi}_y^\lambda(z) - \hat{\chi}_y^j(z) \frac{\partial \alpha}{\partial y_j}(y) + \tilde{\beta}(y).$$

We can compute γ solution of (8.4.6) iff

$$\int_Z (A_2^*(\beta + z_\lambda) + A_3^* \alpha) \, dz = 0$$

which is equivalent to

$$(8.4.11) \qquad -\frac{\partial}{\partial y_i} \int_Z a_{ik}^* \frac{\partial}{\partial z_k}(\beta + z_\lambda) \, dz - \frac{\partial}{\partial y_i} \int_Z a_{ij}^* \, dz \, \frac{\partial \alpha}{\partial y_j} = 0.$$

We replace in (8.4.11) β by its value from (8.4.10); it becomes

(8.4.12)
$$-\frac{\partial}{\partial y_i}\left[\int_Z a_{ij}^*(y,z)\,dz - \int_Z a_{ik}^*\frac{\partial\hat{\chi}_y^j}{\partial z_k}(z)\,dz\right]\frac{\partial\alpha}{\partial y_j}$$

()
$$* -\frac{\partial}{\partial y_i}\left[\int_Z a_{i\lambda}^*\,dz - \int_Z a_{ik}^*\frac{\partial\hat{\chi}_y^j}{\partial z_k}(z)\,dz\right] = 0.$$

But using (8.3.15) ones sees that (8.4.12) is equivalent to

$$\mathcal{A}_1^*\alpha = -\mathcal{A}_1^* y_\lambda.$$

We define $\hat{\chi}^\lambda(y)$ by

(8.4.13)
$$\mathcal{A}_1^*(\hat{\chi}^\lambda - y_\lambda) = 0,\ \hat{\chi}\ Y\text{-periodic.}$$

Then we can take

(8.4.14)
$$\alpha = -\hat{\chi}^\lambda$$

so that (8.4.10) becomes

(8.4.15)
$$\beta = -\hat{\chi}_y^\lambda(z) + \hat{\chi}_y^j(z)\frac{\partial\hat{\chi}^\lambda}{\partial y_j}(y) + \tilde{\beta}(y).$$

It is then possible to solve (8.4.6) for γ. We can next solve (8.4.7) for γ iff

$$\int_Z [A_2^*\gamma + A_3^*(\beta + z_\lambda) + A_4^*\alpha]\,dz = 0$$

and this is an equation in $\tilde{\beta}$. We can then solve for $\tilde{\beta}$ iff

$$\iint_{Y\times Z} [A_2^*\gamma + A_3^*(\beta + z_\lambda) + A_4^*\alpha]\,dy\,dz = 0$$

a condition which is satisfied, so that the calculation is possible. Moreover by (8.4.1) we can assume that

(8.4.16)
$$\alpha,\beta,\gamma,\delta \text{ are } C^\infty \text{ functions in } y \text{ and in } z,$$
$$g_\varepsilon \text{ is also } C^\infty \text{ in } y \text{ and } z.$$

We can now pass to the limit. We know that $\|u_\varepsilon\|_V \le C$. We set, as usual,

$$\xi_i^\varepsilon = a_{ij}\left(\frac{x}{\varepsilon},\frac{x}{\varepsilon^2}\right)\frac{\partial u_\varepsilon}{\partial x_j}$$

and we can assume that

(8.4.17)
$$u_\varepsilon \to u \text{ in } V \text{ weakly, } \xi_i^\varepsilon \to \xi_i \text{ in } L^2(\mathcal{O}) \text{ weakly,}$$

and we have (assuming $a_0 = 0$):

(8.4.18)
$$\left(\xi_i,\frac{\partial v}{\partial x_i}\right) = (f,v)\ \forall v \in V.$$

We take $\phi \in C_0^\infty(\mathcal{O})$ and we choose $v = \phi w_\varepsilon$ in (8.1.26) and we multiply (8.4.2) by ϕu_ε. We obtain

(8.4.19) $\left(\xi_i^\varepsilon,\left(\frac{\partial\phi}{\partial x_i}\right)w_\varepsilon\right) - \int a_{ij}\left(\frac{x}{\varepsilon},\frac{x}{\varepsilon^2}\right)\frac{\partial w_\varepsilon}{\partial x_i}\frac{\partial\phi}{\partial x_j}u_\varepsilon\,dx = (f,\phi w_\varepsilon) - \varepsilon(g_\varepsilon,\phi u_\varepsilon).$

But

$$\frac{\partial w_\varepsilon}{\partial x_i} = \frac{\partial \beta}{\partial y_i} + \frac{\partial \beta}{\partial z_i} + \frac{\partial z_\lambda}{\partial z_i} + \varepsilon\rho_\varepsilon(y,z),$$

$$\rho_\varepsilon(y,z) \in C^\infty \text{ in } y \text{ and } z,$$

and

$$w_\varepsilon \to x_\lambda \text{ in } L^2(\mathcal{O}) \text{ strongly.}$$

Therefore (8.4.19) gives[44]

$$\left(\xi_i, \left(\frac{\partial\phi}{\partial x_i}\right)x_\lambda\right) - \mathcal{M}_{y,z}\left(a_{ij}\left(\frac{\partial\alpha}{\partial y_i} + \frac{\partial\beta}{\partial z_i}\right) + a_{i\lambda}\right)\int_\mathcal{O}\frac{\partial\phi}{\partial x_j}u\,dx = (f,\phi x_\lambda)$$

$$= \left(\xi_i, \frac{\partial(\phi x_\lambda)}{\partial x_i}\right)$$

so that

(8.4.20) $$\xi_\lambda = \mathcal{M}_{y,z}\left(a_{ij}\left(\frac{\partial\alpha}{\partial y_i} + \frac{\partial\beta}{\partial z_i}\right) + a_{i\lambda}\right)ddxux_j.$$

Using (8.4.20) into (8.4.18) gives the desired result provided we verify that (using (8.4.14) and (8.4.15)):

(8.4.21) $$q_{i\lambda} = \mathcal{M}_{y,z}\left(a_{i\lambda} - a_{ij}\frac{\partial\hat\chi_y^\lambda(z)}{\partial z_j} - a_{ij}\frac{\partial\hat\chi^\lambda(y)}{\partial y_j} + a_{ij}\frac{\partial\hat\chi^\lambda}{\partial y_k}\frac{\partial\hat\chi_y^k(z)}{\partial z_j}\right).$$

This identity is a simple computational exercise.

8.5. Correctors. We introduce the first order corrector (using notations of Section 5)

(8.5.1) $$\theta_\varepsilon = (\varepsilon u_1 + \varepsilon^2 u_2)m_\varepsilon(x)$$

i.e.,

(8.5.2) $$\theta_\varepsilon = -\varepsilon\chi^j\left(\frac{x}{\varepsilon}\right)\frac{\partial u}{\partial x_j}(x)m_\varepsilon(x) - \varepsilon^2\chi_\varepsilon^j\left(\frac{x}{\varepsilon^2}\right)\left[\frac{\partial u}{\partial x_j} - \left(\frac{\partial\chi^k}{\partial y_j}\right)\left(\frac{x}{\varepsilon}\right)\frac{\partial u}{\partial x_k}\right]m_\varepsilon.$$

We assume that

$$u \in H^2(\mathcal{O}),$$
$$\chi^j \in W^{2,\infty}(Y),$$
$$\chi_y^j(z) = \chi^j(y,z) \in W^{1,\infty}(Y \times Z).$$

Then, one can verify, by the methods of Section 5, that

(8.5.3) $$u_\varepsilon - u - \theta_\varepsilon \to 0 \text{ in } V \text{ **strongly** as } \varepsilon \to 0.$$

As observed in Section 5, the cut-off functions are not at all indispensable.

---

[44]We use the fact that, for instance, $a_{ij}\left(\frac{x}{\varepsilon},\frac{x}{\varepsilon^2}\right) \to \mathcal{M}_{y,z}(a_{ij}) = \frac{1}{|Y||Z|}\iint_{Y\times Z}a_{ij}(y,z)\,dy\,dz$ in $L^\infty(\mathcal{O})$ weak star.

9. Homogenization of elliptic systems

9.1. Setting of the problem. We now proceed to the homogenization of higher order elliptic operators and, more generally, of some elliptic systems,[45]

NOTATIONS. We consider functions

$$(9.1.1) \qquad a_{\alpha\beta}^{ij}(y) \in L^\infty(\mathbb{R}^n), \text{ real valued, } Y\text{-periodic,}$$

where in (9.1.1) we assume that

$$(9.1.2) \qquad \begin{aligned} &i, j = 1, \dots, N \text{ (we shall consider an } N \times N \text{ system)} \\ &|\alpha| = m_i, \ |\beta| = m_j|, \ \alpha = (\alpha_1, \dots, \alpha_n), \ \beta = (\beta_1, \dots, \beta_n) \\ &m_1, \dots, m_N \text{ given integers } \geq 1. \end{aligned}$$

With the functions $a_{\alpha\beta}^{ij}$ we associate

$$(9.1.3) \qquad {}^\varepsilon a_{\alpha\beta}^{ij}(x) = a_{\alpha\beta}^{ij}(x/\varepsilon),$$

and introduce vector functions

$$u = \{u_1, \dots, u_N\}, \ v = \{v_1, \dots, v_N\}, \ u_i, v_i \text{ defined in } \mathcal{O}.$$

We consider the system

$$(9.1.4) \qquad (-1)^{m_j} D^\beta \left(a_{\alpha\beta}^{ij} \left(\frac{x}{\varepsilon} \right) D^\alpha u_{i\varepsilon} \right) = f_j \text{ in } \mathcal{O}, \qquad j = 1, \dots, N,$$

with $u_\varepsilon = \{u_{i\varepsilon}\}$ subject to appropriate boundary conditions and with a suitable **ellipticity hypothesis**.

We introduce a Hilbert space V such that

$$(9.1.5) \qquad \prod_{i=1}^N H_0^{m_i}(\mathcal{O}) \subseteq V \subseteq \prod_{i=1}^N H^{m_i}(\mathcal{O})$$

where

$$\begin{aligned} H^m(\mathcal{O}) &= \{\phi \mid D^\gamma \phi \in L^2(\mathcal{O}) \ \forall \gamma, |\gamma| \leq m\},{}^{[46]} \\ H_0^m(\mathcal{O}) &= \text{ closure of } C_0^\infty(\mathcal{O}) \text{ in } H^m(\mathcal{O}). \end{aligned}$$

For $u, v \in \prod_{i=1}^N H^{m_i}(\mathcal{O})$, we set

$$(9.1.6) \qquad a^\varepsilon(u, v) = \int {}^\varepsilon a_{\alpha\beta}^{ij}(x) D^\alpha u_i D^\beta v_j \ dx.$$

We shall assume

$$(9.1.7) \qquad a^\varepsilon(v, v) \geq \alpha \|v\|_V^2, \ \alpha > 0, \ \forall v \in V, \ \alpha \text{ independent of } \varepsilon.$$

REMARK 9.1. Condition (9.1.7) will be satisfied if the following conditions hold true:

$$(9.1.8) \qquad a_{\alpha\beta}^{ij}(y) \xi_i^\alpha \xi_j^\beta \geq \alpha_1 \sum_{i=1}^N |\xi_i|_{\mathbb{R}^{m_i}}^2, \text{ a.e. in } y, \ \alpha_1 > 0,$$

---

[45]We do not consider here general elliptic systems in the sense of Agmon-Douglas-Nirenberg [**3**].

[46]Provided with the norm $\left(\sum_{|\gamma| \leq m} |D^\gamma \phi|_{L^2(\mathcal{O})}^2 \right)^{1/2}$.

and if[47]

(9.1.9) $v \in V$ implies $D^\gamma v_j = 0$ on a part Γ_j of $\Gamma = \partial \mathcal{O}$

with positive measure (or capacity) on Γ, $|\gamma| \leq m_j - 1$.

REMARK 9.2. If one has (9.1.8) but not (9.1.9) one would add in (9.1.6) **zero order terms**:

(9.1.10) $$\int {}^\varepsilon a_{00}^{ij}(x) u_i v_j \, dx,$$

where $a_{00}^{ij}(y) = a^{ij}(y)$ is L^∞, Y-periodic, and if

(9.1.11) $a^{ij}(y)\xi_i\xi_j \geq \alpha_0 \xi_i \xi_i$, $\alpha_0 > 0$, a.e. in y.

Let f be given in V', dual of V, and let u_ε be the solution of

(9.1.12) $a^\varepsilon(u_\varepsilon, v) = (f, v), \; \forall v \in V.$

We want to study the behavior of u_ε when $\varepsilon \to 0$.

9.2. Statement of the homogenization procedure. We introduce some more notations. We define

(9.2.1)

$$W(Y) = \left\{ \phi \;\middle|\; \phi \in \prod_{i=1}^N H^{m_i}(Y), \phi_i \text{ periodic in } Y, \text{ i.e. } D^\gamma \phi_i \text{ takes equal values} \right.$$

$$\left. \text{on opposite faces of } Y \text{ for every } \gamma \text{ such that } |\gamma| \leq m_i - 1; \right\}$$

for $\phi, \psi \in \prod_{i=1}^N H^{m_i}(Y)$, we set

(9.2.2) $$a_1(\phi, \psi) = \int_Y a^{ij} D^\alpha \phi_i(y) D^\beta \psi_j(y) \, dy,$$

and we introduce

(9.2.3) $$P_j^\beta(y) = \{0, \ldots, 0, \underbrace{\frac{y^\beta}{\beta}, 0, \ldots, 0}_{j}\}.$$

We then **define**

$$\chi_j^\beta \in W(Y), \text{ such that}$$

(9.2.4) $a_1(\chi_j^\beta - P_j^\beta, \psi) = 0, \; \forall \psi \in W(Y), \; (|\beta| = m_j).$

χ_j^β is defined mod $\mathcal{P}$ where

(9.2.5) $\mathcal{P} = $ set of polynomials $\{P_1, \ldots, P_N\}$,

P_i of degree $\leq m_i - 1$.

Indeed we introduce $W^\circ(Y) = W(Y)/\mathcal{P}$, and we observe that $a_1(\phi, \psi)$ becomes coercive on $W^\circ(Y)$. We then uniquely define

(9.2.6) $$q_{\alpha\beta}^{ij} = \frac{1}{|Y|} a_1(\chi_j^\beta - P_j^\beta, \chi_i^\alpha - P_i^\alpha)$$

---

[47]One can relax this condition. We just present this remark as an example.

and

$$(9.2.7) \qquad \mathcal{A}(u,v) = \int q_{\alpha\beta}^{ij} D^{\alpha} u_i D^{\beta} v_j \, dx.$$

Let $\mathcal{A}$ be the system associated to $\mathcal{A}(u,v)$. **We will show that $\mathcal{A}$ is the homogenized system of A^{ε}** (associated to $a^{\varepsilon}(u,v)$).

REMARK 9.3. The above formulas extend those previously given for the second order scalar operators. We could as well consider systems with coefficients functions of $x, \frac{x}{\varepsilon}, \ldots, \frac{x}{\varepsilon^M}$.

We shall prove in Section 9.3 below the

THEOREM 9.4. *Let u be the solution of*

$$(9.2.8) \qquad u \in V, \ \mathcal{A}(u,v) = (f,v), \ \forall v \in V.$$

Under hypothesis (9.1.7), one has

$$(9.2.9) \qquad u_{\varepsilon} \to u \text{ in } V \text{ weakly, as } \varepsilon \to 0.$$

REMARK 9.5. One can verify that $\mathcal{A}(u,v)$ as defined by (9.2.7) is **coercive on** v.

REMARK 9.6. It is useful to notice that coefficients $q_{\alpha\beta}^{ij}$ can be given an "adjoint" form. We define

$$(9.2.10) \qquad a_1^*(\phi,\psi) = a_1(\psi,\phi), \ \forall \phi, \psi \in \prod H^{m_i}(Y)$$

and we define $\chi_j^{\beta*}$ (mod $\mathcal{P}$) by

$$(9.2.11) \qquad a_1^*(\chi_j^{\beta*} - P_j^{\beta}, \psi) = 0, \ \forall \psi \in W(Y).$$

We have then the formula

$$(9.2.12) \qquad q_{\alpha\beta}^{ij} = \frac{1}{|Y|} a_1^*(\chi_i^{\alpha*} - P_i^{\alpha}, \chi_j^{\beta*} - P_j^{\beta}).$$

Indeed let us denote by $q_{\alpha\beta}^{ij*}$ the right hand side of (9.2.12). We have

$$|Y| q_{\alpha\beta}^{ij} = -a_1(\chi_j^{\beta} - P_j^{\beta}, P_i^{\alpha}),$$
$$|Y| q_{\alpha\beta}^{ij*} = -a_1^*(\chi_i^{\alpha*} - P_i^{\alpha}, P_j^{\beta}),$$

and in order to prove that $q_{\alpha\beta}^{ij} = q_{\alpha\beta}^{ij*}$ it remains only to verify that

$$a_1(\chi_j^{\beta}, P_i^{\alpha}) = a_1^*(\chi_i^{\alpha*}, P_j^{\beta})$$

i.e.,

$$(9.2.13) \qquad a_1^*(P_i^{\alpha}, \chi_j^{\beta}) = a_1(P_j^{\beta}, \chi_i^{\alpha*}).$$

But if we take $\psi = \chi_j^{\beta}$ in (9.2.11) (for $\chi_i^{\alpha*}$ instead of $\chi_j^{\beta*}$) we see that the left hand side of (9.2.13) equals

$$(9.2.14) \qquad a_1^*(\chi_1^{\alpha*}, \chi_j^{\beta}),$$

if we take $\psi = \chi_i^{\alpha*}$ in (9.2.4), we find that the right hand side of (9.2.13) equals

$$(9.2.15) \qquad a_1(\chi_j^{\beta}, \chi_i^{\alpha*})$$

and since the quantities in (9.2.14) and in (9.2.15) are equal, (9.2.13) is proven.

One encounters in a natural fashion the formula (9.2.6) (respectively, (9.2.12)) when using the method of asymptotic expansions (respectively, of energy).

9.3. Proof of the homogenization theorem. We have

$$\|u_\varepsilon\|_V \le C. \tag{9.3.1}$$

If we set

$${}^\varepsilon\xi_\beta^j = {}^\varepsilon a_{\alpha\beta}^{ij} D^\alpha u_{\varepsilon i} \tag{9.3.2}$$

we see that ${}^\varepsilon\xi_\beta^j$ is bounded in $L^2(\mathcal{O})$, therefore we can extract a subsequence, still denoted by u_ε, such that

$$u_\varepsilon \to u \text{ in } V \text{ weakly}, \tag{9.3.3}$$

$${}^\varepsilon\xi_\beta^i \to \xi_\beta^j \text{ in } L^2(\mathcal{O}) \text{ weakly}.$$

Equation (9.1.12) can be rewritten

$$({}^\varepsilon\xi_\beta^j, D^\beta v_j) = (f, v), \tag{9.3.4}$$

(where parentheses denote either the scalar product in $L^2(\mathcal{O})$ or the duality between V' and V). Passing to the limit in (9.3.4), we obtain

$$(\xi_\beta^j, D^\beta v_j) = (f, v), \ \forall v \in V. \tag{9.3.5}$$

We compute now ξ_β^j using "**adjoint functions.**" We introduce:

$$P = \{P_j(y)\}, \ P_j(y) = \text{ homogeneous polynomial of degree } m_j, \tag{9.3.6}$$

and we define w such that

$$A_1^* w = 0 \text{ in } Y, \tag{9.3.7}$$

$$w - P \in W(Y).$$

If we set

$$w - P = -\chi \tag{9.3.8}$$

then (9.3.7) is equivalent to

$$a_1^*(\chi - P, \psi) = 0, \ \forall\psi \in W(Y). \tag{9.3.9}$$

We then introduce

$$w_\varepsilon(x) = \left\{\varepsilon^{m_j} w_j\left(\frac{x}{\varepsilon}\right)\right\}. \tag{9.3.10}$$

We observe that

$$a^{\varepsilon*}w_\varepsilon = 0, \tag{9.3.11}$$

and that

$$w_\varepsilon(x) = p(x) - \left\{\varepsilon^{m_j}\chi_j\left(\frac{x}{\varepsilon}\right)\right\}. \tag{9.3.12}$$

For $\phi \in C_0^\infty(\mathcal{O})$ we set

$$\phi u = \{\phi u_1, \ldots, \phi u_N\}; \tag{9.3.13}$$

we choose

$$v = \phi w_\varepsilon \text{ in } (9.1.12)$$

and we multiply (9.3.11) by ϕu_ε. We obtain:

$$a^\varepsilon(u_\varepsilon, \phi w_\varepsilon) - a^\varepsilon(\phi u_\varepsilon, w_\varepsilon) = (f, \phi w_\varepsilon), \tag{9.3.14}$$

or, more explicitly:

$$(9.3.15) \quad \int_{\mathcal{O}} {}^{\varepsilon}\xi_{\beta}^{j}(D^{\beta}(\phi w_{\varepsilon j}) - \phi D^{\beta} w_{\varepsilon j})\, dx$$

$$- \int_{\mathcal{O}} {}^{\varepsilon}q_{\alpha\beta}^{ij} D^{\beta} w_{\varepsilon j}(D^{\alpha}(\phi u_{\varepsilon i}) - \phi D^{\alpha} u_{\varepsilon i})\, dx = (f, \phi w_{\varepsilon}).$$

But one verifies that (using (9.3.12))

$$(9.3.16) \quad D^{\beta}(\phi w_{\varepsilon j}) - \phi D^{\beta} w_{\varepsilon j} \to D^{\beta}(\phi P_{j}) - \phi D^{\beta} P_{j}, \ |\beta| = m_{j}, \ \text{in } L^{2}(\mathcal{O}) \ \textbf{strongly}$$

and that (compactness of the injection of $H^{\mu}(\mathcal{O})$ into $H^{\mu-1}_{loc}(\mathcal{O})$)

$$(9.3.17) \quad D^{\alpha}(\phi u_{\varepsilon i}) - \phi D^{\alpha} u_{\varepsilon i} \to D^{\alpha}(\phi u_{i}) - \phi D^{\alpha} u_{i}, \ |\alpha| = m_{i}, \ \text{in } L^{2}(\mathcal{O}) \ \textbf{strongly}.$$

One the other hand,

$$(9.3.18) \quad {}^{\varepsilon}a_{\alpha\beta}^{ij} D^{\beta} w_{\varepsilon j} = (a_{\alpha\beta}^{ij} D_{y}^{\beta} w_{j})\left(\frac{x}{\varepsilon}\right) \to (a_{\alpha\beta}^{ij} D_{y}^{\beta} w_{j}) \ \text{in } L^{\infty}(\mathcal{O}) \ \text{weak star},$$

so that (9.3.15) gives
$$(9.3.19)$$
$$\int_{\mathcal{O}} \xi_{\beta}^{j}(D^{\beta}(\phi P_{j}) - \phi D^{\beta} P_{j})\, dx - \mathcal{M}(a_{\alpha\beta}^{ij} D_{y}^{\beta} w_{j}) \int_{\mathcal{O}} (D^{\alpha}(\phi u_{i}) - \phi D^{\alpha} u_{i})\, dx = (f, \phi P).$$

But $\int_{\mathcal{O}} D^{\alpha}(\phi u_{i})\, dx = 0$ and the right hand side of (9.3.19) equals (using (9.3.5)) $(\xi_{\beta}^{j}, D^{\beta}(\phi P_{j}))$; therefore (9.3.19) reduces to

$$- \int_{\mathcal{O}} \xi_{\beta}^{j} D^{\beta} P_{j}\phi\, dx + (a_{\alpha\beta}^{ij} D_{y}^{\beta} w_{j}) \int_{\mathcal{O}} \phi D^{\alpha} u_{i}\, dx = 0,$$

i.e.

$$(9.3.20) \qquad \xi_{\beta}^{j} D^{\beta} P_{j} - \mathcal{M}(a_{\alpha\gamma}^{ik} D_{y}^{\gamma} w_{k}) D^{\alpha} u_{i}.$$

We now take

$$P = P_{j}^{\beta}.$$

Then $w = P_{j}^{\beta} - \chi_{j}^{\beta*}$ and (9.3.20) gives

$$\xi_{\beta}^{j} = \mathcal{M}(a^{ik} D(P_{jk}^{\beta} - \chi_{jk}^{\beta*})) D^{\alpha} u_{i} = \frac{1}{|Y|} a_{1}^{*}(\chi_{j} - P_{j}^{\beta}, -P_{i}^{\alpha}) D^{\alpha} u_{i}$$

so that (using (9.2.12))

$$(\xi_{\beta}^{j}, D^{\beta} v_{j}) = \mathcal{A}(u, v)$$

and (9.3.5) proves the theorem.

9.4. Correctors. We introduce

$$(9.4.1) \qquad \theta_{\varepsilon} = \{\varepsilon^{m_{j}} g_{\varepsilon j}\},$$

$$g_{\varepsilon j} = -m_{\varepsilon}(x)\chi_{kj}^{\beta}\left(\frac{x}{\varepsilon}\right) D^{\beta} u_{k}(x),$$

where $\chi_{k}^{\beta} = \{\chi_{kj}^{\beta}\}$ is defined in (9.2.4) and where m_{ε} is defined in (5.2.2). We have

THEOREM 9.7. *We suppose that the hypotheses of Theorem 9.4 hold true and that $a_1(\phi, \psi)$ is symmetric. We also assume that*

(9.4.2) $$f \in (L^2(\mathcal{O}))^N,$$

(9.4.3) $$\chi_{jk}^\beta \in W^{m_k, \infty}(Y),$$

and that the boundary of $\mathcal{O}$ is smooth enough.

Then if θ_ε is defined by (9.4.1), one has

(9.4.4) $$z_\varepsilon = u_\varepsilon - u - \theta_\varepsilon \to 0 \text{ in } V \text{ strongly};$$

θ_ε *is called the first corrector.*

REMARK 9.8. [Similar to Remark 5.3] It follows from (9.4.2) that u satisfies

(9.4.5) $$u_j \in H^{2m_j}(\mathcal{O}), \ \forall j$$

so that $\theta_\varepsilon \in V$ and therefore

(9.4.6) $$z_\varepsilon \in V.$$

REMARK 9.9. [Similar to Remark 5.4] If $V = \prod H^{m_j}(\mathcal{O})$ (Neumann's problem) one can take $m = 1$ in (9.4.1).

PROOF OF THEOREM 9.7. The method is entirely analogous to the one in Section 5.3. We have

(9.4.7) $$a^\varepsilon(z_\varepsilon) = (f, u_\varepsilon - 2(u + \theta\varepsilon)) + X_\varepsilon,$$

(9.4.8) $$X_\varepsilon = a^\varepsilon(u + \theta_\varepsilon),$$

and all we have to prove is that

(9.4.9) $$X_\varepsilon \to \mathcal{A}(u, u).$$

But

(9.4.10) $$X_\varepsilon = \int_{\mathcal{O}} {}^\varepsilon a_{\alpha\beta}^{ij} D^\alpha(u_i + \theta_{\varepsilon i}) D^\beta(u_j + \theta_{\varepsilon j}) \, dx,$$

and

(9.4.11) $$D^\alpha(u_i + \theta_{\varepsilon i}) = D^\alpha u_i - m_\varepsilon(D_y \chi_{ki}^\beta) D_{uk}^\beta - r_{i\varepsilon}^\alpha,$$

where

(9.4.12) $$r_{i\varepsilon}^\alpha = \varepsilon^{m_i} D^\alpha(m_\varepsilon \chi_{ki}^\beta D^\beta u_k) - m_\varepsilon(D_y^\alpha \chi_{ki}^\beta) D^\beta u_k.$$

By virtue of the construction of m_ε and properties of χ_k^β, we have

(9.4.13) $$r_{i\varepsilon}^\alpha \to 0 \text{ in } L^2(\mathcal{O}).$$

Therefore, if we set

(9.4.14) $$Y_\varepsilon = \int {}^\varepsilon a^{ij}(D^\alpha u_i - m_\varepsilon(D_y^\alpha \chi_{ki}^\gamma) D^\gamma u_k) \cdot (D^\beta u_j - m_\varepsilon(D_y \chi_{\ell j}^\delta) D^\delta u_\ell) \, dx,$$

we have

$$X_\varepsilon - Y_\varepsilon \to 0.$$

But we can pass to the limit in (9.4.14); we obtain

(9.4.15) $$\lim Y_\varepsilon = \int P_{\alpha\beta}^{ij} D^\alpha u_i D^\beta u_j \, dx$$

where

$$P^{ij}_{\alpha\beta} = \mathcal{M}(a^{ij}_{\alpha\beta}) - \mathcal{M}(a^{kj}_{\gamma\beta} D^{\gamma}_{y}\chi^{\alpha}_{ik}) - \mathcal{M}(a^{i\ell}_{\alpha\delta} D^{\delta}_{y}\chi^{\beta}_{j\ell}) + \mathcal{M}(a^{k}_{\gamma\delta} D^{\gamma}\chi^{\alpha}_{ik} D^{\delta}\chi^{\beta}_{j\ell}),$$

i.e.,

$$P^{ij}_{\alpha\beta} = a_1(P^{\alpha}_i, P^{\beta}_j - \chi^{\beta}_j) = a_1(\chi^{\beta}_j - P^{\beta}_j, -P^{\alpha}_i)$$

so that (9.4.15) proves that

$$\lim Y_\varepsilon = \mathcal{A}(u,u)$$

and the theorem is proven.   ∎

REMARK 9.10. The results of this section apply in particular to **the equations of elasticity in composite materials**.

10. Homogenization of the Stokes equation

10.1. Orientation. We give in this section the homogenization of an elliptic system which does not fit into the framework of Section 9 (where, as we said, we did **not** consider the most general elliptic systems). This system is related to the **Stokes equations** and can be used in the modeling of **flows in porous media**.

10.2. Statement of the problem and of the homogenization theorem. In the bounded open set $\mathcal{O}$ of $\mathbb{R}^n$ we consider

$$(10.2.1) \qquad \mathcal{V} = \{\phi \mid \phi \in (C_0^\infty(\mathcal{O}))^n, \ \operatorname{div}\phi = 0\},$$

$$(10.2.2) \qquad H = \text{ closure of } \mathcal{V} \text{ in } (L^2(\mathcal{O}))^n,$$

$$V = \text{ closure of } \mathcal{V} \text{ in } (H^1(\mathcal{O}))^n.$$

One can check that

$$(10.2.3) \qquad V = \{v \mid v = \{v_i\}, v_i \in H_0^1(\Omega), \ \operatorname{div} v = 0\}.$$

We consider the **scalar** operator A^ε defined by

$$(10.2.4) \qquad -\frac{\partial}{\partial x_i}\left(a_{ij}\left(\frac{x}{\varepsilon}\right)\frac{\partial}{\partial x_j}\right),$$

with the hypothesis (1.1.1).[48]

Given u in $(H^1(\mathcal{O}))^n$, we **define** A^ε acting on u in a "diagonal" way, i.e.,

$$(10.2.5) \qquad (A^\varepsilon u)_i = A^\varepsilon u_i, \ i = 1, \ldots, n.$$

For $u, v \in (H^1(\mathcal{O}))^n$, we set

$$(10.2.6) \qquad a^\varepsilon(u,v) = \int_{\mathcal{O}} a_{ij}\left(\frac{x}{\varepsilon}\right)\frac{\partial u_k}{\partial x_j}\frac{\partial v_k}{\partial x_i}\,dx.$$

If $f \in V'$, we denote by (f,v) the scalar product between V' and V, and also the scalar product in $(L^2(\mathcal{O}))^n$.

Given f, **there exists a unique** $u_\varepsilon \in V$ **such that**

$$(10.2.7) \qquad a^\varepsilon(u_\varepsilon, v) = (f, v), \ \forall v \in V.$$

---

[48]We could also consider here coefficients $a_{ij}\left(x, \frac{x}{\varepsilon}, \ldots, \frac{x}{\varepsilon^N}\right)$.

The interpretation of (10.2.7) is: there exists a distribution p_ε in $\mathcal{O}$ (the pressure) such that

(10.2.8) $$A^\varepsilon u_\varepsilon = f - \operatorname{grad} p_\varepsilon \text{ in } \mathcal{O},$$

(10.2.9) $$\operatorname{div} u_\varepsilon = 0 \text{ in } \mathcal{O},$$

$$u_\varepsilon = 0 \text{ on } \Gamma.$$

The **homogenization of this problem** is given as follows. We define

(10.2.10) $$W_{\operatorname{div}}(Y) = \{\phi \mid \phi \in W(Y), \ \operatorname{div}\phi = 0 \text{ in } Y\}$$

(where $W(Y)$ denotes here the subspace of $(H^1(Y))^n$ which consists of periodic functions). For $\phi, \psi \in (H^1(Y))^n$, we set

(10.2.11) $$a_1(\phi, \psi) = \int_Y a_{ij}(y) \frac{\partial \phi_k}{\partial y_j} \frac{\partial \psi_k}{\partial y_i} \, dy,$$

and we introduce (with the notations of Section 9)

(10.2.12) $$P_j(y) = \underbrace{\{0, \ldots, 0, y_\beta, 0 \ldots, 0\}}_{j}, \ j, \beta = 1, \ldots, n.$$

We define χ_j^β (up to the addition of a vector with constant components) by

(10.2.13) $$a_1(\chi_j^\beta - p_j^\beta, \psi) = 0, \ \forall \psi \in W_{\operatorname{div}}(Y),$$

$$\chi_j^\beta \in W_{\operatorname{div}}(Y).$$

Then we set

(10.2.14) $$q_{\alpha\beta}^{ij} = \frac{1}{|Y|} a_1(\chi_j^\beta - p_j^\beta, \chi_i^\alpha - p_i^\alpha)$$

and

(10.2.15) $$\mathcal{A}(u, v) = \int q_{\alpha\beta}^{ij} \frac{\partial u_i}{\partial x_\alpha} \frac{\partial v_j}{\partial x_\beta} \, dx.$$

Let $u \in V$ be the solution of

(10.2.16) $$\mathcal{A}(u, v) = (f, v), \ v \in V;$$

(10.2.16) is **the homogenized problem** of (10.2.9). If $\mathcal{A}$ denotes the operator associated to $\mathcal{A}$, (10.2.16) can be interpreted as

$$\mathcal{A} = f - \operatorname{grad} p,$$

(10.2.17) $$\operatorname{div} u = 0,$$

$$u = 0 \text{ on } \Gamma.$$

REMARK 10.1. Formula (10.2.14) formally coincides with the one of Section 9. But the proof of Section 9 does not apply directly in the present context.

We shall prove in Section 10.3 below the

THEOREM 10.2. *Under the hypothesis (1.1.1), if u_ε (respectively, u) is the solution of (10.2.7) (respectively, (10.2.16)), one has*

(10.2.18) $$u_\varepsilon \to u \text{ in } V \text{ weakly.}$$

We shall give in Section 10.4 (after the "energy" proof of Section 10.3) some indications on the **asymptotic expansion of u_ε.**

10.3. Proof of the homogenization theorem. It immediately follows from (10.2.7) and hypothesis (1.1.1) that

$$(10.3.1) \qquad\qquad \|u_\varepsilon\|_V \leq C.$$

Then $A^\varepsilon u_\varepsilon$ remains bounded in $(H^{-1}(\mathcal{O}))^n$, and therefore by (10.2.9) one has:

$$(10.3.2) \qquad\qquad \operatorname{grad} P_\varepsilon \text{ bounded in } (H^{-1}(\mathcal{O}))^n.$$

Therefore (cf. Deny-Lions [**33**] for instance) one can choose p_ε (which is defined up to an additive constant) such that

$$(10.3.3) \qquad\qquad |p_\varepsilon|_{L^2(\mathcal{O})} \leq C.$$

We introduce

$$(10.3.4) \qquad\qquad \xi_i^\varepsilon = a_{ij}^\varepsilon \frac{\partial u_\varepsilon}{\partial x_j}, \;\; a_{ij}^\varepsilon(x) = a_{ij}\left(\frac{x}{\varepsilon}\right),$$

and by virtue of (10.3.1) one has

$$(10.3.5) \qquad\qquad |\xi_i^\varepsilon|_{(L^2(\mathcal{O}))^n} \leq C.$$

Equation (10.2.7) can be written

$$(10.3.6) \qquad\qquad \left(\xi_i^\varepsilon, \frac{\partial v}{\partial x_i}\right) = (f, v), \;\; \forall v \in V.$$

We can extract a subsequence, still denoted by u_ε, p_ε, such that

$$(10.3.7) \qquad \begin{aligned} &u_\varepsilon \to u \text{ in } V \text{ weakly,} \\ &\xi_i^\varepsilon \to \xi_i \text{ in } (L^2(\mathcal{O}))^n \text{ weakly,} \\ &p_\varepsilon \to p \text{ in } L^2(\mathcal{O}) \text{ weakly;} \end{aligned}$$

in the limit (10.3.6) gives

$$(10.3.8) \qquad\qquad \left(\xi_i, \frac{\partial v}{\partial x_i}\right) = (f, v), \;\; \forall v \in V$$

and (10.2.9) gives

$$(10.3.9) \qquad\qquad -\frac{\partial \xi_i}{\partial x_i} = f - \operatorname{grad} p.$$

The problem is now to compute ξ_i.

We introduce $a_1^*(\phi, \psi) = a_1(\psi, \phi)$ and $p(y)$ such that

$$(10.3.10) \qquad P(y) = \{P_j(y)\}, \; P_j = \text{homogeneous polynomial of first degree.}$$

Given P we define χ as the solution (up to an additive vector with constant components) of

$$(10.3.11) \qquad\qquad a_1^*(\chi - P, \psi) = 0, \; \forall \psi \in W_{\operatorname{div}}(Y), \; \chi \in W_{\operatorname{div}}(Y).$$

The interpretation of (10.3.11) is: there exists π ($\in L^2(Y)$) such that

$$(10.3.12) \qquad \begin{aligned} A_1^* \chi &= a_1^* p - \operatorname{grad} \pi, \\ \operatorname{div} \chi &= 0, \\ \chi \text{ and } \pi &\text{ periodic.} \end{aligned}$$

We now set

$$(10.3.13) \qquad\qquad w = P - \chi$$

and

(10.3.14) $$w_\varepsilon(x) = \varepsilon w(x/\varepsilon), \ \pi_\varepsilon(x) = \pi(x/\varepsilon), \ \chi_\varepsilon = \varepsilon\chi(x/\varepsilon).$$

We have

(10.3.15) $$(A^\varepsilon)^* w_\varepsilon = \operatorname{grad} \pi_\varepsilon.$$

We now use (10.2.9) and (10.3.15). Given $\phi \in C_0^\infty(\mathcal{O})$, we set $\phi v = \{\phi v_j\}$; we multiply (10.2.9) (respectively (10.3.15)) by ϕw_ε (respectively, ϕu_ε); after integrating over $\mathcal{O}$ and subtracting, we obtain

(10.3.16) $\ a^\varepsilon(u_\varepsilon, \phi w_\varepsilon) - a^{\varepsilon*}(w_\varepsilon, \phi u_\varepsilon) = (f, \phi w_\varepsilon) - (\operatorname{grad} p_\varepsilon, \phi w_\varepsilon) - (\operatorname{grad} \pi_\varepsilon, \phi u_\varepsilon).$

Using the fact that $\operatorname{div} \chi_\varepsilon = 0$ and $\operatorname{div} u_\varepsilon = 0$, the right hand side of (10.3.16) equals

(10.3.17) $$(f, \phi w_\varepsilon) + \left(p_\varepsilon, \frac{\partial \phi}{\partial x_i} x_{\varepsilon i} + \phi \operatorname{div} P \right) + \left(\pi_\varepsilon, \frac{\partial \phi}{\partial x_i} u_{\varepsilon i} \right).$$

Since $w_\varepsilon \to p$ in $(L^2(\mathcal{O}))^n$ strongly, and since

$$\pi_\varepsilon \to \mathcal{M}(\pi) \text{ in } L^\infty(\mathcal{O}) \text{ weak star,}$$

and using (10.3.7), one has for the limit of (10.3.17)

(10.3.18) $$(f, \phi P) + \left(p, \frac{\partial \phi}{\partial x_i} p_i + \phi \operatorname{div} P \right) + (\mathcal{M}(pi), \operatorname{div}(\phi u)).$$

But $(\mathcal{M}(\pi), \operatorname{div}(\phi u)) = 0$ and (10.3.18) reduces to

(10.3.19) $$(f, \phi P) + (p, \operatorname{div}(\phi P)).$$

Using (10.3.9), we see that (10.3.19) equals

(10.3.20) $$\left(-\frac{\partial \xi_i}{\partial x_i}, \phi P \right) = \left(\xi_i, \frac{\partial \phi}{\partial x_i} p + \phi \frac{\partial P}{\partial x_i} \right).$$

The left hand side of (10.3.16) equals (this is the usual calculation)

$$\left(\xi_i, \frac{\partial \phi}{\partial x_i} w_\varepsilon \right) - \int_\mathcal{O} \left(a_{ij} \frac{\partial w}{\partial y_i} \right)^\varepsilon \cdot \frac{\partial \phi}{\partial x_j} u_\varepsilon \, dx$$

(where $g^\varepsilon(x) = g(x/\varepsilon)$), and the limit of this term is given by

(10.3.21) $$\left(\zeta_i, \frac{\partial \phi}{\partial x_i} w_\varepsilon \right) - \int_\mathcal{O} \left(a_{ij} \frac{\partial w}{\partial y_i} \right) \cdot \frac{\partial \phi}{\partial x_j} u \, dx.$$

The equality of (10.3.20) and of (10.3.21) gives

$$\left(\xi_i, \phi \frac{\partial P}{\partial x_i} \right) = -\int_\mathcal{O} \mathcal{M} \left(a_{ij} \frac{\partial w}{\partial y_i} \right) \cdot \frac{\partial \phi}{\partial x_j} u \, dx, \ \forall \phi \in C_0^\infty(\mathcal{O}),$$

i.e.,

(10.3.22) $$\xi_i \frac{\partial P}{\partial x_i} = \mathcal{M} \left(a_{k\ell} \frac{\partial w}{\partial y_k} \right) \cdot \frac{\partial u}{\partial x_\ell}.$$

We take now $P = P_j^\beta$. Then $\chi = \chi_j^{\beta*}$ and (10.3.22) gives (if $\xi_\beta = \{\xi_{\beta j}\}$):

(10.3.23) $$\xi_{\beta j} = \mathcal{M} \left(a_{k\ell} \frac{\partial(P_j^\beta - \chi_j^{\beta*})}{\partial y_k} \right) \cdot \frac{\partial u}{\partial x_\ell}$$

which (after a short computation) proves the desired result.

10.4. Asymptotic expansion. We look for $u_\varepsilon, p_\varepsilon$ in the form

(10.4.1)
$$u_\varepsilon = u_0 + \varepsilon u_1 + \varepsilon^2 u_2 + \cdots,$$
$$p_\varepsilon = p_0 + \varepsilon p_1 + \cdots,$$

where $u_k = u_k(x, y)$, $p_k = p_k(x, y)$ are vectors or functions Y-periodic.

We write (10.2.9) in the form

(10.4.2)
$$(\varepsilon^{-2} A_1 + \varepsilon^{-1} A_2 + \varepsilon^0 A_3) u_3 = f - (\varepsilon^{-1} \operatorname{grad}_y + \operatorname{grad}_x) p_\varepsilon$$
$$(\varepsilon^{-1} \operatorname{div}_y + \operatorname{div}_x) u_\varepsilon = 0.$$

We identify the coefficients of the powers $\varepsilon^{-2}, \varepsilon^{-1}, \varepsilon^0$. This gives

(10.4.3) $A_1 u_0 = 0,$

(10.4.4) $A_1 u_1 + A_2 u_0 = - \operatorname{grad}_y P_0,$
$$\operatorname{div}_y u_0 = 0,$$

(10.4.5) $A_1 u_2 + A_2 u_1 + A_3 u_0 = f - \operatorname{grad}_x P_0 - \operatorname{grad}_y P_1,$
$$\operatorname{div}_y u_1 + \operatorname{div}_x u_0 = 0.$$

It follows from (10.4.3) that

(10.4.6) $u_0 = u(x).$

Then (10.4.4) reduces to

$$A_1 u_1 + A_2 u_0 = - \operatorname{grad}_y P_0.$$

The second condition (10.4.5) implies

$$\int_Y (\operatorname{div}_y u_1 + \operatorname{div}_x u_0) \, dy = 0$$

and since this integral equals $|Y| \operatorname{div} u$ one has

(10.4.7) $\operatorname{div} u = 0.$

Then the second condition (10.4.5) is equivalent to $\operatorname{div}_y u_1 = 0$, i.e., finally for u_1

(10.4.8) $A_1 u_1 + A_2 u_0 = - \operatorname{grad}_y P_0,$
$$\operatorname{div}_y u_1 = 0.$$

More explicitly

(10.4.9)
$$A_1 u_1 = \frac{\partial a_{ij}}{\partial y_i} \frac{\partial u_0}{\partial x_j} - \operatorname{grad}_y P_0$$
$$\operatorname{div}_y u_1 = 0.$$

With previous notations, the general solution of (10.4.9) is:

(10.4.10)
$$u_1 = -\chi_j^\beta(y) \frac{\partial u_{0j}}{\partial x_\beta}(x) + \tilde{u}_1(x).$$

The first equation in (10.4.5) can be solved in u_2 if

(10.4.11)
$$\int_Y (A_2 u_1 + A_3 u_0) \, dy = f|Y| - |Y| \operatorname{grad}_x \int_Y P_0 \, dy.$$

Using (10.4.10), (10.4.11) gives the homogenized system (together with (10.4.7)).

REMARK 10.3. Expression (10.4.10) gives the first term in the asymptotic expansion. We have found a technical difficulty in the precise form of the first corrector, since the multiplication by a funciton m_ε does not leave invariant the space of vectors with zero divergence.

REMARK 10.4. One can also use the "adjoint expansion," as in Section 3.3, since we have the Dirichlet's boundary conditions.

REMARK 10.5. One can also consider other boundary value problems, by taking other spaces V. For instance one can take:

$$(10.4.12) \qquad V = \{v \mid v \in (H^1(\mathcal{O}))^n, \ \operatorname{div} v = 0, \ v \cdot \nu = 0 \text{ on } \Gamma\}.$$

As usual, the homogenized operator does not depend on V.

11. Homogenization of equations of Maxwell's type

11.1. Setting of the problem. In $\mathcal{O} \subset \mathbb{R}^3$ we consider two matrices

$$(11.1.1) \qquad a^\varepsilon(x) = [a_{ij}^\varepsilon(x)], \qquad\qquad a_{ij}^\varepsilon(x) = a_{ij}(x/\varepsilon),$$

$$(11.1.2) \qquad a_0^\varepsilon(x) = [a_{0ij}^\varepsilon(x)], \qquad\qquad a_{0ij}^\varepsilon(x) = a_{0ij}(x/\varepsilon),$$

where the functions $a_{ij}(y)$, $a_{0ij}(y)$ are Y-periodic and satisfy the analogs of (1.1.1). Given a vector function on u, we set

$$(11.1.3) \qquad \operatorname{rot} u = \nabla \times u$$

and we consider boundary value problems associated with the equation

$$(11.1.4) \qquad \operatorname{rot}(a^\varepsilon \operatorname{rot} u_\varepsilon) + a_0^\varepsilon u_\varepsilon = f \text{ in } \mathcal{O}.$$

We want to study the behavior of u_ε as $\varepsilon \to 0$.

    11.1.1. *Variational formulation. Functional spaces.* We introduce

$$(11.1.5) \qquad H = (L^2(\mathcal{O}))^3$$

where the scalar product is denoted by (f, q). We introduce

$$(11.1.6) \qquad \mathcal{H}(\mathcal{O}, \operatorname{rot}) = \{v \mid v \in H, \ \operatorname{rot} v \in H\}$$

provided with the norm given by

$$(11.1.7) \qquad \|v\|_{\mathcal{H}(\mathcal{O},\operatorname{rot})}^2 = \|v\|_H^2 + \|\operatorname{rot} v\|_H^2.$$

We introduce next

$$(11.1.8) \qquad \mathcal{H}_0(\mathcal{O}, \operatorname{rot}) = \text{ closure of } (C_0^\infty(\mathcal{O}))^3 \text{ into } \mathcal{H}(\mathcal{O}, \operatorname{rot}).$$

One can verify (cf. for instance Duvaut-Lions [**38**], Chapter 7) that

$$(11.1.9) \qquad \mathcal{H}_0(\mathcal{O}, \operatorname{rot}) = \{v \mid v \in \mathcal{H}(\mathcal{O}, \operatorname{rot}), \ \mathbf{n} \wedge v = 0 \text{ on } \Gamma\}$$

where in (11.1.9) $\mathbf{n} =$ unit normal on Γ directed toward the exterior of $\mathcal{O}$ to fix ideas.

    We introduce V such that

$$(11.1.10) \qquad \mathcal{H}_0(\mathcal{O}, \operatorname{rot}) \subseteq V \subseteq \mathcal{H}(\mathcal{O}, \operatorname{rot}).$$

For $u, v \in \mathcal{H}(\mathcal{O}, \operatorname{rot})$ we set

$$(11.1.11) \qquad a^\varepsilon(u, v) = \int_\mathcal{O} a^\varepsilon \operatorname{rot} u \cdot \operatorname{rot} v \, dx + \int_\mathcal{O} a_0^\varepsilon u \cdot v \, dx.$$

The operator associated to (11.1.11) is

(11.1.12) $$A^\varepsilon = \mathrm{rot}(a^\varepsilon \, \mathrm{rot}) + a_0^\varepsilon.$$

The **boundary value problem** we study is:

(11.1.13) $$a^\varepsilon(u_\varepsilon, v) = (f, v), \ \forall v \in V, \ u_\varepsilon \in V,$$

where $u \to (f, v)$ is a continuous linear form on V $(f \in V')$.

In order to study the asymptotic behavior of u_ε as $\varepsilon \to 0$, we begin with the method of **asymptotic expansions**.

In order to simplify the exposition, we shall consider first the case

(11.1.14) $$a_0^\varepsilon = \lambda I, \ \lambda > 0,$$

and we shall return later on to the more general case (11.1.2).

11.2. Asymptotic expansions. We have

$$A^\varepsilon = \varepsilon^{-2} A_1 + \varepsilon^{-1} A_2 + \varepsilon^0 A_3,$$

(11.2.1)
$$A_1 = \mathrm{rot}_y \, a \, \mathrm{rot}_y,$$
$$A_2 = \mathrm{rot}_y \, a \, \mathrm{rot}_x + \mathrm{rot}_x \, a \, \mathrm{rot}_y,$$
$$A_3 = \mathrm{rot}_x \, a \, \mathrm{rot}_x + \lambda.$$

We look for u_ε in the usual form: $u_\varepsilon = u_0 + \varepsilon u_1 + \cdots$, $u_j = u_j(x, y)$. We obtain

$$A_1 u_0 = 0,$$

(11.2.2)
$$a_1 u_1 + A_2 u_0 = 0,$$
$$A_1 u_2 + A_2 u_1 + A_3 u_0 = f.$$

But $(A_1 u_0, u_0)_Y$[49] equals $(a \, \mathrm{rot}_y \, u_0, \mathrm{rot}_y \, u_0)_Y$ so that the first equation (11.2.2) is equivalent to

$$\mathrm{rot}_y \, u_0 = 0.$$

Then the second equation (11.2.2) reduces to

$$\mathrm{rot}_y(a \, \mathrm{rot}_y \, u_1 + a \, \mathrm{rot}_x \, u_0) = 0.$$

Applying div_y to both sides of the third equation (11.2.2) we obtain

$$\mathrm{div}_y \, \mathrm{rot}_x(a \, \mathrm{rot}_y \, u_1 + a \, \mathrm{rot}_x \, u_0) + \lambda \, \mathrm{div}_y \, u_0 = 0,$$

and since $\mathrm{div}_y \, \mathrm{rot}_x = - \, \mathrm{div}_x \, \mathrm{rot} \, y$, we obtain

$$\mathrm{div}_y \, u_0 = 0.$$

Hence

$$u_0 = u(x)$$

and

(11.2.3) $$\mathrm{rot}_y \, a(\mathrm{rot}_y \, u_1 + \mathrm{rot}_x \, u) = 0.$$

We set

(11.2.4) $$w = a(\mathrm{rot}_y \, u_1 + \mathrm{rot}_x \, u)$$

and

(11.2.5) $$\mathcal{M}(w) = \tilde{w} = \tilde{w}(x).$$

-----
[49]We take the scalar product in $(L^2(Y))^3$.

It follows from (11.2.3) and from the definition of w, that

$$(11.2.6) \qquad \operatorname{rot}_y w = 0,$$

$$(11.2.7) \qquad \operatorname{div}_y(a^{-1}w) = 0.$$

Then $\operatorname{rot}_y(w - \tilde{w}) = 0$, $\mathcal{M}(w - \tilde{w}) = 0$, so that there exists a Y-periodic function $\psi(x, y)$ such that

$$(11.2.8) \qquad w - \tilde{w} = -\operatorname{grad}_y \psi.$$

Using (11.2.8) into (11.2.7) gives

$$(11.2.9) \qquad -\operatorname{div}_y a^{-1} \operatorname{grad}_y \psi = -\operatorname{div}_y a^{-1}\tilde{w}.$$

We introduce, in order to solve (11.2.9), the functions θ^j needed for finding **the homogenized operator associated with** a^{-1}:

$$(11.2.10) \qquad -\operatorname{div}_y(a^{-1}\operatorname{grad}_y \theta^j) = -\operatorname{div}_y(a^{-1}e_j),$$
$$e_1 = (1, 0, 0), \ e_2 = (0, 1, 0), \ e_3 = (0, 0, 1).$$

If we define the matrix $\theta = \{\theta^1, \theta^2, \theta^3\}$, and if we define $\operatorname{grad}_y \theta$ by $\operatorname{grad}_y \theta \cdot \xi = \operatorname{grad}_y(\theta\xi)$, we obtain

$$(11.2.11) \qquad \psi = \theta \cdot \tilde{w}$$

and (11.2.8) gives

$$(11.2.12) \qquad w = (I - \operatorname{grad}_y \theta)\tilde{w}.$$

Applying $\mathcal{M}$ to (11.2.4) after multiplying by a^{-1} gives:

$$\operatorname{rot} u = \mathcal{M}(a^{-1}(I - \operatorname{grad}_y \theta))\tilde{w}.$$

If we define $\mathcal{H}(a)$ to be the homogenized matrix associated with a_{ij}, we obtain

$$(11.2.13) \qquad \operatorname{rot} u = \mathcal{H}(a^{-1})\tilde{w}.$$

Applying $\mathcal{M}$ to (11.1.11) gives

$$\operatorname{rot} \mathcal{M}(w) + \lambda u = f$$

hence, using (11.2.5) and (11.2.13):

$$(11.2.14) \qquad \operatorname{rot} \mathcal{H}(a^{-1})^{-1} \operatorname{rot} u + \lambda u = f.$$

The homogenized operator (obtained in a formal manner, which will be justified below) **associated to** $\operatorname{rot} a^\varepsilon \operatorname{rot} + \lambda I$ **is**

$$\operatorname{rot} \mathcal{H}(a^-1)^{-1} \operatorname{rot} + \lambda.$$

11.2.1. *Asymptotic expansions for the first order system.* Let us write now the equation $A^\varepsilon u_\varepsilon = f$ as the first order system:

$$(11.2.15) \qquad a^\varepsilon \operatorname{rot} u_\varepsilon - v_\varepsilon = 0,$$

$$(11.2.16) \qquad \operatorname{rot} v_\varepsilon + \lambda u_\varepsilon = f.$$

We look for $u_\varepsilon = u_0 + \varepsilon u_1 + \cdots$, $v_\varepsilon = v_0 + \varepsilon v_1 + \cdots$, as usual and we obtain

$$(11.2.17) \qquad a \operatorname{rot}_y u_0 = 0, \qquad \operatorname{rot}_y v_0 = 0,$$

$$(11.2.18) \qquad a \operatorname{rot}_y u_1 + a \operatorname{rot}_x u_0 - v_0 = 0,$$
$$\operatorname{rot}_y v_1 + \operatorname{rot}_x v_0 + \lambda u_0 = f,$$

By the same methods as above, one finds

$$v_0 = (I - \text{grad}_y\,\theta)\tilde{v}_0, \qquad v_0 = \tilde{v}_0,$$

(in fact $v_0 = w$).

11.3. Another asymptotic expansion. Let us prove a simple Lemma:

LEMMA 11.1. *We assume that $a_0^\varepsilon = \lambda I$, $\lambda > 0$, that $\Gamma = \partial\mathcal{O}$ is smooth enough and that*

$$(11.3.1) \qquad\qquad \text{div}\, f \in L^2(\mathcal{O}).$$

Then we do not restrict the generality in (11.1.13) by assuming that

$$(11.3.2) \qquad\qquad \text{div}\, f = 0.$$

PROOF. We introduce the solution z of

$$(11.3.3) \qquad\qquad \lambda\Delta z = \text{div}\, f, \; z = 0 \text{ on } \Gamma,$$

which satisfies

$$(11.3.4) \qquad\qquad z \in H^2(\mathcal{O}) \cap H^1_0(\mathcal{O}).$$

Therefore

$$(11.3.5) \qquad\qquad \tilde{u} = u_\varepsilon - \nabla z \text{ belongs to } V,$$

indeed $\mathbf{n} \wedge \nabla z = 0$ on Γ if $z \in H^2(\mathcal{O}) \cap H^1_0(\mathcal{O})$ so that $\nabla z \in \mathcal{H}_0(\mathcal{O}, \text{rot}) \subseteq V$. Then (11.1.13) is equivalent to

$$a^\varepsilon(\tilde{u}_\varepsilon, v) = (\tilde{f}, v)$$

where $\tilde{f} = f - \lambda\nabla z$ satisfies $\text{div}\,\tilde{f} = 0$.

We now assume that (11.3.2) holds true and therefore $\text{div}(A^\varepsilon u_\varepsilon) = \text{div}\, f = 0$ so that

$$(11.3.6) \qquad\qquad \text{div}\, u_\varepsilon = 0.$$

We look for $u_\varepsilon = u_0 + \varepsilon u_1 + \cdots$, solution of $A^\varepsilon u_\varepsilon = f$ **and** of (11.3.6). We obtain, with the notation (11.2.1):

$$(11.3.7) \qquad\qquad A_1 u_0 = 0, \qquad\qquad \text{div}_y\, u_0 = 0,$$

$$(11.3.8) \qquad\qquad A_1 u_1 + A_2 u_0 = 0, \qquad \text{div}_y\, u_1 + \text{div}_x\, u_0 = 0,$$

$$(11.3.9) \qquad A_1 u_2 + A_2 u_1 + A_3 u_0 = f, \qquad \text{div}_y\, u_2 + \text{div}_x\, u_1 = 0.$$

∎

Now (11.3.7) is equivalent to $u_0 = u(x)$, and $0 = \mathcal{M}(\text{div}_y\, u_1 + \text{div}\, u) = \text{div}\, u$. Therefore (11.3.8) reduces to

$$(11.3.10) \qquad\qquad \text{rot}_y\, a\,\text{rot}\, u_1 = -\text{rot}_y(a\,\text{rot}\, u),$$

$$\text{div}_y\, u_1 = 0, \qquad u_1 \text{ is } Y\text{-periodic}.$$

We now use the vectors $\tilde{\theta}^p(y)$ (**compare to (5.4.28)**):

$$(11.3.11) \qquad\qquad \text{rot}_y\, a\,\text{rot}_y\, \tilde{\theta}^p = \text{rot}_y(ae_p),$$

$$\text{div}_y\, \tilde{\theta}^p = 0, \qquad \tilde{\theta}^p \text{ is } Y\text{-periodic}, \qquad \mathcal{M}(\tilde{\theta}^p) = 0.$$

Then

$$(11.3.12) \qquad\qquad u_1 = -\tilde{\theta}(y)\,\text{rot}\, u + \hat{u}_1(x).$$

Applying $\mathcal{M}$ to be the first equation (11.3.9), we obtain

(11.3.13) $$\mathrm{rot}\,\mathcal{M}(a - a\,\mathrm{rot}_y\,\tilde{\theta})\,\mathrm{rot}\,u + \lambda u = f,$$

(11.3.14) $$\mathrm{div}\,u = 0.$$

REMARK 11.2. Formulas (11.2.14) and (11.3.13) are compatible since

(11.3.15) $$\mathcal{H}(a^{-1})^{-1} = \mathcal{M}(a - a\,\mathrm{rot}_y\,\tilde{\theta});$$

this is formula (5.4.36) with a replaced by a^{-1}.

REMARK 11.3. In the case of (11.3.2), we can formulate (11.1.13) in a different functional setting. We introduce

(11.3.16) $$W = \{v \mid v \in V,\ \mathrm{div}\,v = 0\}.$$

Then

(11.3.17) $$a^\varepsilon(u_\varepsilon, v) = (f, v),\ \forall v \in W,\ u_\varepsilon \in W.$$

There is an important difference between (11.1.13) and (11.3.17). We have

(11.3.18) $$W \to H \textbf{ is locally compact}^{[50]},$$
$$V \to H \textbf{ is not locally compact}.$$

The local compactness is enough to justify in the present situation the expansion obtained in this section, as in Section 9. Therefore we have proven (using Remark 11.2):

THEOREM 11.4. *If we assume $a_0^\varepsilon = \lambda I$, Γ smooth enough and (11.3.1) then $u_\varepsilon \to u$ in V weakly, where u is the solution of*

(11.3.19) $$(\mathcal{H}(a^{-1})^{-1}\,\mathrm{rot}\,u, \mathrm{rot}\,v) + \lambda(u, v) = (f, v),\ \forall v \in V.$$

11.3.1. *Orientation.* In what follows we give a more direct proof of (11.3.19), which does not assume Γ smooth nor (11.3.1); it is based on a result of "compensated compactness" which is of independent interest.

11.4. Compensated compactness. We introduce the space

(11.4.1) $$\mathcal{H}(\mathcal{O}, \mathrm{div}) = \{v \mid v \in H,\ \mathrm{div}\,v \in L^2(\mathcal{O})\},$$

with the norm given by

(11.4.2) $$\|v\|^2_{\mathcal{H}(\mathcal{O},\mathrm{div})} = \|v\|^2_H + \|\,\mathrm{div}\,v\|^2_{L^2(\mathcal{O})}.$$

We have

THEOREM 11.5. *Let u^α and v^α be two sequences of vector functions which satisfy*

(11.4.3) $$u^\alpha \to u \ \text{in } \mathcal{H}(\mathcal{O}, \mathrm{rot}) \ \text{weakly},$$

(11.4.4) $$v^\alpha \to v \ \text{in } \mathcal{H}(\mathcal{O}, \mathrm{div}) \ \text{weakly}.$$

Then

(11.4.5) $$u^\alpha v^\alpha \to uv \ \text{in } \mathcal{D}'(\mathcal{O}).$$

---

[50]By this we mean: if $v_\varepsilon \to v$ in W weakly, then $v\varepsilon \to v$ in $(L^2(\mathcal{O}'))^3$ **strongly** for $\overline{\mathcal{O}}' \subset \mathcal{O}$. We can take $\mathcal{O}' = \mathcal{O}$ with suitable boundary conditions but this is not useful here.

REMARK 11.6. In fact we have a little more than (11.4.5). For every ϕ which is **continuous** with compact support in $\mathcal{O}$, we have

(11.4.6) $$\int_{\mathcal{O}} (u^\alpha v^\alpha)\phi \, dx \to \int_{\mathcal{O}} (uv)\phi \, dx.$$

REMARK 11.7. If we suppose that

$$u^\alpha \to u \text{ in } (H^1(\mathcal{O}))^3 \text{ weakly},$$

$$v^\alpha \to v \text{ in } (L^2(\mathcal{O}))^3 \text{ weakly},$$

then (11.4.5) is immediate because $u^\alpha \to u$ in $(L^2(\mathcal{O}))^3$ **strongly** (compactness of the injection $H^1(\mathcal{O}) \to L^2(\mathcal{O})$ or, if $\partial\mathcal{O}$ is irregular, in $L^2_{local}(\mathcal{O})$). In Theorem 11.5 above we obtain (11.4.5) by using some information on the derivatives of u^α and **some other** information on the derivatives of v^α, hence the terminology for Theorem 11.5; we call it a result of **"compensated" compactness**.

PROOF OF THEOREM 11.5. Let $\phi \in C_0^\infty(\mathcal{O})$. We set $\phi v = \{\phi v_1, \phi v_2, \phi v_3\}$. The mapping $v \to \phi v$ is linear and continuous from $\mathcal{H}(\mathcal{O}, \mathrm{rot})$ (respectively, $\mathcal{H}(\mathcal{O}, \mathrm{div})$) into itself.

We have to show (11.4.6). We take $\psi \in C_0^\infty(\mathcal{O})$ such that $\psi = 1$ on the support of ϕ. Then (11.4.6) amounts to showing that

(11.4.7) $$\int_{\mathcal{O}} (\phi u^\alpha) \cdot (\psi v^\alpha) \, dx \to \int_{\mathcal{O}} (\phi u) \cdot (\psi v) \, dx.$$

But since $v \to \phi v$ (where we extend ϕv by 0 outside $\mathcal{O}$) is a continuous linear mapping from $\mathcal{H}(\mathcal{O}, \mathrm{rot})$ (respectively, $\mathcal{H}(\mathcal{O}, \mathrm{div})$) into $\mathcal{H}(\mathbb{R}^3, \mathrm{rot})$ (respectively, $\mathcal{H}(\mathbb{R}^3, \mathrm{div})$), we are lead to showing the following. We assume that

$$u^\alpha, v^\alpha \text{ have their support in a fixed compact set } K \text{ of } \mathbb{R}^3,$$

(11.4.8) $$u^\alpha \to u \text{ (respectively, } v^\alpha \to v) \text{ in } \mathcal{H}(\mathbb{R}^3, \mathrm{rot}) \text{ weakly},$$

$$\text{(respectively in } \mathcal{H}(\mathbb{R}^3, \mathrm{div}) \text{ weakly)},$$

then

(11.4.9) $$\int_{\mathbb{R}^3} u^\alpha v^\alpha \, dx \to \int_{\mathbb{R}^3} uv \, dx.$$

We can also consider **complex valued functions**. If $\bar{v} = \{\bar{v}_1, \bar{v}_2, \bar{v}_3\}$, $\bar{\phi} = $ complex conjugate of ϕ, we are going to show that

(11.4.10) $$\int_{\mathbb{R}^3} u^\alpha \bar{v}^\alpha \, dx \to \int_{\mathbb{R}^3} u\bar{v} \, dx.$$

We now use the Fourier transform. We define:

(11.4.11) $r^\alpha = \Im(u^\alpha), \qquad s^\alpha = \Im(v^\alpha), \qquad r = \Im(u), \qquad s = \Im(v),$

where in general

$$\Im(v) = \{\Im v_1, \Im v_2, \Im v_3\},$$

$$\Im v_1(\xi) = \int_{\mathbb{R}^3} e^{-2\pi i \xi} v_1(x) \, dx.$$

By virtue of Plancherel's formula, (11.4.10) is equivalent to

(11.4.12) $$I^\alpha = \int_{\mathbb{R}^3} r^\alpha \bar{s}^\alpha \, d\xi \to \int_{\mathbb{R}^3} r\bar{s} \, d\xi.$$

The information (11.4.8) is now "transformed" by Fourier transform. Since the u^α, v^α have their support in K, we have

(11.4.13) $r^\alpha(\xi) \to r(\xi),\ s^\alpha(\xi) \to s(\xi)$ in $\mathbb{R}^3$ uniformly for $\xi \in$ bounded set.

Moreover,

$$(11.4.14) \qquad \xi_j r_k^\alpha \to \xi_j r_k - \xi_k r_j \text{ in } L^2(\mathbb{R}^3) \text{ weakly},$$
$$\xi_j s_j^\alpha \to \xi_j s_j \text{ in } L^2(\mathbb{R}^3) \text{ weakly}.$$

We observe that, $\forall k$:

$$(11.4.15) \qquad \xi_k r^\alpha \bar{s}^\alpha = (\xi_k r_j^\alpha - \xi_j r_k^\alpha)\bar{s}_j^\alpha + r_k^\alpha \xi_j \bar{s}_j^\alpha.$$

Hence it follows, using (11.4.14), that

$$(11.4.16) \qquad |\xi||r^\alpha \bar{s}^\alpha(\xi)| \le g(\xi),$$
$$g(\xi) \ge 0, \qquad g \in L^1(\mathbb{R}^3).$$

We can write

$$(11.4.17) \qquad \left| I^\alpha - \int_{|\xi| \le M} r^\alpha \bar{s}^\alpha \, d\xi \right| \le \int_{|\xi| \ge M} \frac{1}{|\xi|} |\xi||r^\alpha \bar{s}^\alpha(\xi)| \, d\xi$$
$$\le \int_{|\xi| \le M} \frac{1}{|\xi|} g(\xi) \, d\xi \le \frac{C}{M}.$$

This inequality shows that (11.4.12) will be proven if we verify that

$$\int_{|\xi| \le M} r^\alpha \bar{s}^\alpha \, d\xi \to \int_{|\xi| \le M} r\bar{s} \, d\xi$$

which follows from (11.4.13).   ∎

11.5. Homogenization theorem.

THEOREM 11.8. *Let u_ε be the solution of (11.1.13), where $a_0^\varepsilon = \lambda I$. We assume that $f \in V'$ and Γ is not necessarily smooth. Then $u_\varepsilon \to u$ in V weakly, where u is the solution of (11.3.19).*

PROOF. We have $\|u_\varepsilon\|_V \le C$, so that we can extract a subsequence, still denoted by u_ε, so that

$$(11.5.1) \qquad u_\varepsilon \to u \text{ in } V \text{ weakly},\ a^\varepsilon \operatorname{rot} u_\varepsilon \to \xi \text{ in } (L^2(\mathcal{O}))^3 \text{ weakly},$$

and

$$(11.5.2) \qquad f = \operatorname{rot}(a^\varepsilon \operatorname{rot} u_\varepsilon) + \lambda u_\varepsilon \to \operatorname{rot} \xi + \lambda u = f.$$

We introduce

$$(11.5.3) \qquad b^\varepsilon = [a_{ji}^\varepsilon]^{-1} = (a^{\varepsilon *})^{-1};$$

given g in $H^{-1}(\mathcal{O})$ we define v_ε as the solution of

$$(11.5.4) \qquad -\operatorname{div} b^\varepsilon \operatorname{grad} v_\varepsilon = g,\ v_\varepsilon = 0 \text{ on } \Gamma.$$

We know that

$$(11.5.5) \qquad v_\varepsilon \to v \text{ in } H_0^1(\mathcal{O}) \text{ weakly},$$
$$b^\varepsilon \operatorname{grad} v_\varepsilon \to \mathcal{H}(b) \operatorname{grad} v \text{ in } (L^2(\mathcal{O}))^3 \text{ weakly}.$$

We start from the identity

(11.5.6) $a^\varepsilon \operatorname{rot} u_\varepsilon \cdot b^\varepsilon \operatorname{grad} v_\varepsilon = \operatorname{rot} u_\varepsilon \cdot \operatorname{grad} v_\varepsilon.$

By virtue of (11.5.1), (11.5.2), $a^\varepsilon \operatorname{rot} u_\varepsilon \to \xi$ in $\mathcal{H}(\mathcal{O}, \operatorname{rot})$ weakly and by virtue of (11.5.5), (11.5.4), $b^\varepsilon \operatorname{grad} v_\varepsilon \to \mathcal{H}(b) \operatorname{grad} v$ in $\mathcal{H}(\mathcal{O}, \operatorname{div})$ weakly. Using Theorem 11.5, it follows that

(11.5.7) $a^\varepsilon \operatorname{grad} u_\varepsilon \cdot b^\varepsilon \operatorname{grad} v_\varepsilon \to \xi \cdot \mathcal{H}(b) \operatorname{grad} v$ in $\mathcal{D}'(\mathcal{O}).$

Since $\operatorname{div} \operatorname{rot} u_\varepsilon = 0$, $\operatorname{rot} \operatorname{grad} v_\varepsilon = 0$, we have

$$\operatorname{rot} u_\varepsilon \to \operatorname{rot} u \text{ in } \mathcal{H}(\mathcal{O}, \operatorname{div}) \text{ weakly}$$

and

$$\operatorname{grad} v_\varepsilon \to \operatorname{grad} v \text{ in } \mathcal{H}(\mathcal{O}, \operatorname{rot}) \text{ weakly},$$

so that Theorem 11.5 gives

(11.5.8) $\operatorname{rot} u_\varepsilon \cdot \operatorname{grad} v_\varepsilon \to \operatorname{rot} u \cdot \operatorname{grad} v$ in $\mathcal{D}'(\mathcal{O}).$

It follows from (11.5.6), (11.5.7), (11.5.8) that

$$\xi \cdot \mathcal{H}(b) \operatorname{grad} v = \operatorname{rot} u \cdot \operatorname{grad} v$$

i.e.,

(11.5.9) $\mathcal{H}(b) * \xi \cdot \operatorname{grad} v = \operatorname{rot} u \cdot \operatorname{grad} v,$

where v is the solution of $-\operatorname{div} \mathcal{H}(b) \operatorname{grad} v = g$, $v \in H_0^1(\mathcal{O})$; since g is arbitrary in $H^{-1}(\mathcal{O})$, v spans $H_0^1(\mathcal{O})$ and (11.5.9) implies

$$\mathcal{H}(b)^* \xi = \operatorname{rot} u,$$

i.e.,

$$\xi = (\mathcal{H}(a^{*-1})^*)^{-1} \operatorname{rot} u = (\mathcal{H}(a^{-1}))^{-1} \operatorname{rot} u. \qquad \blacksquare$$

REMARK 11.9. We have the analogous result (with the same proof) when $a_{ij}(x) = a_{ij}(x, x/\varepsilon, x/\varepsilon^2, \ldots, x/\varepsilon^N)$.

REMARK 11.10 (Use of the adjoint expansion). We consider problem (11.1.13) with

(11.5.10) $V = \mathcal{H}_0(\mathcal{O}, \operatorname{rot}).$

Then we can use the method of Section 3.3; u_ε is characterized by

(11.5.11) $(u_\varepsilon, A^{\varepsilon *} v_\varepsilon) = (f, v_\varepsilon), \qquad v_\varepsilon \in (C_0^\infty(\mathcal{O}))^3,$ [51]

and by taking enough terms in the asymptotic expansion (this is made more precise below) we can **always** construct (at least if coefficients are smooth enough) a sequence of functions v_ε which belong to V, have compact supports in $\mathcal{O}$ and satisfy

(11.5.12) $v_\varepsilon \to v$ in $(L^2(\mathcal{O}))^3$ **strongly**, $v \in (C_0^\infty(\mathcal{O}))^3$

$$A^{\varepsilon *} v_\varepsilon \to \mathcal{A}^* v \text{ in } (L^2(\mathcal{O}))^3 \text{ strongly.}$$

We can pass to the limit in (11.5.11) and we obtain

$$(u, \mathcal{A}^* v) = (f, v), \ \forall v \in (C_0^\infty(\mathcal{O}))^3,$$

---

[51]Or $v_\varepsilon \in V$ with **compact support**.

which is the desired result.

In order to obtain (11.5.12) we look for v_ε in the form

$$(11.5.13) \qquad v_\varepsilon = v + \varepsilon v_1 + \varepsilon^2 v_2, \;\; v_j = v_j(x, y);$$

two terms are needed here (cf. Remark 3.7).

If $a(y)$ is smooth enough, the identification is possible; we impose

$$(11.5.14) \qquad A_1^* v_1 + A_2^* v = 0,$$
$$A_1^* v_2 + A_2^* v_1 + A_3^* v = \mathcal{A}^* v,$$

so that

$$(11.5.15) \qquad A^{\varepsilon *} v_\varepsilon = \mathcal{A}^* v + \varepsilon (A_2^* v_2 + A_3^* v_1) + \varepsilon^2 A_3^* v_2,$$

and one can then verify (11.5.12).

REMARK 11.11. The result of **compensated compactness** permits us to give a new presentation of the so-called **dual formula** (cf. Remarks 5.10 and 5.11).

We consider simultaneously the problems:

$$(11.5.16) \qquad -\operatorname{div} a^\varepsilon \operatorname{grad} u_\varepsilon + \lambda u_\varepsilon = f,$$

$$(11.5.17) \qquad \operatorname{rot} b^\varepsilon \operatorname{rot} v_\varepsilon + \lambda v_\varepsilon = g, \;\; (\operatorname{div} g = 0),$$

$$\operatorname{div} v_\varepsilon = 0,$$

with appropriate boundary conditions. We choose b^ε such that

$$(11.5.18) \qquad b^{\varepsilon *} a^\varepsilon = I.$$

Then

$$(11.5.19) \qquad a^\varepsilon \operatorname{grad} u_\varepsilon \cdot b^\varepsilon \operatorname{rot} v_\varepsilon = \operatorname{grad} u_\varepsilon \cdot \operatorname{rot} v_\varepsilon.$$

Let us call, as usual, $\mathcal{H}(a)$ the homogenized matrix of a^ε and let us introduce the notation

$$(11.5.20) \qquad \mathcal{H}(b) = \textbf{homogenized matrix for } \boldsymbol{b^\varepsilon} \textbf{ in problem (11.5.17).}$$

We have

$$a^\varepsilon \operatorname{grad} u_\varepsilon \to \mathcal{H}(a) \operatorname{grad} u \text{ in } \mathcal{H}(\mathcal{O}, \operatorname{div}) \text{ weakly},$$
$$b^\varepsilon \operatorname{rot} v_\varepsilon \to \mathcal{H}(b) \operatorname{rot} v \text{ in } \mathcal{H}(\mathcal{O}, \operatorname{rot}) \text{ weakly}$$

so that by virtue of Theorem 11.5:

$$a^\varepsilon \operatorname{grad} u_\varepsilon \cdot b^\varepsilon \operatorname{rot} v_\varepsilon \to \mathcal{H}(a) \operatorname{grad} u \cdot \mathcal{H}(b) \operatorname{rot} v \text{ in } \mathcal{D}'(\mathcal{O}).$$

Since

$$\operatorname{grad} u_\varepsilon \to \operatorname{grad} u \text{ in } \mathcal{H}(\mathcal{O}, \operatorname{rot}) \text{ weakly}$$

and

$$\operatorname{rot} v_\varepsilon \to \operatorname{rot} v \text{ in } \mathcal{H}(\mathcal{O}, \operatorname{div}),$$

Theorem 11.5 shows that

$$\operatorname{grad} u_\varepsilon \cdot \operatorname{rot} v_\varepsilon \to \operatorname{grad} u \cdot \operatorname{rot} v \text{ in } \mathcal{D}'(\mathcal{O}),$$

and therefore (11.5.19) gives in the limit

$$\mathcal{H}(a) \operatorname{grad} u \cdot K(b) \operatorname{rot} v = \operatorname{grad} u \cdot \operatorname{rot} v.$$

Hence it follows that

$$(11.5.21) \qquad K(b)^* \mathcal{H}(a) = I.$$

By as we saw in Section 11.3, the **natural formula** for $K(b)$ is

$$K(b) = \mathcal{M}(a^{*-1}(I - \operatorname{rot}_y \tilde{\chi}))$$

(we use (11.3.13) with a replaced by a^{*-1} so that $\tilde{\theta}$ becomes $\tilde{\chi}$ as defined in (5.4.28)[52]), so that (11.5.21) gives

(11.5.22) $$\mathcal{H}(a) = \mathcal{M}(a^{-1}(I - \operatorname{rot}_y \tilde{\chi}))^{-1}.$$

This gives a new proof of (5.4.36).

11.6. Zero order term.

11.6.1. *Orientation.* We now return to (11.1.13) without the hypothesis (11.1.14). We have

$$u_\varepsilon \to u \text{ in } \mathcal{H}(\mathcal{O}, \operatorname{rot}) \text{ weakly,}$$

and

$$a_0^\varepsilon \to \mathcal{M}(a_0) \text{ in } (L^\infty(\mathcal{O}))^3 \text{ weak star.}$$

Since the injection of $\mathcal{H}(\mathcal{O}, \operatorname{rot})$ into $(L^2(\mathcal{O}))^3$ is **not** compact, **it does not follow** that $a_0^\varepsilon u_\varepsilon \to \mathcal{M}(a_0)u$, in $(L^2(\mathcal{O}))^3$ weakly. We show now how to solve this question.

THEOREM 11.12. *Let u_ε be the solution of (11.1.13). Then $u_\varepsilon \to u$ in V weakly, where u is the solution of*

(11.6.1) $$(\mathcal{H}(a^{-1})^{-1} \operatorname{rot} u, \operatorname{rot} v) + (\mathcal{H}(a_0)u, v) = (f, v), \ \forall v \in V.$$

PROOF. By approximating in V' the function f by a sequence f^j of smooth functions, we do not restrict the generality by assuming that

(11.6.2) $$\operatorname{div} f \in L^2(\mathcal{O}).$$

Given g in $H^{-1}(\mathcal{O})$, we define v_ε as the solution of

(11.6.3) $$- \operatorname{div}(a_0^\varepsilon)^* \operatorname{grad} v_\varepsilon = g, \ v_\varepsilon = 0 \text{ on } \Gamma,$$

and we consider the identity

(11.6.4) $$a_0^\varepsilon u_\varepsilon \cdot \operatorname{grad} v_\varepsilon = u_\varepsilon \cdot a_0^{\varepsilon*} \operatorname{grad} v_\varepsilon.$$

Since $a_0^\varepsilon u_\varepsilon \to \eta$ in $(L^2(\mathcal{O}))^3$ weakly and since

$$\operatorname{div}(a_0^\varepsilon u_\varepsilon) = \operatorname{div} f \in L^2(\mathcal{O}),$$

we have

$$a_0^\varepsilon u_\varepsilon \to \eta \text{ in } \mathcal{H}(\mathcal{O}, \operatorname{div}) \text{ weakly.}$$

Since $\operatorname{rot} \operatorname{grad} v_\varepsilon = 0$, $\operatorname{grad} v_\varepsilon \to \operatorname{grad} v$ in $\mathcal{H}(\mathcal{O}, \operatorname{rot})$ weakly and according to Theorem 11.5

(11.6.5) $$a_0^\varepsilon u_\varepsilon \cdot \operatorname{grad} v_\varepsilon \to \eta \cdot \operatorname{grad} v \text{ in } \mathcal{D}'(\mathcal{O}).$$

We have $u_\varepsilon \to u$ in $\mathcal{H}(\mathcal{O}, \operatorname{rot})$ weakly and according to (11.6.3), $a_0^{\varepsilon*} \operatorname{grad} v_\varepsilon \to \mathcal{H}(a_0^*) \operatorname{grad} v$ in $\mathcal{H}(\mathcal{O}, \operatorname{div})$ weakly, so that according to Theorem 11.5

(11.6.6) $$u_\varepsilon \cdot a_0^{\varepsilon*} \operatorname{grad} v_\varepsilon \to u \cdot \mathcal{H}(a_0^*) \operatorname{grad} v \text{ in } \mathcal{D}'(\mathcal{O}).$$

From (11.6.4), (11.6.5), (11.6.6), it follows that

$$\eta \cdot \operatorname{grad} v = u \cdot \mathcal{H}(a_0^*) \operatorname{grad} v$$

hence we deduce that $\eta = \mathcal{H}(a_0^*)^* u = \mathcal{H}(a_0)u$ and (11.6.1) follows. ∎

---

[52]In fact it should be $\tilde{\chi}^*$ obtained by replacing a by a^* in (5.4.28); it is then $\tilde{\chi}$ which appears in (11.5.22).

REMARK 11.13. A reduction to the case "div $f = 0$" analogous to the one in Lemma 11.1 is possible only if we assume $a_0(y)$ smooth enough (say $a_{0ij}(y) \in W^{1,\infty}(Y)$), an hypothesis which is unnecessary in Theorem 11.12.

11.7. Remark on a regularization method. Let us consider (11.1.3) with

$$(11.7.1) \qquad a_0^\varepsilon = \lambda I, \qquad V = \mathcal{H}(\mathcal{O}, \mathrm{rot}), \qquad \mathrm{div}\, f \in L^2(\mathcal{O}).$$

In that case we saw that we can always reduce the problem to the case when $\mathrm{div}\, f = 0$, $\mathrm{div}\, u = 0$. But there is another possibility of reducing the problem to a situation where we can apply the method of Section 9. **We introduce the regularized problem** defined as follows. We introduce

$$(11.7.2) \qquad \tilde{V} = \{v \mid v \in V, \ \mathrm{div}\, v \in L^2(\mathcal{O})\},$$

and for $\sigma > 0$ we consider

$$(11.7.3) \qquad w_\varepsilon = w_{\varepsilon,\sigma} \in \tilde{V}$$

which is the solution of the (regularized problem)

$$(11.7.4) \qquad a^\varepsilon(w_\varepsilon, v) + \sigma(\mathrm{div}\, w_\varepsilon, \mathrm{div}\, v) = (f, v), \ \forall v \in \tilde{V}.$$

We will verify below that

$$(11.7.5) \qquad \|w_\varepsilon - w_\varepsilon\|_{\mathcal{H}(\mathcal{O},\mathrm{rot})} \le C\sqrt{\sigma}.$$

It follows that we can homogenize (11.7.4) first, (with $\sigma > 0$ **fixed**). Since the injection of $\tilde{V} \to H$ is **locally** compact, we can apply Section 9, and next we let $\sigma \to 0$.

VERIFICATION OF (11.7.5). Given $v \in V$, we have, using (11.1.3) and (11.7.4):

$$(11.7.6) \qquad a^\varepsilon(u_\varepsilon - w_\varepsilon, v) + \lambda(u_\varepsilon - w_\varepsilon, v) = (\mathrm{div}\, w_\varepsilon, \mathrm{div}).$$

Since $\mathrm{div}\, u_\varepsilon = \left(\frac{\mathrm{div}\, f}{\lambda}\right) \in L^2(\mathcal{O})$, we can take $v = u_\varepsilon - w_\varepsilon$ in (11.7.6). Therefore

$$a^\varepsilon(u_\varepsilon - w_\varepsilon, u_\varepsilon - w_\varepsilon) + \lambda(u_\varepsilon - w_\varepsilon)^2 = \sigma|\mathrm{div}\, w_\varepsilon|^2 = \sigma\left(\frac{\mathrm{div}\, f}{\lambda}, \mathrm{div}\, w_\varepsilon\right).$$

Therefore $|\mathrm{div}\, w_\varepsilon| \le \left|\frac{\mathrm{div}\, f}{\lambda}\right|$ and (11.7.5) follows easily. ∎

12. Homogenization with rapidly oscillating potentials

12.1. Orientation. We consider first in this section the equation

$$(12.1.1) \qquad Au_\varepsilon + \frac{1}{\varepsilon}w^\varepsilon u + \mu u_\varepsilon = f \ \text{in}\ \mathcal{O},$$
$$u_\varepsilon = 0 \ \text{on}\ \Gamma,$$

where

$$(12.1.2) \qquad A = -\Delta,$$

$$(12.1.3) \qquad W^\varepsilon(x) = W(x/\varepsilon).$$

W is a Y-periodic L^∞ function such that

$$(12.1.4) \qquad \int_Y W(y)\, dy = 0,$$

and μ is a constant.

We want to study the behavior of u_ε as $\varepsilon \to 0$.

We shall consider next similar questions with other operators A and other boundary conditions.

If we multiply (12.1.1) by u_ε, we find

$$\int \left[|\nabla u_\varepsilon|^2 + \mu |u_\varepsilon|^2 \right] \, dx + \frac{1}{\varepsilon} \int W^\varepsilon u_\varepsilon^2 \, dx = (f, u_\varepsilon).$$

This expression does not give any simple **a priori** estimate, and it is even unclear at this stage if we can solve (12.1.1) for all $\varepsilon \to 0$ with a fixed μ. In order to obtain a priori estimates independent of ε, we shall introduce first the asymptotic expansion of u_ε.

12.2. Asymptotic expansion. We use the notation of Section 2. We have

(12.2.1) $$A = \varepsilon^{-2} A_1 + \varepsilon^{-1} A_2 + \varepsilon^0 A_3,$$

$$A_1 = -\Delta_y, \qquad A_2 = -2 \frac{\partial^2}{\partial x_j \partial y_j}, \qquad A_3 = -\Delta_x.$$

If we look for u_ε in the form

(12.2.2) $$u_\varepsilon = u_0 + \varepsilon u_1 + \varepsilon^2 u_2 + \cdots, \qquad u_j = u_j(x, y),$$

we obtain

(12.2.3) $$A_1 u_0 = 0,$$

(12.2.4) $$A_1 u_1 + (A_2 + W) u_0 = 0,$$

(12.2.5) $$A_1 u_2 + (A_2 + W) u_1 + (A_3 + \mu) u_0 = f.$$

Equation (12.2.3) is equivalent to

$$u_0 = u(x),$$

so that (12.2.4) reduces to

(12.2.6) $$A_1 u_1 + W u = 0.$$

By virtue of (12.1.4), we can find $\chi(y)$ such that

(12.2.7) $$A_1 \chi + W = 0,$$
$$\chi \text{ is } Y\text{-periodic},$$

which defines χ up to an additive constant. It follows that

(12.2.8) $$u_1 = \chi u + \tilde{u}_1(x).$$

We can solve (12.2.5) in u_2, iff

$$\int_Y [(A_2 + W) u_1 + (A_3 + \mu) u_0] \, dy = |Y| f$$

i.e.,

(12.2.9) $$(-\Delta + \mu) u + \mathcal{M}(W_\chi) u = f,$$

where as usual

$$\mathcal{M}(\phi) = \frac{1}{|Y|} \int_Y \phi \, dy.$$

Equation (12.2.9) is the **homogenized equation** of (12.1.1), when

(12.2.10) $\mu + \mathcal{M}(W_\chi) > \lambda_1 = $ first eigenvalue of $-\Delta$

in $\mathcal{O}$ for Dirichlet's boundary condition.

This will be justified in Section 12.3 below.

REMARK 12.1. It follows from (12.2.7) that

$$(12.2.11) \qquad \mathcal{M}(W_\chi) = -\frac{1}{|Y|} \int_Y |\nabla_\chi|^2 \, dy.$$

REMARK 12.2. We justify below (12.2.10) using energy estimates. One could also justify this result directly from the asymptotic expansion, by methods similar to those of Section 2.4.

REMARK 12.3. The **oscillatory potential** W^ε has a "penalty" factor $\frac{1}{\varepsilon}$. Therefore we can also think of the results of this section as results of the **singular perturbations** type, for the equation

$$(12.2.12) \qquad \varepsilon(A + \mu)u_\varepsilon + W^\varepsilon u_\varepsilon = \varepsilon f, \qquad u_\varepsilon = 0 \text{ on } \Gamma.$$

12.3. Estimates for the spectrum and homogenization. We shall prove

THEOREM 12.4. *We assume that (12.1.4) and (12.2.10) hold true. Then, as $\varepsilon \to 0$, equation (12.1.1) admits a unique solution for ε small enough, and*

$$(12.3.1) \qquad u_\varepsilon \to u \text{ in } H_0^1(\mathcal{O}) \text{ weakly,}$$

where u is the solution in $H_0^1(\mathcal{O})$ of (12.2.9).

We shall derive Theorem 12.4 from the following

THEOREM 12.5. *Let $\lambda_1(\varepsilon)$ be the first eigenvalue of the operator $-\Delta + \frac{1}{\varepsilon}W^\varepsilon$ for the Dirichlet boundary condition. Then*

$$(12.3.2) \qquad |\lambda_1(\varepsilon) - (\lambda_1 + \mathcal{M}(W_\chi))| \leq C\varepsilon.$$

PROOF OF THEOREM 12.5. Let us set

$$(12.3.3) \qquad \pi_\varepsilon(v) = \int_\mathcal{O} \left[|\nabla v|^2 + \frac{1}{\varepsilon} W^\varepsilon v^2 \right] dx.$$

Given v **we define** ϕ_ε by

$$(12.3.4) \qquad v = (1 + \varepsilon\chi^\varepsilon)\phi_\varepsilon, \qquad \chi^\varepsilon(x) = \chi(x/\varepsilon),$$

(which indeed uniquely defines ϕ_ε for ε small enough).

We have

$$\pi_\varepsilon(v) = \left(-\Delta v + \frac{1}{\varepsilon} W^\varepsilon v, v \right)$$

and

$$-\Delta v + \frac{1}{\varepsilon} W^\varepsilon v = (1 + \varepsilon\chi^\varepsilon)(-\Delta\phi_\varepsilon) - 2 \left(\frac{\partial\chi}{\partial y_i} \right)^\varepsilon \frac{\partial\phi_\varepsilon}{\partial x_i} + W^\varepsilon \chi^\varepsilon \phi_\varepsilon$$

so that

$$\pi_\varepsilon(v) = (-\Delta\phi_\varepsilon, (1 + \varepsilon\chi^\varepsilon)^2 \phi_\varepsilon) - 2 \int_\mathcal{O} \left(\frac{\partial\chi}{\partial y_i} \right)^\varepsilon (1 + \varepsilon\chi^\varepsilon)\phi_\varepsilon \frac{\partial\phi_\varepsilon}{\partial x_i} \, dx$$

$$+ \int_\mathcal{O} W^\varepsilon \chi^\varepsilon (1 + \varepsilon\chi^\varepsilon)\phi_\varepsilon^2 \, dx,$$

which gives

$$(12.3.5) \qquad \pi_\varepsilon(v) = \int_\mathcal{O} (1 + \varepsilon\chi^\varepsilon)^2 |\nabla\phi_\varepsilon|^2 \, dx + \int_\mathcal{O} W^\varepsilon \chi^\varepsilon (1 + \varepsilon\chi^\varepsilon)\phi_\varepsilon^2 \, dx.$$

Let us introduce $\psi = \psi(y)$ solution of

$$(12.3.6) \qquad -\Delta_y\psi + \chi - \mathcal{M}(W_\chi) = 0, \qquad \psi \ Y\text{-periodic}$$

(ψ exists, and it is defined up to an additive constant). Therefore

$$(12.3.7) \qquad -\varepsilon^2 \Delta\psi^\varepsilon + W^\varepsilon\chi^\varepsilon - \mathcal{M}(W_\chi) = 0.$$

Therefore

$$\int_{\mathcal{O}} W^\varepsilon\chi^\varepsilon(1 + \varepsilon\chi^\varepsilon)\phi_\varepsilon^2\, dx = \int_{\mathcal{O}} (\mathcal{M}(W_\chi) - \varepsilon^2\Delta\psi^\varepsilon)(1 + \varepsilon\chi^\varepsilon)\phi_\varepsilon^2\, dx$$

$$= \mathcal{M}(W_\chi)\int_{\mathcal{O}} (1 + \varepsilon\chi^\varepsilon)\phi_\varepsilon^2\, dx$$

$$- \varepsilon\int_{\mathcal{O}} \left(\frac{\partial\phi}{\partial y_i}\right)^\varepsilon \frac{\partial}{\partial x_i}((1 + \varepsilon\chi^\varepsilon)\phi_\varepsilon^2)\, dx.$$

Consequently (12.3.5) gives:

$$(12.3.8) \quad \pi_\varepsilon(v) = \int_{\mathcal{O}} (1 + \varepsilon\chi^\varepsilon)^2 |\nabla\phi_\varepsilon|^2 |\nabla\phi_\varepsilon|^2\, dx + \mathcal{M}(W_\chi)\int_{\mathcal{O}} (1 + \varepsilon\chi^\varepsilon)\phi_\varepsilon^2\, dx$$

$$- \varepsilon\int_{\mathcal{O}} \left(\frac{\partial\psi}{\partial y_i}\right)^\varepsilon \frac{\partial}{\partial x_i}((1 + \varepsilon\chi^\varepsilon)\phi_\varepsilon^2)\, dx.$$

It follows from (12.3.8) that

$$(12.3.9) \quad \pi_\varepsilon(v) \geq (1 - c_1\varepsilon)\left|\int_{\mathcal{O}} |\nabla\phi_\varepsilon|^2\, dx + \mathcal{M}(W_\chi)\int_{\mathcal{O}} \phi_\varepsilon^2\, dx\right| - c_2\varepsilon\int_{\mathcal{O}} \phi_\varepsilon^2\, dx.$$

But, by definition of λ_1:

$$\int_{\mathcal{O}} |\nabla\phi_\varepsilon|^2\, dx \geq \lambda_1 \int_{\mathcal{O}} \phi_\varepsilon^2\, dx,$$

so that

$$(12.3.10) \qquad \pi_\varepsilon(v) \geq (1 - c_1\varepsilon)[\lambda_1 + \mathcal{M}(W_\chi)]|\phi_\varepsilon|^2 - c_2\varepsilon|\phi_\varepsilon|^2,$$

($|\cdot| = L^2$ norm).
    But $|\phi_\varepsilon|^2 \geq (1 - c_3\varepsilon)|v|^2$, so that

$$(12.3.11) \qquad \frac{\pi_\varepsilon(v)}{|v|^2} \geq (1 - c_4\varepsilon)(\lambda_1 + \mathcal{M}(W_\chi)) - c_5\varepsilon, \ v \in H_0^1(\mathcal{O}),$$

i.e.,

$$(12.3.12) \qquad \lambda_1(\varepsilon) \geq (1 - c_4\varepsilon)(\lambda_1 + \mathcal{M}(W_\chi)) - c_5\varepsilon.$$

To obtain an upper estimate for $\lambda_1(\varepsilon)$, we remark that (12.3.8) implies

$$(12.3.13) \qquad \pi_\varepsilon(v) \leq (1 + c_6\varepsilon)\left[\int_{\mathcal{O}} |\nabla\phi_\varepsilon|^2\, dx + \mathcal{M}(W_\chi)\int_{\mathcal{O}} \phi_\varepsilon^2\, dx\right] - c_7\varepsilon|\phi_\varepsilon|^2.$$

In (12.3.13) we now think of ϕ_ε **given** first ($\phi_\varepsilon = \phi$) and of v **given** by $v = v_\varepsilon = (1_\varepsilon\chi^\varepsilon)\phi$. Then

$$\pi_\varepsilon(v_\varepsilon) \leq (1 + c_6\varepsilon)\left[\sup_{|\phi|=1}\int |\nabla\phi|^2\, dx + \mathcal{M}(W_\chi)\right] - c_7\varepsilon$$

$$= (1 + c_6\varepsilon)(\lambda_1 + \mathcal{M}(W_\chi)) - c_7\varepsilon,$$

this inequality being satisfied, $\forall v_\varepsilon$ such that $|(1 + \varepsilon \chi^\varepsilon)^{-1} v_\varepsilon| = 1$. Therefore

$$\pi_\varepsilon(v) \le (1 + c_8 \varepsilon)[(1 + c_6 \varepsilon)(\lambda_1 + \mathcal{M}(W_\chi)) - c_7 \varepsilon],$$

hence the "reverse" inequality of (12.3.12) follows, and (12.3.2) is proven. ∎

PROOF OF THEOREM 12.4. It follows from Theorem 12.5 that, under condition (12.2.10),

$$(12.3.14) \qquad \left(\left(A + \frac{1}{\varepsilon} W^\varepsilon + \mu \right) v, v \right) \ge c \|v\|^2_{H^1_0(\mathcal{O})}, \ c > 0, \ \forall v \in H^1_0(\mathcal{O}),$$

for ε small enough.

This proves that equation (12.1.1) admits a unique solution for ε small enough and that

$$(12.3.15) \qquad\qquad \|u_\varepsilon\| \le C.$$

We now observe that[53]

$$(12.3.16) \qquad\qquad \left\| \frac{1}{\varepsilon} W^\varepsilon \right\|_{\mathcal{L}(H^1_0(\mathcal{O}); L^2(\mathcal{O}))} \le C.$$

Indeed it follows from (12.2.7) that

$$-\varepsilon \Delta \chi^\varepsilon + \frac{1}{\varepsilon} W^\varepsilon = 0,$$

so that after multiplying by v^2:

$$\left(\frac{1}{\varepsilon} W^\varepsilon v, v \right) = \varepsilon(\Delta \chi^\varepsilon, v^2) = -2 \int_\mathcal{O} \left(\frac{\partial \chi}{\partial y_i} \right)^\varepsilon v \frac{\partial v}{\partial x_i} \, dx$$

so that

$$\left| \left(\frac{1}{\varepsilon} W^\varepsilon v, v \right) \right| \le C |v| \|\nabla v|,$$

hence (12.3.16) follows. It follows from (12.3.15), (12.3.16) that, if we set

$$(12.3.17) \qquad\qquad \eta_\varepsilon = \frac{1}{\varepsilon} W^\varepsilon u_\varepsilon$$

then

$$(12.3.18) \qquad\qquad |\eta_\varepsilon| \le C.$$

We can extract a subsequence, still denoted by u_ε, such that

$$(12.3.19) \qquad\qquad u_\varepsilon \to u \text{ in } H^1_0(\mathcal{O}) \text{ weakly},$$

$$\eta_\varepsilon \to \eta \text{ in } L^2(\mathcal{O}) \text{ weakly},$$

and

$$-\Delta u + \eta u + \mu u = f.$$

In order to compute η, we observe that, if we introduce

$$(12.3.20) \qquad\qquad \theta_\varepsilon = 1 + \varepsilon \chi^\varepsilon$$

then

$$(12.3.21) \qquad\qquad \left(A + \frac{1}{\varepsilon} W^\varepsilon \right) \theta_\varepsilon = W^\varepsilon \chi^\varepsilon.$$

---

[53]It follows from (12.3.16) that $-\Delta + \mu + \frac{1}{\varepsilon} W^\varepsilon$ is an isomorphism from $H^1_0(\mathcal{O}) \to H^{-1}(\mathcal{O})$ for μ large enough, but (12.3.16) does not give the best possible estimates for ν (as given in Theorem 12.5).

Given $\phi \in C_0^\infty(\mathcal{O})$, we multiply (12.1.1) by $\phi\theta_\varepsilon$ and (12.3.21) by ϕu_ε. After integrating and subtracting, we obtain

$$(12.3.22) \quad \mu(u_\varepsilon, \phi\theta_\varepsilon) + \int_\mathcal{O} \frac{\partial u_\varepsilon}{\partial x_i} \frac{\partial \phi}{\partial x_i} \theta_\varepsilon \, dx - \int_\mathcal{O} \frac{\partial \theta_\varepsilon}{\partial x_i} \frac{\partial \phi}{\partial x_i} u_\varepsilon \, dx = (f, \phi) - (W^\varepsilon \chi^\varepsilon, \phi u_\varepsilon),$$

and we obtain

$$\mu(u, \phi) + \int_\mathcal{O} \frac{\partial u}{\partial x_i} \frac{\partial \phi}{\partial x_i} \, dx = (f, \phi) - \mathcal{M}(W_\chi)(u, \phi)$$

i.e., (12.2.9). It follows that

$$(12.3.23) \qquad\qquad\qquad \eta = \mathcal{M}(W_\chi)u.$$

∎

REMARK 12.6. One can avoid the introduction of η_ε in the above proof but η_ε will be needed in Theorem 12.7 below.

12.4. Correctors. We have now a result proving the first order corrector.

THEOREM 12.7. *Under the conditions of Theorem 12.4, one has*

$$(12.4.1) \qquad\qquad z_\varepsilon = u_\varepsilon - (u + \varepsilon\chi^\varepsilon u) \to 0 \ \text{in} \ H_0^1(\mathcal{O}) \ \text{strongly,}$$

where $\chi^\varepsilon(x) = \chi(x/\varepsilon)$, χ defined by (12.2.7).

PROOF. According to Theorem 12.4, we have only to show that

$$(12.4.2) \qquad\qquad \left(\left(A + \frac{1}{\varepsilon} W^\varepsilon + \mu \right) z_\varepsilon, z_\varepsilon \right) \to X_\varepsilon \to 0.$$

But

$$X_\varepsilon = (f, z_\varepsilon) - \left(\left(A + \frac{1}{\varepsilon} W^\varepsilon + \mu \right)(u + \varepsilon\chi^\varepsilon u), u_\varepsilon - (u + \varepsilon\chi^\varepsilon u) \right)$$

$$(12.4.3) \qquad = (f, z_\varepsilon) - \left(\left(A + \frac{1}{\varepsilon} W^\varepsilon + \mu \right) u_\varepsilon, u + \varepsilon\chi^\varepsilon u \right) + Y_\varepsilon,$$

$$Y_\varepsilon = \left(\left(A + \frac{1}{\varepsilon} W^\varepsilon + \mu \right)(u + \varepsilon\chi^\varepsilon u), u + \varepsilon\chi^\varepsilon u \right).$$

Therefore

$$(12.4.4) \qquad\qquad X_\varepsilon = (f, u_\varepsilon - 2(u + \varepsilon\chi^\varepsilon u)) + Y_\varepsilon.$$

But with the notation of the proof of Theorem 12.5, we have

$$Y_\varepsilon = \pi_\varepsilon((1 + \varepsilon\chi^\varepsilon)u) + \mu|u + \varepsilon\chi^\varepsilon u|^2.$$

Hence, using (12.3.5)

$$Y_\varepsilon = \int_\mathcal{O} \left[(1 + \varepsilon\chi^\varepsilon)^2 |\nabla u|^2 + W^\varepsilon \chi^\varepsilon (1 + \varepsilon\chi^\varepsilon)u^2 \right] \, dx + \mu|u + \varepsilon\chi^\varepsilon u|^2$$

so that

$$Y_\varepsilon \to \int_\mathcal{O} \left[|\nabla u|^2 + (\mathcal{M}(W_\chi) + \mu)u^2 \right] \, dx = Y.$$

Therefore

$$X_\varepsilon \to -(f, u) + Y = 0,$$

and (12.4.2) is proven.

∎

12.5. Almost periodic potentials. Theorems 12.4 and 12.5 can be extended to the following situation; (12.1.1) is replaced by

$$(12.5.1) \qquad\qquad Au_\varepsilon + \frac{1}{\varepsilon}W\left(x, \frac{x}{\varepsilon}\right)u_\varepsilon + \mu u_\varepsilon = f \text{ in } \mathcal{O},$$

$$u_\varepsilon = 0 \text{ on } \Gamma,$$

where $W(x,y)$ is defined in $\mathcal{O} \times \mathbb{R}^n$, and satisfies the following properties:

$$(12.5.2) \qquad\qquad W(x,y) = \sum_{m \in \mathbb{Z}^n} e^{ik_m y} a_m(x),$$

$$k_{-m} = -k_m, \qquad \overline{a_m(x)} = a_{-m}(x), \qquad a_0 = 0,$$

$$(12.5.3) \qquad\qquad \sum |a_m(x)| \leq C,$$

$$|k_m| \geq \delta > 0, \text{ if } m \neq 0;$$

$W(x,y)$ is almost periodic in y. We define (compare to (12.1.1)) $\chi(x,y)$ by

$$(12.5.4) \qquad\qquad \chi(x,y) = -\sum_m |k_m|^{-2} e^{ik_m y} a_m(x);$$

χ satisfies (12.2.7) in $\mathbb{R}^n$, x playing the role of a parameter; χ is a-periodic in y.

We introduce the mean-value (in the sense of almost-periodic functions):

$$(12.5.5) \qquad \mathcal{M}_{a.p.}(W_\chi) = \lim_{\lambda \to \infty}\left[\frac{1}{\lambda^n}\int_0^\lambda \cdots \int_0^\lambda W(x,y)\chi(x,y)\, dx\right];$$

one has

$$(12.5.6) \qquad\qquad \mathcal{M}_{a.p.}(W_\chi) = -\sum_m |k_m|^{-2}|a_m(x)|^2$$

(it is now a function of x).

We have the extension of Theorem 12.5:

THEOREM 12.8. *Let $\lambda_1(\varepsilon)$ be the first eigenvalue of the operator $A + \frac{1}{\varepsilon}W(x, x/\varepsilon)$ for the Dirichlet boundary conditions. We assume that (12.5.2), (12.5.3) hold true. Then*

$$(12.5.7) \qquad\qquad |\lambda_1(\varepsilon) - (\lambda_1 + \mathcal{M}_{a.p.}(W_\chi))| \leq C\varepsilon.$$

The proof is analogous to the one of Theorem 12.5, with some more technical difficulties. We refer to Bensoussan, Lions and Papanicolaou [15], [12].

COROLLARY 12.9. *Under the hypothesis of Theorem 12.8, if u_ε denotes the solution of (12.5.1), one has*

$$(12.5.8) \qquad\qquad u_\varepsilon \to u \text{ in } H_0^1(\mathcal{O}) \text{ weakly,}$$

where u is the solution in $H_0^1(\mathcal{O})$ of

$$(12.5.9) \qquad\qquad Au + \mathcal{M}_{a.p.}(W_\chi)u + \mu u = f.$$

REMARK 12.10. One has also a result analogous to the one of Theorem 12.7 for the first order corrector.

12.6. Neumann's problem. We come back to the **periodic** case[54] but with Neumann's boundary condition:

(12.6.1)
$$Au_\varepsilon + \frac{1}{\varepsilon}W^\varepsilon u_\varepsilon + \mu u_\varepsilon = f \text{ in } \mathcal{O},$$

$$\frac{\partial u_\varepsilon}{\partial \nu} = 0 \text{ on } \Gamma,$$

where $\frac{\partial}{\partial \nu}$ = normal derivative to Γ, directed toward the exterior of $\mathcal{O}$.[55]

We have

THEOREM 12.11. *We assume that (12.1.4) holds true. We also assume that Γ does not contain any piece of $(n-1)$ dimensional hyperplane with rational conormal.*[56] *Let $\lambda_1^{Neum}(\varepsilon)$ be the first eigenvalue of the operator $A + \frac{1}{\varepsilon}W^\varepsilon$ for the Neumann's boundary condition. Then*

(12.6.2)
$$|\lambda_1^{Neum}(\varepsilon) - \mathcal{M}(W_\chi)| \le C\varepsilon.$$

REMARK 12.12. The first eigenvalue of $A = -\Delta$ in $\mathcal{O}$ for the Neumann's boundary condition is $\lambda_1^{Neum} = 0$; the analogy between (12.6.2) and (12.3.2) is therefore complete.

REMARK 12.13. The difference between Theorems 12.5 and 12.11 lies in the fact that in Theorem 12.11 we **have** to assume on Γ the special condition in Theorem 12.11. If this condition did not hold, the estimates depend on the manner in which $\varepsilon \to 0$.

PROOF OF THEOREM 12.11. We consider
$$\pi_\varepsilon(v) = \int_{\mathcal{O}} \left[|\nabla v|^2 + \frac{1}{\varepsilon}W^\varepsilon v^2\right] \, dx, \qquad v \in D(A),$$

$D(A)$ = domain of A for Neumann's boundary condition.

With the notation (12.3.4), we find (compare to (12.3.5)) that

(12.6.3)
$$\pi_\varepsilon(v) = -\int_\Gamma \frac{\partial \phi_\varepsilon}{\partial \nu}(1 + \varepsilon\chi^\varepsilon)^2 \phi_\varepsilon \, d\Gamma$$

$$+ \int_{\mathcal{O}} (1 + \varepsilon\chi^\varepsilon)^2 |\nabla \phi_\varepsilon|^2 \, dx$$

$$+ \int_{\mathcal{O}} W^\varepsilon \chi^\varepsilon (1 + \varepsilon\chi^\varepsilon)\phi_\varepsilon^2 \, dx.$$

But if $v \in D(A)$, $\frac{\partial v}{\partial \nu} = 0$ and therefore

(12.6.4)
$$\frac{\partial v}{\partial \nu} = (1 + \varepsilon\chi^\varepsilon)\frac{\partial \phi_\varepsilon}{\partial \nu} + \varepsilon\frac{\partial \chi^\varepsilon}{\partial \nu}\phi_\varepsilon = 0.$$

But $\frac{\partial}{\partial \nu} = \nu_i(x)\frac{\partial}{\partial x_i}$ and therefore (12.6.3) becomes:

(12.6.5)
$$\pi_\varepsilon(v) = \int_\Gamma \nu_i(x)\left(\frac{\partial \chi}{\partial y_i}\right)^\varepsilon (1 + \varepsilon\chi^\varepsilon)\phi_\varepsilon^2 \, d\Gamma$$

$$+ \int_{\mathcal{O}} (1 + \varepsilon\chi^\varepsilon)^2 |\nabla \phi_\varepsilon|^2 \, dx + \int_{\mathcal{O}} W^\varepsilon \chi^\varepsilon (1 + \varepsilon\chi^\varepsilon)\phi_\varepsilon^2 \, dx.$$

---

[54]Results of this section extend to the a.p. case of Section 12.5.
[55]We assume here that the boundary of $\mathcal{O}$ is smooth enough.
[56]i.e., with all cosinus directors rational.

Under the conditions of the Theorem,

$$(12.6.6) \qquad \left(\frac{\partial \chi}{\partial y_i}\right)^\varepsilon \to 0 \text{ in } L^\infty(\Gamma) \text{ weak start.}$$

If $v \in$ bounded set of $H^1(\mathcal{O})$, ϕ_ε belongs also to a bounded set of $H^1(\mathcal{O})$, so that $\phi_\varepsilon \mid_\Gamma \in$ **compact** set of $L^2(\Gamma)$ and therefore
(12.6.7)

$$\int_\Gamma \nu_i(x) \left(\frac{\partial \chi}{\partial y_i}\right)^\varepsilon (1 + \varepsilon \chi^\varepsilon)\phi_\varepsilon^2 \, d\Gamma \to 0 \text{ uniformly for } v \in \text{ bounded set of } H^1(\mathcal{O});$$

(12.6.7) shows that we can neglect in (12.6.5) the surface integral and the proof is completed as in Theorem 12.5 ∎

COROLLARY 12.14. *Under the hypothesis of Theorem 12.11, and if*

$$(12.6.8) \qquad \mu + \mathcal{M}(W_\chi) > 0$$

the solution u_ε of (12.6.1) converges in $H^1(\mathcal{O})$ weakly toward a solution of

$$(12.6.9) \qquad Au + \mathcal{M}(W_\chi) + \mu u = f,$$

$$\frac{\partial u}{\partial \nu} = 0 \text{ on } \Gamma.$$

REMARK 12.15. One can also introduce $\varepsilon \chi u$ as first order corrector.

REMARK 12.16. One can prove similar results for boundary conditions of "mixed" type, i.e., Dirichlet's boundary condition on Γ_0 and Neumann's on $\Gamma - \Gamma_0$.

12.7. Higher order equations. The preceding remarks can be extended in many directions. One of them is obtained by replacing the second order elliptic operator A by a higher order elliptic operator.

We confine ourselves to the following example.

We consider

$$(12.7.1) \qquad A = \Delta^2$$

with Dirichlet boundary conditions and the problem

$$(12.7.2) \qquad Au_\varepsilon + \mu u_\varepsilon + \frac{1}{\varepsilon^2} W^\varepsilon u_\varepsilon = f, \qquad u_\varepsilon = \frac{\partial u_\varepsilon}{\partial \nu} = 0 \text{ on } \Gamma,$$
$$W^\varepsilon(x) = W(x/\varepsilon), \text{ where } W \text{ satisfies (12.1.4).}$$

Let us consider the method of **asymptotic expansions**. We have

$$A = (\varepsilon^{-2}\Delta_y + 2\varepsilon^{-1}\Delta_{yx} + \varepsilon^0\Delta_x)$$

where

$$\Delta_{yx} = \frac{\partial^2}{\partial y_j \partial x_j}.$$

Therefore

$$(12.7.3) \quad A + \mu + \varepsilon^{-2}W^\varepsilon = \varepsilon^{-4}A_1 + \varepsilon^{-3}A_3 + \varepsilon^{-2}(A_3 + W) + \varepsilon^{-1}A_4 + \varepsilon^0(A_5 + \mu)$$

with

$$(12.7.4) \qquad A_1 = \Delta_y^2, \qquad A_2 = 4\Delta_y\Delta_{yx}, \qquad A_3 = 2\Delta_y\Delta_x + 4\Delta_{yx}^2,$$
$$A_4 = 4\Delta_{yx}\Delta_x, \qquad A_5 = \Delta_x^2.$$

If we look for u_ε in the form

(12.7.5) $$u = u_0 + \varepsilon u_1 + \varepsilon^2 u_2 + \cdots,$$

we obtain

(12.7.6) $$A_1 u_0 = 0,$$
(12.7.7) $$A_1 u_1 + A_2 u_0 = 0,$$
(12.7.8) $$A_1 u_2 + A_2 u_1 + (A_3 + W)u_0 = 0,$$
(12.7.9) $$A_1 u_3 + A_2 u_2 + (A_3 + W)u_1 + A_4 u_0 = 0,$$
(12.7.10) $$A_1 u_4 + A_2 u_3 + (A_3 + W)u_2 + A_4 u_1 + A_5 u_0 + \mu u_0 = f.$$

It follows from (12.7.6), (12.7.7), that

$$u_0 = u(x), \qquad u_1 = \tilde{u}_1(x).$$

Then (12.7.8) gives

$$A_1 u_2 + W u_0 = 0.$$

We define χ by

(12.7.11) $$\Delta^2 \chi + W = 0, \qquad \chi \text{ periodic}$$

which defines χ up to an additive constant. Then

(12.7.12) $$u_2 = \chi u + \tilde{u}_2(x).$$

We can also compute u_3 from (12.7.9) but this is not indispensable. Equation (12.7.10) admits a solution u_4, iff

$$\int_Y [A_2 u_3 + (A_3 + W)u_2 + A_4 u_1 + (A_5 + \mu)u_0] \, dy = |Y|f$$

which reduces to

$$\int_Y W u_2 \, dy + |Y|(\Delta^2 + \mu)u = |Y|f$$

and using (12.7.12) it follows that

(12.7.13) $$\left(\Delta^2 + \mu + \frac{1}{|Y|}\left(\int_Y W_\chi \, dy\right)\right) u = f.$$

[We have $\int_Y W_\chi \, dy = -\int_Y (\Delta\chi)^2 \, dy$ which does not depend on the choice of χ.] This procedure can be justified. One can prove: if $\lambda_1(\varepsilon)$ denotes the first eigenvalue of the operator $A + \frac{1}{\varepsilon^2}W^\varepsilon$, and if λ_1 denotes the first eigenvalue of A (for the Dirichlet's boundary conditions), then

(12.7.14) $$|\lambda_1(\varepsilon) - [\lambda_1 + \mathcal{M}(W_\chi)]| \le C\varepsilon$$

and if μ satisfies

(12.7.15) $$\mu + \mathcal{M}(W_\chi) > \lambda_1$$

then

(12.7.16) $$u_\varepsilon \to u \text{ in } H_0^2(\mathcal{O}) \text{ weakly}$$

where u is the solution in $H_0^2(\mathcal{O})$ of

(12.7.17) $$Au + [\mu + \mathcal{M}(W_\chi)]u = f.$$

12.8. Oscillating potential and oscillatory coefficients. We consider now the problem

$$(12.8.1) \qquad A_\varepsilon u_\varepsilon + \frac{1}{\varepsilon} W^\varepsilon u_\varepsilon = f,$$

$$u_\varepsilon = 0 \text{ on } \Gamma,$$

where A^ε is now given by (1.1.3), and W^ε is given by (12.1.2) and (12.1.3). We also assume that

$$(12.8.2) \qquad a_{ij} = a_{ji}, \ \forall i, j.$$

The **asymptotic expansion** proceeds as follows. We use the notations of Section 2. We have

$$(12.8.3) \qquad A^\varepsilon + \frac{1}{\varepsilon} W^\varepsilon = \varepsilon^{-2} A_1 + \varepsilon^{-1}(A_2 + W) + \varepsilon^0 A_3,$$

and the equations,

$$(12.8.4) \qquad A_1 u_0 = 0,$$

$$(12.8.5) \qquad A_1 u_1 + (A_2 + W) u_0 = 0,$$

$$(12.8.6) \qquad A_1 u_2 + (A_2 + W) u_1 + A_3 u_0 = f.$$

It follows from (12.8.4) that $u_0 = u(x)$. If we introduce χ^j as in Section 2:

$$(12.8.7) \qquad A_1(\chi^j - y_j) = 0, \qquad \chi^j \ Y\text{-periodic},$$

and χ by

$$(12.8.8) \qquad A_1(\chi) + W = 0, \qquad \chi \ Y\text{-periodic},$$

then

$$(12.8.9) \qquad u_1 = -\chi^j \frac{\partial u}{\partial x_j} + \chi u + \tilde{u}_1(x).$$

The compatibility condition in (12.8.6) gives:

$$\int_Y (A_2 + W) u_1 \, dy + \int_Y A_3 u \, dy = |Y| f$$

hence

$$(12.8.10) \qquad u + \mathcal{M}(W_\chi) u - \frac{1}{|Y|} \left[\int_Y W \chi^j \, dy + \int_Y a_{jk} \frac{\partial \chi}{\partial y_k} \, dy \right] \frac{\partial u}{\partial x_j} = f.$$

But one easily checks, using (12.8.7), (12.8.8) **and the symmetry (12.8.2) of the coefficients**, that

$$\int_Y W_\chi^j \, dy + \int_Y a_{jk} \frac{\partial \chi}{\partial y_k} \, dy = 0$$

so that (12.8.10) implies

$$(12.8.11) \qquad \mathcal{A} u + \mathcal{M}(W_\chi) u = f.$$

One has the following result:

THEOREM 12.17. *We assume that (1.1.1) holds true and that $a_0 = 0$. Then if λ_1 denotes the first eigenvalue of $\mathcal{A}$ on $\mathcal{O}$, for the Dirichlet boundary condition, and if $\mathcal{M}(W_\chi) > \lambda_1$, then (12.8.1) admits a unique solution for ε small enough and $u_\varepsilon \to u$ in $H_0^1(\mathcal{O})$ weakly, where u is the solution in $H_0^1(\mathcal{O})$ of (12.8.11).*

Let us only sketch the proof.

One verifies first that

$$(12.8.12) \qquad \|u_\varepsilon\| \le c \qquad (\|\cdot\| = \text{ norm in } H_0^1(\mathcal{O})),$$

by introducing $u_\varepsilon = (1 + \varepsilon\chi^\varepsilon)\tilde{u}_\varepsilon$ and arguing as before. Using (12.3.16), it follows that

$$(12.8.13) \qquad \left|\frac{1}{\varepsilon}W^\varepsilon u_\varepsilon\right| \le c \qquad (|\cdot| = L^2(\mathcal{O}) \text{ norm}).$$

Therefore we can extract a subsequence, still denoted by u_ε, such that $u_\varepsilon \to u$ in $H_0^1(\mathcal{O})$ weakly and

$$(12.8.14) \qquad \frac{1}{\varepsilon}W^\varepsilon u_\varepsilon \to \eta \text{ in } L^2(\mathcal{O}) \text{ weakly}.$$

Therefore

$$(12.8.15) \qquad A^\varepsilon u_\varepsilon = f - \frac{1}{\varepsilon}W^\varepsilon u_\varepsilon \to f - \eta \text{ in } L^2(\mathcal{O}) \text{ weakly},$$

and using Remark 3.4, it follows that

$$(12.8.16) \qquad u = f - \eta.$$

We proceed as in (12.3.20). We introduce θ_ε by (12.3.20); we have

$$\left(A^\varepsilon + \frac{1}{\varepsilon}W^\varepsilon\right)\theta_\varepsilon = W^\varepsilon\chi^\varepsilon;$$

if $\xi_i^\varepsilon = a_{ij}^\varepsilon\frac{\partial u_\varepsilon}{\partial x_j}$, we obtain as usually ($\phi \in C_0^\infty(\mathcal{O})$):

$$(12.8.17) \qquad \int_{\mathcal{O}} \xi_i^\varepsilon \frac{\partial\phi}{\partial x_i}\theta_\varepsilon \, dx - \int_{\mathcal{O}} a_{ij}^\varepsilon \frac{\partial\theta_\varepsilon}{\partial x_i}\frac{\partial\phi}{\partial x_j}u_\varepsilon \, dx = (f, \phi\theta_\varepsilon) - (W^\varepsilon\chi^\varepsilon, \phi u_\varepsilon),$$

which gives in the limit

$$\int_{\mathcal{O}} \xi_i \frac{\partial\phi}{\partial x_i} \, dx = (f, \phi) - (\mathcal{M}(W_\chi), \phi u)$$

i.e., $f - \mathcal{M}(W_\chi)u = -\frac{\partial\xi_i}{\partial x_i} = Au$, hence the result follows.

12.9. A phenomenon of uncoupling. We give in this section a simple example where a (rather unsuspected) phenomenon of "uncoupling" arises when passing to the limit in a system of two equations with oscillatory potentials.

We denote by $W_i(y)$, $i = 1, 2$, two functions L^∞ and Y-periodic, such that

$$(12.9.1) \qquad \int_Y W_i(y) \, dy = 0, \qquad i = 1, 2.$$

We denote by $(u_\varepsilon, v_\varepsilon)$ the solution of

$$(12.9.2) \qquad -\Delta u_\varepsilon + \frac{1}{\varepsilon}W_1^\varepsilon v_\varepsilon + \lambda_1 u_\varepsilon = f,$$

$$-\Delta v_\varepsilon + \frac{1}{\varepsilon}W_2^\varepsilon u_\varepsilon + \lambda_2 v_\varepsilon = g,$$

$$(12.9.3) \qquad u_\varepsilon = v_\varepsilon = 0 \text{ on } \Gamma.$$

Using (12.3.16) for W_1^ε, we see that (12.9.2), (12.9.3) admit a unique solution for λ_1, λ_2 large enough.

We introduce χ_i by

(12.9.4) $\qquad\qquad -\Delta\chi_i + W_i = 0, \qquad \chi_i\ Y\text{-periodic}, \qquad i = 1, 2.$

Then, as $\varepsilon \to 0$, one has

(12.9.5) $\qquad\qquad u_\varepsilon \to u, \qquad v_\varepsilon \to v \text{ in } H_0^1(\mathcal{O}) \text{ weakly},$

where

(12.9.6) $\qquad\qquad -\Delta u + \mathcal{M}(W_2\chi_1)u + \lambda_1 u = f,$
$\qquad\qquad\qquad\qquad -\Delta v + \mathcal{M}(W_1\chi_2)v + \lambda_2 v = g.$

REMARK 12.18. The system (12.9.6) is **uncoupled** and consists of two separate equations.

We can prove the above assertion as follows. We introduce $\tilde{u}_\varepsilon$, $\tilde{v}_\varepsilon$ by

$$u_\varepsilon = \tilde{u}_\varepsilon + \varepsilon\chi_1^\varepsilon\tilde{v}_\varepsilon, \qquad v_\varepsilon = \tilde{v}_\varepsilon + \varepsilon\chi_2^\varepsilon\tilde{u}_\varepsilon,$$

and we prove that

$$\|u_\varepsilon\| \leq C, \qquad \|v_\varepsilon\| \leq C.$$

We can therefore assume that, after extracting a subsequence, one has (12.9.5).

Let ϕ and ψ be given in $C_0^\infty(\mathcal{O})$. We multiply the first (respectively, the second) equation (12.9.2) by $\phi + \varepsilon\chi_2^\varepsilon\psi$ (respectively, $\psi + \varepsilon\chi_1^\varepsilon\phi$) and we integrate; adding up, we obtain

$$-\int_\mathcal{O} u_\varepsilon\left(\Delta\phi + \varepsilon\chi^\varepsilon\Delta\psi + 2\left(\frac{\partial\chi_2}{\partial y_i}\right)^\varepsilon\frac{\partial\psi}{\partial x_i} + \frac{1}{\varepsilon}(\Delta\chi_2)^\varepsilon\psi\right)\, dx$$

$$+\int_\mathcal{O}\left[\frac{1}{\varepsilon}W_1^\varepsilon v_\varepsilon\phi + W_1^\varepsilon\chi_2^\varepsilon v_\varepsilon\psi + \lambda_1 u_\varepsilon(\phi + \varepsilon\chi_2^\varepsilon\psi)\right]\, dx$$

$$-\int_\mathcal{O} v_\varepsilon\left(\Delta\psi + \varepsilon\chi_1^\varepsilon\Delta\phi + 2\left(\frac{\partial\chi_1}{\partial y_i}\right)^\varepsilon\frac{\partial\phi}{\partial x_i} + \frac{1}{\varepsilon}(\Delta\chi_1)^\varepsilon\phi\right)\, dx$$

$$+\int_\mathcal{O}\left[\frac{1}{\varepsilon}W_2^\varepsilon u_\varepsilon\psi + W_2^\varepsilon\chi_1^\varepsilon u_\varepsilon\phi + \lambda_2 v_\varepsilon(\psi + \varepsilon\chi_1^\varepsilon\phi)\right]\, dx$$

$$= (f, \phi + \varepsilon\chi_2^\varepsilon\psi) + (g, \psi + \varepsilon\chi_1^\varepsilon\phi).$$

The terms in $\frac{1}{\varepsilon}$ drop out and we obtain in the limit (since $\left(\frac{\partial\chi_1}{\partial y_i}\right)^\varepsilon \to 0$ in $L^\infty(\mathcal{O})$ weak star)

$$-\int_\mathcal{O} u\Delta\phi\, dx + \int_\mathcal{O}\mathcal{M}(W_1\chi_2)v\psi\, dx + \lambda_1\int_\mathcal{O} u\phi\, dx$$

$$-\int_\mathcal{O} v\Delta\psi\, dx + \int_\mathcal{O}\mathcal{M}(W_2\chi_1)u\phi\, dx + \lambda_2\int_\mathcal{O} v\psi\, dx = (f, \phi) + (g, \psi)$$

hence (12.9.6).

13. Study of lower order terms

13.1. Orientation. In this section we consider elliptic operators with rapidly varying coefficients of **all orders**. We confine our study to second order operators, but the methods are general.

Let us consider the equation

(13.1.1) $\qquad\qquad A^\varepsilon u_\varepsilon + B^\varepsilon u_\varepsilon + C^\varepsilon u_\varepsilon + a_0^\varepsilon u_\varepsilon = f \text{ in } \mathcal{O},$

where

(13.1.2) $$A^\varepsilon = -\frac{\partial}{\partial x_i}\left(a_{ij}^\varepsilon \frac{\partial}{\partial x_j}\right), \qquad a_{ij}^\varepsilon(x) = a_{ij}(x/\varepsilon),$$

the a_{ij}'s satisfy (1.1.1),

(13.1.3) $$B^\varepsilon = b_i^\varepsilon \frac{\partial}{\partial x_i}, \qquad b_i^\varepsilon(x) = b_i(x/\varepsilon),$$

(13.1.4) $$C^\varepsilon = \frac{\partial}{\partial x_i}(c_i^\varepsilon), \qquad c_i^\varepsilon(x) = c_i(x/\varepsilon),$$

(13.1.5) $$a_0^\varepsilon(x) = a_0(x/\varepsilon),$$

where the coefficients b_i, c_i, a_0 are all Y-periodic. In (13.1.1) u_ε is subject to appropriate boundary conditions.

From the viewpoint of homogenization, we do not restrict the generality by assuming that

(13.1.6) $$C^\varepsilon = 0.$$

Indeed, **if we obtain the a priori estimate**

(13.1.7) $$\|u_\varepsilon\| \le C \qquad (\|\cdot\| = \text{ norm in } H^1(\mathcal{O}))$$

then, after extracting a subsequence, $u_\varepsilon \to u$ in $H^1(\cdot)$ weakly, hence in $L^2(\mathcal{O})$ strongly, and $c_1^\varepsilon u_\varepsilon \to \mathcal{M}(c_i)u$ so that

(13.1.8) $$C^\varepsilon u_\varepsilon \to \mathcal{M}(c_i)\frac{\partial u}{\partial x_i} \text{ in } L^2(\mathcal{O}) \text{ weakly.}$$

Therefore, we shall only consider the equation

(13.1.9) $$A^\varepsilon u_\varepsilon + B^\varepsilon u_\varepsilon + a_0^\varepsilon u_\varepsilon = f \text{ in } \mathcal{O}.$$

The boundary conditions are as follows. We introduce V as in Section 1. Then we consider the problem

(13.1.10) $$\begin{aligned} &u_\varepsilon \in V, \\ &a^\varepsilon(u_\varepsilon, v) + (B^\varepsilon u_\varepsilon, v) = (f, v), \ \forall v \in V. \end{aligned}$$

We now assume

(13.1.11) $$\begin{aligned} &b_1 \in W^{1,\infty}(Y), \\ &\frac{\partial b_i}{\partial y_i}(y) \le 0, \end{aligned}$$

and

(13.1.12) $$v \in V \Rightarrow v = 0 \text{ if } b_i^\varepsilon n_i \le 0;$$

since in (13.1.12) V should be independent of ε, a simple particular case where (13.1.12) holds true is:

(13.1.13) $$V = H_0^1(\mathcal{O}).$$

Under the conditions (13.1.11), (13.1.12)[57] one has

(13.1.14) $$a^\varepsilon(v, v) + (B^\varepsilon v, v) \ge \min(\alpha_1, \alpha_0)\|v\|$$

---

[57]We emphasize the fact that these conditions are **sufficient** but not necessary in order to obtain (13.1.7). Other sufficient conditions, which do not assume $b_i \in W^{1,\infty}(Y)$ but only $b_i \in L^\infty(Y)$, are obtained by using Cauchy-Schwarz. One has

$$|(B^\varepsilon v, v)| \le c|\nabla v||v|$$

since

$$(B^\varepsilon v, v) = \int_{\mathcal{O}} b_i^\varepsilon \frac{1}{2} \frac{\partial}{\partial x_i} (v^2) \, dx$$

$$= -\frac{1}{2\varepsilon} \int_{\mathcal{O}} \left(\frac{\partial b_i}{\partial x_i} \right)^\varepsilon v^2 \, dx + \frac{1}{2} \int_{\Gamma} (b_i^\varepsilon n_i) v^2 \, d\Gamma$$

$$\geq 0.$$

Consequently, under conditions (13.1.11), (13.1.12), problem (13.1.10) admits a unique solution u_ε, which satisfies (13.1.7).

We want now to study the behavior of u_ε as $\varepsilon \to 0$.

13.2. Asymptotic expansion. Using methods of Section 2, we have

$$(13.2.1) \qquad\qquad B^\varepsilon = \varepsilon^{-1} B_1 + \varepsilon^0 B_2,$$

$$B_1 = b_i \frac{\partial}{\partial y_i}, \qquad B_2 = b_1 \frac{\partial}{\partial x_i}.$$

If we look for $u_\varepsilon = u_0 + \varepsilon u_1 + \cdots$, $u_j = u_j(x, y)$, (13.1.9) becomes:

$$(13.2.2) \qquad\qquad A_1 u_0 = 0,$$

$$(13.2.3) \qquad\qquad A_1 u_1 + (A_2 + B_1) u_0 = 0,$$

$$(13.2.4) \qquad A_1 u_2 + (A_2 + B_1) u_1 + (A_3 + B_2) u_0 = f.$$

Equation (13.2.2) gives $u_0 = u(x)$, and (13.2.3) gives the same result as in Section 2 for u_1; namely,

$$(13.2.5) \qquad\qquad u_1 = -\chi^j \frac{\partial u}{\partial x_j} + \tilde{u}_1(x).$$

Equation (13.2.4) admits a solution in u_2, iff

$$\int_Y (A_2 + B_1) u_1 \, dy + \int_Y (A_3 + B_2) u \, dy = |Y| f.$$

We obtain:

$$(13.2.6) \qquad\qquad \mathcal{A}u + \mathcal{B}u + \mathcal{M}(a_0) u = f,$$

where $\mathcal{A}$ is the homogenized operator of A^ε, and where $\mathcal{B}$ is give by

$$(13.2.7) \qquad \mathcal{B}u = \left(\frac{1}{|Y|} \int_Y \left(b_j - b_k \frac{\partial \chi^j}{\partial y_k} \right) dy \right) \frac{\partial u}{\partial x_j}.$$

This result will be justified in Section 13.3 below.[58]

REMARK 13.1. The operator $\mathcal{B}$ depends of course on B^ε **but also on A^ε. The homogenization of B^ε is "relative to A^ε."** Another example of this type of property will be met again in Section 14 below.

---

and one assumes that

$$\alpha |\nabla v|^2 + \alpha_0 |v|^2 + c |\nabla v||v| \geq \alpha_2 \|v\|^2.$$

[58] One can give a direct justification of this computation for Dirichlet's boundary conditions.

13.3. Energy estimates. We prove now

THEOREM 13.2. *We assume that (13.1.11) and (13.1.12) take place, or the conditions of before (13.1.14).*
Then the solution u_ε of (3.2.4) satisfies

$$(13.3.1) \qquad\qquad u_\varepsilon \to u \ \text{in } V \ \text{weakly},$$

where u is the solution of

$$(13.3.2) \qquad u \in V, \qquad \mathcal{A}(u,v) + (\mathcal{B}u, v) = (f, v), \ \forall v \in V.$$

PROOF. We set

$$(13.3.3) \qquad\qquad \eta_\varepsilon = b_i^\varepsilon \frac{\partial u_\varepsilon}{\partial x_i}.$$

By virtue of (13.1.7), we can extract a subsequence u_ε such that one has (13.3.1) and such that

$$(13.3.4) \qquad\qquad \eta_\varepsilon \to \eta \ \text{in } L^2(\mathcal{O}) \ \text{weakly}.$$

Then $A^\varepsilon u_\varepsilon = f - a_0^\varepsilon u_\varepsilon - \eta_\varepsilon \to f - \mathcal{M}(a_0)u - \eta$ in $L^2(\mathcal{O})$ weakly so that

$$(13.3.5) \qquad\qquad \mathcal{A}u + \eta + \mathcal{M}(a_0)u = f.$$

We now define $\beta(y)$ as the solution (defined up to an additive constant) of

$$(13.3.6) \qquad\qquad A_1^* \beta = \frac{\partial b_j}{\partial y_j}, \qquad \beta \ Y\text{-periodic},$$

i.e.,

$$(13.3.7) \qquad a_1^*(\beta, \psi) = -\left(b_j, \frac{\partial \psi}{\partial y_j}\right)_Y, \ \forall \psi \in W(Y), \ \beta \in W(Y).$$

We define next

$$(13.3.8) \qquad\qquad \gamma_\varepsilon = 1 + c\beta^\varepsilon, \qquad \beta^\varepsilon(x) = \beta(x/\varepsilon).$$

We observe that (we implicitly assume that $a_0 = 0$ to simplify slightly)

$$(13.3.9) \qquad\qquad (A^\varepsilon)^* \gamma_\varepsilon = \frac{\partial b_j^\varepsilon}{\partial x_j}.$$

We multiply (13.1.9) by $\phi\gamma_\varepsilon$, $\phi \in C_0^\infty(\mathcal{O})$, and we multiply (13.3.9) by ϕu_ε. After subtracting, we obtain

$$(13.3.10) \qquad \left(\xi_i^\varepsilon, \frac{\partial \phi}{\partial x_i}\gamma_\varepsilon\right) - \int_\mathcal{O} a_{ij}^\varepsilon \frac{\partial \gamma_\varepsilon}{\partial x_i}\frac{\partial \phi}{\partial x_j} u_\varepsilon \, dx + (\eta_\varepsilon, \phi\gamma_\varepsilon)$$

$$= (f, \phi\gamma_\varepsilon) - \left(b_j^\varepsilon, \frac{\partial}{\partial x_j}(\phi u_\varepsilon)\right)$$

$$= (f, \phi\gamma_\varepsilon) - (\eta_\varepsilon, \phi) - \left(b_j^\varepsilon, u_\varepsilon \frac{\partial \phi}{\partial x_j}\right).$$

Passing to the limit we obtain

$$\left(\xi_i, \frac{\partial \phi}{\partial x_i}\right) - \mathcal{M}\left(a_{ij}\frac{\partial \beta}{\partial y_i}\right)\int_{\mathcal{O}}\frac{\partial \phi}{\partial x_j}u\,dx + (\eta, \phi)$$

$$= (f, \phi) - (\eta, \phi) - \mathcal{M}(b_j)\int_{\mathcal{O}}u\frac{\partial \phi}{\partial x_j}\,dx$$

$$= \left(\xi_i, \frac{\partial \phi}{\partial x_i}\right) - \mathcal{M}(b_j)\int_{\mathcal{O}}u\frac{\partial \phi}{\partial x_j}\,dx.$$

Therefore

$$(13.3.11) \qquad \eta = \mathcal{M}(b_j)\frac{\partial u}{\partial x_j} - \mathcal{M}\left(a_{ij}\frac{\partial \beta}{\partial y_i}\right)\frac{\partial u}{\partial x_j}.$$

It remains to verify that (13.3.11) coincides with (13.2.7). But

$$|Y|\mathcal{M}\left(b_k\frac{\partial \chi^j}{\partial y_k}\right) = (\text{using (13.3.7)}) \ = a_1^*(\beta, \chi_j)$$

$$= a_1(\chi^j, \beta) = \ (\text{by definition of } \chi^j)$$

$$= a_1(y_j, \beta)$$

$$= |Y|\mathcal{M}\left(a_{ij}\frac{\partial \beta}{\partial y_i}\right),$$

hence the result follows.   ∎

REMARK 13.3. One can also understand the homogenization (13.2.7) of B^ε by using correctors, at least in the case when A^ε is symmetric. If $u_\varepsilon + \varepsilon\chi^j\left(\frac{x}{\varepsilon}\right)\frac{\partial u}{\partial x_j}$ converges strongly to u in $H^1_{loc}(\mathcal{O})$, then $B^\varepsilon u_\varepsilon = B^\varepsilon\left(u_\varepsilon + \varepsilon\chi^j\frac{\partial u}{\partial x_j}\right) - b_k\frac{\partial \chi^j}{\partial y_k}\frac{\partial u}{\partial x_j} - \varepsilon\chi^j B^\varepsilon\left(\frac{\partial u}{\partial x_j}\right)$ so that, **locally**, $B^\varepsilon u_\varepsilon$ converges to

$$\mathcal{M}(b_j)\frac{\partial u}{\partial x_j} - \mathcal{M}\left(b_k\frac{\partial \chi^j}{\partial y_k}\right)\frac{\partial u}{\partial x_j},$$

which is formula (13.2.7).

14. Singular perturbations and homogenization

14.1. Orientation. We consider now boundary value problems associated to the equation

$$(14.1.1) \qquad \varepsilon^2 \Delta^2 u_\varepsilon + A^\varepsilon u_\varepsilon = f \text{ in } \mathcal{O},$$

where A is given as in Section 1.

REMARK 14.1. We consider (14.1.1) **as an example**. Very many variants are possible, by the same methods as those given below.

Let us first make precise what are the boundary conditions we consider. We introduce

$$(14.1.2) \qquad H^1(\mathcal{O}; \Delta) = \{v \mid v \in H^1(\mathcal{O}), \ \Delta v \in L^2(\mathcal{O})\},$$

which is a Hilbert space for the norm given by $(|\cdot| = L^2(\mathcal{O})$ norm)

$$\left[\|v\|^2_{H^1(\mathcal{O})} + |\Delta v|^2\right]^{1/2};$$

the closure of $C_0^\infty(\mathcal{O})$ in $H^1(\mathcal{O}; \Delta)$ coincides with $H_0^2(\mathcal{O})$, and we consider V_0, closed subspace of $H^1(\cdot; \Delta)$ such that

$$(14.1.3) \qquad\qquad H_0^2(\mathcal{O}) \subseteq V_0 \subseteq H^1(\mathcal{O}; \Delta).$$

For $u, v \in H^1(\mathcal{O}; \Delta)$ we introduce the bilinear form

$$(14.1.4) \qquad\qquad \pi_\varepsilon(u, v) = \varepsilon^2(\Delta u, \Delta v) + a^\varepsilon(u, v).$$

Under hypothesis (1.1.1) and, when necessary, (1.1.2), we have

$$(14.1.5) \qquad\qquad \pi_\varepsilon(v, v) \geq \varepsilon^2|\Delta v|^2 + c\|v\|^2$$

(where $\|\cdot\|$ denotes $H^1(\mathcal{O})$ norm).

It follows that given f in, say, V_0, there exists a unique u_ε such that

$$(14.1.6) \qquad\qquad \pi_\varepsilon(u_\varepsilon, v) = (f, v), \ \forall v \in V_0,$$
$$u_\varepsilon \in V_0.$$

We want to study u_ε as $\varepsilon \to 0$. **We are going to show that the (weak) limit u of u_ε satisfies an equation**

$$(14.1.7) \qquad\qquad \mathcal{A}^{\Delta^2} u = f$$

where the homogenized operator $\mathcal{A}^{\Delta^2}$ depends of course on A^ε but also on Δ^2. (Compare to Remark 13.1.)

As usual, the best way to obtain the formulas is to use the asymptotic expansion. This is what we do in Section 14.2 below.

14.2. Asymptotic expansion. Using notations of Section 2, we have

$$\Delta = \varepsilon^{-2}\Delta_y + 2\varepsilon^{-1}\Delta_{yx} + \varepsilon^0\Delta_x,$$

where

$$\Delta_{yx} = \frac{\partial^2}{\partial y_j \partial x_j}.$$

Then

$$(14.2.1) \qquad \varepsilon^2\Delta^2 + A^\varepsilon = \varepsilon^{-2}(\Delta_y^2 + A_1) + \varepsilon^{-1}(4\Delta_y\Delta_{yx} + A_2)$$
$$+ \varepsilon^0(2\Delta_y\Delta_x + 4\Delta_{yx}^2 + A_3)$$
$$+ \varepsilon(2\Delta_y\Delta_x + 4\Delta_{yx}^2) + \varepsilon^2\Delta_x^2.$$

If we look for $u_\varepsilon = u_0 + \varepsilon u_1 + \varepsilon^2 u_2 + \cdots$, $u_j = u_j(x, y)$, Y-periodic in y, we obtain:

$$(14.2.2) \qquad\qquad\qquad\qquad\qquad (\Delta_y^2 + A_1)u_0 = 0,$$

$$(14.2.3) \qquad\qquad\qquad (\Delta_y^2 + A_1)u_1 + (4\Delta_y\Delta_{yx} + A_2)u_0 = 0,$$

$$(14.2.4) \quad (\Delta_y^2 + A_1)u_2 + (4\Delta_y\Delta_{yx} + A_2)u_1 + (2\Delta_y\Delta_x + 4\Delta_{yx}^2 + A_3)u_0 = f,$$

$$\vdots$$

Equation (14.2.2) is equivalent to $u_0 = u(x)$. Then (14.2.3) reduces to

$$(14.2.5) \qquad\qquad (\Delta_y^2 + A_1)u_1 = \frac{\partial a_{ij}}{\partial y_j}\frac{\partial u}{\partial x_j}.$$

We introduce ρ^j by

$$(14.2.6) \qquad\qquad (\Delta_y^2 + A_1)\rho^j = A_1 y_j, \qquad \rho^j \ Y\text{-periodic},$$

which defines ρ^j up to an additive constant.

REMARK 14.2. The functions $\rho^j(y)$ "replace" in the present situation the functions $\chi^j(y)$ introduced in Section 2.

Using the functions ρ^j (14.2.5) gives:

$$(14.2.7) \qquad u_1 = -\rho^j(y)\frac{\partial u}{\partial x_j}(x) + \tilde{u}_1(x).$$

Equation (14.2.4) can be solved for u_ε iff

$$\int_Y (4\Delta_y\Delta_{yx} + A_2)u_1 \, dy + \int_Y (2\Delta_y\Delta_x + 4\Delta_{yx}^2 + A_3)u_0 \, dy = \int_Y f \, dy$$

which reduces to

$$(14.2.8) \qquad \int_Y A_2 u_1 \, dy + \int_Y a_{ij}(y) \, dy \frac{\partial^2 u}{\partial x_i \partial x_j} + \int_Y a_0 \, dy \, u = |Y| f.$$

Using (14.2.7), (14.2.8) gives (14.1.7) with

$$(14.2.9) \qquad \mathcal{A}^{\Delta^2} = -\frac{1}{|Y|}\int_Y \left(a_{ij} - a_{ik}\frac{\partial \rho^j}{\partial y_k}\right) dy \frac{\partial^2}{\partial x_i \partial x_j} + \mathcal{M}(a_0).$$

REMARK 14.3. We obtain formulas analogous to those of Section 2, with χ^j replaced by ρ^j.

REMARK 14.4. One can verify, as in Section 2, the **ellipticity** of $\mathcal{A}^{\Delta^2}$. We are going to justify the above procedure.

14.3. Homogenization with respect to Δ^2.

THEOREM 14.5. *We assume that (1.1.1), (1.1.2) hold true. Let u_ε be the solution of (14.2.9). Then, as $\varepsilon \to 0$, one has*

$$(14.3.1) \qquad u_\varepsilon \to u \text{ in } H^1(\mathcal{O}) \text{ weakly,}$$

where u is the solution of

$$(14.3.2) \qquad u \in \bar{V}_0 = \text{ closure of } V_0 \text{ in } H^1(\mathcal{O}),$$

$$\mathcal{A}^{\Delta^2}(u,v) + (\mathcal{M}(a_0)u, v) = (f, v), \ \forall v \in \overline{V}_0,$$

where $\mathcal{A}^{\Delta^2}(u,v)$ is given by

$$(14.3.3) \qquad \mathcal{A}^{\Delta^2}(u,v) = \int_{\mathcal{O}} q_{ij}^{\Delta^2} \frac{\partial u}{\partial x_j}\frac{\partial v}{\partial x_i} \, dx,$$

$$q_{ij}^{\Delta^2} = \frac{1}{|Y|}\int_Y \left(a_{ij} - a_{ik}\frac{\partial \rho^j}{\partial y_k}\right) dy.$$

PROOF. It follows from (14.1.5) that

$$(14.3.4) \qquad \|u_\varepsilon\| \le c, \qquad \varepsilon|\Delta u_\varepsilon| \le c.$$

We set, as usual, $\xi_i^\varepsilon = a_{ij}^\varepsilon \frac{\partial u_\varepsilon}{\partial x_j}$. We can extract a subsequence u_ε such that

$$(14.3.5) \qquad u_\varepsilon \to u \text{ in } H^1(\mathcal{O}) \text{ weakly,}$$

$$\xi_i^\varepsilon \to \xi_i \text{ in } L^2(\mathcal{O}) \text{ weakly,}$$

and since $\varepsilon^2(\Delta u_\varepsilon, \Delta v) \to 0$ (by virtue of (14.3.4)), we obtain

(14.3.6)
$$\left(\xi_i, \frac{\partial v}{\partial x_i}\right) = (f, v), \ \forall v \in V_0,$$

and therefore $\forall v \in \overline{V}_0$.

We also observe that $u \in \overline{V}_0$, so that it only remains to show that

(14.3.7)
$$\xi_i = q_{ij}^{\Delta^2} \frac{\partial u}{\partial x_j}.$$

We introduce $w(y)$ such that

(14.3.8)
$$\begin{aligned} &(\Delta_y^2 + A_1^*)w = 0, \\ &w - P \text{ is } Y\text{-periodic}, \\ &P(y) = \text{ homogeneous polynomial of first degree}. \end{aligned}$$

If we set $w - p = -\rho$, then

(14.3.9)
$$(\Delta_y^2 + A_1^*)\rho = A_1^*\rho, \qquad \rho \text{ is } Y \text{ periodic}.$$

We now introduce

(14.3.10)
$$w_\varepsilon(x) = \varepsilon w\left(\frac{x}{\varepsilon}\right) = P(x) - \varepsilon\rho\left(\frac{x}{\varepsilon}\right).$$

We verify that

(14.3.11)
$$(\varepsilon^2\Delta^2 + (A^\varepsilon)^*)w_\varepsilon = 0.$$

Therefore, multiplying (14.1.1) by ϕw_ε, $\phi \in C_0^\infty(\mathcal{O})$, and (14.3.11) by ϕu_ε, we obtain

$$\varepsilon^2(\Delta u_\varepsilon, \Delta(\phi w_\varepsilon)) - \varepsilon^2(\Delta w_\varepsilon, \Delta(\phi u_\varepsilon)) + \left(\xi_i^\varepsilon, \frac{\partial\phi}{\partial x_i}w_\varepsilon\right) \\ - \int_{\mathcal{O}} a_{ij}^\varepsilon \frac{\partial w_\varepsilon}{\partial x_i}\frac{\partial\phi}{\partial x_j}u_\varepsilon \, dx = (f, \phi w_\varepsilon),$$

which gives in the limit

$$\left(\xi_i, \frac{\partial\phi}{\partial x_i}P\right) - \mathcal{M}\left(a_{ij}\frac{\partial w}{\partial y_i}\right)\int_{\mathcal{O}} \frac{\partial\phi}{\partial x_i}u \, dx = (f, \phi P) = \left(\xi_i, \frac{\partial(\phi P)}{\partial x_i}\right),$$

hence

(14.3.12)
$$\xi_i \frac{\partial P}{\partial x_i} = \mathcal{M}\left(a_{ij}\frac{\partial w}{\partial y_i}\right)\frac{\partial u}{\partial x_j}.$$

Taking $P(y) = y_i$ gives $w = y_i - \rho^i$ so that (14.3.12) gives

(14.3.13)
$$\xi_i = \mathcal{M}\left(a_{kj}\frac{\partial}{\partial y_k}(y_i - \rho^i)\right)\frac{\partial u}{\partial x_j}.$$

If $\tilde{q}_{ij}$ denotes the coefficient of $\frac{\partial u}{\partial x_j}$ in (14.3.13), we have

$$\tilde{q}_{ij} = \mathcal{M}(a_{ij}) - \mathcal{M}\left(a_{kj}\frac{\partial\rho^i}{\partial y_k}\right)$$

and one verifies that this formula coincides with (14.3.3). Hence the theorem follows.   ■

REMARK 14.6 (Correctors). Let us introduce, with the notations of Section 5,

$$(14.3.14) \qquad \theta_\varepsilon = -\varepsilon \rho^j \left(\frac{x}{\varepsilon} \right) m_\varepsilon(x) \frac{\partial u}{\partial x_j}(x).$$

Then, if A is symmetric and if $u \in H^2_{loc}(\mathcal{O})$, we have

$$(14.3.15) \qquad z_\varepsilon = u_\varepsilon - (u + \theta_\varepsilon) \to 0 \text{ in } H^1(\mathcal{O}) \text{ strongly.}$$

For the proof, we compute

$$\pi_\varepsilon(z_\varepsilon) = \pi_\varepsilon(u_\varepsilon, z_\varepsilon) - [\varepsilon^2(\Delta(u + \theta_\varepsilon), \Delta u_\varepsilon) + a^\varepsilon(u + \theta_\varepsilon, u_\varepsilon)]$$
$$+ \pi_\varepsilon(u + \theta_\varepsilon) = (f, z_\varepsilon) - (f, u + \theta_\varepsilon) + \pi_\varepsilon(u + \theta_\varepsilon).$$

We verify that

$$\pi_\varepsilon(u + \theta_\varepsilon) \to \mathcal{A}^{\Delta^2}(u, u),$$

so that

$$\pi_\varepsilon(z_\varepsilon) \to -(f, u) + \mathcal{A}^{\Delta^2}(u, u) = 0,$$

and (14.3.15) follows.

15. Non-local limits

15.1. Setting of the problem. We consider operators A^ε, B^ε defined by

$$(15.1.1) \qquad A^\varepsilon = -\frac{\partial}{\partial x_i} a^\varepsilon_{ij} \frac{\partial}{\partial x_j}, \qquad a^\varepsilon_{ij}(x) = a_{ij}(x/\varepsilon),$$

$$B^\varepsilon = -\frac{\partial}{\partial x_i} b^\varepsilon_{ij} \frac{\partial}{\partial x_j}, \qquad b^\varepsilon_{ij}(x) = b_{ij}(x/\varepsilon),$$

a_{ij}, b_{ij} being Y-periodic functions in $L^\infty(\mathbb{R}^n)$, such that

$$(15.1.2) \qquad a_{ij}(y)\xi_i\xi_j \geq \alpha\xi_i\xi_i,$$
$$b_{ij}(y)\xi_i\xi_j \geq \beta\xi_i\xi_i, \qquad \alpha, \beta > 0.$$

We consider in the cylinder $\mathcal{O} \times \mathbb{R}_\lambda$ the operator

$$(15.1.3) \qquad A^\varepsilon + \Lambda B^\varepsilon, \qquad \Lambda = -\frac{\partial^2}{\partial \lambda^2},$$

and the equation

$$(15.1.4) \qquad (A^\varepsilon + \Lambda B^\varepsilon)u_\varepsilon = f \text{ in } \mathcal{O} \times \mathbb{R}_\lambda.$$

In order to define precisely the boundary conditions, we use a **variational formulation**. We introduce V such that

$$(15.1.5) \qquad H^1_0(\mathcal{O}) \subseteq V \subseteq H^1(\mathcal{O}),$$

and we assume that

$$(15.1.6) \qquad a^\varepsilon(v, v) \geq c\|v\|^2, \ \forall v \in V, \ c > 0,$$
$$b^\varepsilon(v, v) \geq c\|v\|^2, \ \forall v \in V, \ \|\cdot\| = \text{ norm in } V.$$

[We add > 0 zero order terms if $V = H^1(\mathcal{O})$.] In (15.1.6), we have set

$$(15.1.7) \qquad a^\varepsilon(u, v) = \int_\mathcal{O} a^\varepsilon_{ij} \frac{\partial u}{\partial x_j} \frac{\partial v}{\partial x_i} \, dx,$$

$$b^\varepsilon(u, v) = \int_\mathcal{O} b^\varepsilon_{ij} \frac{\partial u}{\partial x_j} \frac{\partial v}{\partial x_i} \, dx.$$

We introduce next

(15.1.8) $\qquad H^1(\mathbb{R}_\lambda, V) = \left\{ v \ \middle| \ v \in L^2(\mathbb{R}_\lambda, V), \dfrac{\partial v}{\partial \lambda} \in L^2(\mathbb{R}_\lambda, V) \right\}$

which is a Hilbert space for the norm given by

(15.1.9) $\qquad \|\|v\|\| = \left(\displaystyle\int_{-\infty}^{+\infty} \left[\|v(\lambda)\|^2 + \left\|\dfrac{\partial v}{\partial \lambda}(\lambda)\right\|^2 \right] d\lambda \right)^{1/2}.$

For $u, v \in H^1(\mathbb{R}_\lambda, V)$, we set

(15.1.10) $\qquad c^\varepsilon(u,v) = \displaystyle\int_{-\infty}^{\infty} \left[a^\varepsilon(u(\lambda), v(\lambda)) + b^\varepsilon\left(\dfrac{\partial u}{\partial \lambda}(\lambda), \dfrac{\partial v}{\partial \lambda}(\lambda)\right) \right] d\lambda.$

The problem we consider is now this: given f such that

(15.1.11) $\qquad f \in H^{-1}(\mathbb{R}_\lambda, V')$ (dual of $H^1(\mathbb{R}_\lambda, V)$),

u_ε is the solution of

(15.1.12) $\qquad c^\varepsilon(u_\varepsilon, v) = \displaystyle\int_{-\infty}^{+\infty} (f, v)\, d\lambda, \ \forall v \in H^1(\mathbb{R}_\lambda, V),$

$\qquad\qquad\qquad u_\varepsilon \in H^1(\mathbb{R}_\lambda, V).$

This equation admits indeed a unique solution, since by virtue of (15.1.6), we have

(15.1.13) $\qquad\qquad\qquad c^\varepsilon(v, v) \geq c\, \|\| v \|\|^2 .$

We want to study the behavior of u_ε as $\varepsilon \to 0$. **We shall show that the limit u of u_ε satisfies an integro-partial differential equation, i.e., a non-local equation.**

REMARK 15.1. The operators A^ε, B^ε could be replaced by other operators such as those introduced before, and the operator Λ could also be replaced by other operators (it suffices then to replace the Fourier transform in λ by spectral diagonalization if Λ is self-adjoint).

15.2. Non-local homogenized operator. We use a Fourier transform in λ; if $g \in L^2(\mathbb{R}_\lambda, V)$ we introduce

$$\mathcal{F}g(\mu) == \hat{g}(\mu) + \int_{-\infty}^{+\infty} e^{-2\pi i \lambda \mu} g(\lambda)\, d\lambda, \qquad \hat{g} \in L^2(\mathbb{R}_\mu, V).$$

Then (15.1.4) is equivalent to

(15.2.1) $\qquad\qquad\qquad (A^\varepsilon + 4\pi^2 \mu^2 B^\varepsilon)\hat{u}_\varepsilon(\mu) = \hat{f}(\mu).$

The variational form is as follows: by Fourier transform in λ, $H^1(\mathbb{R}_\lambda, V)$ becomes

(15.2.2) $\qquad\qquad \mathcal{F}H^1(\mathbb{R}_\mu, V) = \{ v \mid (1 + |\mu|)v \in L^2(\mathbb{R}_\mu, V) \}.$

For $u, v \in \mathcal{F}H^1(\mathbb{R}_\mu, V)$, we define

(15.2.3) $\qquad \gamma(u, v) = \displaystyle\int_{-\infty}^{+\infty} [a^\varepsilon(u(\mu), v(\mu)) + 4\pi^2 \mu^2 b^\varepsilon(u(\mu), v(\mu))]\, d\mu.$

The Fourier transform $\hat{f}$ of f satisfies

(15.2.4) $\qquad\qquad\qquad (1 + |\mu|)^{-1}\hat{f} \in L^2(\mathbb{R}_\mu, V'),$

and $\hat{u}_\varepsilon(\mu)$ is the solution of

$$(15.2.5) \qquad \gamma(\hat{u}_\varepsilon, v) = \int_{-\infty}^{+\infty} (\hat{f}, v)\, d\mu, \ \forall v \in H^1(\mathbb{R}_\mu, V).$$

For each j, we define $\hat{\chi}^j(y, \mu)$ (up to an additive constant chosen, for instance, by adding the condition $\int_Y \hat{\chi}^j(y, u)\, dy = 0$) as the solution of[59]

$$(15.2.6) \qquad (A_1 + 4\pi^2\mu^2 B_1)(\hat{\chi}^j(y, u) - y_j) = 0, \qquad \hat{\chi}^j \ Y\text{-periodic.}$$

We set next (this is the homogenization of $A^\varepsilon + 4\pi^2\mu^2 B^\varepsilon$, i.e., **homogenization with parameter** μ; cf. Section 8.1)

$$(15.2.7) \qquad \hat{q}_{ij}(\mu) = \frac{1}{|Y|}[a_1(\hat{\chi}^j - y_j, \hat{\chi}^i - y_i) + 4\pi^2\mu^2 b_1(\hat{\chi}^j - y_j, \hat{\chi}^i - y_i)],$$

and we define

$$(15.2.8) \qquad \hat{\mathcal{A}}(\mu) = -\hat{q}_{ij}(\mu)\frac{\partial^2}{\partial x_i \partial x_j}.$$

The function $\mu \to \hat{\phi}_{ij}(\mu)$ is bounded by $c(1 + \mu^2)$ (one verifies that $\|\hat{\chi}^j(\mu)\|_{W(Y)} \le c$) so that we can define (cf. L. Schwartz [103] and [104])

$$(15.2.9) \qquad q_{ij} = \mathcal{F}^{-1}(\hat{q}_{ij}) \qquad \text{(a distribution on } \mathbb{R}_\lambda),$$

and (writing distributions as functions)

$$(15.2.10) \qquad \mathcal{A}(\lambda) = -q_{ij}(\lambda)\frac{\partial^2}{\partial x_i \partial x_j}.$$

Then, as we shall show below, the homogenized equation of (15.1.4) is:

$$(15.2.11) \qquad \mathcal{A}(\lambda) *_{(\lambda)} u = f,$$

or

$$(15.2.12) \qquad -\int_{-\infty}^{+\infty} q_{ij}(\lambda - \lambda')\frac{\partial^2 u}{\partial x_i \partial x_j}(x, \lambda')\, d\lambda' = f(x, \lambda).$$

The variational formulation is that

$$\hat{u}(\mu) = \mathcal{F}u(\lambda)$$

satisfies

$$(15.2.13) \qquad \mathcal{A}(\mu; \hat{u}, v) = \int_{-\infty}^{+\infty} (\hat{f}, v)\, d\mu, \ \forall v \in \mathcal{F}H^1(\mathbb{R}_\mu, V),$$
$$\hat{u} \in \mathcal{F}H^1(\mathbb{R}_\mu, V),$$

where we have set

$$(15.2.14) \qquad \mathcal{A}(\mu; u, v) = \int_{-\infty}^{+\infty} \int_{\mathcal{O}} \hat{q}_{ij}(\mu)\frac{\partial u}{\partial x_j}\frac{\partial v}{\partial x_i}\, dx\, d\mu.$$

REMARK 15.2. The functions $\hat{q}_{ij}(\mu)$ are not, in general, polynomials, so that $\mathcal{A}(\lambda)$ is generally not a partial differential operator.

---

[59]A_1 is defined as in Section 2 and $B_1 = -\frac{\partial}{\partial y_i}\left(b_{ij}\frac{\partial}{\partial y_j}\right)$.

15.3. Homogenization theorem.

THEOREM 15.3. *We assume that (15.1.2), (15.1.6) hold true. Let u_ε (respectively, u) be the solution in $H^1(\mathbb{R}_\lambda, V)$ of (15.1.12) (respectively, (15.2.13) via Fourier transform). Then*

$$(15.3.1) \qquad u_\varepsilon \to u \text{ in } H^1(\mathbb{R}_\lambda, V) \text{ weakly.}$$

PROOF. By virtue of (15.1.13) we know that

$$(15.3.2) \qquad \|\|u_\varepsilon\|\| \le c$$

so that, by Fourier transform,

$$(15.3.3) \qquad \hat{u}_\varepsilon(\mu) \text{ remains bounded in } \mathcal{F}H^1(\mathbb{R}_\mu, V);$$

$\hat{u}_\varepsilon(\mu)$ is characterized by (15.2.5); hence a.e. in μ

$$(15.3.4) \qquad a^\varepsilon(\hat{u}_\varepsilon(\mu), v) + 4\pi^2\mu^2 b(\hat{u}_\varepsilon(\mu), v) = (\hat{f}(\mu), v), \; \forall v \in V.$$

It follows from (15.3.3) that, after extracting a subsequence,

$$(15.3.5) \qquad \hat{u}_\varepsilon \to \hat{u} \text{ in } \mathcal{F}H^1(\mathbb{R}_\mu, V) \text{ weakly,}$$

and by the homogenization with parameter, for a.e. μ, $\hat{u}_\varepsilon(\mu)$ converges in V weakly toward $\hat{w}_\mu$ solution of

$$(15.3.6) \qquad \mathcal{A}(\mu; \hat{w}(\mu), v) = (\hat{f}(\mu), v), \; \forall v \in V.$$

But necessarily[60] $\hat{u} = \hat{w}$, hence the theorem follows. $\blacksquare$

REMARK 15.4. We can find the above formulas using asymptotic expansions.

REMARK 15.5. We can also introduce the correctors of first order.

16. Introduction to non-linear problems

16.1. Formal general formulas.
Let us consider a non-linear elliptic equation

$$(16.1.1) \qquad A^\varepsilon u_\varepsilon + B\left(\frac{x}{\varepsilon}, u_\varepsilon, Du_\varepsilon\right) = f \text{ in } \mathcal{O},$$

where

$$(16.1.2) \qquad A^\varepsilon \text{ is given as in Section 1,}$$

$$(16.1.3) \qquad B \text{ is a } \textbf{non-linear perturbation,}$$

$$Du = \nabla u = \operatorname{grad} u.$$

REMARK 16.1. All that follows can be extended to higher order operators A^ε, and accordingly higher order perturbations.

REMARK 16.2. We shall assume that the function

$$(16.1.4) \qquad y \to B(y, u, p) \text{ is } Y \text{ periodic.}$$

(More precise hypothesis will be given below.) One can easily extend the remarks given below to perturbations of the form

$$(16.1.5) \qquad B\left(x, \frac{x}{\varepsilon}, u_\varepsilon, Du_\varepsilon\right).$$

---

[60]We can use the fact that if ϕ_ε is bounded in, say, $L^2(\mathbb{R}_\mu, V)$ and $\phi_\varepsilon(\mu) \to \phi(\mu)$, a.e., $\phi \in L^2(\mathbb{R}_\mu, V)$, then $\phi_\varepsilon \to \phi$ in $L^2(\mathbb{R}_\mu, V)$ weakly.

REMARK 16.3. Boundary conditions are, as it is usual, expressed in variational form:

$$u_\varepsilon \in V,$$

(16.1.6) $$a^\varepsilon(u_\varepsilon, v) + \left(B\left(\frac{x}{\varepsilon}, u_\varepsilon, Du_\varepsilon\right), v\right) = (f, v), \ \forall v \in V.$$

If we look for u_ε in the form

$$u = u_0 + \varepsilon u_1 + \cdots, \qquad u_j = u_j(x, y),$$

we obtain (with the notations of Section 2, and implicitly assuming that $B(y, u, p)$ is "smooth" in u and in p):

(16.1.7) $$A_1 u_0 = 0, \qquad \text{i.e., } u_0 = u(x),$$

(16.1.8) $$A_1 u_1 + A_2 u_0 = 0,$$

(16.1.9)
$$A_1 u_2 + A_2 u_1 + A_3 u_0 + B(y, u_0, D_x u_0 + D_y u_1) = f.$$

Equation (16.1.8) gives, as in the linear case

(16.1.10) $$u_1 = -\chi^j \frac{\partial u}{\partial x_j} + \tilde{u}_1(x).$$

One can solve equation (16.1.9) in u_2, iff

$$\int_Y [A_2 u_1 + A_3 u_0 + B(y, u_0, D_x u_0 + D_y u_1)] \, dy = f|Y|,$$

i.e.,

(16.1.11) $$\mathcal{A}u + \mathcal{B}(u, Du) = f,$$

where

(16.1.12) $$\mathcal{B}(u, Du) = \frac{1}{|Y|} \int_Y B\left(y; u(x), Du(x) - D_y \chi^j(y)\frac{\partial u}{\partial x_j}(x)\right) dy.$$

The variational form of the homogenized equation is:

$$u \in V,$$

(16.1.13) $$\mathcal{A}(u, v) + (\mathcal{B}(u, Du), v) = (f, v), \ \forall v \in V.$$

We now justify this formula on two examples.

16.2. Compact perturbations. Let us consider

(16.2.1) $$A^\varepsilon u_\varepsilon + u_\varepsilon \frac{\partial u_\varepsilon}{\partial x_1} = f, \qquad f \in L^2(\mathcal{O}),$$

(16.2.2) $$u_\varepsilon = 0 \text{ on } \Gamma.$$

The homogenization is straightforward. Since $\left(v\frac{\partial v}{\partial x_i}, v\right) = 0$ if $v = 0$ on Γ, one easily shows the **existence** of (weak) solutions of (16.2.1), (16.2.2), such that

(16.2.3) $$\|u_\varepsilon\| \leq c \qquad (\|\cdot\| = H_0^1(\mathcal{O}) \text{ norm}).$$

(One does not necessarily have uniqueness of the solution u_ε.)

Then u_ε remains bounded in $L^q(\mathcal{O})$, $\frac{1}{q} = \frac{1}{2} - \frac{1}{n}$ (or for every q finite if $n = 2$) and $u_\varepsilon \frac{\partial u_\varepsilon}{\partial x_1}$ remains in a bounded set of $L^r(\mathcal{O})$, $\frac{1}{r} = \frac{1}{q} + \frac{1}{2}$. Then we can extract a subsequence, still denoted by u, such that

$$(16.2.4) \qquad\qquad u_\varepsilon \to u \text{ in } H_0^1(\mathcal{O}) \text{ weakly,}$$

$$A^\varepsilon u_\varepsilon \to g \text{ in } L^r(\mathcal{O}) \text{ weakly.}$$

Necessarily $g = \mathcal{A}u$. On the other hand

$$u_\varepsilon \frac{\partial u_\varepsilon}{\partial x_1} = \frac{1}{2} \frac{\partial}{\partial x_1}(u_\varepsilon^2) \to \frac{1}{2} \frac{\partial}{\partial x_1} u^2 \text{ in } \mathcal{D}'(\mathcal{O}) \text{ (for instance),}$$

(since $u_\varepsilon \to u$ **strongly** in $L^2(\mathcal{O})$) so that u is the solution of

$$(16.2.5) \qquad\qquad \mathcal{A}u + u\frac{\partial u}{\partial x_1} = f.$$

We can apply the general formula (16.1.12). We obtain

$$(16.2.6) \qquad \mathcal{B}(u, Du) = \frac{1}{|Y|} \int_Y u(x) \left(\frac{\partial u}{\partial x_1}(x) - \frac{\partial \chi^j}{\partial y_1}(y) \frac{\partial u}{\partial x_j}(x) \right) dx$$

and since $\int_Y \frac{\partial \chi^j}{\partial y_1}(y)\, dy = 0$, (16.2.6) reduces to

$$\mathcal{B}(u, Du) = u\frac{\partial u}{\partial x_1}.$$

16.3. Non-compact perturbations. We consider now problem (16.1.1) or, more precisely, (16.1.6), with the following hypothesis:

$$(16.3.1) \qquad u_\varepsilon \text{ exists and } \|u_\varepsilon\| \le c, \qquad (\|\cdot\| = V\text{'s norm}),$$

$$(16.3.2) \qquad A^\varepsilon \text{ is symmetric,}$$

$$(16.3.3) \qquad \chi^j \in C^1(\bar{Y}),$$

$$(16.3.4) \qquad \mathcal{B}\left(\frac{x}{\varepsilon}; u_\varepsilon, Du_\varepsilon\right) \text{ remains bounded in } L^2(\mathcal{O}),$$

$$(16.3.5) \qquad |\mathcal{B}(y, u, p) - \mathcal{B}(y, v, q)| \le c(|u - v| + |p - q|).$$

Then $u_\varepsilon \to u$ in V weakly, where u is a solution of (16.1.13).

PROOF. If we set

$$(16.3.6) \qquad\qquad \eta_\varepsilon = \mathcal{B}\left(\frac{x}{\varepsilon}; u_\varepsilon, Du_\varepsilon\right),$$

we can extract a subsequence u_ε, η_ε, such that

$$(16.3.7) \qquad\qquad u_\varepsilon \to u \text{ in } V \text{ weakly,}$$

$$\eta_\varepsilon \to \eta \text{ in } L^2(\mathcal{O}) \text{ weakly.}$$

We introduce (cf. Section 5)

$$(16.3.8) \qquad\qquad \theta_\varepsilon = -\varepsilon m_\varepsilon \chi^j\left(\frac{x}{\varepsilon}\right) \frac{\partial u}{\partial x_j}$$

and a slight variant of Section 5 shows that

$$(16.3.9) \qquad\qquad z_\varepsilon = u_\varepsilon - (u + \theta_\varepsilon) \to 0 \text{ in } V \text{ strongly.}$$

We have

$$(16.3.10) \qquad\qquad \mathcal{A}u \to \eta = f$$

and, in variational form,

(16.3.11) $\qquad \mathcal{A}(u, v) + (\eta, v) = (f, v), \; \forall v \in V.$

It only remains to show that

(16.3.12) $\qquad \eta = \mathcal{B}(u, Du)$ (given in (16.1.12)).

By virtue of (16.3.5), we see that

$$\left| B\left(\frac{x}{\varepsilon}, u + \theta_\varepsilon, D(u + \theta_\varepsilon)\right) - B\left(\frac{x}{\varepsilon}, u, Du - m_\varepsilon(D_y \chi^j)\left(\frac{x}{\varepsilon}\right)\frac{\partial u}{\partial x_j}(x)\right)\right| \le C\varepsilon|\nabla u|$$

so that ζ_ε has the same L^2 limit as

(16.3.13) $\qquad \rho_\varepsilon(x) = B\left(\dfrac{x}{\varepsilon}, u, Du - m_\varepsilon D_y \chi^j\left(\dfrac{x}{\varepsilon}\right)\dfrac{\partial u}{\partial x_j}(x)\right).$

It suffices to show that

(16.3.14) $\qquad \rho_\varepsilon \to \mathcal{B}(u, Du)$ in $L^2(\mathcal{O}')$ weakly,

$$\mathcal{O}' \subseteq \overline{\mathcal{O}}' \subseteq \mathcal{O},$$

and for ε small enough, $m_\varepsilon = 1$ on $\mathcal{O}'$, so that finally it remains to show that, if we set

(16.3.15) $\qquad \sigma_\varepsilon(x) = B\left(\dfrac{x}{\varepsilon}, u(x), Du(x) - (D_y \chi^j)\left(\dfrac{x}{\varepsilon}\right)\dfrac{\partial u}{\partial x_j}(x)\right)$

then

(16.3.16) $\qquad \sigma_\varepsilon \to \mathcal{B}(u, Du)$ in $L^2(\mathcal{O}')$ weakly.

If we set

(16.3.17) $\qquad F(x, y) = B\left(y, u(x), Du(x) - (D_y \chi^j)(y)\dfrac{\partial u}{\partial x_j}(x)\right),$

we have

(16.3.18) $\qquad F \in C^0(\bar{Y}; L^2(\mathcal{O})), \qquad F$ is Y-periodic

so that

$$F(x, x/\varepsilon) \to \frac{1}{|Y|}\int_Y F(x, y)\, dy \text{ in } L^2(\mathcal{O}') \text{ weakly}$$

and (16.3.16) follows. $\qquad\qquad\qquad\qquad\qquad\qquad\qquad\qquad\qquad\blacksquare$

16.4. Non-linearities in the higher derivatives. Let us consider the equation

(16.4.1) $\qquad -\dfrac{\partial}{\partial x_i}\left(a_{ij}\left(\dfrac{x}{\varepsilon}\right)\left|\dfrac{\partial u_\varepsilon}{\partial x_j}\right|^{p-2}\dfrac{\partial u_\varepsilon}{\partial x_j}\right) = f \text{ in } \mathcal{O},$

with $u_\varepsilon = 0$ in Γ, and $p \ne 2$. A formal computation leads to the homogenized equation[61]
(16.4.2)

$$-\dfrac{\partial}{\partial x_i}\left(\mathcal{M}(a_{ij})\left|\dfrac{\partial u}{\partial x_j}\right|^{p-2}\dfrac{\partial u}{\partial x_j}\right) - \dfrac{\partial}{\partial x_i}\left(\left|\dfrac{\partial u}{\partial x_j}\right|^{p-2}\dfrac{1}{|Y|}\int a_{ij}(y)\dfrac{\partial u_1}{\partial y_j}\, dy\right) = f,$$

---

[61]It is actually a **system** in u and u_1.

and

$$(16.4.3) \qquad -\frac{\partial}{\partial y_i}\left(a_{ij}(y)\left|\frac{\partial u}{\partial x_j}\right|^{p-2}\frac{\partial u_1}{\partial y_j}\right) = \left(\frac{\partial a_{ij}}{\partial y_i}\right)\left|\frac{\partial u}{\partial x_j}\right|^{p-2}\frac{\partial u}{\partial x_j},$$

u_1 being Y-periodic in y.

For results along these lines, we refer to L. Tartar [**116**].

17. Homogenization of multi-valued operators

17.1. Orientation. What we have now in mind is the **homogenization of the so-called variational inequalities (V.I.).**

Let A^ε be given as in Section 1 and let K be a subset of V such that

$$(17.1.1) \qquad K \text{ is a closed convex subset of } V \text{ (non-empty).}$$

By a V.I. related to A^ε we mean the problem of finding u_ε such that

$$(17.1.2) \qquad \begin{aligned} &u_\varepsilon \in K, \\ &a^\varepsilon(u_\varepsilon, v - u_\varepsilon) \geq (f, v - u_\varepsilon), \ \forall v \in K. \end{aligned}$$

It is known (Lions-Stampacchia [**73**]) that, under hypothesis (1.1.1), (1.1.2), (17.1.2) **admits a unique solution** u_ε. We want to study u_ε as $\varepsilon \to 0$.

REMARK 17.1. If A^ε is symmetric, (17.1.2) is equivalent to

$$(17.1.3) \qquad \inf_{v \in K}\left(\frac{1}{2}a^\varepsilon(v) - (f, v)\right);$$

the minimum is attained in a unique point u_ε.

REMARK 17.2. One can think of (17.1.2) as an equation for a multivalued operator, introducing the subdifferential of the characteristic function of K.

REMARK 17.3. The problem of the behavior of u_ε becomes trivial if

$$(17.1.4) \qquad K = \text{ compact set in } V.$$

Indeed one has then $\frac{\partial u_\varepsilon}{\partial x_j} \to \frac{\partial u}{\partial x_j}$ in $L^2(\mathcal{O})$ **strongly**, so that

$$a^\varepsilon(u_\varepsilon, v - u_\varepsilon) \to \int_{\mathcal{O}} \mathcal{M}(a_{ij})\frac{\partial u}{\partial x_j}\frac{\partial(v - u)}{\partial x_i}\,dx$$

we have

$$(17.1.5) \qquad u_\varepsilon \to u \text{ in } V,$$

where u is the solution of

$$(17.1.6) \qquad (\mathcal{M}a)(u, v - u) \geq (f, v - u), \ \forall v \in K, \ u \in K.$$

In this case the homogenized operator is simply

$$-\frac{\partial}{\partial x_i}\mathcal{M}(a_{ij})\frac{\partial}{\partial x_j},$$

which is (in general) **different from** $\mathcal{A}$.

REMARK 17.4. If $K = V$, (17.1.2) reduces to the **equation**

$$(17.1.7) \qquad u_\varepsilon \in V, \ a^\varepsilon(u_\varepsilon, v) = (f, v), \ \forall v \in V.$$

Therefore if we **conjecture** the existence of a homogenized problem, **it will depend on K: we** indeed conjecture **the existence of a continuous coercive bilinear form $\mathcal{A}^K(u, v)$ on V, depending on A^ε and on K, such that one has (17.1.5) where u is the solution of**

$$(17.1.8) \qquad u \in K, \ \mathcal{A}^K(u, v - u) \geq (f, v - u), \ \forall v \in K.$$

We give in the next sections some evidence to support this conjecture.

REMARK 17.5. One could also consider in (17.1.2) a family of convex sets K^ε depending on ε, i.e.,

$$(17.1.9) \qquad u_\varepsilon \in K^\varepsilon, \qquad a^\varepsilon(u_\varepsilon, v - u_\varepsilon) \geq (f, v - u_\varepsilon), \ \forall v \in K^\varepsilon.$$

17.2. A formal procedure for the homogenization of problems of the calculus of variations.

NOTATIONS. We denote by K^ε the set of functions $v = v(x)$ on $\mathcal{O}$ such that

$$(17.2.1) \qquad g_1(x) \leq v(x) \leq g_2(x), \text{ where } g_i \text{ are given on } \mathcal{O}$$

and where g_1 (respectively, g_2) can take the

value $-\infty$ (respectively, $+\infty$),

$$(17.2.2) \qquad v = 0 \text{ on } \Gamma_0 \subset \Gamma = \partial\mathcal{O},$$

$$(17.2.3) \qquad |Dv(x)| \leq m(x, x/\varepsilon), \text{ where } m(x, y) \text{ is continuous}$$

on $\overline{\mathcal{O}} \times \mathbb{R}^n$, Y-periodic in y.

Let there be given a function $\phi(x, y, \lambda, \mu)$ which satisfies:

$$(17.2.4) \qquad \phi \text{ is continuous}[62] \text{ on } \overline{\mathcal{O}} \times \mathbb{R}^n \times \mathbb{R} \times \mathbb{R}^n, \ Y\text{-periodic in } y,$$

$$|\phi(x, y, \lambda, \mu) - \phi(x, y, \lambda', \mu')| \leq c(|\lambda - \lambda'| + |\mu - \mu'|).$$

We also restrict the class K^ε to functions which are such that

$$(17.2.5) \qquad \phi(x, x/\varepsilon, v, Dv) \in L^1(\mathcal{O}), \ \forall v \in K^\varepsilon,$$

and we consider the problem

$$(17.2.6) \qquad \inf \int_\mathcal{O} \phi(x, x/\varepsilon, v, Dv) \, dx, \ v \in K^\varepsilon.$$

REMARK 17.6. If we take

$$\phi(x, y, v, Dv) = \frac{1}{2} a_{ij}(y) \frac{\partial v}{\partial x_j} \frac{\partial v}{\partial x_i} + \frac{1}{2} a_0(y) v^2 - f(x) v(x), \ a_{ij} = a_{ji},$$

$$K^\varepsilon = K = \text{ closed convex subset of } V,$$

$$H_0^1(\mathcal{O}) \subseteq V \subseteq H^1(\mathcal{O}),$$

problem (17.2.6) coincides with (17.1.2) (cf. (17.1.3)).

---

[62]This hypothesis can be weakened.

REMARK 17.7. The remarks to follow extend to functionals of the form

$$(17.2.7) \qquad \inf \int_{\mathcal{O}} \phi\left(x, \frac{x}{\varepsilon}, \frac{x}{\varepsilon^2}, v, Dv\right) \, dx,$$

where $\phi(x, y, z, v, Dv)$ is Y periodic in y and Z periodic in z.

We suppose that problem (17.2.6) admits a solution u_ε and we want to define a homogenized problem associated to it.

We consider test functions u_ε of the form

$$(17.2.8) \qquad v_\varepsilon = v(x) + \varepsilon v_1(x, x/\varepsilon)$$

where v is such that

$$(17.2.9) \qquad g_1 \leq v \leq g_2 \text{ on } \mathcal{O}, \qquad v = 0 \text{ on } \Gamma_0,$$

and where $v_1(x, y)$ satisfies

$$(17.2.10) \qquad v_1(x, y) \text{ is } Y\text{-periodic in } y, \ v_1 \text{ is bounded together with } D_x v_1 \text{ and}$$
$$|Dv(x) + D_y v_1(x, y)| \leq m(x, y), \ x \in \mathcal{O}, \ y \in \mathbb{R}^n.$$

Then v_ε does not belong to K^ε but "belongs to K^ε up to ε." Indeed

$$g_1 - \varepsilon M \leq v_\varepsilon \leq g_2 + \varepsilon M, \text{ if } M \text{ is an upper bound for } |v_1|,$$
$$|v_\varepsilon| \leq \varepsilon M \text{ on } \Gamma_0,$$
$$|Dv_\varepsilon(x)| \leq m(x, x/\varepsilon) + M_1 \varepsilon.$$

It is therefore reasonable to think that the problem (17.2.6) has the same limit, when $\varepsilon \to 0$, as

$$(17.2.11) \qquad \inf \int_{\mathcal{O}} \phi\left(x, \frac{x}{\varepsilon}, v, Dv(x) + D_y v_1\left(x, \frac{x}{\varepsilon}\right)\right) \, dx,$$

where v and v_1 satisfy (17.2.9), (17.2.10).

We now define

$$(17.2.12) \qquad \psi(x, y) = v_1(x, y) + y Dv(x).$$

We have $D_y \psi = D_y v_1 + Dv$ so that (17.2.11) can be reformulated:

$$(17.2.13) \qquad \inf \int_{\mathcal{O}} \phi\left(x, \frac{x}{\varepsilon}, v, D_y \psi\right) \, dx,$$

where v satisfies (17.2.9) and where ψ satisfies

$$(17.2.14) \qquad \begin{array}{l} \psi - y \, Dv \text{ is } Y\text{-periodic,} \\ \psi, D_x \psi, D_y \psi \text{ are bounded,} \\ |D_y \psi(x, y)| \leq m(x, y). \end{array}$$

Using the fact that $F\left(x, \frac{x}{\varepsilon}\right) \to \frac{1}{|Y|} \int_Y F(x, y) \, dy$ if **for instance** $F \in C^0(\mathcal{O} \times \mathbb{R}^n)$, F being Y-periodic, it is again reasonable to think that problem (17.2.13) has the same limit as

$$(17.2.15) \qquad \inf \int_{\mathcal{O}} dx \int_Y \phi(x, y, v(x), D_y \psi(x, y)) \, dx,$$

v and ψ being subject to (17.2.9) and (17.2.14).

This leads, in conclusion, to the following general rule.

17.2.1. *First step.* Given $\lambda \in \mathbb{R}$ and $\mu \in \mathbb{R}^n$, we consider

$$(17.2.16) \qquad \inf_\theta \int_Y \phi(x, y, \lambda, D\theta(y)) \, dy = \psi(x, \lambda, \mu)$$

where in (17.2.16), $\theta(y)$ is subject to

$$(17.2.17) \qquad \theta - \mu \cdot y \text{ is } Y\text{-periodic,}$$
$$|D\theta(y)| lem(x, y);$$

ψ **can take the value** $+\infty$ if (17.2.17) defines the empty set.

17.2.2. *Second step.* We consider

$$(17.2.18) \qquad \inf \int_{\mathcal{O}} \psi(x, v, Dv) \, dx,$$

where v is subject to

$$(17.2.19) \qquad g_1 \leq v \leq g_2 \text{ on } \mathcal{O}, \qquad v = 0 \text{ on } \Gamma_0.$$

The problem (17.2.18), (17.2.19) is the homogenized problem associated to (17.2.7).

REMARK 17.8. In all the particular cases where the above conjecture is proved, it gives the same results as those obtained before.

REMARK 17.9. Let us suppose that

$$(17.2.20) \qquad K^\varepsilon = K = \{v \mid v \in H_0^1(\mathcal{O}), \ v \leq 0 \text{ on } \mathcal{O}\}$$

and that ϕ is given as in Remark 17.6 (i.e., $g_1 = -\infty$, $g_2 = 0$, $\Gamma_0 = \Gamma$, $m = +\infty$). Then the problem is

$$(17.2.21) \quad A^\varepsilon u_\varepsilon - f \leq 0, \qquad u_\varepsilon \leq 0, \qquad (A^\varepsilon u_\varepsilon - f)u_\varepsilon = 0 \text{ in } \mathcal{O}, \qquad u_\varepsilon = 0 \text{ on } \Gamma,$$

and the homogenized problem according to the formal rule obtained above is:

$$(17.2.22) \quad \mathcal{A}u - f \leq 0, \qquad u \leq 0, \qquad (\mathcal{A}u - f)u = 0 \text{ in } \mathcal{O}, \qquad u = 0 \text{ on } \Gamma.$$

We are going to **prove** this fact, even with A^ε **not** symmetric.

17.3. Unilateral variational inequalities. We consider problem (17.2.21) with A^ε given as in Section 1; (17.2.21) is equivalent to (17.1.2) when K is given by (17.2.20).

THEOREM 17.10. *We assume that (1.1.1), (1.1.2) hold true. Let $\mathcal{A}$ be the homogenized operator associated to A^ε and let u_ε (respectively, u) be the solution of the unilateral V.I. (17.2.21) (respectively, (17.2.22)). Then, as $\varepsilon \to 0$,*

$$(17.3.1) \qquad u_\varepsilon \to u \text{ in } H_0^1(\mathcal{O}) \text{ weakly.}$$

Several proofs can be given of this result (cf. the Bibliography given in the last section). We give here a proof based on the **penalty method** because of the independent interest of the following **error estimate**.

THEOREM 17.11. *For $\eta > 0$ let $u_{\varepsilon\eta}$ be the solution of*[63]

$$(17.3.2) \qquad A^\varepsilon u_{\varepsilon\eta} + \frac{1}{\eta} u_{\varepsilon\eta}^+ = f \text{ in } \mathcal{O},$$
$$u_{\varepsilon\eta} = 0 \text{ on } \Gamma;$$

---

[63]$\phi^+ = \sup(\phi, 0)$. Equation (17.3.2) admits a unique solution. The term $\frac{1}{\eta} u_{\varepsilon\eta}^+$ is called a **penalty term**.

then

(17.3.3) $$\|u_{\varepsilon\eta} - u_\varepsilon\| \leq C\sqrt{\eta}$$

where the constant C does not depend on ε.

PROOF. We take the scalar product of (17.3.2) with $u_{\varepsilon\eta}^+$. We obtain

(17.3.4) $$a^\varepsilon(u_{\varepsilon\eta}^+) + \frac{1}{\eta}|u_{\varepsilon\eta}^+|^2 = (f, u_{\varepsilon\eta}^+).$$

Hence

(17.3.5) $$|u_{\varepsilon\eta}^+| \leq |f|\eta$$

and $\|u_{\varepsilon\eta}^+\|^2 \leq C|u_{\varepsilon\eta}^+|$, hence

(17.3.6) $$\|u_{\varepsilon\eta}^+\| \leq C\sqrt{\eta}.$$

We write

$$u_\varepsilon - u_{\varepsilon\eta} = u_\varepsilon + u_{\varepsilon\eta}^- u_{\varepsilon\eta}^+$$

and by virtue of (17.3.6), it remains only to prove that

(17.3.7) $$\|u_\varepsilon + u_{\varepsilon\eta}^-\| \leq C\sqrt{\eta}.$$

We choose $v = -u_{\varepsilon\eta}^-$ in (17.1.2) (v is ≤ 0) and we multiply (17.3.2) by $u_\varepsilon + u_{\varepsilon\eta}^-$. We obtain after adding up

$$a^\varepsilon(u_{\varepsilon\eta} - u_\varepsilon, u_\varepsilon + u_{\varepsilon\eta}^-) + \frac{1}{\eta}(u_{\varepsilon\eta}^+, u_\varepsilon + u_{\varepsilon\eta}^-) \geq 0,$$

i.e.,

(17.3.8) $$a^\varepsilon(u_\varepsilon + u_{\varepsilon\eta}^-) + \frac{1}{\varepsilon}(u_{\varepsilon\eta}^+, -u_\varepsilon) \leq a^\varepsilon(u_{\varepsilon\eta}^+, u_\varepsilon + u_{\varepsilon\eta}^-).$$

Since $u_\varepsilon \leq 0$, we have $\frac{1}{\varepsilon}(u_{\varepsilon\eta}^+, -u_\varepsilon) \geq 0$, so that (17.3.8) implies

$$\|u_\varepsilon + u_{\varepsilon\eta}^-\|^2 \leq C\|u_{\varepsilon\eta}^+\|\|u_\varepsilon + u_{\varepsilon\eta}^-\| \leq \sqrt{\eta}\|u_\varepsilon + u_{\varepsilon\eta}^-\|.$$

Hence (17.3.7) follows.   ∎

REMARK 17.12. We have the same estimate (17.3.3) for a family of operators A^ε with coefficients $a_{ij}^\varepsilon \in L^\infty(\mathcal{O})$ and $a_0^\varepsilon \in L^\infty(\mathcal{O})$ such that:

(17.3.9) $$\alpha\xi_i\xi_i \leq a_{ij}^\varepsilon(x)\xi_i\xi_j \leq \beta\xi_i\xi_i, \qquad 0 < \alpha \leq \beta$$
$$0 < \alpha_0 \leq a_0(x) \leq \beta_0.$$

The periodicity of $a_{ij}(y)$, $a_0(y)$ does not play any role.

We can now proceed to the

PROOF OF THEOREM 17.10. Let u_ε be the solution of the penalized problem associated to (17.2.22):

(17.3.10) $$u_\eta + \frac{1}{\eta}u_\eta^+ = f \text{ in } \mathcal{O},$$
$$u_\eta = 0 \text{ on } \Gamma.$$

We have of course

(17.3.11) $$\|u - u_\eta\| \leq C\sqrt{\eta}.$$

Therefore, in order to prove in Theorem, it suffices to prove that **for** $\eta >$ **fixed**, one has

$$(17.3.12) \qquad\qquad u_{\varepsilon\eta} \to u_\eta \text{ in } H_0^1(\mathcal{O}) \text{ weakly, as } \varepsilon \to 0.$$

But (17.3.12) is a simple example of **compact perturbations** as in Section 16.2, and (17.3.12) follows.   ∎

REMARK 17.13. Let us consider the convex set

$$(17.3.13) \qquad\qquad K = \{v \mid v \in H_0^1(\mathcal{O}), v \le g\},$$

the only hypothesis made on g being that K is not empty. Then one can prove (cf. F. Murat [**83**]) that if u_ε is the solution of the V.I. corresponding to K given by (17.3.13) one has (17.3.1) where u is the solution of

$$(17.3.14) \qquad\qquad u \in K, \ \mathcal{A}(u, v - u) \ge f(v, u), \ \forall v \in K.$$

[If g is smooth, this result is immediate by translation.]

REMARK 17.14. The homogenization of quasi-variational inequalities has been considered in Bensoussan-Lions [**9**], and Biroli [**17**].

18. Comments and problems

The problem of homogenization is actually a particular case of the general study of partial differential operators with coefficients in a bounded set of L^∞, and such that they are (in the framework of the present chapter) **uniformly elliptic**.

In the case of **symmetric operators**, this approach is the one followed by E. de Giorgi, Spagnolo and others — and even in a much more general setting of calculus of variations — under the terminology of G-**convergence**. We refer to E. de Giorgi and S. Spagnolo [**30**], S. Spagnolo [**110**], P. Marcellini and C. Sbordone [**76**], [**77**], C. Sbordone [**101**], and to the Bibliography therein.

The special structure studied here — with coefficients which are periodic or connected to periodic or to almost periodic functions — allows us to obtain precise results on the structure of coefficients of the homogenized problem. **Some** quantitative results have been obtained in the general case by Murat [**84**] and Tartar [**117**]. The method of **asymptotic expansions** presented in Section 2 is of the type of the **"multi-scale" methods** — well known in the theory of perturbations. The special way of using this general framework in the present context has been introduced, and used, by the authors in a series of notes and reports; these papers of the authors are quoted in the Bibliography. Its use in the present context was anticipated by J. B. Keller as we mentioned in the Introduction.

Similar problems have been studied by I. Babuska [**8**].

The idea of the proof of Section 3.2 is due to L. Tartar [**116**], together with the method of Section 3.4.

The L^p estimates given in Section 4 are due to N. G. Meyers [**81**] for second order operators. The extension to higher order operators given here, although very simply, seems to be new.

Particular cases of **correctors** (as presented in Section 5) have been given in the notes of the authors.

The results of Section 7 are given here for the first time.

The results of Section 8 were announced in one of the notes of the authors [**15**]. Although the hypothesis made here on the regularity of $a_{ij}(x, y, z)$ improve those of this note, the question of the **most general hypothesis** under which the formulas of Section 8 are valid is unclear.

Formulas of Sections 9, 10, 11 are given here for the first time. Those of Sections 9, 10 were presented in lectures of the authors and those of Section 11 are due to Murat and Tartar. The results of Section 11.4 about compensated compactness follow Murat-Tartar; cf. Murat [84]. Particular cases of Section 9 may be useful in the theory of elasticity.

The theory of Sections 1 to 11 can be extended in several other directions:

(1) One can consider elliptic operators which have singularities or degeneracies (for instance near the boundary). First results in this direction are given in Marcellini-Sbordone [77], [78];

(2) One can consider elliptic operators with coefficients which have a "**stratified periodic structure**." More precisely, let us consider the operator

$$(18.0.15) \qquad A^\varepsilon = -\frac{\partial}{\partial x_i}\left(a_{ij}\left(\frac{\rho(x)}{\varepsilon}\right)\frac{\partial}{\partial x_j}\right)$$

where

$$(18.0.16) \qquad \rho \in C^2(\overline{\mathcal{O}}, \mathbb{R}^m), \ (m \le n),$$

$$(18.0.17) \qquad a_{ij} \in L^\infty(\mathbb{R}^m), \qquad a_{ij} \text{ is } Y\text{-periodic in } \mathbb{R}^m,$$

$$a_{ij}(y)\eta_i\eta_j \ge \alpha\eta_i\eta_j, \ \alpha > 0, \text{ a.e. in } y.$$

Let us assume that

$$(18.0.18) \qquad \sum_{i=1}^n \left(\sum_{k=1}^m \frac{\partial\rho_k(x)}{\partial x_i}\eta_k\right)^2 \ge \beta\sum_{k=1}^m \eta_k^2, \qquad x \in \mathcal{O}.$$

If $\rho(x) = x$, we have the situation of Section 1. If $\rho(x) = |x|$, $(m = 1)$, we have a periodic structure in spheres, and (18.0.18) is value if $0 \notin \overline{\mathcal{O}}$.

The homogenized operator $\mathcal{A}$ of A^ε is constructed as follows. We set

$$(18.0.19) \qquad A_1 = -\frac{\partial}{\partial y_k}\left(a_{ij}(y)\frac{\partial\rho_k}{\partial x_i}(x)\frac{\partial\rho_\ell}{\partial x_j}(x)\frac{\partial}{\partial y_\ell}\right)$$

and we define $\chi^j = \chi^j(x,y)$ as the Y-periodic solution (defined up to an additive constant) of

$$(18.0.20) \qquad A_1\chi^j = -\frac{\partial}{\partial y_k}a_{ij}(y)\frac{\partial\rho_k}{\partial x_i}(x).$$

Then

$$(18.0.21) \qquad \mathcal{A} = -\frac{\partial}{\partial x_i}\left(q_{ij}(x)\frac{\partial}{\partial x_j}\right)$$

where

$$(18.0.22) \qquad q_{ij}(x) = \frac{1}{|Y|}\int_Y \left[a_{ij}(y) - a_{ik}(y)\frac{\partial\chi^j}{\partial y_\ell}(x,y)\frac{\partial\rho_\ell}{\partial x_k}(x)\right]dy;$$

(3) Another extension of the results of Sections 1 to 11 is relative to **quasi-elliptic operators**. For instance let us consider

$$(18.0.23) \qquad A^\varepsilon = -\frac{\partial}{\partial x_1}a\left(\frac{x}{\varepsilon}\right)\frac{\partial}{\partial x_1} + \frac{\partial^2}{\partial x_2^2}b\left(\frac{x}{\varepsilon}\right)\frac{\partial^2}{\partial x_2^2},$$

where $a(y)$, $b(y)$ are L^∞ functions, Y-periodic, $a(y) \geq \alpha > 0$, $b(y) \geq \alpha > 0$. Boundary conditions associated to A^ε are defined by a variational formulation. We consider

$$(18.0.24) \qquad H^{1,2}(\mathcal{O}) = \left\{ v \,\middle|\, v, \frac{\partial v}{\partial x_1}, \frac{\partial v}{\partial x_2}, \frac{\partial^2 v}{\partial x_2^2} \in L^2(\mathcal{O}) \right\}$$

provided with its natural Hilbert space structure. If $H_0^{1,2}(\mathcal{O})$ denotes the closure of $C_0^\infty(\mathcal{O})$ in $H^{1,2}(\mathcal{O})$, we take

$$(18.0.25) \qquad H_0^{1,2}(\mathcal{O}) \subseteq V \subseteq H^{1,2}(\mathcal{O})$$

and we set

$$(18.0.26) \qquad a^\varepsilon(u,v) = \int_{\mathcal{O}} \left(a\left(\frac{x}{\varepsilon}\right) \frac{\partial u}{\partial x_1} \frac{\partial v}{\partial x_1} + b\left(\frac{x}{\varepsilon}\right) \frac{\partial^2 u}{\partial x_2^2} \frac{\partial^2 v}{\partial x_2^2} \right) dx.$$

Then (we add a zero order term if, for instance, $V = H^{1,2}(\mathcal{O})$) u_ε is defined by

$$(18.0.27) \qquad a^\varepsilon(u_\varepsilon, v) = (f, v), \ \forall v \in V, \ u_\varepsilon \in V.$$

When $\varepsilon \to 0$, one has $u_\varepsilon \to u$ in V weakly, where u is the solution of the **homogenized equation**

$$(18.0.28) \qquad \mathcal{A}(u,v) = (f,v), \ \forall v \in V, \ u \in V;$$

in (18.0.28), one has

$$(18.0.29) \qquad \mathcal{A}(u,v) = q_1 \int_{\mathcal{O}} \frac{\partial u}{\partial x_1} \frac{\partial v}{\partial x_1} \, dx + q_2 \int_{\mathcal{O}} \frac{\partial^2 u}{\partial x_2^2} \frac{\partial^2 v}{\partial x_2^2} \, dx,$$

with

$$(18.0.30) \qquad q_1 = \frac{1}{\mathcal{M}_{y_1}\left(1/\mathcal{M}_{y_2}(a)\right)}, \qquad q_2 = \frac{1}{\mathcal{M}\left(\frac{1}{b}\right)},$$

where (18.0.30), $\mathcal{M}_{y_1}\phi = \frac{1}{|Y_1|}\int_{Y_1} \phi(y_1, y_2) \, dy_1$, $\mathcal{M}\phi = \frac{1}{|Y|}\iint \phi \, dy_1 \, dy_2$ and similar notation for $\mathcal{M}_{y_2}$.

(4) One can also consider **non-homogeneous** boundary value problems. For instance, with the notations of Section 1, we can consider the problem

$$(18.0.31) \qquad A^\varepsilon \phi_\varepsilon = 0 \text{ in } \mathcal{O},$$
$$\phi_\varepsilon = g \text{ on } \Gamma,$$

where $g \in H^{1/2}(\Gamma)$. Then

$$(18.0.32) \qquad \phi_\varepsilon \to \phi \text{ in } H^1(\mathcal{O}) \text{ weakly,}$$

where ϕ is the solution of

$$(18.0.33) \qquad \mathcal{A}\phi = 0 \text{ in } \mathcal{O}, \qquad \phi = g \text{ on } \Gamma.$$

Indeed, if we introduce $G \in H^1(\mathcal{O})$ with $G = g$ on Γ, and if we set $u_\varepsilon = \phi_\varepsilon - G$, then

$$(18.0.34) \qquad A^\varepsilon u_\varepsilon = -A^\varepsilon G, \qquad u_\varepsilon \in H_0^1(\mathcal{O}).$$

But $A^\varepsilon G$ remains in a bounded set of $H^{-1}(\mathcal{O})$, so that u_ε remains in a bounded set of $H_0^1(\mathcal{O})$, so that $\|\phi_\varepsilon\|_{H^1(\mathcal{O})} \leq C$. Then we can extract ϕ_ε such that (18.0.32) holds true, and we have (18.0.32).

In connection with (18.0.34), one can more generally consider the equation

(18.0.35) $$A^\varepsilon u_\varepsilon = \varepsilon^{-1} f_\varepsilon + g_\varepsilon$$

where

(18.0.36) $$f_\varepsilon(x) = f\left(x, \frac{x}{\varepsilon}\right), \qquad g_\varepsilon(x) = g\left(x, \frac{x}{\varepsilon}\right),$$

and

(18.0.37) $$f, g \in C^0(\overline{\mathcal{O}}; L^\infty(\mathbb{R}_y^n)), \qquad f, g \text{ being } Y\text{-periodic.}$$

This method of asymptotic expansion (which can be justified) gives the following. We look for u_ε in the for

(18.0.38) $$u = \varepsilon^{-1} u_{-1} + u_0 + \varepsilon u_1 + \cdots, \qquad u_j = u_j(x, y).$$

Identifying the powers in ε, we obtain

(18.0.39) $$A_1 u_{-1} = 0,$$
(18.0.40) $$A_1 u_0 + A_2 u_{-1} = 0,$$
(18.0.41) $$A_1 u_1 + A_2 u_0 + A_3 u - 1 = f(x, y),$$
(18.0.42) $$A_1 u_2 + A_2 u_1 + A_3 u_0 = g(x, y),$$

$$\vdots$$

It follows from (18.0.39) that $u_{-1} = u_{-1}(x)$. Hence (18.0.40) gives

(18.0.43) $$u_0 = -\chi^j(y) \frac{\partial u_{-1}}{\partial x_j} + \tilde{u}_0(x),$$

and (18.0.41) can be solved in u_1, iff

$$\int_Y (A_2 u_0 + A_3 u_{-1})\, dy = \int_Y f(x, y)\, dy$$

hence

(18.0.44) $$\mathcal{A} u_{-1} = \frac{1}{|Y|} \int_Y f(x, y)\, dy.$$

One can then proceed with (18.0.42). If we compute explicitly $A^\varepsilon G$ in (18.0.34), we can find (18.0.35) with

(18.0.45) $$f_\varepsilon = -\frac{\partial a_{ij}}{\partial y_i}(y) \frac{\partial G}{\partial x_j}(x), \qquad y = x/\varepsilon,$$

$$g_\varepsilon = -a_{ij}(y) \frac{\partial^2 G}{\partial x_i \partial x_j}, \qquad y = x/\varepsilon.$$

Therefore $\int_Y f(x, y)\, dy = 0$, so that $u_{-1} = 0$.

The behavior of eigenvalues in the homogenization procedure (for self-adjoint operators) is studied in Kesavan [57] and Planchard (Computer Methods in Applied Mechanics and Engineering, 30 (1982), 75-93).

The homogenization of problems with oblique derivatives in the boundary conditions is not entirely solved and not presented here. Some aspects of that problem are developed in Chapter 3.

One can also study homogenization problems for boundary value problems in media **with "small" holes** having a **periodical structure**. We refer to Cioranescu and Saint Jean Paulin (Journal of Mathematical Analysis and Applications, 71 (1979), 590-607). Asymptotic expansions giving higher order estimates have been introduced in[64], and are currently used in several directions.

The results of Section 12, and others, are given in the notes of the authors. Results not proved here and complements will be given elsewhere. Problems similar to those of Section 12 but with a non-symmetric operator A^ε do not seem to be solved.

Results of Section 13 are taken from Bensoussan, Lions, Papanicolaou [**12**]. In Section 14 we did not make any attempt to study the **boundary layers** arising in these questions.

The results of Section 15 were announced in one of the notes of the authors. They indicate that it could be interesting to consider the general homogenization problem for pseudo-differential operators, a question not considered here.

In Sections 16 and 17 we give some indications on the homogenization of non-linear problems.

The general conjecture presented in Section 17.2 is somewhat similar to the "averaging principle" of Whitham ("Linear and Nonlinear Waves", Wiley, 1974). A proof of the conjecture of Section 17.2 has been announced by L. Carbone [**23**] for the case when $g_1 = -\infty$, $g_2 = +\infty$ (i.e. conditions (17.2.1) are absent) and when in (17.2.3) $m(x,y) = 1$.

There is a "non symmetric" analogue of the conjecture, which can be explained as follows. Let us consider the variational problem[65]

$$(18.0.46) \qquad \int_{\mathcal{O}} a_{ij}(x/\varepsilon) \frac{\partial u_\varepsilon}{\partial x_j} \frac{\partial v}{\partial x_i} \, dx = \int_{\mathcal{O}} fv \, dx \ \forall v \in V.$$

We use the "ansatz" $u_\varepsilon = u_0 + \varepsilon u_1 + \cdots$, $u_j = u_j(x,y)$ and we take $v = v(x,y)$ in (18.0.46). We obtain

$$(18.0.47) \quad \int_{\mathcal{O}} a_{ij}(y) \left(\varepsilon^{-1} \frac{\partial u_0}{\partial y_j} + \varepsilon^0 \left(\frac{\partial u_1}{\partial y_j} + \frac{\partial u_0}{\partial x_j} \right) + \varepsilon \left(\frac{\partial u_2}{\partial y_j} + \frac{\partial u_1}{\partial x_j} \right) + \cdots \right)$$
$$\times \left(\varepsilon^{-1} \frac{\partial v}{\partial y_i} + \frac{\partial v}{\partial x_i} \right) \, dx = \int_{\mathcal{O}} fv \, dx.$$

We identify the various powers of ε. We obtain first

$$\int_{\mathcal{O}} a_{ij}(y) \frac{\partial u_0}{\partial y_j} \frac{\partial v}{\partial y_i} \, dx = 0$$

hence it follows (if we assume that $u_0 \in V$ so that we can take $v = u_0$): $u_0 = u(x)$. The coefficient of ε^{-1} in (18.0.47) gives now:

$$(18.0.48) \qquad \int_{\mathcal{O}} a_{ij}(y) \left(\frac{\partial u_1}{\partial y_j} + \frac{\partial u_0}{\partial x_j} \right) \frac{\partial v}{\partial y_j} \, dx = 0.$$

But assuming that $v = \phi(x)\phi(y)$, (18.0.48) is "approximately" identical to

$$\int_{\mathcal{O}} \phi(x) \, dx \int_Y a_{ij}(y) \left(\frac{\partial u_1}{\partial y_j} + \frac{\partial v}{\partial x_j} \right) \frac{\partial \phi}{\partial y_i}(y) \, dy = 0$$

---

[64] J. L. Lions, Introduction remarks to asymptotic analysis of periodic structures, Symp. on Trends in App. of Math. to Mech., Kozubnik (Poland), Sept. 77.

[65] We emphasize that a_{ij} is not necessarily equal to a_{ji}.

so that

$$\int_Y a_{ij}(y) \left(\frac{\partial u_1}{\partial y_j} + \frac{\partial v}{\partial x_j} \right) \frac{\partial \psi}{\partial y_i}(y) \, dy = 0 \ \forall \psi, \ \psi \text{ periodic.}$$

This gives (with the notations of the text)

$$(18.0.49) \qquad\qquad u_1 = -\chi^j(y) \frac{\partial u}{\partial x_j}(x) + \tilde{u}_1(x).$$

One takes next the coefficient of ε^0 in (18.0.47) and one obtains

$$(18.0.50) \qquad \int_{\mathcal{O}} a_{ij} \left(\frac{\partial u_2}{\partial y_j} + \frac{\partial u_1}{\partial x_j} \right) \frac{\partial v}{\partial y_i} + \left(\frac{\partial u_1}{\partial y_j} + \frac{\partial u_1}{\partial x_j} \right) \frac{\partial v}{\partial x_i} = \int_{\mathcal{O}} fv \, dx.$$

The compatibility condition for solving (18.0.50) in u_2 leads to the usual homogenized equation. The interest of this **formal** procedure (but we conjecture if **can be justified**) is that it can be applied to **variational inequalities for non symmetric operators**.

Another type of homogenization has been considered (for reinforced plates) by energy methods by Artola and Duvaut [5].

We shall return to the problem of **boundary layer terms** in separate papers. In case $\mathcal{O} = \mathbb{R}^n$ the asymptotic expansion constructed in the text gives an approximation of arbitrary high order in ε (in Sobolev's norms).

The homogenization formula given in the text for **periodic coefficients** has been extended by S. M. Kozlov [60] to **almost periodic coefficients** (using the method of asymptotic expansion, the average over the torus being replaced by the average for almost periodic functions and using results of E. Muhamadiev [82] extending the theorems of Favard on the existence of almost periodic solutions of P.D.E.).

A survey of homogenization problems (both for the stationary and for the evolution case) is presented by O. A. Oleinik et. al. Russian Mathematical Surveys, 34 (1979), 65-133.

The results of Section 17.3 were announced in the first of the notes of the authors. Other results along these lines are given in Murat [83], H. Attouch and Y. Konishi [6], L. Boccardo and I. Capuzzo Dolcetta [18], L. Boccardo and P. Marcellini [19]. With a convex K of the type $|Dv(x)| \leq 1$ in $\mathcal{O}$, the problem has been solved only in dimension 1; cf. Carbone [23].

Homogenization formulas can be found — at least for second order operators, but also, in principle, for other cases — by using "**Bloch waves**"; since this theory is naturally connected with wave propagation, remarks along these lines will only be given in Chapter 4.

Numerical computations and numerical problems connected with homogenization of elliptic problems are not studied here. We refer to Bourgat [21], Kesavan [57] and to the reports of I. Babuska.

CHAPTER 2

Evolution Operators

Orientation

The reader mostly interested in Probabilistic Methods (Chapter 3) should read
Section 1 and Sections 2.1 to 2.5, and then proceed with Chapter 3.

The reader mostly interested in High Frequency wave propagation should read
Sections 3.1 to 3.5 and Section 3.8, and the proceed with Chapter 4.

Sections 1 and 2 treat the case of parabolic operators with coefficients rapidly
varying in the space and in the time variable, with different scales. The method of
asymptotic expansion shows that there are three cases to be considered. Proofs of
convergence are given in Section 2; Sections 2.6 and 2.7 give the proof under the
weakest possible hypothesis on the coefficients and are somewhat complicated.

Section 3 studies the operators of hyperbolic type, or of Petrowsky or of
Schrödinger type. Some non-linear problems are considered in Sections 2.12 and
3.7. Section 3.9 gives an example (somewhat connected with Chapter 1, Section 15)
of a situation where local operators of evolution admit for homogenized operator a
non local evolution operator.

Many other situations can be studied using ideas and techniques of Chapters
1 and 3. Some of them are briefly indicated in Section 4 (comments). Another
situation will be studied by these techniques in Chapter 3, Section 5 (since the
motivation for such a study appears in the Probabilistic context).

1. Parabolic operators: Asymptotic expansions

1.1. Notations and orientation. Let $\mathcal{O}$ be a bounded (to fix ideas!) open
set of $\mathbb{R}^n$, with boundary Γ (which will be assumed smooth or **not**, depending on
the situation).

Let $T > 0$ be a given finite number. We set

$$(1.1.1) \qquad \mathcal{G} = \mathcal{O} \times]0, T[, \qquad \Sigma = \Gamma \times]0, T[.$$

We will consider first, equations of the following type:

$$(1.1.2) \qquad \frac{\partial u_\varepsilon}{\partial t} = \frac{\partial}{\partial x_i}\left(a_{ij}^\varepsilon(x,t)\frac{\partial u_\varepsilon}{\partial x_j}\right) = f \text{ in } \mathcal{G},$$

with **initial condition**

$$(1.1.3) \qquad u_\varepsilon(x,0) = u^0(x), \qquad x \in \mathcal{O},$$

and with suitable boundary conditions, defined in Section 1.2 below. On the functions a_{ij} we shall always assume that

$$(1.1.4) \qquad a_{ij}^\varepsilon \text{ remains in a bounded set of } L^\infty(\mathcal{G}),$$

$$a_{ij}^\varepsilon(x,y)\xi_i\xi_j \geq \alpha\xi_i\xi_i, \text{ a.e. in } \mathcal{G}, \alpha > 0, \xi_i \in \mathbb{R}.$$

In order to obtain constructive formulas, we shall assume much more on the structure of functions a_{ij}^ε; namely,

$$(1.1.5) \qquad\qquad a_{ij}^\varepsilon(x,t) = a_{ij}(x/\varepsilon, t/\varepsilon^k),$$

where $k > 0$ is for the moment unspecified, and where:

$$(1.1.6) \qquad
\begin{aligned}
& a_{ij}(y,\tau) \text{ is } Y\text{-}\tau_0 \text{ periodic,}[1] \\
& a_{ij}(y,\tau) \in L^\infty(\mathbb{R}_y^n \times \mathbb{R}_\tau) \\
& a_{ij}(y,\tau)\xi_i\xi_j \geq \alpha\xi_i\xi, \text{ a.e. in } y \text{ and } \tau.
\end{aligned}$$

We want to study the behavior of u_ε as $\varepsilon \to 0$.

We begin by giving a precise definition of u_ε (Section 1.2) and we study next the asymptotic expansion of u_ε. As we shall see the formulas obtained depend on k (which appears in (1.1.5)).

REMARK 1.1. We shall explain in the comments on Section 4 the results one obtains when considering functions a_{ij}^ε of the form

$$(1.1.7) \qquad\qquad a_{ij}^\varepsilon(x,t) = a_{ij}(x, x/\varepsilon, t, t/\varepsilon^k)$$

and more generally

$$(1.1.8) \qquad\qquad a_{ij}^\varepsilon(x,t) = a_{ij}\left(x, \frac{x}{\varepsilon}, \frac{x}{\varepsilon^2}, \ldots, t, \frac{t}{\varepsilon^k}, \frac{t}{\varepsilon^{2k}}, \ldots\right).$$

1.2. Variational formulation. We need some notation. We shall use, as in Chapter 1, spaces V such that

$$(1.2.1) \qquad\qquad H_0^1(\mathcal{O}) \subseteq V \subseteq H^1(\mathcal{O}).$$

Given a Hilbert space $\mathcal{H}$ (or a Banach space!) we denote by $L^p(0,T;\mathcal{H})$ the space of (classes of) functions $t \to v(t)$ from $(0,T) \to \mathcal{H}$ which are measurable and which satisfy

$$(1.2.2) \qquad \|u\|_{L^p(0,T;\mathcal{H})} =
\begin{cases}
\left(\int_0^T \|u(t)\|^p\, dt\right)^{1/p} < \infty, & \text{if } 1 \leq p < \infty, \\
\operatorname{ess\,sup} \|u(t)\|_{\mathcal{H}} < \infty, & \text{if } p = \infty.
\end{cases}$$

Provided with the norm given in (1.2.2), $L^p(0,T;\mathcal{H})$ is a Banach space (and it is a Hilbert space if $p = 2$ and if $\mathcal{H}$ is a Hilbert space).

We shall set

$$(1.2.3) \qquad\qquad H = L^2(\mathcal{O}), \qquad V \subset H \subset V'$$

and we observe that

$$L^2(0,T;H) = L^2(\mathcal{G}).$$

For $u, v \in H^1(\mathcal{O})$, we define:

$$(1.2.4) \qquad\qquad a^\varepsilon(u,v) = \int_{\mathcal{O}} a_{ij}^\varepsilon(x,t)\frac{\partial u}{\partial x_j}\frac{\partial v}{\partial x_i}\, dx.$$

We have to be careful since $a^\varepsilon(u,v)$ depends on t and since, even if the functions a_{ij}'s are smooth (for every ε) the t-derivatives of the a_{ij}'s are **unbounded** — due

---

[1]That is $a_{ij}(y,\tau)$ is Y periodic as a function of y and

$$a_{ij}(y, \tau + \tau_0) = a_{ij}(y,\tau), \text{ a.e. in } y, \tau.$$

to the structure of a_{ij} given in (1.1.5) that, in what follows, the **a priori** estimates should never use $\frac{\partial}{\partial t} a^\varepsilon(u, v)$.

For $u \in V$ we define $A^\varepsilon u \in V'$ by

$$(1.2.5) \qquad (A^\varepsilon u, v) = a^\varepsilon(u, v), \ \forall v \in V.$$

In fact A^ε **depends on** t:

$$(1.2.6) \qquad A^\varepsilon = A^\varepsilon(t).$$

The "abstract" formulation of the problem we want to study is now: let f and u^0 be given, satisfying

$$(1.2.7) \qquad f \in L^2(0, T; V'),$$

$$(1.2.8) \qquad u^0 \in H.$$

Then u_ε is the solution of

$$(1.2.9) \qquad u_\varepsilon \in L^2(0, T; V),$$

$$(1.2.10) \qquad \frac{\partial u_\varepsilon}{\partial t} + A^\varepsilon u_\varepsilon = f \text{ in } (0, T),$$

$$(1.2.11) \qquad u_\varepsilon(0) = u^0.$$

REMARK 1.2. If $v \in L^2(0, T; V)$ then $A^\varepsilon v \in L^2(0, T; V')$ and since a_{ij} remains in a bounded set of $L^\infty(\mathcal{G})$,

$$(1.2.12) \qquad \|A^\varepsilon v\|_{L^2(0,T;V')} \leq c\|v\|_{L^2(0,T;V)}$$

where here and in what follows, the c's denote **various constants which do not depend on** ε.

Therefore (1.2.10) gives:

$$(1.2.13) \qquad \frac{\partial u_\varepsilon}{\partial t} = f - A^\varepsilon u_\varepsilon \in L^2(0, T; V')$$

and

$$(1.2.14) \qquad \left\|\frac{\partial u_\varepsilon}{\partial t}\right\|_{L^2(0,T;V')} \leq c\left(\|u_\varepsilon\|_{L^2(0,T;V)} + 1\right).$$

But it is known (cf. for instance Lions-Magenes [**71**] and [**72**]) that

$$\text{if } v \in L^2(0, T; V) \text{ and } \frac{\partial v}{\partial t} \in L^2(0, T; V'), \text{ then } v$$

$(1.2.15)$ is a.e. equal to a function, still denoted by v,

$$\text{such that } t \to v(t) \text{ is continuous from } [0, T] \to H.$$

Therefore we see that (1.2.9), (1.2.13) imply that (1.2.11) makes sense.

REMARK 1.3. It is proven in Lions [**68**] that the problem (1.2.9), (1.2.10), (1.2.11) **admits a unique solution.**

One has the estimate

$$(1.2.16) \qquad \|u_\varepsilon\|_{L^2(0,T;V)} \leq c\left(\|f\|_{L^2(0,T;V')} + \|u^0\|_H\right).$$

The verification of (1.2.16) is easy. We multiply both sides of (1.2.10) by u_ε. We denote by (f, g) the scalar product in H and we set

$$(1.2.17) \qquad (f, f) = \|f\|_H^2 = |f|^2.$$

We obtain

$$(1.2.18) \qquad \left(\frac{\partial u_\varepsilon}{\partial t}, u_\varepsilon\right) + a^\varepsilon(u_\varepsilon, u_\varepsilon) = (f, u_\varepsilon).$$

But we observe that[2]

$$\left(\frac{\partial \phi}{\partial t}, \phi\right) = \frac{1}{2}\frac{d}{dt}|\phi|^2$$

so that (1.2.18) gives (we write $a^\varepsilon(\phi)$ for $a^\varepsilon(\phi, \phi)$)

$$\frac{1}{2}\frac{d}{dt}|u_\varepsilon(t)|^2 + a^\varepsilon(u_\varepsilon) = (f, u_\varepsilon).$$

Hence

$$(1.2.19) \qquad \frac{1}{2}|u_\varepsilon(t)|^2 + \int_0^t a^\varepsilon(u_\varepsilon(\sigma))\,d\sigma = \frac{1}{2}|u^0|^2 + \int_0^t (f, u_\varepsilon)\,d\sigma.$$

But due to (1.1.4), we can find λ[3] such that

$$(1.2.20) \qquad a^\varepsilon(v) \geq \alpha\|v\|^2 - \lambda|v|^2, \qquad v \in V.$$

We denote by $\|\cdot\|_\bullet$ the norm in V'. We obtain from (1.2.19)

$$\frac{1}{2}|u_\varepsilon(t)|^2 + \alpha\int_0^t \|u_\varepsilon\|^2\,d\sigma \leq \lambda\int_0^t |u_\varepsilon|^2\,d\sigma + \frac{1}{2}|u^0|^2 + \int_0^t \|f\|_\bullet\|u_\varepsilon\|\,d\sigma$$

$$\leq \lambda\int_0^t |u_\varepsilon|^2\,d\sigma + \frac{1}{2}|u^0|^2$$

$$+ \frac{\alpha}{2}\int_0^t \|u_\varepsilon\|^2\,d\sigma + \frac{1}{2\alpha}\int_0^t \|f\|_\bullet^2\,d\sigma.$$

Therefore

$$(1.2.21) \qquad |u_\varepsilon(t)|^2 + \alpha\int_0^t \|u_\varepsilon\|^2\,d\sigma \leq 2\lambda\int_0^t |u_\varepsilon(\sigma)|^2\,d\sigma + |u^0|^2 + \frac{1}{\alpha}\int_0^t \|f\|_\bullet^2\,d\sigma.$$

Using Gronwall's lemma, (1.2.16) follows.

REMARK 1.4. One can replace (1.2.10) by the equivalent formula:

$$(1.2.22) \qquad \left(\frac{\partial u_\varepsilon}{\partial t}, v\right) + a^\varepsilon(u_\varepsilon, v) = (f, v), \ \forall v \in V.$$

EXAMPLE 1.5 ($V = H_0^1(\mathcal{O})$). Then (1.2.9) implies

$$(1.2.23) \qquad u_\varepsilon = 0 \text{ on } \Sigma.$$

It is Dirichlet's boundary value problem.

EXAMPLE 1.6 ($V = H^1(\mathcal{O})$). The one deduces from (1.2.22) that, in a "weak" sense,

$$(1.2.24) \qquad \frac{\partial u_\varepsilon}{\partial \nu_{A^\varepsilon}} = 0 \text{ on } \Sigma$$

---

[2]This is obvious if ϕ is smooth and it can be justified if $\phi \in L^2(0, T; V)$ and $\frac{\partial \phi}{\partial t} \in L^2(0, T; V')$.
[3]One can take $\lambda = 0$ if $V = H_0^1(\mathcal{O})$ or if $V = $ space which does not contain the constants.

where

(1.2.25) $\dfrac{\partial \nu_{A^\varepsilon}}{\partial} = a_{ij}^\varepsilon(x,t)\nu_j\dfrac{\partial}{\partial x_i},$

$\nu = \{\nu_1,\ldots,\nu_n\} = $ normal to Γ directed toward the exterior of $\mathcal{O}$.

It is Neumann's boundary value problem.

REMARK 1.7. By virtue of (1.2.16) we see that we can extract a subsequence, still denoted by u_ε, such that

(1.2.26) $u_\varepsilon \to u$ in $L^2(0,T;V)$ weakly.

Using (1.2.14) (and (1.2.16)), $\frac{\partial u_\varepsilon}{\partial t}$ is bounded in $L^2(0,T;V')$ and therefore we can assume that

(1.2.27) $\dfrac{\partial u_\varepsilon}{\partial t} \to \dfrac{\partial u}{\partial t}$ in $L^2(0,T;V')$ weakly.

It is known that, **when $V \to H$ is compact**, it follows from (1.2.26), (1.2.27) that

(1.2.28) $u_\varepsilon \to u$ in $L^2(0,T;H)$ **strongly**.

If we set

(1.2.29) $$\xi_i^\varepsilon = a_{ij}^\varepsilon\dfrac{\partial u_\varepsilon}{\partial x_j},$$

we can also assume that

(1.2.30) $\xi_i^\varepsilon \to \xi_i$ in $L^2(0,T;H)$ weakly.

One verifies that, if a_{ij}^ε is given by (1.1.5), with (1.1.6), then

(1.2.31) $a_{ij}^\varepsilon \to \mathcal{M}(a_{ij})$ **in $L^\infty(\mathcal{G})$ weak star**,

where in general

(1.2.32) $$\mathcal{M}(b) = \dfrac{1}{|Y|\tau_0}\int_Y\int_0^{\tau_0} b(y,\tau)\,dy\,d\tau.$$

But **it does not follow that** $\xi_i^\varepsilon \to \mathcal{M}(a_{ij})\frac{\partial u}{\partial x_j}$ (in any topology). We shall obtain below the limit of ξ_i^ε (which depends on k). The only thing we can say for the time being is that

(1.2.33) $$\dfrac{\partial u}{\partial t} = \dfrac{\partial \xi_i}{\partial x_i} = f.$$

REMARK 1.8 (The case of coefficients a_{ij}^ε independent of t). Let us assume that

(1.2.34) $a_{ij}^\varepsilon = a_{ij}(x/\varepsilon),$

where the a_{ij}'s are Y-periodic (i.e. with the hypotheses of Chapter 1). Then the answer is immediate. Let $\mathcal{A}$ be the homogenized operator of A^ε, constructed in Chapter 1. **Then**

(1.2.35) $u \in L^2(0,T;V),$

(1.2.36) $\dfrac{\partial u}{\partial t} + \mathcal{A}u = f,$

(1.2.37) $u(0) = u^0,$

defines the limit of u_ε.

Indeed, let ϕ be given in $C_0^\infty(]0, T[)$. For any function $v \in L^p(0, T; \mathcal{H})$, we define

$$(1.2.38) \qquad v(\phi) = \int_0^T v(t)\phi(t)\, dt \in \mathcal{H}.$$

We remark that

$$(1.2.39) \qquad \frac{\partial u_\varepsilon}{\partial t}(\phi) = -u_\varepsilon\left(\frac{d\phi}{dt}\right)$$

so that we deduce from (1.2.10) that

$$(1.2.40) \qquad (A^\varepsilon u_\varepsilon)(\phi) = f(\phi) + u_\varepsilon(\phi'), \qquad \phi' = \frac{d\phi}{dt}.$$

But since A^ε does not depend on t, $(A^\varepsilon u_\varepsilon)(\phi) = A^\varepsilon(u_\varepsilon(\phi))$ and therefore

$$(1.2.41) \qquad A^\varepsilon(u_\varepsilon(\phi)) = f(\phi) + u_\varepsilon(\phi') - f(\phi) + u(\phi') \text{ in } V' \text{ strongly,}$$

(we use (1.2.28)). We also know that

$$(1.2.42) \qquad u_\varepsilon(\phi) \to u(\phi) \text{ in } V \text{ weakly.}$$

It follows from (1.2.41, (1.2.42) and Chapter 1 that

$$\mathcal{A}u(\phi) = f(\phi) + u(\phi'), \ \forall \phi \in C_0^\infty(]0, T[)$$

hence (1.2.36) and the result follows.

1.3. Asymptotic expansions: Preliminary formulas. We shall consider the three most interesting cases for k appearing in (1.1.7). In order to avoid confusion, we shall set:

$$(1.3.1) \qquad P^{\varepsilon, k} = \frac{\partial}{\partial t} + A^\varepsilon\left(\frac{t}{\varepsilon^k}\right)$$

$$A^\varepsilon\left(\frac{t}{\varepsilon^k}\right) = -\frac{\partial}{\partial x_i}\left(a_{ij}\left(\frac{x}{\varepsilon}, \frac{t}{\varepsilon^k}\right)\frac{\partial}{\partial x_j}\right),$$

and we shall consider the values

$$(1.3.2) \qquad k = 1, 2, 3.$$

We shall set $\tau = t/\varepsilon^k$ and

$$(1.3.3) \qquad \begin{aligned} A_1 &= -\frac{\partial}{\partial y_i}\left(a_{ij}(y, \tau)\frac{\partial}{\partial y_j}\right), \\ A_2 &= -\frac{\partial}{\partial y_i}\left(a_{ij}(y, \tau)\frac{\partial}{\partial x_j}\right) - \frac{\partial}{\partial x_i}\left(a_{ij}(y, \tau)\frac{\partial}{\partial y_j}\right) \\ A_3 &= -\frac{\partial}{\partial x_i}\left(a_{ij}(y, \tau)\frac{\partial}{\partial x_j}\right); \end{aligned}$$

all these operators depend on τ parametrically. It follows that

$$(1.3.4) \qquad P^{\varepsilon, 1} = \varepsilon^{-2}Q_1 + \varepsilon^{-1}Q_2 + \varepsilon^0 Q_3,$$

where

$$(1.3.5) \qquad Q_1 = A_1, \qquad Q_2 = \frac{\partial}{\partial \tau} + A_2, \qquad Q_3 = \frac{\partial}{\partial t} + A_3;$$

$$(1.3.6) \qquad P^{\varepsilon, 2} = \varepsilon^{-2}R_1 + \varepsilon^{-1}R_2 + \varepsilon^0 R_3,$$

where

(1.3.7) $$R_1 = \frac{\partial}{\partial \tau} + A_1, \qquad R_2 = A_2, \qquad R_3 = \frac{\partial}{\partial t} + A_3;$$

(1.3.8) $$P^{\varepsilon,3} = \varepsilon^{-3} S_1 + \varepsilon^{-2} S_2 + \varepsilon^{-1} S_3 + \varepsilon^0 S_4,$$

where

(1.3.9) $$S_1 = \frac{\partial}{\partial \tau}, \qquad S_2 = A_1, \qquad S_3 = A_2, \qquad S_4 = \frac{\partial}{\partial \tau} + A_3.$$

We use now these formulas to find an asymptotic expansion of u_ε. We distinguish the three cases $k = 1, 2, 3.$[4] We look for u_ε in the form

(1.3.10) $$u_\varepsilon = u_0 + \varepsilon u_1 + \varepsilon^2 u_2 + \cdots, \qquad u_j = u_j(x, y, t, \tau),$$

$$u_j \text{ being } Y \text{ periodic in } y \text{ and } \tau_0 \text{ periodic in } \tau.$$

1.4. Asymptotic expansions: The case $k = 1$. By identifying powers of ε in the expansion of $p^{\varepsilon,1} u_\varepsilon = f$, we obtain:

(1.4.1) $$Q_1 u_0 = 0,$$

(1.4.2) $$Q_1 u_1 + Q_2 u_0 = 0,$$

(1.4.3) $$Q_1 u_2 + Q_2 u_1 + Q_3 u_0 = f,$$

$$\vdots$$

In (1.4.1), x, t, τ are parameters, so that (1.4.1) is equivalent to

(1.4.4) $$u_0 \text{ does not depend on } y.$$

We denote by $\mathcal{M}_y$ (respectively, $\mathcal{M}_\tau$) the average in y (respectively, τ) so that $\mathcal{M}_y \cdot \mathcal{M}_\tau = \mathcal{M}_\tau \cdot \mathcal{M}_y = \mathcal{M}$. By taking $\mathcal{M}_y$ of (1.4.2), we obtain

$$\mathcal{M}_y \frac{\partial u_0}{\partial \tau} = 0,$$

i.e., since we have (1.4.4), $\frac{\partial u_0}{\partial \tau} = 0$ and therefore

(1.4.5) $$u_0 = u(x, t).$$

Then (1.4.2) reduces to:

(1.4.6) $$A_1 u_1 = -A_2 u = +\frac{\partial}{\partial y_i} a_{ij}(y, \tau) \frac{\partial u}{\partial x_j}(x, t).$$

We can write the general solution of (1.4.6). We introduce $\chi^j = \chi^j(y, \tau)$ as the Y-periodic solution, defined up to an additive constant, of

(1.4.7) $$A_1(\chi^j(y, \tau) - y_j) = 0,$$

and we choose the constants such that

(1.4.8) $$\chi^j(y, \tau) \text{ is } \tau_0 \text{ \textbf{periodic}};$$

(for instance we can take $\mathcal{M}_y(\chi^j(y, \tau)) = 0, \forall \tau$).

Then (1.4.6) gives

(1.4.9) $$u_1 = -\chi^j(y, \tau) \frac{\partial u}{\partial x_j}(x, t) + \hat{u}_1(x, t, \tau).$$

---

[4] Another ansatz will be necessary when k is not an integer.

We can solve (1.4.3) for u_2 iff

$$\mathcal{M}_y(Q_2 u_1 + Q_3 u) = f,$$

i.e.,

(1.4.10) $$\frac{\partial \hat{u}_1}{\partial \tau} + \mathcal{M}_y Q_2 \left(-chi^j \frac{\partial u}{\partial x_j}\right) + \mathcal{M}_y(Q_3 u) = f.$$

We can solve (1.4.10) for $\hat{u}_1$ (with a τ_0 periodic function of τ) iff

$$\mathcal{M}_\tau \left[\mathcal{M}_y Q_2 \left(-\chi^j \frac{\partial u}{\partial x_j}\right) + \mathcal{M}_y(Q_3 u)\right] = f$$

i.e.,

$$\mathcal{M} Q_2 \left(-\chi^j \frac{\partial u}{\partial x_j}\right) + \mathcal{M} Q_3 u = f.$$

We observe that

$$\mathcal{M}\left(\frac{\partial}{\partial \tau}\left(-\chi^j \frac{\partial u}{\partial x_j}\right)\right) = 0$$

so that we obtain

(1.4.11) $$\frac{\partial u}{\partial t} = \mathcal{M}\left(a_{ij} - a_{ik}\frac{\partial \chi^j}{\partial y_k}\right)\frac{\partial^2 u}{\partial x_i \partial x_j} = f.$$

This is the **homogenized operator** (as we shall justify below).

1.5. Asymptotic expansions: The case $k = 2$. We use now (1.3.6), (1.3.7). We obtain

(1.5.1) $$R_1 u_0 = 0,$$

(1.5.2) $$R_1 u_1 + R_2 u_0 = 0,$$

(1.5.3) $$R_1 u_2 + R_2 u_1 + R_3 u_0 = f.$$

We multiply (1.5.1) by u_0 and we integrate over Y. Let us set

(1.5.4) $$a_1(\phi, \psi) = \int_Y a_{ij}(y, \tau)\frac{\partial \phi}{\partial y_j}\frac{\partial \psi}{\partial y_i}\, dy, \qquad (\phi, \psi)_Y = \int_Y \phi\psi\, dy.$$

We obtain

(1.5.5) $$\left(\frac{\partial u_0}{\partial \tau}, u_0\right)_Y + A_1(u_0, u_0) = 0.$$

Since u_0 is τ_0 periodic, $\int_0^{\tau_0} \left(\frac{\partial u_0}{\partial \tau}, u_0\right)_Y d\tau = 0$, so that (1.5.5) implies $\frac{\partial u_0}{\partial y_i} = 0 \; \forall i$. Then $A_1 u_0 = 0$ and $\frac{\partial u_0}{\partial \tau} + A_1 u_0 = \frac{\partial u_0}{\partial \tau} = 0$, so that

(1.5.6) $$u_0 = u(x, t).$$

Then (1.5.2) reduces to

$$\frac{\partial u_1}{\partial \tau} + A_1 u_1 = \frac{\partial a_{ij}}{\partial y_i}\frac{\partial u}{\partial x_j}.$$

We introduce

$$\theta^j(y, \tau), \qquad Y\text{-}\tau_0 \text{ periodic, solution of}$$

(1.5.7) $$\frac{\partial \theta^j}{\partial \tau} + A_1(\theta^j - y_j) = 0,$$

(θ^j is defined up to an additive constant). Then

$$(1.5.8) \qquad u_1 = -\theta^j(y,\tau)\frac{\partial u}{\partial x_j} + \hat{u}_1(x,t).$$

We can solve (1.5.3) for u_2 iff[5]

$$\mathcal{M}(R_2 u_1 + R_3 u_0) = f,$$

i.e.,

$$(1.5.9) \qquad \frac{\partial u}{\partial t} - \mathcal{M}\left(a_{ij} - a_{ik}\frac{\partial\theta^j}{\partial y_k}\right)\frac{\partial^2 u}{\partial x_i \partial x_j} = f.$$

This is the homogenized operator of $P^{\varepsilon,2}$ (as we shall justify below). It has of course the same structure as the one obtained in (1.4.11) for $P^{\varepsilon,1}$, but coefficients are different.

1.6. Asymptotic expansions: The case $k = 3$. We now use (1.3.8), (1.3.9). We obtain:

$$(1.6.1) \qquad\qquad S_1 u_0 = 0,$$
$$(1.6.2) \qquad\qquad S_1 u_1 + S_2 u_0 = 0,$$
$$(1.6.3) \qquad\qquad S_1 u_2 + S_2 u_1 + S_3 u_0 = 0,$$
$$(1.6.4) \qquad S_1 u_3 + S_2 u_2 + S_3 u_1 + S_4 u_0 = f,$$

$$\vdots$$

It follows from (1.6.1) that u_0 does not depend on τ; (1.6.2) admits a τ_0-periodic solution in u_1 iff

$$\mathcal{M}_\tau S_2 u_0 = 0,$$

i.e.,

$$(1.6.5) \qquad -\frac{\partial}{\partial y_i}(\mathcal{M}_\tau a_{ij})\frac{\partial u_0}{\partial y_j} = 0.$$

We define

$$(1.6.6) \qquad \bar{A}_1 = \mathcal{M}_\tau(A_1) = -\frac{\partial}{\partial y_i}\left(\mathcal{M}_\tau(a_{ij})\frac{\partial}{\partial y_j}\right).$$

It is of course an elliptic operator in Y, so that (1.6.5) is equivalent to $\frac{\partial u_0}{\partial y_j} = 0$, $\forall j$ and therefore $u_0 = u(x,t)$. Then (1.6.2) reduces to $\frac{\partial u_1}{\partial \tau} = 0$, i.e.

$$(1.6.7) \qquad u_1 \text{ does not depend on } \tau.$$

We can solve (1.6.3) for u_2 iff

$$\mathcal{M}_\tau(S_2 u_1 + S_3 u_0) = 0,$$

i.e.

$$\bar{A}_1 u_1 = \frac{\partial}{\partial y_i}\mathcal{M}_\tau(a_{ij})\frac{\partial u}{\partial x_j}.$$

---

[5] $R_1\phi = P$ admits a Y-τ_0 periodic solution iff $\mathcal{M}(F) = 0$; (observe that $R_1^*\psi = 0$ admits the constants as only Y-τ_0 periodic solution).

If we introduce $\phi^j(y)$ by:

$$\phi^j(y) \text{ is } Y\text{-periodic,}$$

(1.6.8)
$$\bar{A}_1(\phi^j - y_j) = 0,$$

(which defines ϕ^j up to an additive constant), we see that

(1.6.9)
$$u_1 = -\phi^j(y)\frac{\partial u}{\partial x_j} + \hat{u}_1(x,t).$$

Then (1.6.3) can be solved for u_2; if

(1.6.10)
$$u_2^0 = -\int_0^\tau (S_2 u_1 + S_3 u_0)\, d\sigma,$$

we have

(1.6.11)
$$u_2 = u_2^0 + \hat{u}_2(x,y,t).$$

Equation (1.6.4) can be solved for u_3 iff

(1.6.12)
$$\mathcal{M}_\tau(S_2 u_2 + S_3 u_1 + S_4 u_0) = f$$

and (1.6.12) can be solved for $\hat{u}_2$ iff

(1.6.13)
$$\mathcal{M}_y \mathcal{M}_\tau(S_2 u_2 + S_3 u_1 + S_4 u_0) = f.$$

But $\mathcal{M}_y \mathcal{M}_\tau(S_2 u_2) = \mathcal{M}_\tau \mathcal{M}_y(S_2 u_2) = 0$, so that (1.6.13) reduces to

(1.6.14)
$$\frac{\partial u}{\partial t} - \mathcal{M}\left(a_{ij} - a_{ik}\frac{\partial \phi^j}{\partial y_k}\right)\frac{\partial^2 u}{\partial x_i \partial x_j} = f.$$

We have again the same structure for the homogenized operator, with again different formulas.

1.7. Other form of homogenization formulas. Let us denote by $q_{ij}^k = $ homogenized coefficient of $\left(\frac{\partial^2 u}{\partial x_i \partial x_j}\right)$, for $k = 1, 2, 3$. Let us check the following formulas:

(1.7.1)
$$q_{ij}^1 = \frac{1}{|Y|\tau_0}\int_0^{\tau_0} a_1(\chi^j - y_j, \chi^i - y_i)\, d\tau,$$

(1.7.2)
$$q_{ij}^2 = \frac{1}{|Y|\tau_0}\int_0^{\tau_0} a_1(\theta^j - y_j, \theta^i - y_i)\, d\tau,$$

(1.7.3)
$$q_{ij}^3 = \frac{1}{|Y|}\bar{a}_1(\phi^j - y_j, \phi^i - y_i),$$

where a_1 is defined in (1.5.4) and where

(1.7.4)
$$\bar{a}_1(\phi, \psi) = \int_Y \mathcal{M}_\tau(a_{ij})\frac{\partial \phi}{\partial y_j}\frac{\partial \psi}{\partial y_i}\, dy.$$

For (1.7.2), we will actually prove that

(1.7.5)
$$q_{ij}^2 = \frac{1}{|Y|\tau_0}\left[\int_0^{\tau_0} a_1(\theta^j - y_j, \theta^i - y_i)\, d\tau + \int_0^{\tau_0}\left(\frac{\partial \theta^j}{\partial \tau}, \theta^i\right)_Y d\tau\right].$$

This is sufficient since

$$\sum_{i,j} \frac{1}{|Y|\tau_0} \left(\int_0^{\tau_0} \left(\frac{\partial \theta^j}{\partial \tau}, \theta^i \right)_Y d\tau \right) \frac{\partial^2 u}{\partial x_i \partial x_j} = \sum_{i,j} \frac{1}{|Y|\tau_0} \frac{\partial^2 u}{\partial x_i \partial x_j}$$

$$\cdot \left[\frac{1}{2} \int_0^{\tau_0} \left(\frac{\partial \theta^j}{\partial \tau}, \theta^i \right)_Y d\tau + \frac{1}{2} \int_0^{\tau_0} \left(\frac{\partial \theta^i}{\partial \tau}, \theta^j \right)_Y d\tau \right] = 0$$

taking into account that

$$\int_0^{\tau_0} \left(\frac{\partial \theta^j}{\partial \tau}, \theta^i \right)_Y d\tau + \int_0^{\tau_0} \left(\frac{\partial \theta^i}{\partial \tau}, \theta^j \right)_Y d\tau = \int_0^{\tau_0} \frac{d}{d\tau}(\theta^i, \theta^j)_Y \, d\tau = 0.$$

Therefore, by comparing the expressions (1.4.11), (1.5.9), (1.6.14) with (1.7.1), (1.7.5), (1.7.3), we have to verify that

$$(1.7.6) \qquad\qquad \int_0^{\tau_0} a_1(\chi^j - y_j, \chi^i) \, d\tau = 0,$$

$$(1.7.7) \qquad\qquad \int_0^{\tau_0} \left[a_1(\theta^j - y_j, \theta^i) + \left(\frac{\partial \theta^j}{\partial \tau}, \theta^i \right)_Y \right] d\tau = 0,$$

$$(1.7.8) \qquad\qquad \bar{a}_1(\phi^j - y_j, \phi^i) = 0.$$

This is straightforward. Indeed (1.4.7) implies that

$$a_1(\chi^j - y_j, \chi^i) = 0 \text{ for every } \tau,$$

hence (1.7.6) follows.

If we multiply (1.5.7) by θ^i and integrate in y and in τ, we obtain (1.7.7) and (1.7.8) follows from (1.6.8).

Formulas (1.7.1), (1.7.2), (1.7.3) imply the ellipticity of

$$(1.7.9) \qquad\qquad \mathcal{A}^k = -q_{ij}^k \frac{\partial^2}{\partial x_i \partial x_j}.$$

Indeed, if we set:

$$\phi = \xi_i(\chi^j - y_j), \qquad \psi = \xi_j(\theta^j - y_j), \qquad \rho = \xi_j \phi^j,$$

we have

$$(1.7.10) \qquad\qquad q_{ij}^1 \xi_i \xi_j = \frac{1}{|Y|\tau_0} \int_0^{\tau_0} a_1(\phi, \phi) \, d\tau,$$

$$(1.7.11) \qquad\qquad q_{ij}^2 \xi_i \xi_j = \frac{1}{|Y|\tau_0} \int_0^{\tau_0} a_1(\phi, \psi), \, d\tau, {}^6$$

$$(1.7.12) \qquad\qquad q_{ij}^3 \xi_i \xi_j = \frac{1}{|Y|} \bar{a}_1(\rho, \rho).$$

Hence the result follows.

---

[6]Since $\int_0^{\tau_0} \left(\frac{\partial \psi}{\partial \tau}, \psi \right)_Y d\tau = 0.$

1.8. The role of k. Let us consider now $k > 0$ not necessarily an integer. We set

(1.8.1) $u_\varepsilon^k = $ solution of (1.2.9), (1.2.10), (1.2.11) relative to $A^\varepsilon = A(t/\varepsilon^k)$,

and we denote by u^k, $k = 1, 2, 3$, the solution of

(1.8.2) $$\frac{\partial u^k}{\partial t} + \mathcal{A}^k u^k = f,$$

(1.8.3) $$u^k \in L^2(0, T; V),$$

(1.8.4) $$u^k(0) = u^0.$$

Then we shall see that the limit of u_ε^k when $k < 2$ (respectively, $k > 2$) **does not depend on** k, i.e. that $u_\varepsilon^k \to u^1$ (in the appropriate topology) when $k < 2$ (respectively, $\to u^3$ when $k > 2$). But the asymptotic expansion depends on k, and actually new "ansatz" will be necessary for the cases when k is not an integer.

2. Convergence of the homogenization of parabolic equations

2.1. Statement of the homogenization result. We shall prove in several steps the

THEOREM 2.1. *We assume that (1.1.5), (1.1.6) hold true. We denote by u_ε^k the solution of (1.2.9), (1.2.10), (1.2.11) relative to $A^\varepsilon = A(t/\varepsilon^k)$ and by u^k the solution of (1.8.2), (1.8.3), (1.8.4), $k = 1, 2, 3$, where $\mathcal{A}^1$ (respectively, $\mathcal{A}^2$, respectively, $\mathcal{A}^3$) is given by (1.4.11) (respectively, (1.5.9), respectively, (1.6.14)). Also assume that there exists $p > 2$ such that $f \in L^p(0, T; W^{-1,p}(\mathcal{O}))$ and that $u^0 \in W_0^{1,p}(\mathcal{O})$, or at least that f and u_0 belong locally to these spaces.*[7] *Then, as $\varepsilon \to 0$, one has*

(2.1.1) $$u_\varepsilon^k \to u^1 \text{ in } L^2(0, T; V) \text{ weakly if } k < 2,$$

(2.1.2) $$u_\varepsilon^2 \to u^2 \text{ in } L^2(0, T; V) \text{ weakly if } k = 2,$$

(2.1.3) $$u_\varepsilon^k \to u^3 \text{ in } L^2(0, T; V) \text{ weakly if } k > 2.$$

We begin by the case (2.1.2) which is the simplest; we shall next proceed with (2.1.1) and (2.1.3).

2.2. Proof of the homogenization when $k = 2$. Given $P(y) =$ homogeneous polynomial in y of first degree, we define $w(y, \tau)$ as the solution (defined up to an additive constant) of

(2.2.1) $$-\frac{\partial w}{\partial \tau} + A_1^* w = 0, \qquad w - P \text{ is } Y - \tau_0 \text{ periodic.}$$

We write:

(2.2.2) $$A^\varepsilon(t/\varepsilon^2) = A^\varepsilon, \qquad u_\varepsilon^2 = u_\varepsilon.$$

If we set

(2.2.3) $$w - P = -\theta^*$$

then

(2.2.4) $$-\frac{\partial \theta^*}{\partial \tau} + A_1^*(\theta^* - P) = 0, \qquad \theta^* \text{ being } Y - \tau_0 \text{ periodic.}$$

---

[7]i.e. if $\mathcal{O}' \subset \overline{\mathcal{O}} \subset \mathcal{O}$, $f \in L^p(0, T; W^{-1,p}(\mathcal{O}'))$, $u^0 \in W^{1,p}(\mathcal{O}')$. This hypothesis is not necessary if $k = 2$ or if $k \neq 1$ and $a_{ij} \in C^0$ (cf. (2.2.19) below).

We now define w_ε by

(2.2.5)
$$w_\varepsilon(x,t) = \varepsilon w(x/\varepsilon, t/\varepsilon^2).$$

We observe that

(2.2.6)
$$-\frac{\partial w_\varepsilon}{\partial t} + A^{\varepsilon*} w_\varepsilon = 0.$$

For any $\phi \in C_0^\infty(\mathcal{G})$, $\mathcal{G} = \mathcal{O} \times]0, T[$, we multiply (1.2.10) by ϕw_ε and (2.2.6) by ϕu_ε. We integrate over $\mathcal{G}$; we obtain (compare with Chapter 1, Section 2):

(2.2.7)
$$\int_\mathcal{G} \left[\left(\frac{\partial u_\varepsilon}{\partial t} \right) \phi w_\varepsilon + \left(\frac{\partial w_\varepsilon}{\partial t} \right) \phi u_\varepsilon \right] dx\, dt + \int_\mathcal{G} \left[\xi_i^\varepsilon \frac{\partial \phi}{\partial x_i} w_\varepsilon - a_{ij} \frac{\partial w_\varepsilon}{\partial x_i} \frac{\partial \phi}{\partial x_j} u_\varepsilon \right] dx\, dt$$
$$= \int_\mathcal{G} f \phi w_\varepsilon\, dx\, dt.$$

The first term in (2.2.7) equals

(2.2.8)
$$\int_\mathcal{G} \phi \frac{\partial}{\partial t}(u_\varepsilon w_\varepsilon)\, dx\, dt = -\int_\mathcal{G} u_\varepsilon w_\varepsilon \frac{\partial \phi}{\partial t}\, dx\, dt.$$

Since $w_\varepsilon \to P$ in $L^2(\mathcal{G})$ strongly and since we have (1.2.28), we can pass to the limit in all terms of (2.2.7), (2.2.8). We obtain

(2.2.9)
$$-\int_\mathcal{G} uP \frac{\partial \phi}{\partial t}\, dx\, dt + \int_\mathcal{G} \left[\xi_i \frac{\partial \phi}{\partial x_i} P - \mathcal{M}\left(a_{ij} \frac{\partial w}{\partial y_i} \right) \frac{\partial \phi}{\partial x_j} u \right] dx\, dt = \int_\mathcal{G} f \phi P\, dx\, dt.$$

Using (1.2.33) we see that

$$\int_\mathcal{G} f \phi P\, dx\, dt = -\int_\mathcal{G} uP \frac{\partial \phi}{\partial t}\, dx\, dt + \int_\mathcal{G} \xi_i \frac{\partial}{\partial x_i}(\phi P)\, dx\, dt,$$

so that (2.2.9) reduces to

$$\int_Q \xi_i \frac{\partial P}{\partial x_i} \phi\, dx\, dt = \mathcal{M}\left(a_{kj} \frac{\partial w}{\partial y_k} \right) \int_\mathcal{G} \phi \frac{\partial u}{\partial x_j}\, dx\, dt, \quad \forall \phi \in C_0^\infty(\mathcal{G}).$$

Therefore

(2.2.10)
$$\xi_i \frac{\partial P}{\partial x_i} = \mathcal{M}\left(a_{kj} \frac{\partial w}{\partial y_k} \right) \frac{\partial u}{\partial x_j}.$$

Let us take

$$P(y) = y_i.$$

We define θ^{i*} by

(2.2.11)
$$-\frac{\partial \theta^{i*}}{\partial \tau} + A_1^*(\theta^{i*} - y_i) = 0,$$
$$\theta^{i*} \text{ being } Y - \tau_0 \text{ periodic.}$$

Then $w = -(\theta^{i*} - y_i)$ and (2.2.10) gives:

$$\xi_i = \mathcal{M}\left(a_{kj} \frac{\partial(y_i - \theta^{i*})}{\partial y_k} \right) \frac{\partial u}{\partial x_j},$$

or

(2.2.12)
$$\xi_i = \mathcal{M}\left(a_{ij} - a_{kj} \frac{\partial \theta^{i*}}{\partial y_k} \right) \frac{\partial u}{\partial x_j}.$$

The variational formulation of the problem is:

$$\left(\frac{\partial u_\varepsilon}{\partial t}, v\right) + \left(\xi_i^\varepsilon, \frac{\partial v}{\partial x_i}\right) = (f, v), \ \forall v \in V,$$

and passing to the limit we find

$$(2.2.13) \qquad \left(\frac{\partial u}{\partial t}, v\right) + \left(\xi_i, \frac{\partial v}{\partial x_i}\right) = (f, v), \ \forall v \in V.$$

This proves the Theorem (for $k = 2$) if we verify that we obtain the same formula as in (1.5.9), i.e. that

$$(2.2.14) \qquad \mathcal{M}\left(a_{ik}\frac{\partial \theta^j}{\partial y_k}\right) = \mathcal{M}\left(a_{kj}\frac{\partial \theta^{i*}}{\partial y_k}\right).$$

This is equivalent to proving that

$$(2.2.15) \qquad \int_0^{\tau_0} a_1(\theta^j, y_i)\, d\tau = \int_0^{\tau_0} a_1^*(\theta^{i*}, y_j)\, d\tau$$

or that

$$(2.2.16) \qquad \int_0^{\tau_0} a_1^*(y_i, \theta^j)\, d\tau = \int_0^{\tau_0} a_1(y_j, \theta^{i*})\, d\tau.$$

We multiply (2.2.11) by θ^j, (1.5.7) by θ^{i*}. This shows that the left hand side of (2.2.16) equals

$$(2.2.17) \qquad \int_0^{\tau_0} \left[a_1^*(\theta^{i*}, \theta^j) - \left(\frac{\partial \theta^{i*}}{\partial \tau}, \xi^j\right)\right] d\tau$$

and that the right hand side of (2.2.16) equals

$$(2.2.18) \qquad \int_0^{\tau_0} \left[a_1(\theta^j, \theta^{i*}) + \left(\frac{\partial \theta^j}{\partial \tau}, \theta^{i*}\right)\right] d\tau$$

and the two expressions (2.2.17), (2.2.18) are indeed equal.

2.2.1. *Orientation.* We consider now the cases when $k \neq 2$. We begin with the study of this question **under the extra hypothesis**

$$(2.2.19) \qquad a_{ij} \in C^0(\bar{Y} \times [0, \tau_0]), \qquad \text{periodic},$$

2.3. Reduction to the smooth case. We are going to show

LEMMA 2.2. *Under the supplementary hypothesis (2.2.19), it suffices to prove Theorem 2.1 under the hypothesis*

$$(2.3.1) \qquad a_{ij} \in C^\infty(\bar{Y} \times [0, \tau_0]), \text{ periodic together with all its derivatives.}$$

PROOF. We denote by u_ε the solution u_ε^k of (1.2.22), k **fixed** arbitrarily. We approximate a_{ij} in $C^\infty(\bar{Y} \times [0, \tau_0])$ by $a_{ij}^\beta \in C^\infty(\bar{Y} \times [0, \tau_0])$, a_{ij}^β being periodic together with all its derivatives and satisfying

$$(2.3.2) \qquad a_{ij}^\beta \xi_i \xi_j \geq \alpha_1 \xi_i \xi_i, \qquad \alpha_1 > 0, \ \forall y, \tau.$$

We denote by u_ε^β (not to be confused with u_ε^k) the solution in $L^2(0, T; V)$ of

$$(2.3.3) \qquad \left(\frac{\partial u_\varepsilon^\beta}{\partial t}, v\right) + a^{\varepsilon\beta}(u_\varepsilon^\beta, v) = (f, v), \ \forall v \in V$$

where

(2.3.4)
$$a^{\varepsilon\beta}(u,v) = \int_{\mathcal{O}} a_{ij}^{\beta}\left(\frac{x}{\varepsilon},\frac{t}{\varepsilon^k}\right)\frac{\partial u}{\partial x_j}\frac{\partial v}{\partial x_i}\,dx.$$

We shall check below that

(2.3.5)
$$\|u_{\varepsilon}^{\beta} - u_{\varepsilon}\|_{L^2(0,T;V)} \le C\sup_{i,j}\|a_{ij}^{\beta} - a_{ij}\|_{C^0(\bar{Y}\times[0,\tau_0])}.$$

Let us consider now the case $k=1$ to fix ideas. We define $\chi^{j,\beta}$ by
$$A_1(\chi^{j,\beta} - y_j) = 0, \qquad \chi^{j,\beta}\ Y - \tau_0 \text{ periodic,}$$
where $A_1^{\beta} = -\frac{\partial}{\partial y_i}\left(a_{ij}^{\beta}(y,\tau)\frac{\partial}{\partial y_j}\right)$, and we define
$$q_{ij}^{1\beta} = \frac{1}{|Y|\tau_0}\int_0^{\tau_0} a_1(\chi^{j\beta} - y_j, \chi^{i\beta} - y_i)\,d\tau.$$

One shows easily that (normalizing χ^j, $\chi^{j,\beta}$ in the same manner, for instance $\mathcal{M}_y(\chi^j) = 0$, $\mathcal{M}_Y(\chi^{j,\beta}) = 0$)
$$\|\chi^{j,\beta} - \chi^j\|_{H^1(Y)} \le C\sup_{i,j}\|a_{ij}^{\beta} - a_{ij}\|_{C^0(\bar{Y}\times[0,\tau_0])}$$

so that

(2.3.6)
$$|q_{ij}^{1\beta} - q_{ij}^1| \le C\sup_{i,j}\|a_{ij}^{\beta} - a_{ij}\|_{C^0(\bar{Y}\times[0,\tau_0])}.$$

The same proof as (2.3.5) will give

(2.3.7)
$$\|u^{1,\beta} - u^1\|_{L^2(0,T;V)} \le C\sup_{i,j}|q_{ij}^{1\beta} - q_{ij}^1|$$

where $u^{1,\beta}$ is the solution of
$$\frac{\partial u^{1,\beta}}{\partial t} + \mathcal{A}^{1,\beta}u^{1,\beta} = f, \qquad u^{1,\beta} \in L^2(0,T;V),$$
$$u^{1,\beta}(0) = u^0, \qquad \mathcal{A}^{1,\beta} = -q_{ij}^{1\beta}\frac{\partial^2}{\partial x_i\partial x_j}.$$

It follows from (2.3.5), (2.3.6), (2.3.7) that it is enough to prove that $u_{\varepsilon}^{1,\beta} \to u^{1,\beta}$ in $L^2(0,T;V)$ weakly when $\varepsilon \to 0$, β **fixed**.

The same remarks apply to all values of k, so that it remains only to prove (2.3.5), which is standard. Indeed if we set
$$m_{\varepsilon} = u_{\varepsilon}^{\beta} - u_{\varepsilon}$$

we have
$$\left(\frac{\partial m_{\varepsilon}}{\partial t}, v\right) + a^{\varepsilon\beta}(m_{\varepsilon}, v) = a^{\varepsilon\beta}(u_{\varepsilon}, v) - a^{\varepsilon}(u_{\varepsilon}, v) = (f_{\varepsilon}, v),$$
$$m_{\varepsilon}(0) = 0,$$

where
$$|(f_{\varepsilon}, v)| = \left|\int_{\mathcal{O}}\left[a_{ij}^{\beta}\left(\frac{x}{\varepsilon},\frac{t}{\varepsilon^k}\right) - a_{ij}\left(\frac{x}{\varepsilon},\frac{t}{\varepsilon^k}\right)\right]\frac{\partial u_{\varepsilon}}{\partial x_j}\frac{\partial v}{\partial x_i}\,dx\right|$$
$$\le C\sup\|a_{ij}^{\beta} - a_{ij}\|_{C^0(\bar{Y}\times[0,\tau_0])}\|u_{\varepsilon}\|\|v\|$$

so that
$$\|f_{\varepsilon}\|_* \le C\sup\|a_{ij}^{\beta} - a_{ij}\|_{C^0(\bar{Y}\times[0,\tau_0])}\|u_{\varepsilon}\|$$

and therefore
$$\|f_\varepsilon\|_{L^2(0,T;V')} \le C \sup \|a_{ij}^\beta - a_{ij}\|_{C^0(\bar{Y}\times[0,\tau_0])}.$$
Thus (2.3.5) follows from (1.2.16).   ∎

2.4. Proof of the homogenization when $0 < k < 2$. We now prove (2.1.1)
under the assumption (2.2.19), by Lemma 2.2. We can therefore assume that
(2.3.1) holds true.

2.4.1. *The case* $0 < k \le 1$. Let $P(y)$ be given as in Section 2.2. Let w be the
solution (defined up to an additive constant) of

$$(2.4.1) \qquad A_1^* w = 0, \qquad w - P(y) \text{ is } Y - \tau_0 \text{ periodic.}$$

We use notation (2.2.2) and we set $u_\varepsilon^k = u_\varepsilon$. If we set

$$w - P = -\chi^*$$

then

$$(2.4.2) \qquad A_1^*(\chi^* - P) = 0, \qquad \chi^* \text{ being } Y - \tau_0 \text{ periodic.}$$

We now define w_ε by

$$w_\varepsilon(x,t) = w\left(\frac{x}{\varepsilon}, \frac{t}{\varepsilon^k}\right).$$

We observe that

$$A^{\varepsilon *} w_\varepsilon = 0.$$

The same procedure as for (2.2.7) leads to

$$(2.4.3) \quad \int_{\mathcal{G}} \frac{\partial u_\varepsilon}{\partial t} \phi w_\varepsilon \, dx \, dt + \int_{\mathcal{G}} \left[\xi_i^\varepsilon \frac{\partial \phi}{\partial x_i} w_\varepsilon - a_{ij} \frac{\partial w_\varepsilon}{\partial x_i} \frac{\partial \phi}{\partial x_j} u_\varepsilon \right] dx \, dt = \int_{\mathcal{G}} f \phi w_\varepsilon \, dx \, dt.$$

The first term in (2.4.3) equals

$$(2.4.4) \qquad -\int_{\mathcal{G}} u_\varepsilon w_\varepsilon \frac{\partial \phi}{\partial t} \, dx \, dt - \int_{\mathcal{G}} u_\varepsilon \phi \frac{\partial w_\varepsilon}{\partial t} \, dx \, dt.$$

But $\frac{\partial w_\varepsilon}{\partial t} = \varepsilon^{1-k} \left(\frac{\partial w}{\partial \tau}\right)^\varepsilon$ where $\phi^\varepsilon(x,t) = \phi\left(\frac{x}{\varepsilon}, \frac{t}{\varepsilon^k}\right)$.

Wince we have assumed the coefficients a_{ij} to be **smooth**, it follows from
(2.4.1) that

$$A_1^*\left(\frac{\partial w}{\partial \tau}\right) + \left(\frac{\partial A_1^*}{\partial \tau}\right) w = 0$$

so that we have **in particular**

$$\left|\frac{\partial w}{\partial \tau}(y,\tau)\right| \le C.$$

Therefore $\left|\frac{\partial w_\varepsilon}{\partial t}\right| \le C\varepsilon^{1-k}$ and

$$\left|\int_{\mathcal{G}} u_\varepsilon \phi \frac{\partial w_\varepsilon}{\partial t} \, dx \, dt\right| \le C\varepsilon^{1-k} \to 0, \text{ if } k < 1.$$

If $k = 1$, $\frac{\partial w_\varepsilon}{\partial \tau} = \left(\frac{\partial w}{\partial \tau}\right)^\varepsilon$ and $\frac{\partial w_\varepsilon}{\partial \tau} \to \mathcal{M}\left(\frac{\partial w}{\partial \tau}\right) = 0$ in $L^\infty(\mathcal{G})$ weak star so that

$$\int_{\mathcal{G}} u_\varepsilon \phi \frac{\partial w_\varepsilon}{\partial t} \, dx \, dt \to 0.$$

In all cases, the limit of (2.4.4) is

$$-\int_{\mathcal{G}} u P \frac{\partial \phi}{\partial t} \, dx \, dt,$$

so that (2.4.12) gives in the limit

$$-\int_{\mathcal{G}} uP\frac{\partial\phi}{\partial t}\,dx\,dt + \int_{\mathcal{G}}\left[\xi_i\frac{\partial\phi}{\partial x_i}P - \mathcal{M}\left(a_{ij}\frac{\partial w}{\partial y_i}\right)\frac{\partial\phi}{\partial x_j}u\right]dx\,dt = \int_{\mathcal{G}} f\phi P\,dx\,dt.$$

This is the same formula as (2.2.9), **with a different** w. We obtain (2.2.10) and if we define χ^{i*} by

(2.4.5) $$A_1^*(\chi^{i*} - y_i) = 0, \qquad \chi^{i*}\text{ being }Y\text{-periodic,}$$

then

(2.4.6) $$\xi_i = \mathcal{M}\left(a_{ij} - a_{kj}\frac{\partial\chi^{i*}}{\partial y_k}\right)\frac{\partial u}{\partial x_j}.$$

The Theorem is proven, provided we verify that (2.4.6) furnishes the same formula as in (1.4.11), i.e. that

$$\mathcal{M}\left(a_{kj}\frac{\partial\chi^{i*}}{\partial y_k}\right) = \mathcal{M}\left(a_{ij}\frac{\partial\chi^j}{\partial y_k}\right).$$

The verification is made along similar lines as for (2.2.14).

2.4.2. *The case* $1 < k < 3/2$. We prove now the result for $1 < i < 3/2$ and we shall indicate next how to proceed when $k \to 2$.

We begin by **general considerations on the asymptotic expansions.**

With the notation (1.3.1) we have

(2.4.7) $$P^{\varepsilon,k} = \varepsilon^{-1}A_1 + \varepsilon^{-1}A_2 + \varepsilon^0\left(A_3 + \frac{\partial}{\partial\tau}\right) + \varepsilon^{-k}\frac{\partial}{\partial\tau}.$$

We are looking for an asymptotic expansion of u_ε where $P^{\varepsilon,k}u_\varepsilon = f$.

The usual expansion using terms $\varepsilon^j u_j(x, y, t, \tau)$ is impossible **without additional terms**, since there is otherwise no term to "compensate" the powers ε^{j-k} which would appear.[8]

Therefore it is natural to look for u in the form:

$$u_\varepsilon = u_0 + \varepsilon u_1 + \varepsilon^2 u_2 + \cdots + \varepsilon^{2-k}v_0 + \varepsilon^{3-k}v_1 + \cdots,$$

where all functions depend on x, y, t, τ and are periodic in y and in τ_0. But if we use the above "ansatz" for u_ε, we do not have τ derivatives coming in the identification for u_0, u_1, u_2 and one easily checks that the computations are impossible. Therefore one should add to the above "ansatz," terms of the type $\varepsilon^k w_0 + \varepsilon^{k+1}w_1 + \cdots$, so that finally we look for u_ε in the form

(2.4.8) $$u_\varepsilon = u_0 + \varepsilon u_1 + \varepsilon^2 u_2 + \cdots + \varepsilon^{2-k}v_0 + \varepsilon^{3-k}v_1 + \cdots + \varepsilon^k w_0 + \varepsilon^{k+1}w_1 + \cdots.$$

Identifying the coefficients of ε^{-2}, ε^{-1}, ε^0 gives

(2.4.9) $$A_1 u_0 = 0,$$

(2.4.10) $$A_1 u_1 + A_2 u_0 = 0,$$

(2.4.11) $$A_1 u_2 + A_2 u_1 + A_3 u_0 + \frac{\partial u_0}{\partial t} + \frac{\partial w_0}{\partial\tau} = f.$$

---

[8]Except if the functions u_j do not depend on τ, but then the identification is impossible.

Identifying next the coefficients of ε^{-k}, ε^{1-k}, ε^{-k-2}, ε^{-k-1}, we obtain

(2.4.12)
$$\frac{\partial u_0}{\partial \tau} + A_1 v_0 = 0,$$

(2.4.13)
$$\frac{\partial u_1}{\partial \tau} + A_1 v_1 + A_2 v_0 = 0,$$

(2.4.14)
$$A_1 w_0 = 0,$$

(2.4.15)
$$A_1 w_1 + A_2 w_0 = 0.$$

It follows from (2.4.9) that u_0 does not depend on y; then, by integrating (2.4.12) in y, we obtain $\frac{\partial u_0}{\partial \tau} = 0$ and therefore

$$u_0 = u(x,t).$$

Then (2.4.12) reduces to $A_1 v_0 = 0$ and v_0 does not depend on y.

If we use the functions $\chi^j(y,\tau)$ satisfying (1.4.7), (1.4.8), we have from (2.4.10)

$$u_1 = -\chi^j \frac{\partial u}{\partial x_j} + \tilde{u}_1(x,t,\tau).$$

In order to be able to solve (2.4.13) for v_1 we need

$$\int_Y \left(\frac{\partial u_1}{\partial \tau} + A_2 v_0 \right) dy = 0.$$

But since v_0 does not depend on y, $\int_Y A_2 v_0 \, dy = 0$, so that the only way to satisfy this condition is to choose χ^j such that

$$\int_Y \chi^j(y,\tau) \, dy = 0 \text{ for every } \tau,$$

and $\tilde{u}_1$ independent of τ.

Then we can compute v_1 no matter how we choose v_0. Therefore we choose

$$v_0 = 0$$

and v_1 is defined (up to the addition of a function of x, t, τ) by

$$A_1 v_1 - \frac{\partial \chi^j}{\partial \tau} \frac{\partial u}{\partial x_j} = 0.$$

We see from (2.4.14) that w_0 is independent of y and then w_1 can be computed by (2.4.15).

It only remains to satisfy (2.4.11) (observe that this identity is impossible to satisfy if there is no w_0!). By integration **in y and in τ** we find that u satisfies (1.4.11). Summing up: **with the "ansatz"**

(2.4.16) $u_\varepsilon = u + \varepsilon u_1 + \varepsilon^2 u_2 + \cdots + \varepsilon^{3-k} v_1 + \cdots + \varepsilon^k w_0 + \varepsilon^{k+1} w_1 + \cdots,$

where the coefficients are computed above, one has

$$P^{\varepsilon,k} u_\varepsilon = f + \text{ terms of order } \geq k - 1 \text{ in } \varepsilon.$$

REMARK 2.3.   (1) We have introduced the above ansatz on a rather "algebraic" basis; another motivation will be obtained in Chapter 3 (probabilistic methods).

(2) One sees from the above computation that the limit of u_ε does not depend on k (when $1 < k < 3/2$)[9] but that the expansion itself **depends on k**.

---

[9]Of course once the above (formal) proof is **justified**.

(3) For k satisfying

$$\frac{2p-1}{2p} < k < \frac{2p+1}{p+1}, \quad (p \text{ integer} \geq 1),$$

the "ansatz" for the asymptotic expansion will be:

$$u_\varepsilon = u + \varepsilon u_1 + \varepsilon^2 u_2 + \varepsilon^{2p+1-pk} v_0 + \cdots + \varepsilon^k w_0 + \cdots .$$

We can now proceed with the proof of the Theorem.

We construct a function w_ε such that

$$(2.4.17) \qquad \left(\frac{\partial}{\partial t} + A^\varepsilon\right)^* w = \varepsilon^{k-1} g_\varepsilon,$$

where g_ε is bounded in $L^2(\mathcal{G})$ as $\varepsilon \to 0$ and where $w_\varepsilon \to y_i$ in $L^2(Y \times (0, \tau_0))$ as $\varepsilon \to 0$.

One looks for w_ε in the form[10]

$$(2.4.18) \qquad w_\varepsilon = \varepsilon(\alpha(y, \tau) - y_i) + \varepsilon^2 \beta(y, \tau) + \varepsilon^{3-k} \lambda(y, \tau).$$

One finds:

$$(2.4.19) \qquad A_1^*(\alpha - y_i) = 0, \qquad A_1^* \beta + A_2^*(\alpha - y_i) = 0, \qquad -\frac{\partial \alpha}{\partial \tau} + A_1^* \lambda = 0.$$

One defines χ^{i*} by (2.4.5) **and** $\int_Y \chi^{i*}(y, \tau)\, dy = 0$, $\forall \tau$. Then one can take $\alpha = \chi^{i*}$ and one can compute λ from the third equation (2.4.19). One computes β using the second equation (2.4.19) and one obtains the result. The proof is then completed along the usual lines.

2.5. Proof of the homogenization when $k > 2$.

REMARK 2.4.   (1) We are going to prove the theorem only for $k = 3$. The proof for the general case proceeds along the lines of Section 2.4.2.

(2) We assume again that (2.2.19) holds true. See Section 2.6 for the general case.

According to Lemma 2.2 we assume (2.3.1). We are going to construct a function w_ε satisfying

$$(2.5.1) \qquad \left(\frac{\partial}{\partial t} + A^\varepsilon\right)^* w_\varepsilon = \varepsilon g_\varepsilon,$$

such that as $\varepsilon \to 0$

$$(2.5.2) \qquad g_\varepsilon \text{ is bounded in } L^2(\mathcal{G}),$$

$$(2.5.3) \qquad w_\varepsilon \to y_i = P(y) \text{ in } L^2(\mathcal{G}).$$

We use asymptotic expansions. We look for w_ε in the form

$$(2.5.4) \qquad w_\varepsilon = \varepsilon(\alpha(y) - y_i) + \varepsilon^2 \beta(y, \tau) + \varepsilon^3 \gamma(y, \tau), \qquad y = x/\varepsilon, \ \tau = t/\varepsilon^3,$$
$$\alpha \text{ being } Y\text{-periodic and } \beta \text{ and } \tau \text{ being } Y - \tau_0 \text{ periodic.}$$

Using the notations (1.3.8), (1.3.9) we have

$$(2.5.5) \qquad \left(\frac{\partial}{\partial t} + A^\varepsilon\right)^* = \varepsilon^{-3} S_1^* + \varepsilon^{-2} S_2^* + \varepsilon^{-1} S_3^* + \varepsilon^0 S_4^*.$$

---

[10]We do not need in this expansion terms in ε^3 or in ε^k.

Since α does not depend on τ, the coefficient of ε^{-2} is zero. The coefficients of ε^{-1} and of ε^0 are zero iff

(2.5.6) $$S_1^* \beta + S_2^*(\alpha - y_i) = 0,$$

(2.5.7) $$S_1^* \gamma + S_2^* \beta + S_3^*(\alpha - y_i) = 0.$$

Then we find (2.5.1) (assuming we can solve (2.5.6), (2.5.7)) with

$$g_\varepsilon = h + \varepsilon \ell + \varepsilon^2 m,$$

(2.5.8)
$$h = S_2^* \gamma + S_3^* \beta + S_4^*(\alpha - y_i),$$
$$\ell = S_3^* \gamma + S_4^* \beta,$$
$$m = S_4^* \gamma.$$

We can solve (2.5.6) for β iff $\left(S_1^* = -\frac{\partial}{\partial \tau}\right)$:

(2.5.9) $$\mathcal{M}_\tau S_2^*(\alpha - y_i) = 0.$$

But $S_2^* = A_1^*$ and since α does not depend on τ, (2.5.9) is equivalent to

$$\left(\mathcal{M}_\tau A_1^*\right)(\alpha - y_i) = 0.$$

Using the notation (1.6.6), this means

(2.5.10) $$\bar{A}_1^*(\alpha - y_i) = 0.$$

We denote by ϕ^{i*} (compare to (1.6.8)) the solution which is Y-periodic (defined up to an additive constant) of (2.5.10). Then

(2.5.11) $$\alpha \phi^{i*}.$$

Now we can solve (2.5.6); if we set

(2.5.12) $$\beta_0 = \int_0^\tau S_2^*(\sigma)\, d\sigma (\alpha - y_i),$$

then

(2.5.13) $$\beta = \beta_0 + \tilde{\beta}(x, y, t).$$

We can solve (2.5.7) for γ iff

$$\mathcal{M}_\tau(S_2^* \beta + S_3^*(\alpha - y_i)) = 0,$$

i.e.,

(2.5.14) $$\bar{A}_1^* \tilde{\beta} + \mathcal{M}_\tau(S_2^* \beta_0 + S_3^*(\alpha - y_i)) = 0, \ \forall \tilde{\beta} \ Y\text{-periodic},$$

which defines (up to an additive constant) the function $\tilde{\beta}(y)$.

All the functions introduced are C^∞ in $\tilde{Y}$ or in $\bar{Y} \times [0, \tau_0]$.

Then we can solve (2.5.7) for γ. Functions α, β, γ are C^∞ so that in particular

$$|h(y, \tau)| + |k(y, \tau)| + |m(y, \tau)| \leq C$$

and

$$\|g_\varepsilon\|_{L^\infty(\mathcal{G})} \leq C.$$

We now use (2.5.1). We take $\phi \in C_0^\infty(\mathcal{G})$ and, as before, we multiply the equation for u_ε by ϕw_ε and (2.5.1) by ϕu_ε. We find

$$(2.5.15) \qquad -\int_{\mathcal{G}} u_\varepsilon w_\varepsilon \frac{\partial \phi}{\partial t}\, dx\, dt + \int_{\mathcal{G}} \left[\xi_i^\varepsilon \frac{\partial \phi}{\partial x_i} w_\varepsilon - a_{ij}^\varepsilon \frac{\partial w_\varepsilon}{\partial x_i} \frac{\partial \phi}{\partial x_j} u_\varepsilon \right]\, dx\, dt$$

$$= \int_{\mathcal{G}} f\phi w_\varepsilon\, dx\, dt - \varepsilon \int_{\mathcal{G}} g_\varepsilon \phi u_\varepsilon\, dx\, dt.$$

We can pass to the limit and we find

$$-\int_{\mathcal{G}} uw \frac{\partial \phi}{\partial t}\, dx\, dt + \int_{\mathcal{G}} \xi_i \frac{\partial \phi}{\partial x_i} P\, dx\, dt - \mathcal{M}\left(a_{kj} \frac{\partial(\alpha - y_i)}{\partial y_k} \right) \int_{\mathcal{G}} \frac{\partial \phi}{\partial x_j} u\, dx\, dt$$

$$= \int_{\mathcal{G}} f\phi P\, dx\, dt,$$

and finally

$$(2.5.16) \qquad \xi_i = \mathcal{M}\left(a_{ij} - a_{kj} \frac{\partial \phi^{i*}}{\partial y_k} \right) \frac{\partial u}{\partial x_j}.$$

It remains only to verify that this formula coincides with what we have obtained in (1.6.14), which is left to the reader.

2.6. Proof of the homogenization formulas when $a_{ij} \in L^\infty(\mathbb{R}_y^n \times \mathbb{R}_\tau)$ using L^p estimates. We show now that the assumption (2.2.19) **is not necessary** for Theorem 2.1 to be true. The hypothesis $a_{ij} \in L^\infty(\mathbb{R}_y^n \times \mathbb{R}_\tau)$ is sufficient. We assume for a moment the following estimate: for the Dirichlet's boundary conditions there exists $p > 2$, independent of ε, such that

$$(2.6.1) \qquad \left\| \frac{\partial u_\varepsilon}{\partial x_i} \right\|_{L^p(\mathcal{G})} \leq C.$$

The proof of this result is given in Section 2.7 below.

We introduce q such that

$$(2.6.2) \qquad \frac{1}{p} + \frac{1}{q} = \frac{1}{2}.$$

We can find a sequence of functions a_{ij}^β such that

$(2.6.3)$

$a_{ij}^\beta \in C^\infty(\bar{Y} \times [0, \tau_0])$, a_{ij}^β is periodic together with all its derivatives, and

$$\sup_{i,j} \|a_{ij}^\beta - a_{ij}\|_{L^q(Y \times [0,\tau_0])} = \rho(\beta) \to 0 \text{ as } \beta \to \infty.$$

With the notations of Section 2.3 we show that

$$(2.6.4) \qquad \|u_\varepsilon^\beta - u_\varepsilon\|_{L^2(0,T;V)} \leq C\rho(\beta).$$

Then Lemma 2.2 is still valid and the Theorem is proven.

With the notations of the end of Section 2.3 we observe that

$$|(f_\varepsilon, v)| \leq C\rho(\beta)\|\nabla u_\varepsilon\|_{(L^p(\mathcal{O}))^n}\|v\|$$

so that

$$(2.6.5) \qquad \|f_\varepsilon\|_* \leq C\rho(\beta)\|\nabla u_\varepsilon\|_{(L^p(\mathcal{O}))^n}.$$

By virtue of (2.6.1) it follows that

(2.6.6) $$\|f_\varepsilon\|_{L^p(0,T;V')} \le C\rho(\beta)$$

so that in particular

$$\|f_\varepsilon\|_{L^2(0,T;V')} \le C\rho(\beta)$$

and (2.6.4) follows.

For other boundary conditions than Dirichlet, the analogue of (2.6.1) is valid **locally**, i.e. for $\mathcal{G}' \subset \mathcal{O}' \times]0, T[$, $\overline{\mathcal{O}}' \subset \mathcal{O}$, and the analogue of (2.6.4) is valid locally, i.e. in

$$L^2(0,T;H^1(\mathcal{O}')).$$

2.7. The L^p estimates. We now prove the L^p estimates used in Section 2.6 above. They are of independent interest. The method of proof is analogous to the one used in Chapter 1, Section 4, for elliptic problems. The notations are those of this Section 4, Chapter 1; in particular, we use norms (4.1.5), (4.1.7), (4.1.8), (4.1.11) of that section. We define

(2.7.1) $$X_p = L^p(0,T;W_0^{1,p}(\mathcal{O}))$$

provided with the norm

(2.7.2) $$\left(\int_0^T \|v(t)\|_{W_0^{1,p}(\mathcal{O})}^p \, dt \right)^{1/p} = \|\nabla v\|_{L^p(\mathcal{G})^n}.$$

We introduce

(2.7.3) $$Y_p = L^p(0,T;W^{-1,p}(\mathcal{O}))$$

and we observe that the mapping $v \to \operatorname{div}_x v$ maps $(L^p(\mathcal{G}))^n$ onto $L^p(0,T;W^{-1,p}(\mathcal{O}))$. We then provide Y_p with the norm

(2.7.4) $$\|f\|_{Y_p} = \inf_{\operatorname{div}_x g = f} \|g\|_{(L^p(\mathcal{G}))^n}.$$

We consider a **family** of functions a_{ij} such that

(2.7.5)
$$a_{ij} \in L^\infty(\mathcal{G}),$$
$$a_{ij}(x,t)\xi_i\xi_j \ge \alpha\xi_i\xi_i, \ \alpha > 0, \ \text{a.e. in } \mathcal{G}, \ \xi_i \in \mathbb{R}.$$

We denote by $[a_{ij}(x,t)]$ the $n \times n$ matrix with entries $a_{ij}(x,t)$ and we set

(2.7.6) $$\operatorname*{ess\,sup}_{x,t \in \mathcal{G}} \|[a_{ij}(x,t)]\|_{\mathcal{L}(\mathbb{R}^n)} = \beta.$$

We set

(2.7.7) $$A = A(t) = -\frac{\partial}{\partial x_i}\left(a_{ij}(x,t)\frac{\partial}{\partial x_j} \right).$$

THEOREM 2.5. *Let f and u^0 be given such that $f \in L^2(0,T;H^{-1}(\mathcal{O}))$ and $u^0 \in H = L^2(\mathcal{O})$. Let u be the solution of*

(2.7.8) $$\frac{\partial u}{\partial t} + Au = f \ in \ \mathcal{G},$$

(2.7.9) $$u \in L^2(0,T;H_0^1(\mathcal{O})),$$

(2.7.10) $$u(0) = u^0.$$

Then assuming Γ to be smooth enough there exists $p > 2$, depending only on α, β and $\mathcal{O}$, such that if

$$(2.7.11) \qquad\qquad f \in L^p(0, T; W^{-1,p}(\mathcal{O}))$$

and if[11]

$$(2.7.12) \qquad\qquad u^0 \in W_0^{1,p}(\mathcal{O}),$$

then $u \in L^p(0, T; W_0^{1,p}(\mathcal{O}))$ and remains in a bounded set of this space when f and u^0 remain bounded in $L^p(0, T; W^{-1,p}(\mathcal{O}))$ and in $W_0^{1,p}(\mathcal{O})$.

REMARK 2.6. We can reduce the problem to the case when $u^0 = 0$. Indeed we can find $\phi \in L^p(0, T; W_0^{1,p}(\mathcal{O}))$, $\frac{\partial \phi}{\partial t} \in L^p(0, T; W_0^{1,p}(\mathcal{O}))$, ϕ depending continuously on u^0 in the corresponding topology, so that $\phi(0) = u^0$ and then we consider $u - \phi$ instead of u.

PROOF OF THEOREM 2.5. As in Section 4.2 of Chapter 1, we can reduce the problem to

$$(2.7.13) \qquad\qquad \frac{1}{(\beta + c)} \frac{\partial u}{\partial t} + (A_1 + A_2)u + \frac{1}{\beta + c} f$$

where

$$(2.7.14) \qquad\qquad A_k = -\frac{\partial}{\partial x_i}\left(a_{ij}^k(x, t)\frac{\partial}{\partial x_j}\right), \qquad k = 1, 2,$$

$$(2.7.15) \qquad\qquad a_{ij}^1 = a_{ji}^1, \qquad a_{ij}^1 \xi_i \xi_j \geq \mu \xi_i \xi_i, \qquad 0 < \mu < 1,$$

$$(2.7.16) \qquad\qquad \|[a_{ij}^2(x, t)]\|_{\mathcal{L}(\mathbb{R}^n)} \leq \nu, \text{ a.e. in } \mathcal{G},$$

$$(2.7.17) \qquad\qquad \nu < \mu.$$

By changing the scale of time, we see that we are reduced to proving the theorem for

$$(2.7.18) \qquad \frac{\partial u}{\partial t} + A_1 u + A_2 u = f, \qquad u(0) = 0, \qquad f \text{ satisfying } (2.7.11),$$
$$u = 0 \text{ on } \Sigma = \Gamma \times]0, T[,$$

and with $(2.7.14), \ldots, (2.7.17)$.

We set

$$(2.7.19) \qquad\qquad P = \frac{\partial}{\partial t} = \Delta.$$

It is known that, given $F \in Y_p$, (**where p is arbitrary for the time being**), there exists a unique u such that

$$(2.7.20) \qquad\qquad Pu = F \text{ in } \mathcal{G},$$
$$u \in X_p, \qquad u(0) = 0.$$

Equation $(2.7.18)$ is equivalent to

$$(2.7.21) \qquad\qquad Pu + (A_1 + \Delta)u + A_2 u = f,$$

and if the solution u of $(2.7.20)$ is denoted by

$$(2.7.22) \qquad\qquad u = P^{-1}F$$

---

[11]This hypothesis can be very much improved.

then (2.7.21) is equivalent to

$$(2.7.23) \qquad u + (P^{-1}(A_1 + \Delta) + P^{-1}A_2)u = P^{-1}f.$$

We introduce

$$(2.7.24) \qquad k(p) = \|P^{-1}(A_1 + \Delta) + P^{-1}A_2\|_{\mathcal{L}(X_p;X_p)}$$

and the Theorem will be proven if we verify that

$$(2.7.25) \qquad \textbf{one can find } p > 2 \textbf{ such that } k(p) < 1.$$

We have

$$(2.7.26) \qquad k(p) \leq \|P^{-1}\|_{\mathcal{L}(Y_p;X_p)}\|A_1 + \Delta + A_2\|_{\mathcal{L}(X_p;Y_p)}.$$

But exactly as in Chapter 1, Section 4.3 (t plays the role of a parameter) we have

$$\|A_1 + \Delta\|_{\mathcal{L}(X_p;Y_p)} \leq 1 - \mu,$$
$$\|A_2\|_{\mathcal{L}(X_p;Y_p)} \leq \nu$$

so that

$$(2.7.27) \qquad k(p) \leq (1 - \mu + \nu)\|P^{-1}\|_{\mathcal{L}(Y_p;X_p)}.$$

It is straightforward to verify that $\|P^{-1}\|_{\mathcal{L}(Y_2;X_2)} = 1$ so that, since $1 - \mu + \nu < 1$, we shall have (2.7.25) if we can prove that

$$(2.7.28) \qquad \text{there exists a function } \rho(p) \text{ which is continuous of } p \geq 2,$$
$$\text{such that } \|P^{-1}\|_{\mathcal{L}(Y_p;X_p)} \leq \rho(p) \text{ and } \rho(2) = 1.$$

Indeed we have then $k(p) < \rho(p)$ and we can find $p_0 > 2$ such that $k(p_0) < 1$.

PROOF OF (2.7.28). We take $F = \operatorname{div} g$, $g \in (L^p(\mathcal{G}))^n$, in (2.7.20) and we consider the mapping

$$(2.7.29) \qquad g \xrightarrow{\pi} \operatorname{grad}_x u,$$

which maps continuously

$$(2.7.30) \qquad (L^{P_0}(\mathcal{G}))^n \to (L^{P_0}(\mathcal{G}))^n, \qquad \text{norm } \omega_0,$$
$$(2.7.31) \qquad (L^2(\mathcal{G}))^n \to (L^2(\mathcal{G}))^n, \qquad \text{norm } 1.$$

Using Riesz-Thorin's theorem it follows that

$$(2.7.32) \qquad \|\operatorname{grad}_x u\|_{(L^p(\mathcal{G}))^n} \leq \omega_0^{1-\theta(p)}\|g\|_{(L^p(\mathcal{G}))^n},$$
$$\frac{1}{p} = \frac{1 - \theta(p)}{p_0} + \frac{\theta(p)}{2}.$$

Therefore, according to (2.7.2) and to (2.7.4), we have

$$\|u\|_{X_p} \leq \omega_0^{1-\theta(p)} \inf_{\operatorname{div}_x g = F} \|g\|_{(L^p(\mathcal{G}))^n}$$
$$= \omega_0^{1-\theta(p)}\|F\|_{Y_p},$$

so that we can take $\rho(p) = \omega_0^{1-\theta(p)}$ and the result follows.   ∎

∎

REMARK 2.7. We have, **for arbitrary boundary conditions**, a local result (as in Theorem 4.5, Chapter 1): there exists $p > 2$ (and not too large: $p \leq \frac{2n}{n-2}$) such that if f and u^0 satisfy **locally** (2.7.11) and (2.7.12), then

$$(2.7.33) \qquad \phi u \in L^p(0, T; W_0^{1,p}(\mathcal{O})).$$

REMARK 2.8. It follows of course from these estimates that, if f and u_0 satisfy conditions of the type (2.7.11), (2.7.12), then

$$(2.7.34) \quad \frac{\partial u_\varepsilon}{\partial x_i} \to \frac{\partial u}{\partial x_i} \text{ in } L^p(\mathcal{G}) \text{ weakly or in } L^p(0, T; L^p(\mathcal{O}')) \text{ weakly, for } \overline{\mathcal{O}}' \subset \mathcal{O}.$$

2.8. The adjoint expansion. Let us remark that convergence can be proven for **Dirichlet's boundary conditions** and when coefficients a_{ij} are smooth enough (cf. Section 2.3) using the method of the adjoint expansion as in Chapter 1, Section 3.3. We present this briefly, since the idea is **exactly analogous to the one in the elliptic case**. We can always assume that $u^0 = 0$, and we write the equation in the equivalent form

$$(2.8.1) \qquad \left(u_\varepsilon, \left(-\frac{\partial}{\partial t} + A^{\varepsilon *} \right) v_\varepsilon \right) = (f, v_\varepsilon)$$

where here $(f, v) = \int_{\mathcal{G}} fv \, dx \, dt$.

Given $v \in \mathcal{D}(\mathcal{G})$, we construct — **using asymptotic expansions** — a sequence of functions v_ε such that $v_\varepsilon \to v$ in $L^2(\mathcal{G})$ weakly (say), and such that

$$(2.8.2) \qquad \left(-\frac{\partial}{\partial t} + A^{\varepsilon *} \right) v_\varepsilon \to \left(-\frac{\partial}{\partial t} + \mathcal{A}^* \right) v \text{ in } L^2(\mathcal{G}) \text{ weakly,}$$

where $\mathcal{A}^* =$ adjoint of the operator $\mathcal{A}$ constructed in Section 1, i.e. $\mathcal{A} = \mathcal{A}^1$, $\mathcal{A}^2$ or $\mathcal{A}^3$ according to the case $k < 1$, $k = 2$, $k > 3$. Then

$$\left(u, \left(-\frac{\partial}{\partial t} + \mathcal{A}^* \right) v \right) = (f, v)$$

and the result follows.

2.9. Use of the maximum principle. Assuming **all data to be very smooth** and for the **Dirichlet's boundary conditions**, we have

$$(2.9.1) \qquad \|u_\varepsilon - u\|_{L^\infty(\mathcal{G})} \leq C\varepsilon.$$

Indeed let us write

$$(2.9.2) \qquad P^\varepsilon = \frac{\partial}{\partial t} + A^\varepsilon = \varepsilon^{-2} P_1 + \varepsilon^{-1} P_2 + \varepsilon^0 P_3,$$

valid (with **different** P_i's) for $k = 1$ and 2. (For $k = 3$ we use the same method and add an $\varepsilon^3 u_3$ term in the proof below.) We introduce (with notations of Section 1),

$$(2.9.3) \qquad z_\varepsilon = u_\varepsilon - (u + \varepsilon u_1 + \varepsilon^2 u_2)$$

and we observe that

$$(2.9.4) \qquad P^\varepsilon z_\varepsilon = -\varepsilon(P_2 u_2 + P_3 u_1) - \varepsilon^2 P_3 u_2.$$

If all data are C^∞, the right hand side of (2.9.4) is $O(\varepsilon)$ in the $L^\infty(\mathcal{G})$ norm. But

$$\|z_\varepsilon\|_{L^\infty(\Sigma)} = \|\varepsilon u_1 + \varepsilon^2 u_2\|_{L^\infty(\Sigma)} = O(\varepsilon)$$

and

$$\|z_\varepsilon(0)\|_{L^\infty(\mathcal{O})} = \|\varepsilon u_1(0) + \varepsilon^2 u_2(0)\|_{L^\infty(\mathcal{O})} = O(\varepsilon)$$

so that, **by using the maximum principle**

$$\|z_\varepsilon\|_{L^\infty(\mathcal{G})} = O(\varepsilon)$$

hence (2.9.1) follows.

2.10. Higher order equations and systems. Except in Section 2.9 above, we never used the maximum principle. All results obtained so far extend to higher order equations and to systems, i.e. to the "parabolic analogue" of the problems considered in Chapter 1, Sections 9 and 10. The results of Section 6, Chapter 1 also extend easily. Let us consider an equation of the type

$$(2.10.1) \qquad \frac{\partial u_\varepsilon}{\partial t} - \frac{\partial}{\partial x_i}\left(a_{ij}\left(x, \frac{x}{\varepsilon}, t, \frac{t}{\varepsilon^k} \right) \frac{\partial u_\varepsilon}{\partial x_j} \right) = f,$$

where we suppose that

$$(2.10.2) \qquad a_{ij}(x, y, t, \tau) \in C^0(\overline{\mathcal{G}}; L_p^\infty(\mathbb{R}_y^n \times \mathbb{R}_\tau))$$

where $L_p^\infty(\mathbb{R}_y^n \times \mathbb{R}_\tau)$ = space of L^∞ functions of y and τ which are $Y \times \tau_0$ periodic. We also assume that

$$(2.10.3) \qquad a_{ij}(x, y, t, \tau)\xi_i\xi_j \geq \alpha \xi_i\xi_i, \qquad \alpha > 0, \qquad \xi_i \in \mathbb{R}.$$

Then we denote by $\mathcal{A}^k(x, t)$ the homogenized operator

$$(2.10.4) \qquad \mathcal{A}^k(x, t) = -\frac{\partial}{\partial x_i} q_{ij}^k(x, t) \frac{\partial}{\partial x_j}$$

where the coefficients $q_{ij}^k(x, t)$ are computed as the q_{ij}'s in Section 1, with x and t playing the role of parameters. If u^k denotes the solution of

$$(2.10.5) \qquad \begin{aligned} &u^k \in L^2(0, T; V), \\ &\frac{\partial u^k}{\partial t} + \mathcal{A}^k(x, t)u^k = f, \\ &u^k(0) = u^0, \end{aligned}$$

then

$$(2.10.6) \qquad u_\varepsilon \to u^k \text{ in } L^2(0, T; V) \text{ weakly as } \varepsilon \to 0.$$

One can also extend, always with the same type of method, the results of Chapter 1, Section 8.

Let us consider now briefly the "parabolic analogue" of the situation of Section 11, Chapter 1. We assume that the space dimension equals 3, and we denote by $a^\varepsilon(x, t)$ a 3×3 matrix. We assume that

$$(2.10.7) \qquad a^\varepsilon(x, t) = a(x/\varepsilon, t/\varepsilon^k), \qquad a(y, \tau) = \|a_{ij}(y, \tau)\|,$$

$a_{ij}(y, \tau)$ having the same properties as in (1.1.6).

We consider, with the notations of Chapter 1, Section 11:

$$(2.10.8) \qquad \mathcal{H}_0(\mathcal{O}, \text{rot}) \subseteq V \subseteq \mathcal{H}(\mathcal{O}, \text{rot}),$$

and we observe that there exists a unique function u_ε such that

(2.10.9)
$$\left(\frac{\partial u_\varepsilon}{\partial t}, v\right) + (a^\varepsilon, \operatorname{rot} u_\varepsilon, \operatorname{rot} v) = (f, v), \qquad \forall v \in V,$$

$$u_\varepsilon \in L^2(0, T; V), \qquad u_\varepsilon(0) = u^0,$$

where $f \in L^2(0, T; V')$, $u^0 \in H = (L^2(\mathcal{O}))^3$.

2.10.1. *Asymptotic expansions.* We give only **the results** of the asymptotic expansion; one has to consider three cases.

2.10.2. *Case $k = 1$.* We define $\tilde{\chi}^p(y, \tau)$ as a $Y - \tau_0$ periodic solution of

(2.10.10)
$$\operatorname{rot}_y a(y, \tau)(\operatorname{rot}_y \tilde{\chi}^p - e_p) = 0,$$

$$\operatorname{div}_y \tilde{\chi}^p = 0.$$

Then

(2.10.11)
$$\mathcal{A}^1(u, v) = (\mathcal{M}(a - a \operatorname{rot}_y \tilde{\chi}) \operatorname{rot} u, \operatorname{rot} v)$$

and the **homogenized equation** is

(2.10.12)
$$\left(\frac{\partial u}{\partial t}, v\right) + \mathcal{A}^1(u, v) = (f, v), \qquad \forall v \in V.$$

2.10.3. *Case $k = 2$.* We define $\tilde{\theta}^p(y, \tau)$ as a $Y - \tau_0$ periodic solution of

(2.10.13)
$$\frac{\partial \tilde{\theta}^p}{\partial \tau} + \operatorname{rot}_y a(y, \tau)(\operatorname{rot}_y \tilde{\theta}^p - e_p) = 0,$$

$$\operatorname{div}_y \tilde{\theta}^p(y, \tau) = 0.$$

Then

(2.10.14)
$$\mathcal{A}^2(u, v) = (\mathcal{M}(a - a \operatorname{rot}_y \tilde{\theta}) \operatorname{rot} u, \operatorname{rot} v)$$

and the homogenized equation is the analogue of (2.10.12) with $\mathcal{A}^2$ instead of $\mathcal{A}^1$.

2.10.4. *Case $k = 3$.* We define

(2.10.15)
$$\bar{a}(y) = \mathcal{M}_\tau a(y, \tau)$$

and we define $\tilde{\phi}^p(y)$ as a Y-periodic solution of

(2.10.16)
$$\operatorname{rot}_y \bar{a}(\operatorname{rot}_y \tilde{\phi}^p - e_p) = 0,$$

$$\operatorname{div}_y \phi^p = 0.$$

Then

(2.10.17)
$$\mathcal{A}^3(u, v) = (\mathcal{M}_y(\bar{a} - \bar{a} \operatorname{rot}_y \tilde{\phi}) \operatorname{rot} u, \operatorname{rot} v)$$

and the homogenized equation is the analogue of (2.10.12) with $\mathcal{A}^3$ instead of $\mathcal{A}^1$.

The **justification** of these formulas is simple if we assume:

(2.10.18)
$$\operatorname{div}_x f \in L^2(\mathcal{G}), \qquad \operatorname{div} u^0 \in L^2(\mathcal{O}).$$

Indeed, we define z^0 as the solution of

(2.10.19)
$$\Delta z^0 = \operatorname{div} u^0, \qquad z^0 \in H_0^1(\mathcal{O})$$

and **assuming the boundary Γ smooth enough**, we have

(2.10.20)
$$z^0 \in H^2(\mathcal{O}) \cap H_0^1(\mathcal{O}).$$

We then define z as the solution of

(2.10.21)
$$\frac{\partial}{\partial t}\Delta z = \operatorname{div}_x f, \qquad z(0) = z^0, \qquad z = 0 \text{ on } \Sigma.$$

(This is an elliptic problem: $\Delta z = \int_0^t \operatorname{div}_x f(x,\sigma)\, d\sigma + \Delta z^0$); we have

$$z \in L^2(0,T; H^2(\mathcal{O}) \cap H_0^1(\mathcal{O}))$$

and $\frac{\partial z}{\partial t} \in L^2(0,T; H^2(\mathcal{O}) \cap H_0^1(\mathcal{O}))$. Then if we introduce

$$\tilde{u}_\varepsilon = u_\varepsilon - \nabla z$$

we see that

$$\left(\frac{\partial \tilde{u}_\varepsilon}{\partial t}, v\right) + (a\operatorname{rot}\tilde{u}_\varepsilon, \operatorname{rot} v) = \left(f - \frac{\partial}{\partial t}\nabla z, v\right) = (\tilde{f}, v),$$

where $\tilde{f} = f - \frac{\partial}{\partial t}\nabla z$, so that $\operatorname{div}_x \tilde{f} = 0$. Since $\operatorname{div}\tilde{u}_\varepsilon(0) = \operatorname{div} u^0 - \Delta z^0 = 0$, we see that $\operatorname{div}\tilde{u}_\varepsilon = 0$.

Therefore we can assume (if (2.10.8) holds true) that

$$\operatorname{div}_x f = 0, \qquad \operatorname{div}_x u^0 = 0, \qquad \operatorname{div}_x u_\varepsilon = 0.$$

Then one can prove the above formulas by arguments similar to those of Sections 2.2, 2.4, 2.5.

REMARK 2.9. A proof along the lines of Chapter 1, Sections 11.4, 11.5, does not seem to work in the present situation. The homogenization problem seems to be open if we do not assume (2.10.18).

2.11. Correctors.

2.11.1. *Orientation.* We now return to the situation of (1.2.9), (1.2.10), (1.2.11). We want to introduce first order correctors θ_ε such that

(2.11.1) $u_\varepsilon - (u + \theta_\varepsilon) \to 0$ in $L^2(0,T; H^1(\mathcal{O}))$ **strongly**.

REMARK 2.10. By using cut-off functions $m_\varepsilon(x)$, as in Chapter 1, Section 5.1, we can have correctors giving an approximation in $L^2(0,T;V)$ strongly.

REMARK 2.11. We did not make any attempt to find the analogous results to those of Chapter 1, Section 18.

NOTATIONS. We shall denote by u_1 the function:

(2.11.2)
$$u_1 = \begin{cases} -\chi^j \frac{\partial u}{\partial x_j} & \text{if } k = 1, \\ -\theta^j \frac{\partial u}{\partial x_j} & \text{if } k = 2, \\ -\phi^j \frac{\partial u}{\partial x_j} & \text{if } k = 3. \end{cases}$$

We shall also denote by the **same** notation $u_2, u_3, \ldots$, the next terms of the asymptotic expansion, for $k = 1, 2, 3$. We shall use

(2.11.3)
$$\tilde{u}_\varepsilon = \begin{cases} u + \varepsilon u_1 & \text{if } k = 1, 2, \\ u + \varepsilon u_1 + \varepsilon^2 u_2 & \text{if } k = 3. \end{cases}$$

THEOREM 2.12. *We assume the hypothesis of Theorem 2.1 to hold true and we assume all data smooth enough.*[12] *Then*

(2.11.4) $u_\varepsilon - (u + \varepsilon u_1) \to 0$ in $L^2(0,T;V)$ strongly.

---

[12]This is made more precise in the proof below.

PROOF. We introduce

$$(2.11.5) \qquad \zeta_\varepsilon = u_\varepsilon - \tilde{u}_\varepsilon$$

and we consider

$$(2.11.6) \qquad X_\varepsilon = \int_0^T (T-t)[(\zeta_\varepsilon', \zeta_\varepsilon) + a^\varepsilon(\zeta_\varepsilon)]\, dt.$$

We observe that

$$X_\varepsilon = -\frac{T}{2}|\zeta_\varepsilon(0)|^2 + \frac{1}{2}\int_0^T |\zeta_\varepsilon|^2\, dt + \int_0^T (T-t)a^\varepsilon(\zeta_\varepsilon)\, dt.$$

Since $|\zeta_\varepsilon(0)|^2 = \varepsilon^2|u_1(0)|^2$ or $\varepsilon^2|u_1(0) + \varepsilon u_2(0)|^2$, we see that, if we prove that $X_\varepsilon \to 0$ as $\varepsilon \to 0$, then we shall have (2.11.4) with T replaced by $T-\eta$, $\eta > 0$, but since T does not play any role, this gives the general result.

We can write

$$X_\varepsilon = \int_0^T (T-t)[(u_\varepsilon', u_\varepsilon) + a^\varepsilon(u_\varepsilon)]\, dt$$
$$+ \int_0^T (T-t)[(\tilde{u}_\varepsilon', \tilde{u}_\varepsilon) + a^\varepsilon(\tilde{u}_\varepsilon)]\, dt - Y_\varepsilon - Z_\varepsilon,$$

$$Y_\varepsilon = \int_0^T (T-t)[(\tilde{u}_\varepsilon', u_\varepsilon) + a^\varepsilon(u_\varepsilon, \tilde{u}_\varepsilon)]\, dt,$$

$$Z_\varepsilon = \int_0^T (T-t)[(\tilde{u}_\varepsilon', u_\varepsilon) + a^\varepsilon(\tilde{u}_\varepsilon, u_\varepsilon)]\, dt.$$

The first term in X_ε equals

$$\int_0^T (T-t)(f, u_\varepsilon)\, dt \to \int_0^T (T-t)(f, u)\, dt = \int_0^T (T-t)[(u', u) + \mathcal{A}(u, u)]\, dt.^{[13]}$$

The second term in X_ε equals

$$-\frac{T}{2}|\tilde{u}_\varepsilon(0)|^2 + \int_0^T (T-t)a^\varepsilon(\tilde{u}_\varepsilon)\, dt.$$

One verifies, exactly as in Chapter 1, Section 5, (5.3.4), that $\int_0^T (T-t)a^\varepsilon(\tilde{u}_\varepsilon)\, dt \to \int_0^T (T_t)\mathcal{A}(u, u)\, dt$, so that the second term tends to $-\frac{T}{2}|u^0|^2 + \int_0^T (T-t)\mathcal{A}(u, u)\, dt = \int_0^T (T-t)[(u'.u) + \mathcal{A}(u, u)]\, dt$. Therefore, if we show that

$$(2.11.7) \qquad Y_\varepsilon \to \int_0^T (T-t)[(u', u) + \mathcal{A}(u, u)]\, dt,$$

$$(2.11.8) \qquad Z_\varepsilon \to \int_0^T (T-t)[(u', u) + \mathcal{A}(u, u)]\, dt,$$

it will follows that $X_\varepsilon \to 0$, hence the Theorem follows.

---

[13]We set $\mathcal{A} = \mathcal{A}^k$, $k = 1, 2, 3$ and $\mathcal{A}(u, v) = (\mathcal{A}u, v)$.

Let ϕ be a given function in $C_0^\infty(\mathcal{O})$, $\phi = 1$ except on a set E (near Γ) of measure $|E|$. We define (compare to Chapter 1, (5.3.11), (5.3.12))

$$Y_{\varepsilon\phi} = \int_0^T (T - t)[(u_\varepsilon', \tilde{u}_\varepsilon\phi) + a_\phi^\varepsilon(u_\varepsilon, \tilde{u}_\varepsilon)]\, dt,$$

$$z_{\varepsilon\phi} = \int_0^T (T - t)[(\tilde{u}_\varepsilon', u_\varepsilon\phi) + a_\phi^\varepsilon(\tilde{u}_\varepsilon, u_\varepsilon)]\, dt.$$

We see exactly as in Chapter 1, Section 5.3, that (2.11.7), (2.11.8) will be proven if we verify that, for **fixed** ϕ,

$$(2.11.9) \qquad Y_{\varepsilon\phi} \to \int_0^T (T - t)[(u', u\phi) + \mathcal{A}_\phi(u, u)]\, dt,$$

$$Z_{\varepsilon\phi} \to \int_0^T (T - t)[(u', u\phi) + \mathcal{A}_\phi(u, u)]\, dt.$$

But

$$Y_{\varepsilon\phi} = \int_{\mathcal{G}} (T - t)\left(\frac{\partial u_\varepsilon}{\partial t} + A^\varepsilon u_\varepsilon\right) \tilde{u}_\varepsilon\phi\, dx\, dt - \int_{\mathcal{G}} (T - t)a_{ij}^\varepsilon \frac{\partial u_\varepsilon}{\partial x_j} \tilde{u}_\varepsilon \frac{\partial \phi}{\partial x_i}\, dx\, dt.$$

If we denote by q_{ij} the coefficients of $\mathcal{A}^k$, $k = 1, 2, 3$, we have

$$Y_{\varepsilon\phi} \to \int_{\mathcal{G}} (T - t)fu\phi\, dx\, dt - \int_{\mathcal{G}} (T - t)q_{ij} \frac{\partial u}{\partial x_j} u \frac{\partial \phi}{\partial x_i}\, dx\, dt$$

$$= \int_{\mathcal{G}} (T - t)\left[(u', u) + \left(q_{ij}\frac{\partial u}{\partial x_j}, \frac{\partial(u\phi)}{\partial x_i} - u\frac{\partial \phi}{\partial x_i}\right)\right] dt$$

hence (2.11.9) follows for $Y_{\varepsilon\phi}$. We observe next that

$$(2.11.10)$$

$$Z_{\varepsilon\phi} = \int_0^T (T - t)\left[(\tilde{u}_\varepsilon', u_\varepsilon\phi) + \left(a_{ij}^\varepsilon\frac{\partial \tilde{u}_\varepsilon}{\partial x_j}, \frac{\partial}{\partial x_i}(u_\varepsilon\phi) - u_\varepsilon\frac{\partial \phi}{\partial x_i}\right)\right] dt$$

$$= \int_0^T (T - t)\left(\frac{\partial \tilde{u}_\varepsilon}{\partial t_\varepsilon} + A^\varepsilon \tilde{u}_\varepsilon, u_\varepsilon\phi\right) dt - \int_0^T (T - t)\left(a_{ij}^\varepsilon\frac{\partial \tilde{u}_\varepsilon}{\partial x_j}, u_\varepsilon\frac{\partial \phi}{\partial x_i}\right) dt.$$

But from the definition of $\tilde{u}_\varepsilon$ it follows that $\frac{\partial \tilde{u}_\varepsilon}{\partial t} + A^\varepsilon \tilde{u}_\varepsilon \to \frac{\partial u}{\partial t} + \mathcal{A}u$ in $L^2(\mathcal{G})$ weakly, so that (2.11.10) gives

$$(2.11.11) \qquad Z_{\varepsilon\phi} \to \int_0^T (T - t)\left[(u', u\phi) + \mathcal{A}(u, u\phi) - \left(q_{ij}\frac{\partial u}{\partial x_j}, u\frac{\partial \phi}{\partial x_i}\right)\right] dt$$

hence the result follows.   ∎

REMARK 2.13. We can also write the equation as a first order system and develop arguments along the lines of Chapter 1, Section 5.5. For **Dirichlet's boundary conditions** we obtain in this manner error estimates which do not rely on the maximum principle.

2.12. Non-linear problems.

2.12.1. *Orientation.* We confine ourselves to **one** aspect of the homogenization of non-linear parabolic equations (or inequalities), namely to Variational Inequalities (V.I.) of evolution. Many other cases can be treated by using the **methods** introduced below.

2.12.2. *Setting of the problem.* We consider operators A^ε as in Section 1.1; $A^\varepsilon = A(x/\varepsilon, t/\varepsilon^k)$. **Formally** we want to consider V.I. of the following types:

$$(2.12.1) \qquad \frac{\partial u_\varepsilon}{\partial t} + A^\varepsilon u_\varepsilon - f \le 0, \qquad u_\varepsilon \le 0,$$

$$\left(\frac{\partial u_\varepsilon}{\partial t} + A^\varepsilon u_\varepsilon - f \right) u_\varepsilon = 0 \text{ in } \mathcal{G},$$

u_ε being subject to usual boundary and initial conditions. To simplify a little, we shall take:

$$(2.12.2) \qquad u_\varepsilon(x, 0) = 0.$$

The variational form of (2.12.1) is as follows: let f be given in $L62(0, T; H)$, then we are looking for u_ε such that:

$$u_\varepsilon \in L^2(0, T; V),$$
$$(2.12.3) \qquad \frac{\partial u_\varepsilon}{\partial t} \in L^2(0, T; V'), \qquad u_\varepsilon \le 0, \qquad \text{a.e. in } \mathcal{G},$$

and

$$(2.12.4) \qquad \int_0^T \left[\left(\frac{\partial u_\varepsilon}{\partial t}, v - u_\varepsilon \right) + a^\varepsilon(u_\varepsilon, v - u_\varepsilon) - (f, v - u_\varepsilon) \right] dt \ge 0,$$
$$\forall v \in L^2(0, T; V), \qquad v \le 0 \text{ in } \mathcal{G},$$

and such that (2.12.2) holds true.

We shall show below that **this problem admits a unique solution.** The main result we want to prove is

THEOREM 2.14. *We assume that the hypotheses of Theorem 2.1 hold true. We denote by $\mathcal{A}$ the homogenized operator found in Theorem 2.1 (i.e. $\mathcal{A} = \mathcal{A}^k$, $k = 1, 2, 3$) and by $\mathcal{A}(u, v)$ the corresponding bilinear form on V. Let u be the solution of the V.I. of evolution*

$$(2.12.5) \qquad u \in L^2(0, T; V), \qquad \frac{\partial u}{\partial t} \in L^2(0, T; V'), \qquad u \le 0, \text{ a.e. in } \mathcal{G},$$

$$(2.12.6) \qquad \int_0^T \left[\left(\frac{\partial u}{\partial t}, v - u \right) + \mathcal{A}(u, v - u) - (f, v - u) \right] dt \ge 0,$$
$$\forall v \in L^2(0, T; V), \qquad v \le 0 \text{ a.e.,}$$

and

$$(2.12.7) \qquad u(x, 0) = 0.$$

Then when $\varepsilon \to 0$,

$$(2.12.8) \qquad u_\varepsilon \to u \text{ in } L^2(0, T; V) \text{ weakly,}$$

$$\frac{\partial u_\varepsilon}{\partial t} \to \frac{\partial u}{\partial t} \text{ in } L^2(0, T; V') \text{ weakly.}$$

PROOF. The proof is divided in several steps.

We first introduce the **penalized equation** associated to the V.I. (2.12.4). For $\eta > 0$, we denote by $u_{\varepsilon\eta}$ **the** solution of the nonlinear equation

(2.12.9)
$$\frac{\partial u_{\varepsilon\eta}}{\partial t} + A^\varepsilon u_{\varepsilon\eta} + \frac{1}{\eta} u_{\varepsilon\eta}^+ = f,$$

$$u_{\varepsilon\eta}(x,0) = 0, \qquad u_{\varepsilon\eta} \in L^2(0,T;V)$$

(and as usual, the boundary conditions which correspond to the variational formulation). The equation (2.12.9) is called the **penalized equation** associated to (2.12.4). It admits a unique solution, for which we now obtain **a priori estimates**.

If we multiply (2.12.9) by $u_{\varepsilon\eta}^+$, we obtain (since $a^\varepsilon(v,v^+) = a^\varepsilon(v^+,v^+) = a^\varepsilon(v^+)$):

$$\left(\frac{\partial u_{\varepsilon\eta}^+}{\partial t}, u_{\varepsilon\eta}^+ \right) + a^\varepsilon(u_{\varepsilon\eta}^+) + \frac{1}{\eta} |u_{\varepsilon\eta}^+|^2 = (f, u_{\varepsilon\eta}^+),$$

hence

(2.12.10)
$$\frac{1}{2} \frac{d}{dt} |u_{\varepsilon\eta}^+|^2 + a^\varepsilon(u_{\varepsilon\eta}^+) + \frac{1}{\eta} |u_{\varepsilon\eta}^+|^2 \le |f| |u_{\varepsilon\eta}^+|.$$

We deduce from (2.12.10) that

(2.12.11)
$$\frac{1}{\eta} u_{\varepsilon\eta}^+ \text{ is bounded in } L^2(\mathcal{G}), \ \forall\varepsilon \text{ and } \forall\eta,$$

(2.12.12)
$$\|u_{\varepsilon\eta}^+\|_{L^2(0,T;V)} + \|u_{\varepsilon\eta}^+\|_{L^\infty(0,T;H)} \le C\sqrt{\eta}.$$

It follows from (2.12.11) that

$$\frac{\partial u_{\varepsilon\eta}}{\partial t} + A^\varepsilon u_{\varepsilon\eta} = f - \frac{1}{\eta} u_{\varepsilon\eta}^+ \in \text{ bounded set of } L^2(\mathcal{G}),$$

(and u_ε satisfies initial and boundary conditions) so that the usual a priori estimates are valid:

(2.12.13)
$$\|u_{\varepsilon\eta}\|_{L^2(0,T;V)} + \left\| \frac{\partial u_{\varepsilon\eta}}{\partial t} \right\|_{L^2(0,T;V')} \le C.$$

If we let $\eta \to 0$, we can extract a subsequence, still denoted by $u_{\varepsilon\eta}$, such that

(2.12.14)
$$u_{\varepsilon\eta} \to u_\varepsilon \text{ in } L^2(0,T;V) \text{ weakly},$$

(2.12.15)
$$\frac{\partial u_{\varepsilon\eta}}{\partial t} \to \frac{\partial u_\varepsilon}{\partial t} \text{ in } L^2(0,T;V') \text{ weakly},$$

(2.12.16)
$$u_{\varepsilon\eta}^+ \to 0 \text{ in } L^2(\mathcal{G}).$$

It follows from (2.12.14) that $u_{\varepsilon\eta} \to u_\varepsilon$ in $L^2(\mathcal{G})$ **strongly**, so that $u_{\varepsilon\eta}^+ \to u_\varepsilon^+$ in $L^2(\mathcal{G})$ and by comparison with (2.12.16) we obtain $u_\varepsilon^+ = 0$, i.e. $u_\varepsilon \le 0$. Therefore u_ε satisfies (2.12.4) and (2.12.2). Let v be in $L^2(0,T;V)$, $v \le 0$. We multiply (2.12.9) by $v - u_{\varepsilon\eta}$; we obtain:

$$\left(\frac{\partial u_{\varepsilon\eta}}{\partial t} - f, v - u_{\varepsilon\eta} \right) + a^\varepsilon(u_{\varepsilon\eta}, v - u_{\varepsilon\eta}) = \left(-\frac{1}{\eta} u_{\varepsilon\eta}^+, v - u_{\varepsilon\eta} \right) \ge 0,$$

hence

$$\int_0^T \left[\left(\frac{\partial u_{\varepsilon\eta}}{\partial t} - f, v - u_{\varepsilon\eta} \right) + a^\varepsilon(u_{\varepsilon\eta}, v - u_{\varepsilon\eta}) \right] dt \ge 0$$

or
(2.12.17)
$$\int_0^T \left[\left(\frac{\partial u_{\varepsilon\eta}}{\partial t}, v \right) - (f, v - u_{\varepsilon\eta}) + a^\varepsilon(u_{\varepsilon\eta}, v) \right] dt \geq \frac{1}{2} |u_{\varepsilon\eta}(T)|^2 + \int_0^T a^\varepsilon(u_{\varepsilon\eta}, u_{\varepsilon\eta}) \, dt.$$

Since
$$\liminf_{\eta \to 0} \left[\frac{1}{2} |u_{\varepsilon\eta}(T)|^2 + \int_0^T a^\varepsilon(u_{\varepsilon\eta}, u_{\varepsilon\eta}) \, dt \right] \geq \frac{1}{2} |u_\varepsilon(T)|^2 + \int_0^T a^\varepsilon(u_\varepsilon, u_\varepsilon) \, dt$$

it follows from (2.12.17) that u_ε satisfies (2.12.3). Therefore we have shown the existence of u_ε, satisfying (2.12.3), (2.12.4), (2.12.2) and such that

$$(2.12.18) \qquad \|u_\varepsilon\|_{L^2(0,T;V)} + \left\| \frac{\partial u_\varepsilon}{\partial t} \right\|_{L^2(0,T;V')} \leq C.$$

The **uniqueness** of u_ε is straightforward.

We now establish an error estimate of independent interest:

$$(2.12.19) \qquad \|u_{\varepsilon\eta} - u_\varepsilon\|_{L^2(0,T;V)} \leq C\eta^{1/4}.$$

PROOF. We have $u_\varepsilon - u_{\varepsilon\eta} = u_\varepsilon + u_{\varepsilon\eta}^- - u_{\varepsilon\eta}^+$ and by virtue of (2.12.12) we have only to show that

$$(2.12.20) \qquad \|u_\varepsilon + u_{\varepsilon\eta}^-\|_{L^2(0,T;V)} \leq C\eta^{1/4}.$$

We take $v = -u_{\varepsilon\eta}^-$ in (2.12.4) and we multiply (2.12.9) by $u_\varepsilon + u_{\varepsilon\eta}^-$. Adding up, we obtain

$$-\left(\frac{\partial(u_\varepsilon + u_{\varepsilon\eta}^-)}{\partial t}, u_\varepsilon + u_{\varepsilon\eta}^- \right) + \left(\frac{\partial(u_{\varepsilon\eta}^+)}{\partial t}, u_\varepsilon + u_{\varepsilon\eta}^- \right) - a^\varepsilon(u_\varepsilon + u_{\varepsilon\eta}^-)$$
$$+ a^\varepsilon(u_{\varepsilon\eta}^+, u_\varepsilon + u_{\varepsilon\eta}^-) + \frac{1}{\eta}(u_{\varepsilon\eta}^+, u_\varepsilon + u_{\varepsilon\eta}^-) \geq 0,$$

i.e.

$$\frac{1}{2} \frac{d}{dt} |u_\varepsilon + u_{\varepsilon\eta}^-|^2 + a^\varepsilon(u_\varepsilon + u_{\varepsilon\eta}^-) + \frac{1}{\eta}(u_{\varepsilon\eta}^+, -u_\varepsilon) \leq \left(\frac{\partial(u_{\varepsilon\eta}^+)}{\partial t}, u_\varepsilon \right) + a^\varepsilon(u_{\varepsilon\eta}^+, u_\varepsilon + u_{\varepsilon\eta}^-)$$

and integrating over $(0,t)$, we obtain (since $\frac{1}{\eta}(u_{\varepsilon\eta}^+, -u_\varepsilon) \geq 0$)

$$(2.12.21) \quad \frac{1}{2} |u_\varepsilon(t) + u_{\varepsilon\eta}^-(t)|^2 + \int_0^t a^\varepsilon(u_\varepsilon + u_{\varepsilon\eta}^-) \, dx$$
$$\leq (u_{\varepsilon\eta}^+(t), u_\varepsilon(t)) - \int_0^t \left(u_{\varepsilon\eta}^+, \frac{\partial u_\varepsilon}{\partial t} \right) ds + \int_0^t a^\varepsilon(u_{\varepsilon\eta}^+, u_\varepsilon) \, dx.$$

Using (2.12.12) and (2.12.18), the right hand side is bounded by $C\sqrt{\eta}$, hence (2.12.20) follows.   ∎

We have also proved that

$$(2.12.22) \qquad \|u_\varepsilon + u_{\varepsilon\eta}^-\|_{L^\infty(0,T;H)} \leq C\eta^{1/4}.$$

It follows from (2.12.19) that it suffices to study the limit of $u_{\varepsilon\eta}$ as $\varepsilon \to 0$ **for η fixed**. But this is immediate. By virtue of (2.12.13), we can suppose, by extracting a subsequence, still denoted by $u_{\varepsilon\eta}$, that

(2.12.23) $$u_{\varepsilon\eta} \to u_\eta \text{ in } L^2(0, T; V) \text{ weakly,}$$

$$\frac{\partial u_{\varepsilon\eta}}{\partial t} \to \frac{\partial u_\eta}{\partial t} \text{ in } L^2(0, T; V') \text{ weakly.}$$

Then $u_{\varepsilon\eta} \to u_\eta$ in $L^2(\mathcal{G})$ **strongly**, so that $f - \frac{1}{\eta} u_{\varepsilon\eta}^+ \to f - \frac{1}{\eta} u_\eta^+$ in $L^2(\mathcal{G})$ and then

$$\left(\frac{\partial u_\eta}{\partial t}, v \right) + \mathcal{A}(u_\eta, v) = \left(f - \frac{1}{\eta} u_\eta^+, v \right), \ \forall v \in V,$$

where $\mathcal{A}(u, v)$ is defined as in the statement of the Theorem. The proof is completed. ∎

REMARK 2.15. Theorem 2.14 gives, **in particular**, the homogenization of the Stefan's free boundary problem for a "granular" medium — i.e. when coefficients are functions of x/ε: $a_{ij}(x/\varepsilon)$.[14]

REMARK 2.16. The result of Theorem 2.14 extends to the case of coefficients of the form $a_{ij}\left(x, \frac{x}{\varepsilon}, t, \frac{t}{\varepsilon^k} \right)$.

REMARK 2.17. Theorem 2.14 can be extended **to obstacles** of the type

$$\psi_1(x, t) \le v \le \psi_2(x, t),$$

when ψ_1 and ψ_2 are smooth enough. The case of the "most general" obstacles ψ_i under which the analogue of Theorem 2.14 still holds does not seem to be known.

REMARK 2.18. The homogenization of V.I. of evolution for "arbitrary constraints" $v \in K(t)$ is largely open.

2.13. Remarks on averaging.

2.13.1. *Setting of the problem.* We consider now the case $T = +\infty$:

(2.13.1) $$\mathcal{G} = \mathcal{O} \times]0, +\infty[$$

and we consider the equation

$$\frac{\partial u_\varepsilon}{\partial t} + A^\varepsilon u_\varepsilon = f \text{ in } \mathcal{G},$$

(2.13.2) $$A^\varepsilon v = -\frac{\partial}{\partial x_i} \left(a_{ij}^\varepsilon(x, t) \frac{\partial v}{\partial x_j} \right) + b_j(x, t) \frac{\partial v}{\partial x_j}.$$

For $u, v \in H^1(\mathcal{O})$, we set now

(2.13.3) $$a^\varepsilon(u, v) = \int_{\mathcal{O}} a_{ij}^\varepsilon(x, t) \frac{\partial u}{\partial x_j} \frac{\partial v}{\partial x_i} \, dx + \int_{\mathcal{O}} b_j^\varepsilon \frac{\partial u}{\partial x_j} v \, dx.$$

The precise problem we consider is then:

$$u_\varepsilon \in L^2(0, T; V), \ \forall T \text{ finite,}$$

(2.13.4) $$\left(\frac{\partial u_\varepsilon}{\partial t}, v \right) + a^\varepsilon(u_\varepsilon, v) = (f, v), \ \forall v \in V, \ t > 0,$$

$$u^\varepsilon(0) = u^0.$$

We assume the following:

(2.13.5) $$a_{ij}^\varepsilon, \ \frac{\partial a_{ij}^\varepsilon}{\partial x_k}, \ b_j^\varepsilon \text{ remain in a bounded set of } L^\infty(\mathcal{G})$$

---

[14]This is because the Stefan's free boundary problem can be transformed into a V.I. of evolution.

and of course

$$(2.13.6) \qquad a_{ij}^\varepsilon(x,t)\xi_i\xi_j \geq \alpha\xi_i\xi_i, \qquad \alpha > 0.$$

EXAMPLE 2.19. We take:

$$(2.13.7) \qquad a_{ij}^\varepsilon(x,t) = a_{ij}(x,t/\varepsilon), \qquad b_j^\varepsilon(x,t) = b_j(x,t/\varepsilon),$$

where

$$(2.13.8) \qquad a_{ij}(x,\tau), \qquad b_j(x,\tau), \qquad \frac{\partial}{\partial x_k}a_{ij}(x,\tau)$$

belong to $L^\infty(\mathcal{O} \times \mathbb{R}_\tau)$ and are almost periodic in τ.

We extract a subsequence such that

$$(2.13.9) \qquad a_{ij}^\varepsilon \to \hat{a}_{ij}, \quad \frac{\partial a_{ij}^\varepsilon}{\partial x_k} \to \frac{\partial \hat{a}_{ij}}{\partial x_k}, \quad b_j^\varepsilon \to \hat{b}_j \text{ in } L^\infty(\mathcal{G}) \text{ weak star.}$$

We want to study the behavior of u_ε as $\varepsilon \to 0$.
In the case of Example 2.19 we have

$$(2.13.10) \qquad \hat{a}_{ij} = \mathcal{M}_\tau(a_{ij}), \qquad \hat{b}_j = \mathcal{M}_\tau(b_j),$$

where we have set here

$$(2.13.11) \qquad \mathcal{M}_\tau(\phi) = \lim_{\lambda \to +\infty} \frac{1}{\lambda}\int_0^\lambda \phi(x,\sigma)\,d\sigma.$$

THEOREM 2.20. *We assume that (2.13.5), (2.13.6) hold true and that*[15]

$$(2.13.12) \qquad V = H^1(\mathcal{O}) \text{ or } V = H_0^1(\mathcal{O}).$$

We also assume that

$$(2.13.13) \qquad f, \frac{\partial f}{\partial x_i} \in L^2(\mathcal{O}\times]0,T[), \ \forall T \text{ finite,}$$

$$u^0 \in V.$$

Then as $\varepsilon \to 0$, one has

$$(2.13.14) \qquad u_\varepsilon \to u \text{ in } L^2(0,T;V) \text{ weakly for every } T \text{ finite,}$$

where u is the solution of

$$(2.13.15) \qquad \left(\frac{\partial u}{\partial t},v\right) + \hat{a}(u,v) = (f,v), \ \forall v \in V,$$

$$u(0) = u^0,$$

where

$$(2.13.16) \qquad \hat{a}(u,v) = \int_\mathcal{O} \hat{a}_{ij}\frac{\partial u}{\partial x_j}\frac{\partial v}{\partial x_i}\,dx + \int_\mathcal{O} \hat{b}_j\frac{\partial u}{\partial x_j}v\,dx.$$

REMARK 2.21. In fact we shall prove more: namely,

$$(2.13.17) \qquad u_\varepsilon \to u \text{ in } L^2(0,T;H^2(\mathcal{O})) \text{ weakly, } \forall T < \infty,$$

$$(2.13.18) \qquad \frac{\partial u_\varepsilon}{\partial t} \to \frac{\partial u}{\partial t} \text{ in } L^2(0,T;L^2(\mathcal{O})) \text{ weakly, } \forall T < \infty.$$

---

[15]This hypothesis is by no means indispensable. If V consists of the functions of $H^1(\mathcal{O})$ which are zero on a subset Γ_0 of Γ, the result is still valid with a slightly more complicated proof.

REMARK 2.22. In case of Example 2.19 (formulas (2.13.7)) the theorem holds true (**without** (2.13.17), (2.13.18)) **without conditions on** $\frac{\partial}{\partial x_k} a_{ij}$. One uses these methods along the lines of those of Section 2.2.

PROOF OF THEOREM 2.20. The whole problem is reduced to proving that, $\forall T$ finite,

$$(2.13.19) \qquad \|u_\varepsilon\|_{L^2(0,T;H^2(\mathcal{O}))} + \left\|\frac{\partial u_\varepsilon}{\partial t}\right\|_{L^2(0,T;L^2(\mathcal{O}))} \le C.$$

Indeed, if we admit (2.13.19), then one can extract a subsequence, still denoted by u_ε, such that one has (2.13.14), (2.13.17) and (2.13.18). It follows that, $\forall T$ finite

$$\frac{\partial u_\varepsilon}{\partial x_j} \to \frac{\partial u}{\partial x_j} \text{ in } L^2(\mathcal{O}\times]0,T[) \text{ strongly}$$

and therefore

$$a_{ij}^\varepsilon \frac{\partial u_\varepsilon}{\partial x_j} \to \hat{a}_{ij} \frac{\partial u}{\partial x_j}, \qquad b_j^\varepsilon \frac{\partial u_\varepsilon}{\partial x_j} \to \hat{b}_j \frac{\partial u}{\partial x_j} \text{ in } L^2(\mathcal{O}\times]0,T[) \text{ weakly}$$

and the result follows.

Estimates (2.13.19) are obtained by using first local mappings, to reduce the problem to the same question when

$$(2.13.20) \qquad \mathcal{O} = \{x \mid x_n > 0\}.$$

Then one uses the method of **translations parallel to the boundary.**[16]

We suppress the index "ε" to simplify the writing. Given $h = \{h_1, \ldots, h_{n-1}, 0\}$, we define

$$v_h(x) = v(x_1 - h_1, \ldots, x_{n-1} - h_{n-1}, x_n).$$

We have

$$\frac{\partial u}{\partial t} + Au = f \text{ in } \mathcal{G},$$

and with the obvious notations

$$\frac{\partial u_h}{\partial t} + A_h u_h = f_h \text{ in } \mathcal{G}.$$

Therefore

$$(2.13.21) \qquad \frac{\partial(u_h - u)}{\partial t} + A_h(u_h - u) + (A_h - A)u = f_h - f.$$

It follows from (2.13.21) and from the hypothesis that

$$\|u_h - u\|_{L^2(0,T;V)} \le C|h|$$

so that by letting $|h| \to 0$, we see that when $\varepsilon \to 0$
(2.13.22)
$$\frac{\partial^2 u}{\partial x_i' \partial x_j}$$ remains in a bounded set of $L^2(\mathcal{O}\times]0,T[)$, where $x_i' = x_i$, $i \le n-1$.

We now return to the equation, which we write

$$(2.13.23) \qquad \frac{\partial u}{\partial t} - a_{nn} \frac{\partial^2 u}{\partial x_n^2} = g,$$

---

[16]This is a classical method, introduced by L. Nirenberg [**87**] for the study of the regularity of elliptic equations. When V does not satisfy (2.13.12) but consists of functions which are zero on $\Gamma_0 \subset \Gamma$, then there is a difficulty because V is not "translation invariant." One can then use a compensation model, due to Aronszajn and Smith, and reported in Lions [**69**].

where g remains in a bounded set of $L^2(\mathcal{O}\times]0, T[)$ (we use (2.13.22)); multiplying (2.13.23) by $\frac{1}{a_{nn}}\frac{\partial u}{\partial t}$ we obtain (since $a_{nn} \geq \gamma > 0$)

$$\int \frac{1}{a_{nn}}\left(\frac{\partial u}{\partial t}\right)^2 dx + \frac{1}{2}\frac{d}{dt}\int \left(\frac{\partial u}{\partial x_n}\right)^2 dx = \left(g, \frac{1}{a_{nn}}\frac{\partial u}{\partial t}\right)$$

(since $u = 0$ or $\frac{\partial u}{\partial x_n} = 0$ on Γ) hence it follows that

(2.13.24) $\dfrac{\partial u}{\partial t}$ remains in a bounded set of $L^2(\mathcal{O}\times]0, T[)$.

But then (2.13.23), which can be written:

$$\frac{\partial^2 u}{\partial x_n^2} = \frac{1}{a_{nn}}\left(\frac{\partial u}{\partial t} - g\right),$$

implies

$$\frac{\partial^2 u}{\partial x_n^2} \in \text{ bounded set of } L^2(\mathcal{O}\times]0, T[)$$

and (2.13.19) is proven.   ∎

REMARK 2.23. In the case of almost periodic coefficients, Theorem 2.20 is related to the classical theory of averaging of ordinary differential equations with almost periodic coefficients. Cf. Bogoliubov and Mitropolsky [20] and the Bibliography therein.

REMARK 2.24. We confine ourselves to remarks made in the comments of the last section of this chapter for the case of coefficients $a_{ij}(x/\varepsilon, t/\varepsilon^k)$, where the a_{ij}'s are almost periodic in τ (and periodic in y).

3. Evolution operators of hyperbolic, Petrowsky, or Schrödinger type

3.1. Orientation. We present now some very simple remarks connected with the homogenization problem for **hyperbolic operators**. Further results, using some different ideas and techniques, are given in Chapter 4 of this book. We also give in this section some results for operators of Petrowsky or of Schrödinger type, the hyperbolicity not being used in an essential manner. We give next some examples of "non-local" operators which appear in the homogenization process.

3.2. Linear operators with coefficients which are regular in t. Let us consider the operator

(3.2.1) $$A^\varepsilon v = -\frac{\partial}{\partial x_i}\left(a_{ij}\left(\frac{x}{\varepsilon}\right)\frac{\partial v}{\partial x_j}\right),$$

where the a_{ij}'s satisfy:

(3.2.2) $a_{ij} = a_{ji}, \ \forall i, j,$

$a_{ij}(y) \in L^\infty(\mathbb{R}_y^n), \qquad a_{ij}$ is Y-periodic,

(3.2.3) $a_{ij}(y)\xi_i\xi_j \geq \alpha\xi_i\xi_i, \qquad \alpha > 0, \qquad \text{a.e. in } y.$

We consider V as usual ($H_0^1(\mathcal{O}) \subseteq V \subseteq H^1(\mathcal{O})$) and with the usual notations we consider the following problem:

(3.2.4) $u_\varepsilon \in L^\infty(0, T; V), \qquad \dfrac{\partial u_\varepsilon}{\partial t} \in L^\infty(0, T; H),$

$$(3.2.5) \qquad \left(\frac{\partial^2 u_\varepsilon}{\partial \partial t^2}, v\right) + a^\varepsilon(u_\varepsilon, v) = (f, v), \ \forall v \in V,$$

$$(3.2.6) \qquad u_\varepsilon(0) = u^0, \qquad \frac{\partial u_\varepsilon}{\partial t}(0) = u^1,$$

where f, u^0, u^1 are given such that

$$(3.2.7) \qquad f \in L^2(0, T; H), \qquad u^0 \in V, \qquad u^1 \in H.$$

It is known (cf. for instance Lions [68], Lions-Magenes [71] and [72]) that **problem (3.2.4), (3.2.5), (3.2.6) admits a unique solution**.

3.2.1. *A priori estimates.* We shall write ϕ', ϕ'' for $\frac{\partial \phi}{\partial t}$, $\frac{\partial^2 \phi}{\partial t^2}$. Taking $v = u'_\varepsilon$ in (3.2.5) gives

$$(3.2.8) \qquad \frac{1}{2}\frac{d}{dt}\left[|u'_\varepsilon|^2 + a^\varepsilon(u_\varepsilon)\right] = (f, u'_\varepsilon).$$

We emphasize that in order to obtain (3.2.8) we used in an essential manner the fact that $a^\varepsilon(u, v)$ is **symmetric and does not depend on** t. It follows immediately from (3.2.8) that

$$(3.2.9) \qquad \|u_\varepsilon\|_{L^\infty(0,T;V)} + \|u'_\varepsilon\|_{L^\infty(0,T;H)} \leq C.$$

The behavior of u_ε **as** $\varepsilon \to 0$ is very simple to study. It follows from (3.2.9) that we can extract a subsequence, still denoted by u_ε, such that

$$(3.2.10) \qquad u_\varepsilon \to u \text{ in } L^\infty(0, T; V) \text{ weak star},$$
$$u'_\varepsilon \to u' \text{ in } L^\infty(0, T; H) \text{ weak star}.$$

We use next the method of Remark 1.8 which is entirely general. With the notation (1.2.38) one has, $\forall \phi \in C_0^\infty(]0, T[)$:

$$(3.2.11) \qquad a^\varepsilon(u_\varepsilon(\phi), v) = (f(\phi) - u_\varepsilon(\phi''), v), \ \forall v \in V.$$

Since $u_\varepsilon(\phi'') \to u(\phi'')$ in H strongly, one has

$$u_\varepsilon(\phi) \to u(\phi) \text{ in } V \text{ weakly, where}$$
$$\mathcal{A}(u(\phi), v) = (f(\phi) - u(\phi''), v), \ \forall v \in V,$$

where $\mathcal{A}(u, v) = $ homogenized form associated to $a^\varepsilon(u, v)$. Therefore u is the solution of

$$(3.2.12) \qquad \left(\frac{\partial^2 u}{\partial t^2}, v\right) + \mathcal{A}(u, v) = (f, v), \ \forall v \in V,$$

$$(3.2.13) \qquad u \in L^\infty(0, T; V), \qquad u' \in L^\infty(0, T; H),$$

$$(3.2.14) \qquad u(0) = u^0, \qquad u'(0) = u^1.$$

Let us consider now the case when

$$(3.2.15) \qquad A^\varepsilon(t)v = -\frac{\partial}{\partial x_i}\left(a_{ij}\left(x, \frac{x}{\varepsilon}, t\right)\frac{\partial v}{\partial x_j}\right).$$

We assume that

(3.2.16) $$a_{ij} = a_{ji},$$

$$a_{ij}(x, y, t) \text{ is } Y\text{-periodic}, \forall x, t,$$

(3.2.17) $$a_{ij} \in C^0(\overline{\mathcal{O}} \times [0, T]; L^\infty(\mathbb{R}_y^n)),$$

$$\frac{\partial a_{ij}}{\partial t} \in L^\infty(\mathcal{O} \times (0, T) \times \mathbb{R}_y^n).$$

We set, $\forall u, v \in H^1(\mathcal{O})$:

(3.2.18) $$a^\varepsilon(t; u, v) = \int_{\mathcal{O}} a_{ij}\left(x, \frac{x}{\varepsilon}, t\right) \frac{\partial u}{\partial x_j} \frac{\partial v}{\partial x_i} \, dx.$$

We can define

(3.2.19) $$\dot{a}^\varepsilon(t; u, v) = \int_{\mathcal{O}} \frac{\partial a_{ij}}{\partial t}\left(x, \frac{x}{\varepsilon}, t\right) \frac{\partial u}{\partial x_j} \frac{\partial v}{\partial x_i} \, dx.$$

We consider again (3.2.4), (3.2.5), (3.2.6) (**with** $a^\varepsilon(u, v)$ **replaced by** $a^\varepsilon(t; u, v)$), **and we have existence and uniqueness of the solution.**

The a priori estimates are now the following. Equation (3.2.8) is replaced by

(3.2.20) $$\frac{1}{2} \frac{d}{dt} \left[|u'_\varepsilon|^2 + a^\varepsilon(u_\varepsilon) \right] - \frac{1}{2} \dot{a}^\varepsilon(t, u_\varepsilon) = (f, u'_\varepsilon)$$

and by virtue of the last condition in (3.2.17) **we obtain the same a priori estimates (3.2.9)**.

By the same technique as in Chapter 1, Section 6, one proves that u_ε satisfies (3.2.10) where u is the solution of (3.2.13), (3.2.14) and

(3.2.21) $$\left(\frac{\partial^2 u}{\partial t^2}, v\right) + \mathcal{A}(t; u, v) = (f, v),$$

(3.2.22) $$\mathcal{A}(t; u, v) = (\mathcal{A}(t)u, v) \text{ if } u, v \in C_0^\infty(\mathcal{O}),$$

where $\mathcal{A}(t)$ is the homogenized operator of $A^\varepsilon(t)$ **for fixed** t.

REMARK 3.1. The operator $\frac{\partial^2}{\partial t^2} + A^\varepsilon$ is a second order **hyperbolic** operator. But the results indicated so far really **do not** depend at all on the hyperbolicity. All that we said remains unchanged (except obvious modifications) if one considers

(3.2.23)

$$a^\varepsilon(t; u, v) = \int_{\mathcal{O}} a_{\alpha\beta}\left(x, \frac{x}{\varepsilon}, t\right) D^\alpha u D^\beta v \, dx, \qquad |\alpha| = |\beta| = m, \qquad a_{\alpha\beta} = a_{\beta\alpha},$$

$a_{\alpha\beta}(x, y, t)$ being Y-periodic in y,

(3.2.24)

$$a_{\alpha\beta} \in C^0(\overline{\mathcal{O}} \times [0, T]; L^\infty(\mathbb{R}_y^n)),$$

$$\frac{\partial a_{\alpha\beta}}{\partial t} \in L^\infty(\mathcal{O} \times (0, T) \times \mathbb{R}_y^n).$$

The homogenized operator is computed for every fixed t according to the results of Chapter 1, Section 9.

We remark that the corresponding operator is **not** of hyperbolic type when $m > 1$; it is an operator of Petrowsky type.

REMARK 3.2. One can also add to the operator A^ε lower order terms. We return to that in Section 3.3 which follows.

3.3. Linear operators with coefficients which are irregular in t. A natural problem is the following: Instead of the situation (3.2.1) we consider functions $a_{ij}(y, \tau)$ which satisfy (3.2.2) and

(3.3.1) $\qquad\qquad a_{ij}(y, \tau) \in L^\infty(\mathbb{R}^n_y \times \mathbb{R}_\tau), \qquad Y - \tau_0 \text{ periodic}$

and the usual ellipticity condition. We then consider the operator A^ε defined by

$$(3.3.2) \qquad\qquad A^\varepsilon v = -\frac{\partial}{\partial x_i} a_{ij}\left(\frac{x}{\varepsilon}, \frac{t}{\varepsilon}\right) \frac{\partial v}{\partial x_j}.$$

If we assume that

$$(3.3.3) \qquad\qquad \frac{\partial}{\partial \tau} a_{ij} \in L^\infty(\mathbb{R}^n_y \times \mathbb{R}_\tau),$$

then problem (3.2.4), (3.2.5), (3.2.6) **admits a unique solution. But there are no[17] a priori estimates independent of ε.** Indeed if we define

$$(3.3.4) \qquad \dot{a}^\varepsilon(t; u, v) = \int \left(\frac{\partial a_{ij}}{\partial t}\right)\left(x, \frac{x}{\varepsilon}, \frac{t}{\varepsilon}\right) \frac{\partial u}{\partial x_j} \frac{\partial v}{\partial x_i}\, dx$$

then (3.2.20) is replaced by

$$(3.3.5) \qquad \frac{1}{2}\frac{d}{dt}\left[|u'|^2 + a^\varepsilon(u_\varepsilon)\right] - \frac{1}{2\varepsilon}\dot{a}^\varepsilon(t, u_\varepsilon) = (f, u'_\varepsilon),$$

which does not give estimates independent of ε. **The behavior of u_ε as $\varepsilon \to 0$ does not seem to be known in general.** We shall give in Section 3.4 below some formal computations (which can be justified in **some** cases).

It is possible to consider **lower order terms** which have coefficients irregular in t. Let us consider

(3.3.6) $\qquad\qquad A^\varepsilon \text{ given by (3.2.1)},$

$$B^\varepsilon = b_j\left(\frac{x}{\varepsilon}, \frac{t}{\varepsilon^k}\right)\frac{\partial}{\partial x_j}$$

where

(3.3.7) $\qquad b_j(y, \tau) \text{ is } Y - \tau_0 \text{ periodic}, \qquad b_j \in L^\infty(\mathbb{R}^n_y \times \mathbb{R}_\tau).$

We have then: **there exists a function u_ε and only one which satisfies (3.2.4), (3.2.6) and**

(3.3.8) $\qquad (u''_\varepsilon, v) + a^\varepsilon(u_\varepsilon, v) + (B^\varepsilon u_\varepsilon, v) = (f, v), \ \forall v \in V.$

Then a priori estimates are as follows. Taking $v = u'_\varepsilon$ in (3.3.8), we obtain

$$(3.3.9) \qquad \frac{1}{2}\frac{d}{dt}\left[|u'_\varepsilon|^2 + a^\varepsilon(u_\varepsilon)\right] + (B^\varepsilon u_\varepsilon, u'_\varepsilon) = (f, u'_\varepsilon).$$

But

(3.3.10) $\qquad\qquad\qquad |B^\varepsilon v| \leq C\|v\|$

so that (3.3.9) gives:

$$(3.3.11) \quad |u'_\varepsilon(t)|^2 + a^\varepsilon(u_\varepsilon(t)) \leq |u^1|^2 + a^\varepsilon(u^0) + 2C\int_0^t \|u_\varepsilon\|\|u'_\varepsilon\|\, d\sigma + 2\int_0^t |f||u'_\varepsilon|\, d\sigma,$$

---

[17]At least there do not seem to be known a priori estimates strong enough to pass to the limit in ε.

and we obtain (3.2.9).

Therefore one can extract a subsequence, still denoted by u_ε, such that one has (3.2.10) and

$$(3.3.12) \qquad\qquad B^\varepsilon u_\varepsilon \to \eta \text{ in } L^\infty(0,T;H) \text{ weak star.}$$

Then (3.3.8) gives, $\forall \phi \in C_0^\infty(]0,T[)$

$$(3.3.13) \qquad a^\varepsilon(u_\varepsilon(\phi), v) = (f(\phi) - (B^\varepsilon u_\varepsilon)(\phi) - u_\varepsilon(\phi''), v)$$

and $(B^\varepsilon u_\varepsilon)(\phi) \to \eta(\phi)$ in H weakly, therefore in V' strongly and it follows from (3.3.13) that, $\mathcal{A}(u,v)$ being the homogenized form associated to a^ε

$$\mathcal{A}(u(\phi), v) = (f(\phi) - \eta(\phi) - u(\phi''), v), \ \forall v \in V.$$

Therefore

$$(3.3.14) \qquad\qquad (u'', v) + \mathcal{A}(u, v) + (\eta, v) = (f, v),$$

and the problem is now **to compute** η. We study this question in the following sections.

3.4. Asymptotic expansions (I). Let us consider first — **in a purely formal fashion** — the problem (3.2.4), (3.2.5), (3.2.6) for A^ε given by (3.3.2). We have with the usual notations:

(3.4.1)

$$\frac{\partial^2}{\partial t^2} + A^\varepsilon = \varepsilon^{-2}Q_1 + \varepsilon^{-1}Q_2 + \varepsilon^0 Q_3,$$

$$Q_1 = \frac{\partial^2}{\partial \tau^2} + A_1, \qquad\qquad A_1 = -\frac{\partial}{\partial y_i} a_{ij}(y,\tau) \frac{\partial}{\partial y_j},$$

$$Q_2 = 2\frac{\partial^2}{\partial t \partial \tau} + A_2, \qquad A_2 = -\frac{\partial}{\partial y_i}\left(a_{ij}\frac{\partial}{\partial x_j}\right) - \frac{\partial}{\partial x_i}\left(a_{ij}\frac{\partial}{\partial y_j}\right),$$

$$Q_3 = \frac{\partial^2}{\partial t^2} + A_3, \qquad\qquad A_3 = -\frac{\partial}{\partial x_i}\left(a_{ij}\partial x_j\right).$$

Then, if we look for $u_\varepsilon = u_0 + \varepsilon u_1 + \cdots$, we find as usual:

$$(3.4.2) \qquad\qquad Q_1 u_0 = 0,$$
$$(3.4.3) \qquad\qquad Q_1 u_1 + Q_2 u_0 = 0,$$
$$(3.4.4) \qquad\qquad Q_1 u_2 + Q_2 u_1 + Q_3 u_0 = f.$$

We are thus led to the following question: given a function F such that

$$(3.4.5) \qquad\qquad F \in L^2(Y \times (0, \tau_0))$$

is it possible to find ϕ which is $Y - \tau_0$ periodic and satisfies

$$(3.4.6) \qquad\qquad \frac{\partial^2 \phi}{\partial \tau^2} + A_1\phi = F,$$

i.e., ϕ takes equal values on opposite faces of Y and

$$\phi(y, 0) = \phi(y, \tau_0), \qquad \frac{\partial \phi}{\partial \tau}(y, 0) = \frac{\partial \phi}{\partial \tau}(y, \tau_0)?$$

A necessary condition for ϕ to exist is

$$(3.4.7) \qquad\qquad \mathcal{M}(F) = 0.$$

Let us assume that this condition is sufficient (so that (3.4.6) defines ϕ up to an additive constant). It can be proven that this is "generically" the case. Then (3.4.2) is equivalent to $u_0 = u(x,t)$ and (3.4.3) reduces to

$$Q_1 u_1 + A_2 u = 0.$$

If we introduce χ^j as some $Y - \tau_0$ periodic solution of

$$(3.4.8) \qquad \frac{\partial^2 \chi^j}{\partial \tau^2} + A_1(\chi^j - y_j) = 0,$$

then

$$(3.4.9) \qquad u_1 = -\chi^j(y,\tau) \frac{\partial u}{\partial x_j} + \hat{u}_1(x,t)$$

and (3.4.4) can be solved for u_2 iff

$$\mathcal{M}(Q_2 u_1 + Q_3 u_0) = f,$$

i.e.,

$$(3.4.10) \qquad \frac{\partial^2 u}{\partial t^2} - \frac{\partial}{\partial x_i} \mathcal{M}\left(a_{ij} - a_{ik}\frac{\partial \chi^j}{\partial y_k}\right) \frac{\partial u}{\partial x_j} = f.$$

Of course we add the appropriate initial and boundary conditions. In which sense u is an "approximation" of u_ε is an open question.

3.5. Asymptotic expansions (II). We now consider the situation (3.3.6) and we take

$$(3.5.1) \qquad\qquad k = 1.$$

Then

$$(3.5.2) \qquad \frac{\partial^2}{\partial t^2} + A^\varepsilon + B^\varepsilon = \varepsilon^{-2} R_1 + \varepsilon^{-1} R_2 + \varepsilon^0 R_3,$$

$$R_1 = -\frac{\partial^2}{\partial \tau^2} + A_1,$$

$$R_2 = 2\frac{\partial^2}{\partial t \partial \tau} + A_2 + B_2, \qquad B_2 = -b_j \frac{\partial}{\partial y_j},$$

$$R_3 = \frac{\partial^2}{\partial t^2} + A_3 + B_3, \qquad B_3 = -b_j \frac{\partial}{\partial x_j}.$$

We find

$$(3.5.3) \qquad\qquad R_1 u_0 = 0,$$
$$(3.5.4) \qquad\qquad R_1 u_1 + R_2 u_0 = 0,$$
$$(3.5.5) \qquad\qquad R_1 u_2 + R_2 u_1 + R_3 u_0 = f.$$

We are led again to (3.4.6) **but this time with A_1 independent of τ**. It follows from de Simon [**31**] that for almost every τ_0, condition (3.4.7) is necessary and sufficient for (3.4.6) to admit a $Y - \tau_0$ periodic solution. We therefore **assume**

$$(3.5.6) \qquad \text{condition (3.4.7) is necessary and sufficient for the}$$
$$\text{existence of } \phi \text{ which is a } Y - \tau_0 \text{ periodic solution of (3.4.6).}$$

Then (3.5.4) gives u_1 by formula (3.4.9), χ^j being defined by (3.4.8); but since A_1 does not depend on τ, we have

$$\chi^j = \chi^j(y),$$

(3.5.7) $$A_1(\chi^j - y_j) = 0, \text{ and}$$

(3.5.8) $$u_1 = -\chi^j(y)\frac{\partial u}{\partial x_j}(x,t) + \hat{u}_1(x,t).$$

Then (3.5.5) admits a solution for u_2 iff

$$\mathcal{M}(R_2u_1 + R_3u_0) = f,$$

i.e.

(3.5.9) $$\frac{\partial^2 u}{\partial t^2} + \mathcal{A}u + \mathcal{M}\left(b_i - b_k\frac{\partial \chi^i}{\partial y_k}\right) = f.$$

This can actually be proven:

THEOREM 3.3. *We assume that A^ε is given by (3.2.1), (3.2.2), (3.2.3) and that B^ε is given by (3.3.6) with $k = 1$. We assume that (3.5.6) holds true and that $b_j \in C^1(\bar{Y} \times [0,\tau_0])$. Then u_ε satisfies (3.2.10) where u is the solution of (3.3.14) with η given by*

(3.5.10) $$\eta = \left(b_i - b_k\frac{\partial \chi^i}{\partial y_k}\right).$$

SKETCH OF PROOF. We introduce $B^{\varepsilon*} = $ adjoint of B in the sense of distributions, i.e., $B^{\varepsilon*}v = -\frac{\partial}{\partial x_j}(b_j^\varepsilon(x,t)v)$. We construct a family of functions w_ε such that

(3.5.11) $$\left(\frac{\partial^2}{\partial t^2} + A^\varepsilon + B^{\varepsilon*}\right)w_\varepsilon = \varepsilon g_\varepsilon, \qquad g_\varepsilon \text{ bounded in } L^2(\mathcal{G}),$$

$$w_\varepsilon + 1 \to 0 \text{ in } L^2(\mathcal{G}) \text{ as } \varepsilon \to 0.$$

For the construction of w_ε we need all coefficients a_{ij} and b_j to be smooth — we can always reduce the problem to this case by approximation. We look for w_ε in the form:

(3.5.12) $$w_\varepsilon = -1 + \varepsilon\alpha(y,\tau) + \varepsilon^2\beta(y,\tau), \qquad \alpha \text{ and } \beta \text{ being } Y - \tau_0 \text{ periodic.}$$

We find, using (3.5.2) and since $R_1^* = R_1$:

(3.5.13) $$R_1\alpha + R_2^*(-1) = 0,$$

(3.5.14) $$R_1\beta + R_2^*\alpha + R_3^*(-1) = 0,$$

and

$$g_\varepsilon = R_2^*\beta.$$

We define α as a $Y - \tau_0$ periodic solution of (3.5.13), i.e.

(3.5.15) $$R_1\alpha = -\frac{\partial b_j}{\partial y_j}(y,\tau).$$

Then (3.5.14) gives (using the fact that $R_3^*(-1) = 0$):

$$R_1\beta = -R_2^*\alpha$$

which admits a $Y - \tau_0$ periodic solution by virtue of (3.5.6). Then we have (3.5.11).

We now use $\phi \in C_0^\infty(\mathcal{G})$ and we take $v = \phi w_\varepsilon$ in (3.2.22). We multiply (3.5.11) by ϕu_ε. We obtain, by subtracting

(3.5.16)
$$\int_0^T \left[\left(\frac{\partial^2 u_\varepsilon}{\partial t^2}, \phi w_\varepsilon \right) - \left(\frac{\partial^2 w_\varepsilon}{\partial t^2}, \phi u_\varepsilon \right) \right] dt + \int_0^T \left[a^\varepsilon(u_\varepsilon, \phi w_\varepsilon) - a^\varepsilon(w_\varepsilon, \phi u_\varepsilon) \right] dt$$
$$+ \int_\mathcal{G} \left[b_j^\varepsilon \frac{\partial u_\varepsilon}{\partial x_j} \phi w_\varepsilon + \frac{\partial}{\partial x_j}(b_j^\varepsilon w_\varepsilon) \phi u_\varepsilon \right] dx \, dt = \int_0^T \left[(f, \phi w_\varepsilon) - \varepsilon(g_\varepsilon, \phi u_\varepsilon) \right] dt.$$

The first term in (3.5.16) equals

(3.5.17)
$$2 \int_0^T \left(u_\varepsilon, \frac{\partial \phi}{\partial t} \frac{\partial w_\varepsilon}{\partial t} \right) dt + \int_0^T \left(u_\varepsilon, w_\varepsilon \phi'' \right) dt.$$

In (3.5.17), the first term converges to 0, since $u_\varepsilon \to u$ in $L^2(\mathcal{G})$ **strongly** and $\frac{\partial w_\varepsilon}{\partial t} = \frac{\partial \alpha}{\partial \tau} + \varepsilon \frac{\partial \beta}{\partial \tau} \to \mathcal{M}\left(\frac{\partial \alpha}{\partial \tau} \right) = 0$ in $L^2(\mathcal{G})$ weakly. Therefore (3.5.17) converges to $-\int_0^T (u, \phi'') \, dt$.

If we set $\xi_i^\varepsilon = a_{ij}^\varepsilon \frac{\partial u_\varepsilon}{\partial x_j}$, the second term in (3.5.16) equals

$$\int_\mathcal{G} \left[\xi_i^\varepsilon \frac{\partial \phi}{\partial x_i} w_\varepsilon - a_{ij}^\varepsilon \frac{\partial w_\varepsilon}{\partial x_i} \frac{\partial \phi}{\partial x_j} u_\varepsilon \right] dx \, dt \to - \int_\mathcal{G} \xi_i \frac{\partial \phi}{\partial x_i} \, dx \, dt$$
$$- \mathcal{M}\left(a_{ij} \frac{\partial \phi}{\partial y_i} \right) \int_\mathcal{G} \frac{\partial \phi}{\partial x_j} \, dx \, dt.$$

The third term in (3.5.16) equals

$$- \int_\mathcal{G} b_j^\varepsilon u_\varepsilon w_\varepsilon \frac{\partial \phi}{\partial x_j} \, dx \, dt \to \mathcal{M}(b_j) \int_\mathcal{G} u \frac{\partial \phi}{\partial x_j} \, dx \, dt.$$

The right hand side in (3.5.16) converges to $-\int_0^T (f, \phi) \, dt = $ (using (3.3.14)) $= -\int_0^T (u'', \phi) \, dt - \int_0^T \mathcal{A}(u, \phi) \, dt = \int_0^T (\eta, \phi) \, dt$. Since $\left(\xi_i, \frac{\partial \phi}{\partial x_i} \right) = \mathcal{A}(u, \phi)$, it follows that

(3.5.18)
$$\eta = \mathcal{M}\left(b_i - a_{ij} \frac{\partial \alpha}{\partial y_j} \right) \frac{\partial u}{\partial x_i}.$$

It remains to show the identity of (3.5.18) with (3.5.10), i.e. that

(3.5.19)
$$\mathcal{M}\left(a_{ij} \frac{\partial \alpha}{\partial y_j} \right) = \mathcal{M}\left(b_k \frac{\partial \chi^i}{\partial y_k} \right).$$

We multiply (3.5.15) by χ^i. It follows that

$$\mathcal{M}\left(b_k \frac{\partial \chi^i}{\partial y_k} \right) = (R_1 \alpha, \chi^i) = (\alpha, R_1 \chi^i) = (\alpha, A_1 \chi^i) = (\alpha, A_1 y_i)$$

hence the result follows.   ∎

3.6. Remarks on correctors. We present here only a very preliminary result on the question of correctors. We consider the case (3.2.1). We have

(3.6.1)
$$u_1 = -\chi^j \frac{\partial u}{\partial x_j}, \qquad \chi^j(y) \text{ being solution of } A_1(\chi^j - y_j) = 0 \text{ and being } Y\text{-periodic,}$$

and we use cut-off functions $m_\varepsilon(x)$ as in Chapter 1, Section 5.2. We set

(3.6.2)
$$\tilde{u}_\varepsilon = u + \varepsilon m_\varepsilon u_1.$$

More generally we suppose that u_ε is the solution of

(3.6.3)
$$(u_\varepsilon'', v) + a^\varepsilon(u_\varepsilon, v) = (f_\varepsilon, v)$$

where

(3.6.4) $f_\varepsilon, f \in L^2(0, T; H)$, $f_\varepsilon \to f$ in $L^1(0, T; H)$ strongly.

We have again (same proof), (3.2.10), (3.2.12). **If we assume that**

(3.6.5) $u \in L^2(0, T; H^2(\mathcal{O}))$, $\dfrac{\partial u}{\partial t} \in L^2(0, T; V)$,

then

(3.6.6) $z_\varepsilon = u_\varepsilon - \tilde{u}_\varepsilon \to 0$ in $L^2(0, T; V)$ **strongly,**

$z_\varepsilon' \to 0$ in $L^2(0, T; H)$ **strongly.**

The proof of this result relies on the following identity. We set $\phi(t) = \frac{1}{2}(T - t)^2$. We have

$$X^\varepsilon = \int_0^T \left(\frac{T - t}{2}\right) \left[a^\varepsilon(z_\varepsilon) + |z_\varepsilon'|^2\right] \, dt$$

$$= \int_0^T \left\{ \phi(f_\varepsilon, u_\varepsilon') - \frac{1}{2}\phi' \left[a^\varepsilon(\tilde{u}_\varepsilon) + |u_\varepsilon'|^2\right] \right\} \, dt$$

$$+ \int_0^T \left[\phi'(f_\varepsilon, \tilde{u}_\varepsilon) + 2\phi'(u_\varepsilon', (\tilde{u}_\varepsilon)') + \phi''(u_\varepsilon, \tilde{u}_\varepsilon)\right] \, dt.$$

To prove this identity, we can assume that f_ε and f are smooth and with values in H (by approximation of these functions). It is then obtained by integration by parts.

Since T does not play any role here, the result will be proved if we verify that $X^\varepsilon \to 0$.

One verifies that $\int_0^T \phi' a^\varepsilon(\tilde{u}_\varepsilon) \, dt \to \int_0^T \phi' \mathcal{A}(u) \, dt$ and one can pass to the limit directly in all other terms. The result follows.

3.7. Remarks on nonlinear problems. Let us give a simple example. We consider the situation of Section 3.6 and we set

(3.7.1)
$$\phi(v) = \int_{\mathcal{O}} |\operatorname{grad} v|^2 \, dx.$$

Let us consider **the nonlinear equation**

(3.7.2)
$$(u_\varepsilon'', v) + a^\varepsilon(u_\varepsilon, v) + \phi(u_\varepsilon)(u_\varepsilon', v) = (f, v),$$
$$u_\varepsilon \in L^\infty(0, T; V), \qquad u_\varepsilon' \in L^\infty(0, T; H),$$
$$u_\varepsilon(0) = 0, \qquad u_\varepsilon'(0) = 0.$$

We assume that

(3.7.3)
$$f, \frac{\partial f}{\partial t} \in L^2(0, T; H).$$

Then **there exists a unique solution of (3.7.2) which satisfies**

(3.7.4) $u_\varepsilon' \in L^\infty(0, T; V)$, $u_\varepsilon'' \in L^\infty(0, T; H)$.

The proof of this fact is standard. One obtains the a priori estimates

(3.7.5) $$\|u_\varepsilon\|_{L^\infty(0,T;V)} + \|u'_\varepsilon\|_{L^\infty(0,T;V)} + \|u''_\varepsilon\|_{L^\infty(0,T;H)} \le C.$$

Therefore we can extract a subsequence, still denoted by u_ε, and such that

(3.7.6) $$u_\varepsilon \to u \text{ and } u'_\varepsilon \to u' \text{ in } L^\infty(0,T;V) \text{ weak star,}$$
$$u''_\varepsilon \to u'' \text{ in } L^\infty(0,T;H) \text{ weak star.}$$

Since grad u_ε is bounded in $L^\infty(0,T;(L^2(\mathcal{O}))^n)$, $\phi(u_\varepsilon)$ is bounded in $L^\infty(0,T)$ and we can assume that

(3.7.7) $$\phi(u_\varepsilon) \to \zeta \text{ in } L^\infty(0,T) \text{ weak star;}$$

since $u'_\varepsilon \to u'$ in $L^2(\mathcal{G})$ strongly, we have $\phi(u_\varepsilon)u'_\varepsilon \to \zeta u'$ in $L^2(\mathcal{G})$ weakly and if we write

(3.7.8) $$(u''_\varepsilon, v) + a^\varepsilon(u_\varepsilon, v) = (f - \phi(u_\varepsilon)u'_\varepsilon, v) = (f_\varepsilon, v),$$

we have $f_\varepsilon \to f - \zeta u'$ in $L^2(\mathcal{G})$ weakly. But one can improve this result; indeed

$$\frac{d}{dt}\phi(u_\varepsilon) = 2\int_\mathcal{O} \text{grad } u_\varepsilon \cdot \text{grad } u'_\varepsilon \, dx$$

remains in a bounded set of $L^\infty(0,T)$ so that we can assume that $\phi(u_\varepsilon) - \zeta$ in $L^\infty(0,T)$ **strongly** and therefore

(3.7.9) $$f_\varepsilon \to f - \zeta u' \text{ in } L^2(\mathcal{G}) \text{ strongly.}$$

Therefore, using (3.6.6), we see that

$$\phi(u_\varepsilon) - \phi(\tilde{u}_\varepsilon) \to 0 \text{ in } L^1(0,T) \text{ for instance.}$$

But

$$\phi(\tilde{u}_\varepsilon) = \int_\mathcal{O} \frac{\partial}{\partial x_i}(u + \varepsilon m_\varepsilon u_1)\frac{\partial}{\partial x_i}(u + \varepsilon m_\varepsilon u_1)\, dx,$$

$$\frac{\partial}{\partial x_i}(u + \varepsilon m_\varepsilon u_1) = \frac{\partial u}{\partial x_i} - m_\varepsilon\frac{\partial \chi^j}{\partial y_i}\frac{\partial u}{\partial x_j} - \varepsilon m_\varepsilon \chi^j\frac{\partial^2 u}{\partial x_i \partial x_j} - \varepsilon u_1\frac{\partial m_\varepsilon}{\partial x_i},$$

so that
(3.7.10)

$$\phi(\tilde{u}_\varepsilon) \to \psi(u) = \int_\mathcal{O} \left(\frac{\partial u}{\partial x_i} - \frac{\partial \chi^j}{\partial y_i}\frac{\partial u}{\partial x_j}\right)\left(\frac{\partial u}{\partial x_i} - \frac{\partial \chi^k}{\partial y_i}\frac{\partial u}{\partial x_k}\right)\, dx \text{ in } L^1(0,T).$$

Therefore u is the solution of

(3.7.11) $$(u'', v) + \mathcal{A}(u, v) + \psi(u)(u', v) = (f, v),$$

where $\psi(u)$ is given in (3.7.10).

REMARK 3.4 (Homogenization of variational inequalities). The general question of the homogenization in variational inequalities of "hyperbolic type" is largely open. We give only a simple result along the following lines. We set

(3.7.12) $$j(v) = g\int_\mathcal{O} |v|\, dx, \qquad g > 0$$

and, always with the conditions of Section 3.6, we consider the solution u_ε of the V.I.

$$(3.7.13) \qquad (u_\varepsilon'', v - u_\varepsilon') + a^\varepsilon(u_\varepsilon, v - u_\varepsilon') + j(v) - j(u_\varepsilon') \geq (f, v - u_\varepsilon'), \ \forall v \in V,$$

$$u_\varepsilon, u_\varepsilon' \in L^\infty(0, T; V),$$

$$u_\varepsilon'' \in L^\infty(0, T; H),$$

$$u_\varepsilon(0) = 0, \qquad u_\varepsilon'(0) = 0.$$

We assume that f satisfies (3.7.3). It is known (cf. for instance Duvaut-Lions [**38**]) that (3.7.13) admits a unique solution. Then one can prove that (3.7.6) holds true, with u being the solution of the **homogenized** V.I.

$$(3.7.14) \qquad (u'', v - u') + \mathcal{A}(u, v - u') + j(v) - j(u') \geq (f, v - u')$$

$$\forall v \in V, \qquad u(0) = 0, \qquad u'(0) = 0.$$

REMARK 3.5. If in the V.I. of the preceding Remark, we replace $j(v)$ given by (3.7.12) by

$$(3.7.15) \qquad j_1(v) = g \int_{\mathcal{O}} |\operatorname{grad} v| \, dx$$

it is likely — but it is not proved — that one obtains for the limit a V.I. of type (3.7.14) with j replaced by

$$(3.7.16) \qquad \tilde{j}_1(v) = g \int_{\mathcal{O}} \left[\left(\frac{\partial u}{\partial x_i} - \frac{\partial \chi^j}{\partial y_i} \frac{\partial u}{\partial x_j} \right) \left(\frac{\partial u}{\partial x_i} - \frac{\partial \chi^k}{\partial y_i} \frac{\partial u}{\partial x_k} \right) \right]^{1/2} \, dx.$$

3.8. Remarks on Schrödinger type equations. We can consider, **now with complex valued functions**, the equation (of the Schrödinger type)[18]

$$(3.8.1) \qquad \sqrt{-1} \left(\frac{\partial u_\varepsilon}{\partial t}, v \right) + a^\varepsilon(u_\varepsilon, v) = (f, v), \ \forall v \in V, \ u_\varepsilon(0) = 0,$$

where a^ε is given as in (3.2.1), (3.2.2), (3.2.3). [We can also consider other cases considered in the preceding sections for a^ε.] If we assume (3.7.3), we can obtain the following a priori estimate: taking $v = u_\varepsilon'$ in (3.8.1) and taking the real part

$$\frac{1}{2} \frac{d}{dt} a^\varepsilon(u_\varepsilon) = \Re(f, u_\varepsilon')$$

hence

$$a^\varepsilon(u_\varepsilon(t)) = 2\Re(f(t), u_\varepsilon(t)) - 2\Re \int_0^t (f', u_\varepsilon) \, d\sigma.$$

Hence it follows that (one obtains a preliminary estimate by taking $v = u_\varepsilon$ in (3.8.1))

$$(3.8.2) \qquad \|u_\varepsilon\|_{L^\infty(0,T;V)} \leq C.$$

Then we can extract a subsequence, still denoted by u_ε, such that

$$(3.8.3) \qquad u_\varepsilon \to u \text{ in } L^\infty(0, T; V) \text{ weak star}$$

---

[18]Of course one has now:

$$(f, v) = \int_{\mathcal{O}} f \bar{v} \, dx, \qquad a^\varepsilon(u, v) = \int_{\mathcal{O}} a_{ij}^\varepsilon \frac{\partial u}{\partial x_j} \frac{\partial \bar{v}}{\partial x_i} \, dx.$$

and

$$(3.8.4) \qquad \sqrt{-1}\left(\frac{\partial u}{\partial t}, v\right) + \mathcal{A}(u, v) = (f, v), \ \forall v \in V, \ u(0) = 0.$$

3.9. Nonlocal operators.
3.9.1. *Setting of the problem.* We consider:

$$(3.9.1) \qquad\qquad A^\varepsilon \text{ given with } (3.2.1), (3.2.3)^{19}$$

$$
B^\varepsilon = -\frac{\partial}{\partial x_i}\left(b_{ij}\left(\frac{x}{\varepsilon}\right)\frac{\partial}{\partial x_j}\right),
$$
$$(3.9.2) \qquad b_{ij} = b_{ji}, \qquad b_{ij} \in L^\infty(\mathbb{R}^n_y), \qquad b_{ij}\ Y\text{-periodic},$$
$$
b_{ij}(y)\xi_i\xi_j \geq \beta\xi_i\xi_i, \qquad \beta > 0
$$

and we take[20]

$$(3.9.3) \qquad\qquad\qquad V = H^1_0(\mathcal{O}).$$

We set

$$(3.9.4) \qquad\qquad b^\varepsilon(u, v) = \int_{\mathcal{O}} b^\varepsilon_{ij}(x)\frac{\partial u}{\partial x_j}\frac{\partial v}{\partial x_i}\, dx$$

and we consider the equation

$$(3.9.5) \qquad\qquad b^\varepsilon(u'_\varepsilon, v) + a^\varepsilon(u_\varepsilon, v) = (f, v),$$
$$u_\varepsilon(0) = 0.$$

REMARK 3.6. The ellipticity of A^ε is not necessary in what follows.

The problem (3.9.5) admits a unique solution $u_\varepsilon \in L^2(0, T; V)$, if

$$(3.9.6) \qquad\qquad\qquad f \in L^2(0, T; V'),$$

and

$$(3.9.7) \qquad\qquad\qquad \|u_\varepsilon\|_{L^\infty(0,T;V)} \leq C.$$

Indeed taking $v = u_\varepsilon$ in (3.9.5) we have

$$\frac{1}{2}\frac{d}{dt}b^\varepsilon(u_\varepsilon) + a^\varepsilon(u_\varepsilon) = (f, u_\varepsilon)$$

hence (3.9.7) follows. Therefore one can extract a subsequence, still denoted by u_ε, such that

$$(3.9.8) \qquad\qquad u_\varepsilon \to u \text{ in } L^\infty(0, T; V) \text{ weak star.}$$

The problem is now to see what is the "homogenized" equation which characterizes u. We are going to show it is not a partial differential equation but an equation with pseudo-differential operator.

We take the **Laplace transform** in t of (3.9.5). Let us set

$$(3.9.9) \qquad\qquad \hat{u}_\varepsilon(p) = \int_0^\infty e^{-pt} u_\varepsilon(t)\, dt, \qquad \Re p > 0.$$

---

[19] The symmetry is now necessary.

[20] We can take $H^1_0(\mathcal{O}) \subseteq V \subseteq H^1(\mathcal{O})$ if $1 \notin V$. If $V = H^1(\mathcal{O})$, what we are going to say is valid if one replaces B^ε by $B^\varepsilon + b_0(x/\varepsilon)$, $b_0(y) \geq \beta_0 > 0$.

We do not restrict the generality by assuming that (3.9.5) is taken for every $t > 0$ and with $f \in L^2(0, \infty; H)$ with compact support. We can also consider (3.9.5) as an equation over $\mathbb{R}_t$, with u and f zero for $t < 0$. We then set

$$\hat{f}(p) = \int_0^\infty e^{-pt} f(t)\, dt$$

and we write (3.9.5)

$$(3.9.10) \qquad pb^\varepsilon(u_\varepsilon(p), v) + a^\varepsilon(u_\varepsilon(p), v) = (\hat{f}(p), v).$$

This leads to "homogenization with parameters" as in Chapter 1, Sections 8.1 and 15. We define $\chi^j(p)$ as a Y-periodic solution of

$$(3.9.11) \qquad (pB_1 + A_1)(\chi^j(p) - y_j) = 0$$

normalized by, say, $\int_Y \chi^j(p)\, dy = 0$; in (3.9.11) we use the standard notation for A_1 and $B_1 = -\frac{\partial}{\partial y_i}\left(b_{ij}\frac{\partial}{\partial y_j}\right)$. Then we define

$$(3.9.12) \qquad q_{ij}(p) = \frac{1}{|Y|}\left[pb_1(\chi^j(p) - y_j, \chi^i(p) - y_i) + a_1(\chi^j(p) - y_j, \chi^i(p) - y_i)\right]$$

and

$$(3.9.13) \qquad Q(p) = -q_{ij}(p)\frac{\partial^2}{\partial x_i \partial x_j}.$$

We can take the inverse Laplace transform of $q_{ij}(p)$ and set

$$(3.9.14) \qquad Q = \mathcal{L}^{-1}Q(p), \qquad \mathcal{L}^{-1} = \text{ inverse Laplace transform.}$$

Then **the homogenized equation is**

$$(3.9.15) \qquad u \in L^2(\mathbb{R}; V), \qquad u = 0 \text{ for } t < 0, \qquad Q *_{(t)} u = f.$$

The proof proceeds by methods similar to those of Chapter 1, Section 15.

REMARK 3.7. The operator $Q*_{(t)}$ is not, in general, a partial differential operator.

REMARK 3.8. Let us consider materials which have a composite periodical structure and with a long memory (cf. for this last point, for instance, Duvaut-Lions [**38**], Chapter 3, Section 7). We consider a simple model. These general cases will be treated with similar methods. We are given functions $a_{ij}(y)$ satisfying (3.2.1), (3.2.2), (3.2.3). We are also given functions $b_{ij}(y, t)$ such that

$$(3.9.16) \qquad \begin{aligned} &b_{ij}, \quad \frac{\partial}{\partial t}b_{ij}(y, t) \in L^\infty(\mathbb{R}^n_y \times \mathbb{R}_t), \\ &\forall t, \quad b_{ij}(y, t) \text{ is } Y\text{-periodic}, \\ &b_{ij} = b_{ji}, \ \forall i, j. \end{aligned}$$

(No ellipticity is assumed.) Given $v \in L^2 loc(\mathbb{R}_t; V)$, we set

$$(3.9.17) \qquad B^\varepsilon(t) *_{(t)} v = -\frac{\partial}{\partial x_i}\int_0^t b_{ij}\left(\frac{x}{\varepsilon}, t - s\right)\frac{\partial v}{\partial x_j}(x, s)\, ds$$

and we consider the problem:

$$u_\varepsilon \in L_{loc}(\mathbb{R}; V), \qquad \frac{\partial u_\varepsilon}{\partial t} \in L^\infty(\mathbb{R}_t; H),$$

(3.9.18)
$$\frac{\partial^2 u_\varepsilon}{\partial t^2} + A^\varepsilon u_\varepsilon + B^\varepsilon(t) *_{(t)} u_\varepsilon = f$$

$$u_\varepsilon = 0 \text{ for } t < 0,$$

where

(3.9.19) $$f \in L^2_{loc}(\mathbb{R}_t; H), \qquad f = 0 \text{ for } t < 0.$$

This problem admits a unique solution. One has the following a priori estimates:

(3.9.20) $$\|u_\varepsilon\|_{L^\infty(0,T;V)} + \|u'_\varepsilon\|_{L^\infty(0,T;H)} \leq C.$$

Indeed, if we set

$$b^\varepsilon(t; u, v) = \int_{\mathcal{O}} b_{ij}\left(\frac{x}{\varepsilon}, t\right) \frac{\partial u}{\partial x_j} \frac{\partial v}{\partial x_i} \, dx,$$

$$\dot{b}^\varepsilon(t; u, v) = \int_{\mathcal{O}} \frac{\partial b_{ij}}{\partial t}\left(\frac{x}{\varepsilon}, t\right) \frac{\partial u}{\partial x_j} \frac{\partial v}{\partial x_i} \, dx,$$

we obtain

(3.9.21) $$\frac{1}{2} \frac{d}{dt}\left[|u'_\varepsilon|^2 + a^\varepsilon(u_\varepsilon)\right] + \int_0^t b^\varepsilon(t - s; u_\varepsilon(s), u'_\varepsilon(t)) \, ds = (f, u'_\varepsilon)$$

and

$$\frac{d}{dt} \int_0^t b^\varepsilon(t - s; u_\varepsilon(s), u'_\varepsilon(t)) \, ds = \int_0^t b^\varepsilon(t - s; u_\varepsilon(s), u'_\varepsilon(t)) \, ds$$
$$+ \int_0^t \dot{b}^\varepsilon(t - s; u_\varepsilon(s), u_\varepsilon(t)) \, ds + b^\varepsilon(0; u_\varepsilon(t), u_\varepsilon(t))$$

so that (3.9.21) gives:

$$\frac{d}{dt}\left[\frac{1}{2}|u'_\varepsilon|^2 + \frac{1}{2}a^\varepsilon(u_\varepsilon) + \int_0^t b^\varepsilon(t - s; u_\varepsilon(s), u_\varepsilon(t)) \, ds\right]$$
$$- \int_0^t \dot{b}^\varepsilon(t - s; u_\varepsilon(s), u_\varepsilon(t)) \, ds - b^\varepsilon(0; u_\varepsilon(t), u_\varepsilon(t)) = (f, u'_\varepsilon)$$

and therefore, for $t \leq T$ finite (arbitrary)

$$\frac{1}{2}|u'_\varepsilon(t)|^2 + \frac{1}{2}a^\varepsilon(u_\varepsilon(t)) + \int_0^t b^\varepsilon(t - s; u_\varepsilon(s), u_\varepsilon(t)) \, ds$$
$$\leq \int_0^t |f||u'_\varepsilon| \, ds + C \int_0^t \|u_\varepsilon\|^2 \, ds$$

hence

$$|u'_\varepsilon(t)|^2 + a^\varepsilon(u_\varepsilon(t)) \leq C \int_0^t |f|^2 \, ds + C \int_0^t \left[\|u_\varepsilon\|^2 + |u'_\varepsilon|^2\right] \, ds$$

hence (3.9.20) follows.

Therefore we can extract a subsequence, still denoted by u_ε, that satisfies (3.2.10). **We show now that u is characterized by the solution of a pseudo differential operator equation.**

We define, for $\Re p > 0$

(3.9.22)
$$\hat{b}_{ij}(y;p) = \int_0^\infty e^{-pt} b_{ij}(y,t)\,dt,$$

$$\hat{B}^\varepsilon(p) = -\frac{\partial}{\partial x_i}\left(\hat{b}_{ij}\left(\frac{x}{\varepsilon};p\right)\frac{\partial}{\partial x_j}\right).$$

Then (3.9.18) can be written

(3.9.23)
$$(A^\varepsilon + \hat{B}^\varepsilon(p) + p^2 I)\hat{u}_\varepsilon(p) = \hat{f}(p).$$

(We can always assume that f admits a Laplace transform in t.)

We use now "homogenization with parameter." We define $\chi^j(p)$ as the Y-periodic solution (normalized by $\int_Y \chi^j(p)\,dy = 0$) of

(3.9.24)
$$(A_1 + \hat{B}_1(p))(\chi^j - y_j) = 0;$$

we have set

(3.9.25)
$$\hat{B}_1(p) = -\frac{\partial}{\partial y_i}\left(\hat{b}_{ij}(y;p)\frac{\partial}{\partial y_j}\right).$$

We observe that $|\hat{b}_{ij}(y;p)| \le \frac{C}{\Re p}$ so that the operator

(3.9.26) $\qquad A_1 + \hat{B}_1(p)$ **is elliptic for $\Re p$ large enough**.

Therefore (3.9.24) defines $\chi^j(y;p)$. We define

(3.9.27) $\quad q_{ij}(p) = \dfrac{1}{|Y|}\left[a_1(\chi^j(p) - y_j, \chi^i(p) - y_i) + b_1(p; \chi^j(p) - y_i, \chi^i(p) - y_i) \right]$

$$b_1(p;\phi,\psi) = \int_Y \hat{b}_{ij}(y,p)\frac{\partial\phi}{\partial y_j}\frac{\partial\psi}{\partial y_i}\,dy,$$

and $Q(p)$ and Q as in (3.9.13), (3.9.14).

Then **the homogenized equation** is

(3.9.28)
$$\frac{\partial^2 u}{\partial t^2} + Q *_{(t)} u = f,$$

$$u \in L^2_{loc}(\mathbb{R}_t; V), \qquad u = 0 \text{ for } t < 0.$$

4. Comments and problems

Most of the results presented in this chapter are given for the first time. Preliminary announcements were made in notes by the authors (cf. Bibliography of Chapter 1).

The L^p estimates of Section 2.7 were obtained, in the symmetric case, by Pulvirenti [95], [96]. The method given here extends to parabolic equations of higher order or to systems.

Given a family of elliptic operators, say A^ε, one can say that A^ε "G converges" to $\mathcal{A}$ if $\forall f$, $(A^\varepsilon)^{-1}f \to \mathcal{A}^{-1}f$ in a weak Sobolev space, where the inverses are taken for a given set of homogeneous boundary conditions. If one considers next a family of elliptic operators depending on t, say $A^\varepsilon(t)$, one can say that A^ε "$P - G$ converges" to $\mathcal{A}$ (where "P" stands for "parabolic") if $\forall f$, $\left(\frac{\partial}{\partial t} + A^\varepsilon\right)^{-1}f \to \left(\frac{\partial}{\partial t} + \mathcal{A}\right)^{-1}f$ in a weak Sobolev space (where again the inverses are taken for a given set of homogeneous boundary conditions). Then one can compare the "$P - G$ convergence" with the "G-convergence $\forall t$" (in general these are different notions!)

Cf. F. Colombini and S. Spagnolo [**28**]. Cf. also for this general approach, Spagnolo [**111**], [**110**], Sbordone [**102**].

In case the coefficients $a_{ij}(y, \tau)$ are Y-periodic in y and **almost periodic in** τ, one obtains formulas similar to those of the text, with a difficulty for $k = 2$. In the case $k = 1$, one considers $\chi^j(y, \tau)$ as the solution of (1.4.7) which satisfies, say $\int_Y \chi^j(y, \tau)\, dy = 0$. Then χ^j **is almost periodic in** τ (τ plays the role of a parameter) and one obtains

$$(4.0.29) \qquad \mathcal{A}^1 = -q_{ij}^1 \frac{\partial^2}{\partial x_i \partial x_j}$$

where

$$(4.0.30) \qquad q_{ij}^1 = \lim_{\lambda \to \infty} \frac{1}{|Y|\lambda} \int_Y \int_0^\lambda \left(a_{ij} - a_{ik} \frac{\partial \chi^j}{\partial y_k} \right)(y, \sigma)\, d\sigma\, dy.$$

In case $k = 3$, one introduces (compare to (1.6.6))

$$(4.0.31) \qquad \bar{A}_1 = -\frac{\partial}{\partial y_i} \left(\left(\lim_{\lambda \to \infty} \frac{1}{\lambda} \int_0^\lambda a_{ij}(y, \tau)\, d\tau \right) \frac{\partial}{\partial y_j} \right)$$

and $\mathcal{A}^3$ is given again as in (1.6.14) where

$$\phi = \lim_{\lambda \to \infty} \frac{1}{|Y|\lambda} \int_Y \int_0^\lambda \phi(y, \sigma)\, d\sigma\, dy.$$

In case $k = 2$, one has to consider the τ-almost periodic solution of

$$(4.0.32) \qquad \frac{\partial \theta^j}{\partial \tau} + A_1(\theta^j - y_j) = 0$$

where is **periodic** in y, and many problems remain open in this direction.

One can of course in the τ-almost periodic case consider **correctors**, as in the case of periodic coefficients.

Let us given now some indications on possible extensions of the results given in the text. Let $b(y, \tau)$ be given satisfying

$$(4.0.33) \qquad b \text{ is } Y - \tau_0 \text{ periodic}, \qquad b(y, \tau) \geq \beta > 0, \qquad b \in L^\infty(\mathbb{R}_y^n \times \mathbb{R}_\tau),$$

and let us consider the operator

$$(4.0.34) \quad b^\varepsilon \frac{\partial}{\partial t} - \frac{\partial}{\partial x_i} \left(a_{ij}^\varepsilon \frac{\partial}{\partial x_j} \right), \qquad b^\varepsilon = b(x/\varepsilon, t/\varepsilon^k), \qquad a_{ij}^\varepsilon = a_{ij}(x/\varepsilon, t/\varepsilon^k).$$

(If one considers the case $b^\varepsilon(x, t) = b(x/\varepsilon)$ when b does not depend on τ, the homogenization is immediate and is given by

$$\mathcal{M}_y(b) \frac{\partial}{\partial t} + \mathcal{A}^k, \qquad k = 1, 2, 3.)$$

Then general case (4.0.34) leads to a number of interesting questions. In case $k = 1$, the asymptotic expansion, with usual notations, leads to

$$(4.0.35) \qquad\qquad\qquad\qquad A_1 u_0 = 0,$$

$$(4.0.36) \qquad\qquad\qquad A_1 u_1 + b \frac{\partial u_0}{\partial \tau} + A_2 u_0 = 0,$$

$$(4.0.37) \qquad A_1 u_2 + \left(b \frac{\partial}{\partial \tau} + A_2 \right) u_1 + \left(b \frac{\partial}{\partial t} + A_3 \right) u_0 = f.$$

It follows from (4.0.35) that u_0 does not depend on y. Taking $\mathcal{M}_y$ of (4.0.36) gives: $\mathcal{M}_y(b)\frac{\partial u_0}{\partial \tau} = 0$, so that $u_0 = u(x,t)$ and with the notation (1.4.7), one will have:

$$(4.0.38) \qquad u_1 = -\chi^j(y,\tau)\frac{\partial u}{\partial x_j} + \hat{u}_1(x,t,\tau).$$

We use (4.0.38) into (4.0.37). Taking $\mathcal{M}_y$ of the result, we obtain:

$$\mathcal{M}_y(b)\frac{\partial \hat{u}_1}{\partial \tau} - \mathcal{M}\left(b\frac{\partial \chi^j}{\partial \tau}\right)\frac{\partial u}{\partial x_j} + \mathcal{M}_y(b)\frac{\partial u}{\partial t} - \frac{\partial}{\partial x_i}\mathcal{M}_y\left(a_{ij} - a_{ik}\frac{\partial \chi^j}{\partial y_k}\right)\frac{\partial u}{\partial x_j} = f.$$

We then multiply by $\frac{1}{\mathcal{M}_y(b)}$ and we take the average in τ:

$$(4.0.39) \quad \frac{\partial u}{\partial t} - \mathcal{M}_\tau\left(\frac{1}{\mathcal{M}_y(b)}\mathcal{M}_y\left(b\frac{\partial \chi^j}{\partial \tau}\right)\right)\frac{\partial u}{\partial x_j}$$

$$- \frac{\partial}{\partial x_i}\mathcal{M}_\tau\left(\frac{1}{\mathcal{M}_y(b)}\mathcal{M}_y\left(a_{ij} - a_{ik}\frac{\partial \chi^j}{\partial y_k}\right)\right)\frac{\partial u}{\partial x_j} = \mathcal{M}_\tau\left(\frac{1}{\mathcal{M}_y(b)}\right)f.$$

One sees that appearance of **first order derivatives** in x in the homogenization process. One also sees that one needs neither regularity on the a_{ij}'s of the type: $\frac{\partial}{\partial \tau}a_{ij} \in L^\infty(\mathbb{R}^n_y \times \mathbb{R}_\tau)$, in order $\frac{\partial \chi^j}{\partial \tau}$ to make sense nor regularity of b: $\frac{\partial b}{\partial \tau} \in L^\infty(\mathbb{R}^n_y \times \mathbb{R}_\tau)$ (since one can integrate by parts in τ in the coefficients of $\frac{\partial u}{\partial x_j}$). The justification can then be made along the lines of Section 2. If $k = 2$, one runs into the difficulty of finding $Y - \tau_0$ periodic solutions of

$$(4.0.40) \qquad b(y,\tau)\frac{\partial \phi}{\partial \tau} + A_1\phi = P$$

a question whose general study does not seem to have been made (the name is a fortiori the case for the **almost periodic** situation).

In the case $k = 3$, one defines:

$$(4.0.41) \qquad \bar{A}_1 = -\frac{\partial}{\partial y_i}\mathcal{M}_\tau\left(\frac{a_{ij}}{b}\right)\frac{\partial}{\partial y_j}$$

and the functions ϕ^j by (1.6.8) (with the new $\bar{A}_1$). Then the homogenized equation is

$$(4.0.42) \quad \frac{\partial u}{\partial t} = \mathcal{M}_\tau\left(\frac{1}{\mathcal{M}_y(b)}\mathcal{M}_y\left(a_{ij} - a_{ik}\frac{\partial \phi^j}{\partial y_k}\right)\right)\frac{\partial^2 u}{\partial x_i \partial x_j} = \mathcal{M}_\tau\left(\frac{1}{\mathcal{M}_y(b)}\right)f.$$

4.1. Singular perturbation and homogenization. Let us consider first the problem

$$(4.1.1) \qquad \frac{\partial u_\varepsilon}{\partial t} + (\varepsilon^2 \Delta^2 + A^\varepsilon)u_\varepsilon = f,$$

u_ε being subject, say, to Dirichlet's boundary conditions (with usual boundary conditions). Then the homogenized equation is obtained by formulas analogous to those for homogenizing $\frac{\partial}{\partial t} + A^\varepsilon$ but **where A_1 is replaced by $A_1 + \Delta^2_y$. For instance** if $k = 2$, one considers the solution $\chi^j(y,\tau)$ which is $Y - \tau_0$ periodic, of

$$(4.1.2) \qquad \frac{\partial \chi^j}{\partial \tau} + (A_1 + \Delta^2_y)(\chi^j - y_j) = 0,$$

and $\mathcal{A}^2 = \mathcal{A}^{2,\Delta^2}$ is now given by

(4.1.3)
$$\mathcal{A}^{2,\Delta^2} = -q_{ij}^{2,\Delta^2} \frac{\partial^2}{\partial x_i \partial x_j},$$

$$q_{ij}^{2,\Delta^2} = \mathcal{M}\left(a_{ij} - a_{ik}\frac{\partial \chi^j}{\partial y_k}\right).$$

One can also consider the problem:

(4.1.4)
$$\varepsilon\frac{\partial u_\varepsilon}{\partial t} + A^\varepsilon(t/\varepsilon^k)u_\varepsilon = f,$$

with usual boundary conditions and with $u_\varepsilon(0) = 0$.

If $k = 2$, one has

(4.1.5)
$$u_\varepsilon \to u \text{ in } L^2(0,T;V) \text{ weakly}$$

where u is the solution of

(4.1.6)
$$\mathcal{A}^1 u = f = f(x,t), \qquad (t \text{ plays the role of a parameter}).$$

(More precisely,

(4.1.7)
$$u \in V, \qquad \mathcal{A}^1(u,v) = (f,v), \ \forall v \in V.)$$

If $k = 3$ **or** $k = 4$, one has again (4.1.5), with u given by

(4.1.8)
$$\mathcal{A}^2(u,v) = (f,v), \ \forall v \in V, \text{ if } k = 3,$$

(4.1.9)
$$\mathcal{A}^3(u,v) = (f,v), \ \forall v \in V, \text{ if } k = 4.$$

Actually the proof of (4.1.8) has been obtained only for the Dirichlet's boundary condition by the method of the adjoint expansion, and assuming all data to be smooth. Proofs of (4.1.7), (4.1.9) proceed along the lines of Section 2.

There is one more a priori estimate one can obtain when

(4.1.10)
$$k = 1, \qquad a_{ij} = a_{ji}, \qquad \frac{\partial a_{ij}}{\partial \tau} \in L^\infty(\mathbb{R}_y^n \times \mathbb{R}_\tau).$$

[One will obtain the same estimates for the equation

(4.1.11)
$$\varepsilon^k \frac{\partial u_\varepsilon}{\partial t} + A(t/\varepsilon^k)u_\varepsilon = f.]$$

Indeed if we multiply (4.1.4), for $k = 1$, by $A^\varepsilon(t/\varepsilon)u_\varepsilon$, we obtain (we use the symmetry):

(4.1.12)
$$\varepsilon a^\varepsilon\left(\frac{\partial u_\varepsilon}{\partial t}, u_\varepsilon\right) + |A^\varepsilon(t/\varepsilon)u_\varepsilon|^2 = (f, A^\varepsilon(t/\varepsilon)u_\varepsilon).$$

Hence

(4.1.13)
$$\frac{\varepsilon}{2}\frac{d}{dt}a^\varepsilon(u_\varepsilon) - (\dot{A})^\varepsilon(u_\varepsilon) + |A^\varepsilon(t/\varepsilon)u_\varepsilon|^2 = (f, A^\varepsilon(t/\varepsilon)u_\varepsilon)$$

in (4.1.13), $(\dot{a})^\varepsilon(u,v) = \int_\mathcal{O}\left(\frac{\partial a_{ij}}{\partial \tau}\right)(x/\varepsilon,t/\varepsilon)\frac{\partial u}{\partial x_j}\frac{\partial v}{\partial x_i}\,dx$ and $(\dot{a})^\varepsilon(u_\varepsilon) = (\dot{a})^\varepsilon(u_\varepsilon, u_\varepsilon)$. It follows from (4.1.13) that

(4.1.14)
$$A^\varepsilon(t/\varepsilon)u_\varepsilon \text{ remains in a bounded set of } L^2(0,T;H).$$

Using (4.1.4) and (4.1.14) we see that

(4.1.15)
$$\varepsilon\frac{\partial u_\varepsilon}{\partial t} \text{ remains in a bounded set of } L^2(0,T;H).$$

The behavior of the solution of (4.1.4) for $k = 1$ is unsettled.

4.2. Reiteration. Let us consider for instance the equation

$$(4.2.1) \qquad \frac{\partial u_\varepsilon}{\partial t} - \frac{\partial}{\partial x_i} a_{ij} \left(\frac{x}{\varepsilon}, \frac{x}{\varepsilon^2}, \frac{t}{\varepsilon^{k_1}}, \frac{t}{\varepsilon^{k_2}}, \frac{t}{\varepsilon^{k_3}} \right) \frac{\partial u_\varepsilon}{\partial x_j} = f,$$

with the usual boundary conditions and initial conditions, where the a_{ij}'s satisfy the following conditions:

$$(4.2.2) \qquad a_{ij}(y, z, \tau_1, \tau_2, \tau_3) \text{ is } Y - Z \text{ periodic in } y, z \text{ and admits}$$

the period τ_j^0 in the variable τ_j, $j = 1, 2, 3$;

$$(4.2.3) \qquad a_{ij} \in C^0(\bar{Y} \times \bar{Z} \times [0, \tau_1^0] \times [0, \tau_2^0] \times [0, \tau_3^0])^{21}$$

and the usual hypothesis

$$a_{ij} \xi_i \xi_j \geq \alpha \xi_i \xi_i, \qquad \alpha > 0.$$

In (4.2.1) we assume that

$$(4.2.4) \qquad 0 < k_1 < k_2 < k_3.$$

We have then the "usual" result that $u_\varepsilon \to u$ in $L^2(0, T; V)$ weakly, where u is the solution of $\frac{\partial u}{\partial t} + \mathcal{A}u = f$ subject to appropriate boundary conditions and where $\mathcal{A}$ is computed as follows.

One considers the operator with coefficients

$$(4.2.5) \qquad a_{ij} \left(\lambda, \frac{x}{\varepsilon}, \mu, \nu, \frac{t}{\varepsilon^k} \right)$$

where $k = k_3/2$. One homogenizes the corresponding operator according to the rules of Section 1; therefore we have three cases to consider: $k < 2$, $k = 2$ or $k > 2$, i.e.

$$(4.2.6) \qquad k_3 < 4, \qquad k_3 = 4 \qquad \text{or} \qquad k_3 > 4.$$

Let us denote by $b_{ij}(\lambda, \mu, \nu)$ the coefficients of the homogenized operator obtained in this way. We then consider the operator with coefficients

$$(4.2.7) \qquad b_{ij} \left(\frac{x}{\varepsilon}, \mu, \frac{t}{\varepsilon^{k_2}} \right)$$

and we homogenize the corresponding operator; therefore there are again three cases to consider:

$$(4.2.8) \qquad k_2 < 2, \qquad k_2 = 2 \qquad \text{or} \qquad k_2 > 2.$$

Let us denote by $c_{ij}(\mu)$ the coefficients of the homogenized operator obtained in this way. The coefficients of the operator we are looking for are then given by

$$(4.2.9) \qquad \frac{1}{\tau_1^0} \int_0^{\tau_1^0} c_{ij}(\tau_1) \, d\tau_1 = d_{ij}.$$

The proof is technically long but the ideas are those given in Chapters 1 and 2.

In case of coefficients of the form

$$(4.2.10) \qquad a_{ij} \left(x, \frac{x}{\varepsilon}, \frac{x}{\varepsilon^2}, t, \frac{t}{\varepsilon^{k_1}}, \frac{t}{\varepsilon^{k_2}}, \frac{t}{\varepsilon^{k_3}} \right)$$

---

[21] One could weaken this hypothesis but the weakest hypothesis is not known.

one will apply the same rule to $a_{ij}(x_0, y, z, t_0, \tau_1, \tau_2, \tau_3)$ and one will obtain coefficients $d_{ij}(x_0, t_0)$. Then the homogenized operator is $d_{ij}(x, t)$.

4.3. Homogenization with rapidly oscillating potentials. One can extend the considerations of Chapter 1, Section 12, to the evolution problems, and this for parabolic and hyperbolic equations, and also for Schrödinger equations. For instance let u_ε be the solution of

$$(4.3.1) \qquad \frac{\partial u_\varepsilon}{\partial t} = \Delta u_\varepsilon + \frac{1}{\varepsilon} W^\varepsilon u_\varepsilon = f_0 \text{ in } \mathcal{G} = \mathcal{O} \times]0, T[,$$

$$u_\varepsilon = 0 \text{ on } \Sigma, \qquad u_\varepsilon(x, 0) = 0.$$

We assume that $W^\varepsilon = W(x/\varepsilon, t/\varepsilon^2)$ (one can also consider $W(x/\varepsilon, t/\varepsilon^k)$, $k \neq 2$, and a function $W(x, t, x/\varepsilon, t/\varepsilon^2)$) is almost periodic. Then one considers the almost periodic solution χ (a.p. in y and in τ) of

$$(4.3.2) \qquad -\frac{\partial \chi}{\partial \tau} - \Delta \chi + W(y, \tau) = 0,$$

and we set $\bar{W}$ = average in y and τ of W_χ

$$\bar{W}_\chi = \lim_{\lambda \to +\infty} \frac{1}{(2\lambda)^{n+1}} \int_{-\lambda}^{\lambda} \cdots \int_{-\lambda}^{\lambda} W(y, \tau) \chi(y, \tau) \, dy \, d\tau.$$

Then

$$(4.3.3) \qquad u_\varepsilon \to u \text{ in } L^2(0, T; H_0^1(\mathcal{O})) \text{ weakly,}$$

where u is the solution of

$$(4.3.4) \qquad \frac{\partial u}{\partial t} - \Delta u - \bar{W}u = f, \qquad u = 0 \text{ on } \Sigma, \qquad u(x, 0) = 0;$$

cf. Bensoussan, Lions, Papanicolaou [**16**] (where one will find analogous formulas also for the hyperbolic case).

4.4. Homogenization and penalty. Let us consider

$$(4.4.1) \qquad A_1^\varepsilon = -\frac{\partial}{\partial x_i} \left(a_{ij1} \left(\frac{x}{\varepsilon}, \frac{t}{\varepsilon^k} \right) \frac{\partial}{\partial x_j} \right),$$

$$A_2 = -\frac{\partial}{\partial x_i} \left(a_{ij2} \left(\frac{x}{\varepsilon}, \frac{t}{\varepsilon^k} \right) \frac{\partial}{\partial x_j} \right),$$

where the $a_{ijk}(y, \tau)$ satisfy the usual hypotheses. We consider the system:

$$(4.4.2) \qquad \frac{\partial u_{\varepsilon 1}}{\partial t} + A_1^\varepsilon u_{\varepsilon 1} + \frac{1}{\varepsilon^2}(u_{\varepsilon 1} - u_{\varepsilon 2}) = f_1,$$

$$\frac{\partial u_{\varepsilon 2}}{\partial t} + A_2^\varepsilon u_{\varepsilon 2} - \frac{1}{\varepsilon^2}(u_{\varepsilon 1} - u_{\varepsilon 2}) = f_2,$$

with boundary conditions, say the Dirichlet's boundary conditions but this is without importance,

$$(4.4.3) \qquad u_{\varepsilon 1} = u_{\varepsilon 2} = 0 \text{ on } \Sigma$$

and

$$(4.4.4) \qquad u_{\varepsilon 1}(x, 0) = u_{\varepsilon 2}(x, 0) = 0.$$

This problem admits a unique solution. Multiplying the first (respectively, second) equation (4.4.2) by $u_{\varepsilon 1}$ (respectively, $u_{\varepsilon 2}$) one obtains:

$$\frac{1}{2}\frac{d}{dt}\left[|u_{\varepsilon 1}|^2 + |u_{\varepsilon 2}|^2\right] + a_1^\varepsilon(u_{\varepsilon 1}) + a_2^\varepsilon(u_{\varepsilon 2}) + \frac{1}{\varepsilon^2}|u_{\varepsilon 1} - u_{\varepsilon 2}|^2 = (f, u_{\varepsilon 1}) + (f_2, u_{\varepsilon 2}).$$

Hence it follows that

(4.4.5) $$\|u_{\varepsilon i}\|_{L^2(0,T;V)} + \|u_{\varepsilon i}\|_{L^\infty(0,T;H)} \le C$$

and

(4.4.6) $$\frac{1}{\varepsilon}(u_{\varepsilon 1} - u_{\varepsilon 2}) \text{ is bounded in } L^2(0,T;H).$$

Estimate (4.4.6) comes from **the penalty terms** $\pm\frac{1}{\varepsilon^2}(u_{\varepsilon 1} - u_{\varepsilon 2})$ in (4.2.6). It follows from (4.2.9), (4.2.10) that one can extract a subsequence still denoted by $u_{\varepsilon i}$ such that

(4.4.7) $\quad u_{\varepsilon i} \to u$ **in** $L^2(0,T;V)$ **weakly**

(where we have the same limit for $u_{\varepsilon 1}$ and for $u_{\varepsilon 2}$).

The limit u is the solution of the following homogenized problem. We give the formulas only for the case $k = 2$. One defines A_{1j}, A_{2j} by

$$A_1^\varepsilon = \varepsilon^{-2}A_{11} + \varepsilon^{-1}A_{12} + \varepsilon^0 A_{13},$$
$$A_2^\varepsilon = \varepsilon^{-2}A_{21} + \varepsilon^{-1}A_{22} + \varepsilon^0 A_{23},$$

and $\{\chi_1^j, \chi_2^j\}$ is the solution which is $Y - \tau_0$ periodic of

(4.4.8) $$\frac{\partial}{\partial\tau}\begin{pmatrix}\chi_1^j\\\chi_2^j\end{pmatrix} + \begin{pmatrix}A_{11}+I & 0\\0 & A_{21}+I\end{pmatrix}\begin{pmatrix}\chi_1^j - y_j\\\chi_2^j - y_j\end{pmatrix} = 0.$$

Then the equation for u is

(4.4.9) $$2\frac{\partial u}{\partial t} - q_{ij}\frac{\partial^2 u}{\partial x_i \partial x_j} = f_1 + f_2,$$

where

(4.4.10) $$q_{ij} = \mathcal{M}\left(a_{ik1}\frac{\partial\chi_1^j}{\partial y_k} + a_{ij2} - a_{ik2}\frac{\partial\chi_2^j}{\partial y_k}\right).$$

The proof proceeds along the lines of Section 2.

In (4.4.9) the boundary condition is

(4.4.11) $$u = 0 \text{ on } \Sigma.$$

If in (4.4.2) one takes boundary conditions which are the same for $u_{\varepsilon 1}$ and for $u_{\varepsilon 2}$, that is if we take a variational formulation with a space

(4.4.12) $$V = W \times W, \qquad H_0^1(\mathcal{O}) \subseteq W \subseteq H^1(\mathcal{O}),$$

(with the same W), the result extends: u will be characterized by $u(t) \in W$ and

(4.4.13) $$\left(2\frac{\partial u}{\partial t}, v\right) + \int_{\mathcal{O}} q_{ij}\frac{\partial u}{\partial x_j}\frac{\partial v}{\partial x_i}\, dx = (f_1 + f_2, v), \ \forall v \in W.$$

In case we take $V = V_1 \times V_2$ with **different spaces V_1 and V_2** one obtains

(4.4.14) $$u(t) \in V_1 \cap V_2, \text{ and (4.4.13) } \forall v \in V_1 \cap V_2.$$

Problems where one has rapidly oscillating coefficients in t and a penalty term are studied by Simonenko [107], [108].

One can also study the evolution problems with **degenerate** elliptic part and also with domains with "**periodic holes.**"

4.5. Homogenization and regularization. It is known that the solution of a parabolic equation can be obtained as the limit of the solution of elliptic equations, via the elliptic regularization. Applying this idea to operators of the type $\frac{\partial}{\partial t} + A^\varepsilon$ leads to the behavior as $\varepsilon \to 0$ of equations of the type

$$(4.5.1) \qquad -\varepsilon^m \frac{\partial^2 u_\varepsilon}{\partial t^2} + \frac{\partial u_\varepsilon}{\partial t} + A^\varepsilon u_\varepsilon = f,$$

$$u_\varepsilon(0) = 0, \qquad \frac{\partial u_\varepsilon}{\partial \tau}(T) = 0.$$

The behavior of u_ε will depend on m and on k, if A^ε is with coefficients $a_{ij}(x/\varepsilon, t/\varepsilon^k)$.

For instance if we consider

$$(4.5.2) \qquad -\varepsilon^2 \frac{\partial^2 u_\varepsilon}{\partial t^2} + \frac{\partial u_\varepsilon}{\partial t} - \frac{\partial}{\partial x_i} a_{ij}\left(\frac{x}{\varepsilon}, \frac{t}{\varepsilon^2}\right) \frac{\partial u_\varepsilon}{\partial x_j} = f,$$

then the limit value u of u_ε is the solution of a new parabolic problem obtained in the following manner. One defines $\psi^j(y, \tau)$ as the $Y - \tau_0$ periodic solution of

$$(4.5.3) \qquad \left(-\frac{\partial^2}{\partial \tau^2} + \frac{\partial}{\partial \tau}\right) \psi^j + A_1(\psi^j - y_j) = 0.$$

Then

$$(4.5.4) \qquad \frac{\partial u}{\partial t} + \mathcal{B}u = f,$$

where $\mathcal{B}u = -\mathcal{M}\left(a_{ij} - a_{ik}\frac{\partial \psi^j}{\partial y_k}\right) \frac{\partial^2 u}{\partial x_i \partial x_j}$.

The results presented for the **homogenization** of V.I. are sketchy. Other results are announced in the first note of the authors in the Bibliography of this chapter.

The problem studied in Section 3.9 corresponds to operators which have been studied (without homogenization) by Showalter and Ting [106]; cf. also other equations of this type, or which present somewhat analogous properties, in Carroll-Showalter [24]. Some results for the homogenization of V.I. connected with these operators are given in the first note of the authors in the Bibliography. Many questions arise for these equations. It would be interesting to understand "intrinsically" when the homogenization of partial differential operators leads to **non-local operators in the limit** (this question could also have a physical interest). For instance let us consider the equation

$$(4.5.5) \qquad B^\varepsilon \frac{\partial u_\varepsilon}{\partial t} - \frac{\partial}{\partial x_i}\left(a_{ij}\left(\frac{x}{\varepsilon}, \frac{t}{\varepsilon}\right) \frac{\partial u_\varepsilon}{\partial x_j}\right) = f,$$

B_ε given as in Section 3.9 and the a_{ij}'s not necessarily elliptic. We set: $B_1 = -\frac{\partial}{\partial y_i}\left(b_{ij}(y)\frac{\partial}{\partial y_j}\right)$ and let us **assume** that there exists a function $\chi^j(y, \tau)$ which is $Y - \tau_0$ periodic and which satisfies

$$(4.5.6) \qquad B_1 \frac{\partial \chi^j}{\partial \tau} = -\frac{\partial a_{ij}}{\partial y_i}$$

(i.e., $\int_0^{\tau_0} \frac{\partial a_{ij}}{\partial y_i}(y,\tau)\,d\tau = 0$). Then the homogenized equation (found by asymptotic expansion) is formally

$$-\mathcal{M}\left(b_{ij} - b_{ik}\frac{\partial \chi^j}{\partial y_k}\right)\frac{\partial^2}{\partial x_i \partial x_j}\frac{\partial u}{\partial t} - \mathcal{M}\left(a_{ij} - a_{ik}\frac{\partial \chi^j}{\partial y_k}\right)\frac{\partial^2 u}{\partial x_i \partial x_j} = f.$$

For the homogenization of first order hyperbolic systems, we refer to Chapter 4. One can also study a case when one has simultaneously "homogenization" and "penalty":

(4.5.7)
$$\frac{\partial u_{\varepsilon 1}}{\partial t} + \frac{a^\varepsilon}{\varepsilon}\frac{\partial u_{\varepsilon 1}}{\partial x} + \frac{b^\varepsilon}{\varepsilon^2}(u_{\varepsilon 1} - u_{\varepsilon 2}) = f_1,$$
$$\frac{\partial u_{\varepsilon 2}}{\partial t} - \frac{a^\varepsilon}{\varepsilon}\frac{\partial u_{\varepsilon 2}}{\partial x} - \frac{b^\varepsilon}{\varepsilon^2}(u_{\varepsilon 1} - u_{\varepsilon 2}) = f_2,$$
$$x \in (0,1), \qquad u_{\varepsilon 1}(0,t) = 0, \qquad u_{\varepsilon 2}(1,t) = 0, \qquad u_{\varepsilon i}(x,0) = 0,$$
$$a^\varepsilon, b^\varepsilon \in L^\infty(0,1), \qquad a^\varepsilon, b^\varepsilon \geq \alpha > 0,$$

(4.5.8)
$$\frac{1}{a^\varepsilon} \to \mu \text{ in } L^\infty(0,1) \text{ weak star,}$$
$$\frac{b^\varepsilon}{a^\varepsilon} \to \nu \text{ in } L(0,1) \text{ weak star.}$$

Then, assuming that $f_i, \frac{\partial f_i}{\partial t} \in L^2((0,1) \times (0,T))$, one has

$$u_{\varepsilon i} \to u \text{ in } L^\infty(0,T; L^2(0,1)) \text{ weak star,}$$

(same limit u for $i = 1,2$) and where u is the solution of the **parabolic equation**

(4.5.9)
$$2\mu\frac{\partial u}{\partial t} - \frac{\partial}{\partial x}\left(\frac{1}{\nu}\frac{\partial u}{\partial x}\right) = \mu(f_1 + f_2)$$
$$u(0,t) = u(1,t) = 0, \qquad u(x,0) = 0.$$

In (4.5.7) we have only (4.5.8), **without periodic, or almost periodic, structures**. For problems of this type for parabolic equations, we refer to Markov and Oleinik [**79**].

Very many questions arise in connection with the homogenization of second order hyperbolic operators. Some of them are indicated in the text. We shall return to this question in Chapter 4. For a recent work on the regularity of the solution of hyperbolic equations with irregular coefficients we refer to Colombini [**27**].

One can also study the homogenization of **coupled hyperbolic parabolic systems**; cf. Bensoussan, Lions, Papanicolaou [**14**].

Another problem is the following. Consider the equation

(4.5.10)
$$k_2^\varepsilon \frac{\partial^2 u_\varepsilon}{\partial t^2} + k_1^\varepsilon \frac{\partial u_\varepsilon}{\partial t} + Au_\varepsilon = f,$$

where $k_1^\varepsilon \in L^\infty(\mathcal{O})$,

(4.5.11)
$$k_1^\varepsilon \to \hat{k}_1 \text{ in } L^\infty(\mathcal{O}) \text{ weak star.}$$

We assume that

(4.5.12)
$$k_1^\varepsilon \geq \beta > 0$$

and that

(4.5.13)
$$k_2^\varepsilon \geq 0 \text{ (but } k_2^\varepsilon \text{ can be zero).}$$

In (4.5.10) A is a fixed second order elliptic operator. Then one can add to (4.5.10) standard boundary conditions on u_ε and for the **initial conditions**

$$(4.5.14) \qquad u_\varepsilon(x,0) = u^0(x),$$

$$(4.5.15) \qquad k_2^\varepsilon \frac{\partial u_\varepsilon}{\partial t}(x,0) = v^1(x),$$

where u^0 and v^1 are given; (4.5.15) gives a condition on $\frac{\partial u_\varepsilon}{\partial \tau}(x,0)$ only on the set where $k_2^\varepsilon \neq 0$. This is an equation of "**hyperbolic-parabolic**" type. Cf. V. N. Vragov [122] for the case of **fixed** coefficients k_1, k_2. One can show the following (cf. Bensoussan, Lions, Papanicolaou [11]); we do not restrict the generality in assuming that

$$(4.5.16) \qquad \sqrt{k_2^\varepsilon} \to \chi \text{ in } L^\infty(\mathcal{O}) \text{ weak star}.$$

Then $u_\varepsilon \to u$ in $L^\infty(0,T;V)$ weak star, $u'_\varepsilon \to u'$ in $L^2(0,T;H)$ weak star, where u is the solution of

$$(4.5.17) \qquad \hat{k}_2 u'' + \hat{k}_1 u' + Au = f,$$

$$(4.5.18) \qquad u(0) = u^0,$$

$$(4.5.19) \qquad u'(0) = \frac{\chi}{\hat{k}_2} v^1.$$

We remark that in (4.5.19), $u'(x,0) = u'(0)$ is given over **the whole domain** $\mathcal{O}$. Comparing with (4.5.15) we see that there is, here, some sort of "increase" in the initial data.

Other evolution equations are studied from the new point of homogenization in separated papers by the authors. For the transport equations, cf. already the authors' [13] and the Bibliography therein. For other physical aspects of these questions, cf. Sánchez-Palencia [100].

We also refer to the survey of O. A. Oleinik et. al. Russian Mathematical Surveys, 34 (1979), 65-133.

We also mention applications of the asymptotic expansion method to problems in turbulence.[22]

---

[22] P. Perrier and O. Pironneau, C.R. Acad. Sc. Paris, 1978.

CHAPTER 3

Probabilistic Problems and Methods

Orientation

In this chapter we will develop the probabilistic approach to homogenization. This approach is based upon the probabilistic interpretation of the solution of elliptic and parabolic PDE, as averages of functionals of the trajectory of a diffusion process. Therefore, the problem is reduced to the study of the behavior of diffusion processes with rapidly varying drift and diffusion terms. In paragraphs 1 and 2, we give a brief survey of the theory of stochastic differential equations and their connections with PDE. To keep the survey as brief as possible, no proofs are given. This survey should not be considered as complete, even for the reading of the chapter, where some other notions will be used and briefly recalled when necessary. But for the reader with no probabilistic background, the survey will give a quick overview of the basic results and concepts. Paragraph 1 gives the elements of stochastic differential and integral calculus, and the theory of strong solutions of stochastic differential equations. This material is quite standard and the reader may use for example the following references: Gikhman-Skorokhod [**47**], Friedman [**43**], Itô-McKean [**52**]. Paragraph 2 presents the weak formulation of stochastic differential equations and the martingale formulation, which turns out to be instrumental in the sequel. The martingale formulation is due to Stroock-Varadhan [**112**] and [**113**]. Other basic references are Watanabe-Yamada [**125**], Yamada-Watanabe [**128**], Meyer [**80**], Doleans-Dade-Meyer [**34**], Girsanov [**48**]. The study of the behavior of stochastic differential equations (S.D.E) with rapidly varying coefficients is very much connected to ergodic theory. We consider it in Section 3 and we give all results necessary for the sequel (with proofs). The reader may consult Doob [**36**], Neveu [**86**] for other results and technical details. In the applications we have in mind, we are mainly concerned with the ergodic properties of diffusions on the torus. An important analytic counterpart is the Fredholm alternative for differential operators on the torus, which is also a basic tool in the study of homogenization. In Section 4, we consider the theory of homogenization where the limit operator has constant coefficients. It is then possible to use directly ergodic theory. Also since the limit operator has constant coefficients, the limit process is a Gaussian diffusion, and therefore direct identification of the limit finite dimensional distributions is possible. Here we do not use the martingale formulation of diffusions.

The probabilistic approach naturally leads to the study of operators which are not written in divergence form. The analytic approach is therefore not included in Chapters 1 and 2. This is the reason why we give in Section 5 the analytic approach for the case of operators which are not written in divergence form. A serious drawback of all these theories is the need of quite strong regularity assumptions on the coefficients. This is also the reason why the results of Chapters 1 and 2 (where minimal regularity assumptions were made) are not a particular case of what is considered in this chapter. On the other hand, if the regularity of the coefficients is introduced, then the results of Chapters 1 and 2 become a particular case of those of the present chapter, where we give more general homogenization formulas.

In Section 6, we consider the situation of operators with non-uniformly oscillating coefficients. Some of the methods developed in the previous paragraphs for uniformly oscillating coefficients cannot be applied anymore. The methods we use here require in general more regularity on the coefficients. Since the limit process is a general diffusion, the use of the martingale approach is very important to identify the limit process. It turns out that the martingale approach is related to the ideas of the "adjoint expansion" considered in Chapter 1. We also give the analytic approach to the problem of homogenization with nonuniformly oscillating coefficients. It requires some nontrivial generalizations of the "uniformly oscillating" case.

In Section 7, we consider the problem of reiterated homogenization for operators not written in divergence form. Only the probabilistic approach is given completely, to avoid too lengthy developments. Tedious calculations are necessary, but the general structure of the homogenization formulas is interesting, and it cannot be easily guessed from what has been done in Chapter 1 (divergence form operators). Paragraph 8 is devoted to the study of problems with potentials. Up to Section 8, only the Dirichlet problem has been considered. In Section 9, we give some results for the Neumann problem. This involves the study of reflected diffusion processes with rapidly varying coefficients. However we always assume here that the direction of reflection is not rapidly varying. Therefore we cannot recover the general results of Chapter 1, even if we assume regularity of the coefficients.

In Section 10, we turn to evolution problems. Our purpose is to give the probabilistic counterpart of Chapter 2 (Sections 1 and 2). We recover the three cases, depending on how the coefficients depend on time. However we do not consider necessarily the case of operators in divergence form. Paragraph 11 is devoted to the problem of averaging, where some specific methods can be developed. Also we give a general framework, that we call generalized averaging which includes most of the problems considered before. The generality is paid by the complication of the arguments. So we do not give all the developments, but the framework is useful as a general reference.

1. Stochastic differential equations and connections with partial differential equations

1.1. Stochastic integrals. Let $(\Omega, \mathcal{A}, P)$ be a probability space, and $\mathcal{F}^t$ be an increasing family of sub σ-algebras of $\mathcal{A}$ ($\mathcal{F}^{t_1} \subset \mathcal{F}^{t_2}$, if $t_1 \leq t_2$). A stochastic process $w(t)$ with values in $\mathbb{R}^n$, which is **continuous** and satisfies

$$(1.1.1) \qquad\qquad w(0) = 0$$

$$(1.1.2) \qquad\qquad E[w(t) \mid \mathcal{F}^s] = w(s), \ \forall s \leq t$$

$$(1.1.3) \qquad E[(w(t) - w(s))(w(t) - w(s))^* \mid \mathcal{F}^s] = I(t - s), \ s \leq t,$$

where I is the identity matrix, is called a **Wiener process with respect to** $\mathcal{F}^t$. Here $*$ stands for transpose.

Such a process is a **Gaussian process**, i.e., $\forall t_1, t_2, \ldots, t_k$, the random vector $w(t_1), \ldots, w(t_k)$ is Gaussian. The property (1.1.2) expresses the fact that $w(t)$ is an $\mathcal{F}^t$ martingale.

Wiener processes will be the basic process in the following. We now introduce the spaces

(1.1.4) $L_{\mathcal{F}}^p(0, T; \mathbb{R}^n)$

(1.1.5) $\Big\{ \phi(t; \omega) \mid \phi$ is measurable, $\forall t, \phi(t) = \phi(t, \cdot)$ is $\mathcal{F}^t$ measurable,

$$\text{a.s.} \int_0^T |\phi(t)|^p \, dt < \infty \Big\}, \ p \geq 1.$$

(1.1.6) $M_{\mathcal{F}}^p(0, T; \mathbb{R}^n) = \Big\{ \phi \in L_{\mathcal{F}}^p(0, T; \mathbb{R}^n) \ \Big| \ E \int_0^T |\phi(t)|^p \, dt < \infty \Big\}.$

The space $M_{\mathcal{F}}^p$ is isomorphic to a Banach subspace of $L^p((0, T) \times \Omega, dt \otimes dP; \mathbb{R}^n)$. In particular $M_{\mathcal{F}}^2$ is a Hilbert space.

If $\phi \in M_{\mathcal{F}}^2$ is a step process, i.e. if there exists a deterministic splitting of $(0, T)$

$$t_0 = 0 < t_1 < t_2 < \cdots < t_N = T$$

such that

$$\phi(t) = \phi^n \text{ in } [t_n, t_{n+1}[, \qquad n = 0, \ldots, N - 1,$$

then one defines the **stochastic integral** as follows

(1.1.7) $$I = I(\phi) = \sum_{n=0}^{N-1} \phi^n \cdot (w(t_{n+1}) - w(t_n))$$

and the following properties hold true

(1.1.8) $$E I(\phi) = 0,$$

(1.1.9) $$E I(\phi) I(\psi) = E \int_0^T \phi(t) \cdot \psi(t) \, dt.$$

The relation (1.1.9) permits to extend the stochastic integral to the whole space $M_{\mathcal{F}}^2$, and the mapping I is an isometry from $M_{\mathcal{F}}^2$ to $L^2(\Omega, \mathcal{A}, P; \mathbb{R}^n)$. The notation

$$I(\phi) = \int_0^T \phi(t) \cdot dw(t)$$

is standard.

REMARK 1.1. The stochastic integral extends to integrands ϕ which are random matrices.

REMARK 1.2. The stochastic integral extends to integrands ϕ which belong only to $L_{\mathcal{F}}^2$.

It is clear that one can define for any $t \in [0, T]$

(1.1.10) $$I(t) = \int_0^t \phi(s) \cdot dw(s).$$

One can show that the process $I(t)$ is **continuous**.[1] Moreover $I(t)$ is an $\mathcal{F}^t$ martingale.

---

[1]Up to a stochastic equivalence.

1.2. Itô's formula. The stochastic integral leads to a stochastic differential calculus, which differs from the usual differential calculus. Let us consider

(1.2.1) $$a \in L^1_{\mathcal{F}}(0,T;\mathbb{R}^n), \qquad b \in L^2_{\mathcal{F}}(0,T;\mathcal{L}(\mathbb{R}^m;\mathbb{R}^n))$$

$\xi(0)$ random variable with values in $\mathbb{R}^n$, $\mathcal{F}^0$ measurable.

We define

(1.2.2) $$\xi(t) = \xi(0) + \int_0^t a(s)\,ds + \int_0^t b(s) \cdot dw(s)$$

which is a continuous, adapted process. Let now $\psi(x,t)$ be a function on $\mathbb{R}^n \times [0,T]$, belonging to $C^{2,1}(\mathbb{R}^n \times [0,T])$; then the following formula holds:

(1.2.3) $\psi(\xi(t),t) = \psi(\xi(0),0)+$

$$\int_0^t \left(\frac{\partial\psi}{\partial t} + \frac{\partial\psi}{\partial x} \cdot a(s) + \frac{1}{2}\operatorname{tr}\frac{\partial^2\psi}{\partial x^2}bb^*(s) \right)(\xi(s),s)\,ds$$

$$+ \int_0^t \frac{\partial\psi}{\partial x}(\xi(s),s) \cdot b(s)\,dw(s), \ \forall t \in [0,t], \ \text{a.s.}$$

The formula (1.2.3) is called Itô's formula.

A standard notation for (1.2.2) is:

(1.2.4) $$d\xi(t) = a(t)\,dt + b(t)\,dw(t)$$

and one says that the right hand side of (1.2.4) is the stochastic differential of the process $\xi(t)$. Then (1.2.3) expresses the fact that the process $\psi(\xi(t),t)$ has also a stochastic differential whose integral form is (1.2.3).

1.3. Strong formulation of stochastic differential equations. Let $(\Omega, \mathcal{A}, P, \mathcal{F}^t, w(t))$ be as above, and let $g(x,t)$, $\sigma(x,t)$ be two functions such that

(1.3.1)

$g : \mathbb{R}^n \times [0,T] \to \mathbb{R}^n, \qquad \sigma : \mathbb{R}^n \times [0,T] \to \mathcal{L}(\mathbb{R}^m;\mathbb{R}^n)$ are measurable and

$$|g(x,t) - g(x',t)| + |\sigma(x,t) - \sigma(x',t)| \leq K|x - x'|,$$

(1.3.2) $$|g(x,t)|^2 + |\sigma(x,t)|^2 \leq K_0^2(1 + |x|^2).^2$$

We define the following problem: for x,t given $t \in [0,T]$, to find a process $y(s) = y_{xt}(s)$, $s \in [t,T]$ such that

(1.3.3) $$y(s) \text{ is adapted and continuous,}$$

(1.3.4) $$y(s) = x + \int_t^s g(y(\lambda),\lambda)\,d\lambda + \int_t^s \sigma(y(\lambda),\lambda)\,dw(\lambda), \ s \in [t,T], \ \text{a.s.}$$

Equation (1.3.4) is called a stochastic differential equation.

Under the assumption (1.3.1) and (1.3.2) there exists one and only one process satisfying (1.3.3), (1.3.4). The uniqueness is understood in the following sense: if $y_1(s)$, $y_2(s)$ are two processes satisfying (1.3.4) then one has

$$y_1(s) = y_2(s), \qquad s \in [t,T], \ \text{a.s.}$$

$^2\sigma^2 = \operatorname{tr}\sigma\sigma^*.$

REMARK 1.3. The formulation (1.3.4) is a natural extension of the formulation of ordinary differential equations. We will usually write (1.3.4) in the more intuitive way

$$(1.3.5) \qquad dy = g(y(s), s) \, ds + \sigma(y(s), s) \, dw(s), \qquad s \in]t, T[, \qquad y(t) = x.$$

A **stopping time** τ with respect to $\mathcal{F}^t$ is a nonnegative random variable such that

$$\{\tau \leq t\} \subset \mathcal{F}^t, \ \forall t.$$

To a stopping time τ we associate the σ-algebra

$$\mathcal{F}^\tau = \{A \in \mathcal{A} \mid A \cap \{\tau \leq t\} \subset \mathcal{F}^t, \ \forall t\},$$

which is interpreted as the σ-algebra of events which take place before the random time τ.

Let τ_1, τ_2 be two stopping times such that

$$0 \leq \tau_1 \leq \tau_2 \leq T \text{ a.s.}$$

One can define the stochastic integral in the interval $[\tau_1, \tau_2]$, as

$$\int_{\tau_1}^{\tau_2} \phi(t) \cdot dw(t) = \int_0^T \phi(t)(\chi_{\tau_2}(t) - \chi_{\tau_1}(t)) \cdot dw(t)$$

where

$$\chi_\tau(t) = \begin{cases} 1 & \text{if } t < \tau, \\ 0 & \text{if } t \geq \tau. \end{cases}$$

It is then possible to solve a stochastic differential equation like (1.3.5) with stochastic initial conditions, i.e., the pair (x, t) in (1.3.5) can be changed into a pair (ξ, τ), where τ is a stopping time with respect to $\mathcal{F}^t$, and ξ is $\mathcal{F}^\tau$ measurable.

1.4. Connections with partial differential equations. Let $\mathcal{O}$ be an open bounded subset of $\mathbb{R}^n$, whose boundary Γ is of class C^2. In $\mathcal{F}$ functions $a_{ij}(x)$, g_i, $a_0(x)$ are given such that

$$(1.4.1) \qquad a_{ij}, g_i \in W^{1,\infty}(\mathcal{O}), \qquad a_0 \in C^0(\overline{\mathcal{O}}), \qquad a_0 \geq 0,$$

$$a_{ij} = a_{ji}, \qquad \sum_{i,j} a_{ij} \xi_j \xi_i \geq \alpha \sum_i \xi_i^2, \ \forall \alpha > 0.$$

We denote by A the second order differential operator

$$(1.4.2) \qquad A = -a_{ij}(x) \frac{\partial^2}{\partial x_i \partial x_j} - g_i(x) \frac{\partial}{\partial x_i} + a_0(x).$$

Let f be such that

$$(1.4.3) \qquad f \in C^0(\overline{\mathcal{O}}).$$

We consider the Dirichlet problem

$$(1.4.4) \qquad Au = f \text{ in } \mathcal{O}, \qquad u \mid_\Gamma = 0,$$

$$u \in W^{2,p}(\mathcal{O}) \cap C^1(\overline{\mathcal{O}}), \qquad p \geq 1,$$

which has one and only one solution. One can give an explicit formula for $u(x)$, in terms of a function of the solution of a stochastic differential equation.

Ones defines an n by n matrix $\sigma(x)$ which is symmetric, positive definite and Lipschitz such that $a = \sigma^2/2$. An explicit formula for $\sigma(x)$ is given by

$$(1.4.5) \qquad \sigma(x) = \frac{1}{\pi} \int_0^\infty \lambda^{-1/2}(2a(x) + \lambda)^{-1} 2a \; d\lambda.$$

Given a system $(\Omega, \mathcal{A}, P, \mathcal{F}^t, w(t))$, one can solve the stochastic differential equation

$$(1.4.6) \qquad dy = g(y(t)) \, dt + \sigma(y(t)) \, dw(t), \qquad y(0) = x.$$

Let now τ_x be the exit time from $\mathcal{O}$, i.e.,

$$(1.4.7) \qquad \tau_x = \inf\{t \mid y_x(t) \notin \mathcal{O}\},$$

which is an $\mathcal{F}^t$ stopping time. Then the following formula holds

$$(1.4.8) \qquad u(x) = E \int_0^{\tau_x} f(y_x(t)) \left(\exp - \int_0^t a_0(y_x(s)) \, ds\right) dt.$$

REMARK 1.4. In (1.4.6), g, σ must be defined on $\mathbb{R}^n$. This can be done by extending g, σ, defined on $\overline{\mathcal{O}}$, so that they remain globally bounded and Lipschitz.

Let us now turn to the evolution case. We set $\Omega = \mathcal{O} \times]0, T[$, $\Sigma = \Gamma \times]0, T[$ and we consider the family $A(t)$ of operators

$$(1.4.9) \qquad A(t) = -a_{ij}(x, t)\frac{\partial^2}{\partial x_i \partial x_j} - g_i(x, t)\frac{\partial}{\partial x_i}.$$

We will assume that a_{ij} is symmetric, strongly elliptic, and regular enough. At this level, we do not make precise the regularity assumptions, since the weakest assumptions will be given later on.

Let f be defined on $\overline{\Omega}$, regular enough and such that $\bar{u} \mid_\Gamma = 0$. Then one can find one and only one $u \in C^{2,1}(\Omega)$, $u \in C^0(\overline{\Omega})$ such that

$$(1.4.10) \qquad -\frac{\partial u}{\partial t} + A(t)u = f \text{ in } \Omega, \qquad u \mid_\Sigma = 0, \qquad u(x, T) = \bar{u}(x).$$

One can also give an explicit formula for $u(x, t)$. If $\sigma(x, t)$ is again the square root of $a(x, t)$, the one can solve the equation

$$(1.4.11) \qquad dy = g(y, x) \, dx + \sigma(y, s) \, dw(s), \qquad y(t) = x.$$

We set

$$(1.4.12) \qquad \tau_{xt} = \inf\{s \geq t \mid y(s) \notin \mathcal{O}\}.$$

Then we have

$$(1.4.13) \qquad u(x, t) = E \int_t^{T \wedge \tau_{xt}} f(y_{xt}(s), s) \, dx + E\bar{u}(y_{xt}(T))\chi_T \leq \tau_{xt}.$$

REMARK 1.5. One can consider a zero order term in $A(t)$, and modify the relation (1.4.13) along the lines of (1.4.8).

From the explicit formulas (1.4.8), (1.4.13) and standard estimates in PDE, one obtains the following probabilistic estimates,

$$(1.4.14) \qquad \left| E \int_0^{\tau_x} \psi(y_x(s)) \, ds \right| \leq C\|\psi\|_{L^p(\mathcal{O})}, \qquad p > \frac{n}{2},$$

$$(1.4.15) \qquad \left| E \int_t^{\tau_{xt} \wedge T} \psi(y_{xt}(s), s) \, ds \right| \leq C\|\psi\|_{L^p(\Omega)}, \qquad p > \frac{n}{2} + 1,$$

where C does not depend on ψ. The estimates (1.4.14), (1.4.15) prove that the mathematical expectations on the left hand side of (1.4.14), (1.4.15) make sense for functions ψ not necessarily continuous, but in L^p, with p large enough. This permits to give an integrated form of Itô's formula, which holds true for functions $\psi(x,t)$ which are not necessarily in $C^{2,1}(\overline{\Omega})$, with $\Omega = \mathbb{R}^n \times]0,T[$. Indeed let ψ be such that

(1.4.16) $\quad \psi \in C^0(\overline{\Omega}), \quad \dfrac{\partial \psi}{\partial t} - A(t)\psi \in L^p(0,T; L^p_{loc}(\mathbb{R}^n)), \quad p > \dfrac{n}{2} + 1.$

Let $0 < \theta \le \theta' \le T$, be two stopping times and let τ be the exit time from a regular bounded domain $\mathcal{O}$; then one has
(1.4.17)

$$E\psi(y(\theta' \wedge \tau), \theta' \wedge \tau) = E\psi(y(\theta \wedge \tau), \theta \wedge \tau) + E \int_{\theta \wedge \tau}^{\theta' \wedge \tau} \left(\dfrac{\partial \psi}{\partial t} - A(t)\psi \right)(y(s),s)\, ds.$$

The result (1.4.17) is very useful in giving a probabilistic interpretation of the solution of parabolic equations, with the weakest possible assumptions. We will do this in the next paragraph.

2. Martingale formulation of stochastic differential equations

2.1. Martingale problem. We suppose that two functions

(2.1.1) $\qquad\qquad g(x,t) : \mathbb{R}^n \times [0,\infty[\to \mathbb{R}^n;$

$\qquad\qquad\qquad a(x,t) : \mathbb{R}^n \times [0,\infty[\to \mathcal{L}(\mathbb{R}^n; \mathbb{R}^n),$

are given. We assume that

(2.1.2) $\qquad\qquad g,a$ are Borel measurable and bounded;

$\qquad\qquad\qquad a$ is symmetric non-negative definite.

We now set

(2.1.3) $\quad \Omega = C([0,\infty[; \mathbb{R}^n)$; it is a Fréchet space for the topology

$\qquad\qquad\qquad$ of uniform convergence on compact subsets of $[0,\infty[$,

(2.1.4) $\qquad \mathcal{F}^s_t = \sigma - $ algebra generated by $\omega(\lambda), \ \lambda \in [t,s], \ \mathcal{F}^\infty_t = \mathcal{F}_t.$

One can identify $\mathcal{F}^\infty_0$ with the Borel σ-algebra on Ω.

The family $(\Omega, \mathcal{F}_t, \mathcal{F}^s_t)$ will be called the **canonical space**. The **canonical process** is then defined as follows

(2.1.5) $\qquad\qquad\qquad y(s,\omega) = \omega(s).$

We will say that a probability measure $P \equiv P^{xt}$ on $(\Omega, \mathcal{F}_t)$ is a solution of the martingale problem, starting from (x,t) if

(2.1.6) $\qquad\qquad\qquad P(y(t) = x) = 1,$

(2.1.7) $\quad \forall \phi \in C^\infty_0(\mathbb{R}^n), \quad \phi(y(s)) + \displaystyle\int_t^s A(\lambda)\phi(y(\lambda))\, d\lambda$ is a P-martingale.

In (2.1.7) we have set

(2.1.8) $\qquad\qquad\qquad A(t) = -a_{ij} \dfrac{\partial^2}{\partial x_i \partial x_j} - g_i \dfrac{\partial}{\partial x_i}$

where $a = (a_{ij})$.

REMARK 2.1. Many other formulations of the martingale problem are possible. The one we have chosen is the most convenient for our purpose.

The most important result which we will use about the martingale problem is the following. **If besides (2.1.2) we assume that**

(2.1.9) a **is continuous,** a^{-1} **exists and it is continuous and bounded,**

then there exists one and only one solution of the martingale problem.

REMARK 2.2. The existence of a^{-1} is instrumental for obtaining the uniqueness of the solution of the martingale problem. If we have only (2.1.2) and

(2.1.10) a continuous, g continuous,

then there exists a solution of the martingale problem, but this solution is not necessarily unique. The invertibility of a is also useful in removing the assumption of continuity of g.

2.2. Weak formulation of stochastic differential equations. Let us assume now that there exists a factorization of the matrix a such that

$$(2.2.1)\quad a(x,t) = \frac{1}{2}\sigma(x,t)\sigma^*(x,t), \qquad \sigma \in \mathcal{L}(\mathbb{R}^n; \mathbb{R}^n), \text{ measurable and bounded.}$$

We consider a set $\mathcal{E} = (\Omega_0, \mathcal{G}_t, \mathcal{G}_t^2, w(s), y(s), \pi)$, where Ω_0 is a set, $\mathcal{G}_t$, $\mathcal{G}_t^s$ are σ-algebras on Ω_0. $\mathcal{G}_t^{s_1} \subset \mathcal{G}_t^{s_2}$ if $s_1 \leq s_2$, $\mathcal{G}_t = \mathcal{G}_t^\infty$, π is a probability measure on $(\Omega_0, \mathcal{G}_t)$, $w(s)$ is a Wiener process with respect to $\mathcal{G}_t^s$, with values in $\mathbb{R}^n$, and $y(s)$ is a continuous process, adapted to $\mathcal{G}_t^s$, such that

(2.2.2) $y(s) = x,$

$$y(s) = x + \int_t^s g(y(\lambda), \lambda)\, d\lambda + \int_t^s \sigma(y(\lambda), \lambda)\, dw(\lambda).$$

A set $\mathcal{E}$ satisfying all the above conditions is called a weak solution of the stochastic differential equation (2.2.2), with initial conditions (x, t).

REMARK 2.3. If one can choose arbitrarily the subset $(\Omega_0, \mathcal{G}_t, \mathcal{G}_t^s, w(s), \pi)$ then $\mathcal{E}$ reduces to $y(s)$, in which case we recover the concept of strong solution.

As far as the uniqueness of weak solutions is concerned, two types of uniqueness are considered.

2.2.1. *Pointwise uniqueness.* There is pointwise uniqueness of the weak solutions of (2.2.2) if for any two weak solutions $\mathcal{E}_1$, $\mathcal{E}_2$ such that

$$\mathcal{E}_1 = (\Omega_0, \mathcal{G}_t, \mathcal{G}_t^s, w(s), y_1(s), \pi),$$
$$\mathcal{E}_2 = (\Omega_0, \mathcal{G}_t, \mathcal{G}_t^s, w(s), y_2(s), \pi),$$

then one has

$$\text{a.s. } y_1(s) = y_2(s), \ \forall s \geq t.$$

2.2.2. *Uniqueness in law.* There is uniqueness in law, if for any weak solutions

$$\mathcal{E}_1 = (\Omega_0^1, \mathcal{G}_t^1, \mathcal{G}_t^{1s}, w_1(s), y_1(s), \pi_1),$$
$$\mathcal{E}_2 = (\Omega_0^2, \mathcal{G}_t^2, \mathcal{G}_t^{2s}, w_2(s), y_2(s), \pi_2),$$

then y_1 and y_2 have the same probability law (i.e., define the same image probability on $C([0, \infty[; \mathbb{R}^n))$.

The relation between the martingale formulation and the concept of weak solutions is given by the following statement.

If (2.1.2) and (2.2.1) hold true, then the martingale problem has one and only one solution, if and only if there is existence and uniqueness in law of the weak solutions of (2.2.2). It is important to know that under (2.1.2) and (2.2.1), then

- **pointwise uniqueness implies uniqueness in law;**
- **existence of a weak solution and pointwise uniqueness imply existence of a strong solution.**

2.3. Connections with PDE. We can now give the probabilistic interpretations of weak solutions of PDE in terms of functionals of weak solutions of stochastic differential equations or solutions of the martingale problem.

Let $\mathcal{O}$ be an open bounded subset of $\mathbb{R}^n$, whose boundary Γ is of class C^2. Let g and σ be such that

$$(2.3.1) \quad \begin{array}{l} g : \mathbb{R}^n \to \mathbb{R}^n \text{ continuous and bounded,} \\ \sigma : \mathbb{R}^n \to \mathcal{L}(\mathbb{R}^n; \mathbb{R}^n), \quad \sigma \text{ continuously differentiable and bounded,} \\ a = \dfrac{\sigma\sigma^*}{2} \geq \beta I, \quad \beta > 0.^3 \end{array}$$

Let also a_0 be given satisfying

$$(2.3.2) \qquad a_0(x) \geq 0, \text{ continuous and bounded.}$$

For $f \in L^p(\mathcal{O})$, $p > 1$, there exists one and only one solution of

$$(2.3.3) \quad a_{ij}(x)\frac{\partial^2 u}{\partial x_i \partial x_j} - g_i \frac{\partial u}{\partial x_i} + a_0(u) = f, \quad u\mid_\Gamma = 0, \quad u \in W^{2,p}(\mathcal{O}).$$

Now under the assumption (2.3.1), we know that there exists one and only one solution of the martingale problem (2.1.6), (2.1.7), $P \equiv P^{x0}$ (starting from 0), or one and only one (in law) weak solution of the stochastic differential equation

$$(2.3.4) \qquad \qquad dy = g(y)\,dt + \sigma(y)\,dw,$$
$$y(0) = x.$$

Then for $p > \frac{n}{2}$, $u(x)$ is continuous and is given explicitly by

$$(2.3.5) \qquad u(x) = E^x \int_0^\tau f(y(t)) \left(\exp - \int_0^t a_0(y(s))\,ds \right) dt.$$

A similar result holds true for parabolic equations. We assume that

$$(2.3.6) \quad \begin{array}{l} g : \mathbb{R}^n \times]0, T[\to \mathbb{R}^n, \text{ is continuous bounded,} \\ \sigma : \mathbb{R}^n \times]0, T[\to \mathcal{L}(\mathbb{R}^n; \mathbb{R}^n), \quad \sigma \text{ is continuous on } \mathbb{R}^n \times [0, T] \text{ and bounded,} \\ a = \dfrac{\sigma\sigma^*}{2} \geq \beta I, \quad \beta > 0. \end{array}$$

Let also a_0 be given satisfying

$$(2.3.7) \qquad a_0(x, t) \text{ continuous and bounded.}$$

---

[3]Actually only the values on $\mathcal{O}$ will play a role.

For $f \in L^p(Q)$, $Q = \mathcal{O} \times]0, T[$, $p \geq 1$, and $\bar{u} \in W^{2,p}(\mathcal{O}) \cap W_0^{1,p}(\mathcal{O})$ there exists one and only one solution of

$$(2.3.8) \qquad -\frac{\partial u}{\partial t} - a_{ij}(x,t)\frac{\partial^2 u}{\partial x_i \partial x_j} - g_i(x,t)\frac{\partial u}{\partial x_i} + a_0(x,t)u = f,$$

$$u\,|_\Sigma = 0, \qquad u(x,T) = \bar{u},$$

$$u \in \mathcal{W}^{2,1,p}(Q).^4$$

Under the assumption (2.3.6) we also know that there exists one and only one solution of the martingale problem (2.1.6), (2.1.7), $P \equiv P^{xt}$ (starting from x at time t), or one and only one (in law) weak solution of the stochastic differential equation

$$(2.3.9) \qquad dy = g(y,s)\,ds + \sigma(y,s)\,dw(s)$$

$$y(t) = x.$$

Then, for $p > \frac{n}{2} + 1$, $u(x,t)$ is continuous and is given explicitly by
(2.3.10)

$$u(x,t) = E^{xt}\left[\int_t^{\tau \wedge T} f(y(s),s)\left(\exp - \int_t^s a_0(y(\lambda),\lambda)\,d\lambda \right)\,ds + \bar{u}(y(T))\chi_{T \leq \tau} \right].$$

REMARK 2.4. We will be interested particularly in PDE written in variational form (of Chapters 1 and 2), i.e. instead of (2.3.3), or (2.3.8) we will have

$$(2.3.11) \qquad -\frac{\partial}{\partial x_i}a_{ij}(x)\frac{\partial u}{\partial x_j} - a_i\frac{\partial u}{\partial x_i} + a_0 u = f,$$

$$(2.3.12) \qquad -\frac{\partial u}{\partial t} - \frac{\partial}{\partial x_i}a_{ij}(x,t)\frac{\partial u}{\partial x_j} - a_i(x,t)\frac{\partial u}{\partial x_i} + a_0 u = f.$$

We will then assume that besides (2.3.1), or (2.3.6) that

$$(2.3.13) \qquad a_{ij} \in W^{1,\infty}(\mathcal{O}) \text{ (or } W^{1,\infty}(Q)), \ a_i \in L^\infty(\mathcal{O}) \text{ (or } L^\infty(Q)).$$

In that case the forms (2.3.11), (2.3.12) are equivalent to (2.3.3), (2.3.8) provided that we set

$$g_i = a_i + \frac{\partial}{\partial x_j}a_{ji}.$$

3. Some results from ergodic theory

3.1. General results. Let (S, Σ) be a probability space, on which a family $P(x,E)$, $x \in S$, $E \in \Sigma$, of **probability measures** is given. We assume that

$$(3.1.1) \qquad x \to P(x,E) \text{ is measurable, } \forall E \in \Sigma.$$

An **invariant (probability) measure** $\pi(E)$ is a probability on (S, Σ) such that

$$(3.1.2) \qquad \int_S \pi(dx)P(x,E) = \pi(E), \ \forall E \in \Sigma.$$

$^4 \mathcal{W}^{2,1,p}(Q) = \left\{ z \in L^p(Q) \ \middle| \ \frac{\partial z}{\partial t}, \frac{\partial z}{\partial x_i}, \frac{\partial^2 z}{\partial x_i \partial x_j} \in L^p(Q) \right\}.$

To $P(x, E)$ corresponds a linear bounded operator on $B(S)$ (set of bounded measurable functions on S) which is a Banach space for the sup norm. This operator, denoted by P, is defined by

$$(3.1.3) \qquad P\phi(x) = \int_S P(x, dz)\phi(z).$$

Clearly

$$(3.1.4) \qquad \|P\| \leq 1.$$

Of course, we have

$$P\chi_E(x) = P(x, E),$$

where χ_E is the characteristic function of E.

One of the main objectives of ergodic theory is to study the limit of the sequence P^n as $n \to +\infty$. We will give only the result which we are going to use in homogenization. We assume that

(3.1.5) $\qquad S$ is a compact metric space, and Σ is the Borel σ-algebra,

(3.1.6)

there exists a probability measure μ on (S, Σ) such that $P(x, E) = \int_E p(x, y)\mu(dy)$,

$$(3.1.7) \qquad p(x, y) : S \times S \to \mathbb{R}^+ \text{ is continuous},$$

(3.1.8)

there exists a ball U_0 such that $\mu(U_0) > 0$ and $p(x, y) > 0$, $x \in S$, $\forall y \in U_0$.

Since S is compact we can replace (3.1.8) without loss of generality by

(3.1.9) $\qquad$ there exists a ball U_0 such that $\mu(U_0) > 0$ and a positive number

$\delta > 0$ (depending on U_0) such that $p(x, y) \geq \delta$, $x \in S$, $\forall y \in U_0$.

The next result is basic for the following. Its proof can be found in Doob [36]. We give it for the convenience of the reader.

THEOREM 3.1. *Under the assumptions (3.1.5), (3.1.6), (3.1.7), (3.1.8), there exists one and only one invariant probability measure on (S, Σ) and one has*

$$(3.1.10) \qquad \left| P^n\phi(x) - \int \phi\pi(dx) \right| \leq K\|\phi\|e^{-\rho n}, \ \forall \phi \in B(S),$$

where $\rho > 0$, $K > 0$ are independent of ϕ.

PROOF OF THEOREM 3.1. If there exists a probability π for which (3.1.10) holds true, then π is necessarily invariant. Indeed taking in (3.1.10) $\phi = \mathcal{Z}_E$, we can state that

$$\pi(E) = \lim_{n \to \infty} P^n\mathcal{Z}_E(x), \ \forall x.$$

Obviously also

$$\pi(E) = \lim_{n \to \infty} P^n P\chi_E(x).$$

From (3.1.10) again with $\phi = P\chi_E$, we have

$$P^n P\chi_E(x) \to \int P\chi_E(x)\pi(dx) = \int \pi(dx)P(x, E).$$

Hence
$$\pi(E) = \int \pi(dx)P(x,E),$$
which proves that π is invariant.

Let us prove that the invariant measure is unique. Indeed, if $\tilde{\pi}(dx)$ is an invariant probability measure, then

(3.1.11)
$$\int \tilde{\pi}(dx)P\phi(x) = \int \tilde{\pi}(dx) \int P(x,dz)\phi(z).$$

We can use Fubini theorem, since
$$\tilde{\pi}(dx)P(x,dz) = p(x,z)\tilde{\pi}(dx)\mu(dz).$$

Therefore
$$\int \tilde{\pi}(dx)P\phi(x) = \int \mu(dz)\phi(z) \int p(x,z)\tilde{\pi}(d\tilde{x}) = \int \phi(z)\tilde{\pi}(dz).$$

Also
$$\int \tilde{\pi}(dz)P^2\phi(x) = \int \tilde{\pi}(dx)P\phi(x) = \int \tilde{\pi}(dx)\phi(x).$$

More generally
$$\int \tilde{\pi}(dx)P^n\phi(x) = \int \tilde{\pi}(dx)\phi(x).$$

Since $P^n\phi(x) \to \int \phi\pi(dx)$ (which is a constant), and is bounded in the sup norm, it follows from Lebesgue's theorem that
$$\int \phi\pi(dx) = \int \phi\tilde{\pi}(dx), \ \forall\phi,$$
hence $\pi = \tilde{\pi}$.

Let us prove now (3.1.10). For any $E \in \Sigma$, we set

(3.1.12)
$$m_n(E) = \int_x P^n\chi_E(x),$$
$$M_n(E) = \sup_x P^n\chi_E(x).$$

Obviously we have

(3.1.13)
$$m_1(E) \le m_2(E) \le \cdots \le m_n(E) \le \cdots,$$
$$M_1(E) \ge M_2(E) \ge \cdots \ge M_n(E) \ge \cdots,$$
$$m_n(E) \le M_n(E).$$

For x,y fixed, we set

(3.1.14)
$$\lambda(E) = \lambda_{xy}(E) = P\chi_E(x) - P\chi_E(y).$$

Now λ is a measure on (S,Σ) (not necessarily positive), and $\lambda(E)$ is finite $\forall E$. Therefore by the Hahn decomposition theorem, there exists S_0 such that
$$\lambda^+(E) = \lambda(ES_0), \qquad \lambda^-(E) = -\lambda(ES_0'),$$
(where $S_0' = S - S_0$).

Moreover
$$\lambda(s) = \lambda(S_0) + \lambda(S_0') = 0.$$

Now we have

$$\lambda(S_0) = P\chi_{S_0}(x) - P\chi_{S_0}(y)$$
$$= 1 - P\chi_{S_0'}(x) - P\chi_{S_0}(y)$$
$$\leq 1 - P\chi_{S_0'U_0}(x) - P\chi_{S_0U_0}(y).$$

But

$$P\chi_{S_0U_0}(y) = \int_{S_0U_0} p(y,z)\mu(dz) \geq \delta\mu(S_0U_0).$$

Similarly

$$P\chi_{S_0'U_0}(x) \geq \delta\mu(S_0'U_0).$$

Therefore, collecting results we obtain

(3.1.15) $$\lambda(S_0) \leq 1 - \delta\mu(U_0).$$

Next we have

$$M_{n+1}(E) - m_{n+1}(E) = \sup_{x,y}(P^{n+1}\chi_E(x) - P^{n+1}\chi_E(y))$$
$$= \sup_{x,y}[P(P^n\chi_E)(x) - P(P^n\chi_E)(y)]$$
$$= \sup_{x,y}\left|\int_S p(x,z)P^n\chi_E(z)\mu(dz) - \int_S p(y,z)P^n\chi_E(z)\mu(dz)\right|$$
$$= \sup_{x,y}\left|\int_S P^n\chi_E(z)\lambda_{xy}(dz)\right|$$
$$\leq \sup_{x,y}\left|\int_{S_0} M_n(E)\lambda_{xy}(dz) + \int_{S_0'} m_n(E)\lambda_{xy}(dz)\right|$$
$$= \sup_{x,y}(M_n(E) - m_n(E))\lambda_{xy}(S_0)$$
$$\leq (M_n(E) - m_n(E))(1 - \delta\mu(U_0)).$$

Therefore we have

(3.1.16) $$M_n(E) - m_n(E) \leq (1 - \delta\mu(U_0))^{n-1}, \ n \geq 1.$$

From (3.1.13) and (3.1.16) it follows that

(3.1.17) $$M_n(E) \downarrow \pi(E), \qquad m_n(E) \uparrow \pi(E),$$

and

$$m_n(E) - M_n(E) \leq P^n\chi_E(x) - \pi(E) \leq M_n(E) - m_n(E).$$

Therefore we obtain

(3.1.18) $$|P^n\chi_E(x) - \pi(E)| \leq M_n(E) - m_n(E)$$
$$\leq c_1(1 - c_2)^n, \ c_2 \in]0,1[.$$

Clearly π is a probability measure on (S, Σ). For $\phi \in B(S)$, we have,

(3.1.19) $$\left|P^n\phi(x) - \int \phi\pi(dx)\right| \leq \sup_x\left|P^n\phi(x) - \int \phi\pi(dx)\right|$$
$$\leq \|P^n - \pi\|\|\phi\|$$

where $P^n - \pi$ here denotes the measure $P^n\chi_E - \pi(E)$, $E \in \Sigma$. But

$$\|P^n - \pi\| = (P^n(S) - \pi(S))^+ + (P^n(S) - \pi(S))^-,$$

and

$$(P^n(S) - \pi(S))^+ = \sup_E (P^n(E) - \pi(E)) \le c_1(1 - c_2)^n.$$

Similarly,

$$(P^n(S) - \pi(S))^- = -\inf_E (P^n(E) - \pi(E)) \le c_1(1 - c_2)^n.$$

Hence from (3.1.19) and the above estimates, we obtain (3.1.10), with

$$K = 2c_1 = \frac{2}{1 - \delta\mu(U_0)}, \qquad \rho = \log \frac{1}{1 - \delta\mu(U_0)}. \qquad \blacksquare$$

3.2. Ergodic properties of diffusions on the torus. Let us consider functions $a_{ij}(x)$, $b_i(x)$, $i, j = 1, \ldots, n$ on $\mathbb{R}^n$ such that

(3.2.1) $a_{ij} \in C^1(\mathbb{R}^n)$ with holderian derivatives, $\dfrac{\partial^2 a_{ij}}{\partial x_i \partial x_j} \in L^\infty(\mathbb{R}^n)$,

  $b_i \in C^1(\mathbb{R}^n)$, with holderian derivatives,

(3.2.2) $a_{ij} = a_{ji}$, $a_{ij}\xi_i\xi_j \ge \beta|\xi|^2$, $\beta > 0$, $\xi \in \mathbb{R}^n$,

(3.2.3) a_{ij}, b_i are periodic in all variables with period 1.[5]

We define $\sigma(x)$ such that

(3.2.4) $a = \dfrac{\sigma^2}{2}.$

By virtue of the regularity assumptions, σ and b are lipschitz functions, and bounded. Therefore we can solve on an arbitrary probability space and for an arbitrary Wiener process, the stochastic differential equation

(3.2.5) $dy = b(y)\, dt + \sigma(y)\, dw(t)$, $y(0) = x.$

We recall that $Y =]0, 1[^n$, and we denote by $\tilde{Y}$ the torus obtained by identifying the opposite faces of Y. We can also write

$$\tilde{Y} = \mathbb{R}^n / \mathbb{Z}^n.$$

Then $\tilde{Y}$ is a compact metric space. The Borel σ-algebra on $\tilde{Y}$ can be identified with the sub σ-algebra of periodic Borel sets of $\mathbb{R}^n$ (i.e., whose characteristic function is periodic), by the formula

$$E = \bigcup_{k_1, \ldots, k_n \in \mathbb{Z}} \dot{E} + k_i e_i$$

where $\dot{E}$ is a Borel subset of $\tilde{Y}$, and E is a periodic Borel subset of $\mathbb{R}^n$ ($e_1, \ldots, e_n$ are the unit coordinate vectors). Let us set

(3.2.6) $A = -a_{ij}\dfrac{\partial^2}{\partial x_i \partial x_j} - b_i\dfrac{\partial}{\partial x_i}.$

---

[5]We will consider homogenization only in the case when the period is 1. This is only for simplicity, the general case is obtained by an easy extension. With the notation of preceding chapters, $Y =]0, 1[^n$, (cf. Section 1.1).

We will use the fact (cf., A Friedman [42] for instance) that there exists a unique Green's function $p(x, t, y)$ from $\mathbb{R}^n \times]0, \infty[\times \mathbb{R}^n \to \mathbb{R}$ such that

(3.2.7) $p(x, t, y) > 0$, p is continuous on $\{t > 0, x \in \mathbb{R}^n, y \in \mathbb{R}^n\}$,

p is C^2 in x, C^1 in t,

as a function of x, t, p satisfies the relations $\dfrac{\partial p}{\partial t} + Ap = 0$

(3.2.8) $\forall x, \displaystyle\int p(x, t, y) f(y) \, dy \to f(x)$ as $t \to 0$, $\forall f$ continuous and bounded.[6]

Moreover we have the representation formula

(3.2.9) $Ef(y_x(t)) = \displaystyle\int_{\mathbb{R}^n} f(y) p(x, t, y) \, dy,$[7] f measurable and bounded.

Suppose now that $f(y)$ is periodic (i.e., f is a bounded Borel function on the torus); then we have

$$\int_{\mathbb{R}^n} f(y) p(x, t, y) \, dy = \sum_{k_1, \dots, k_n \in \mathbb{Z}} \int_Y p\left(x, t, y + \sum k_i e_i\right) f(y) \, dy.$$

A point in the torus will be represented by $\dot{x}$, where $x \in \mathbb{R}^n$ is a representative of the equivalence class $\dot{x}$.

Since b, σ are periodic

$$y_{x + \sum k_i e_i}(t) = y_x(t) + \sum k_i e_i$$

hence $y_x(t)$ can be identified with a process on the torus denoted by $\dot{y}_x(t)$; moreover the function $p(x, t, y)$ is periodic in x, hence depends only on $\dot{x}$. We can thus rewrite (3.2.9) as follows (for f periodic)

(3.2.10) $Ef(\dot{y}_{\dot{x}}(t)) = \displaystyle\int_{\tilde{Y}} p_0(\dot{x}, t, \dot{y}) f(\dot{y}) \, d\dot{y} = Ef(y_x(t))$

where

(3.2.11) $p_0(x, t, y) = \displaystyle\sum_{k_1, \dots, k_n \in \mathbb{Z}} p\left(x, t, y + \sum k_i e_i\right),$[8]

and $d\dot{y}$ is the Lebesgue measure on the torus (which is a probability on the torus).

As a consequence of the general results of the preceding paragraph, we can state the

THEOREM 3.2. *Under the assumptions (3.2.1), (3.2.2), (3.2.3), there exists one and only one invariant probability measure on the torus $\tilde{Y}$, denoted by π, such that for any f periodic bounded Borel function on $\mathbb{R}^n$ (i.e., f is Borel bounded on $\tilde{Y}$), one has*

(3.2.12) $\displaystyle\sup_x \left| Ef(y_x(t)) - \int_{\dot{y}} f \pi(d\dot{y}) \right| \le K \|f\| e^{-\rho t}$

where K, ρ are positive constants depending only on β and the bounds of a_{ij}, b_i.

---

[6]The periodicity does not play any role here.

[7]The function $u(x, t) = \int f(y) p(x, t, y) \, dy$ is a solution of $\frac{\partial u}{\partial t} + Au = 0$, $u(x, 0) = f(x)$.

[8]Note that the right hand side of (3.2.11) is a function of x, y convergent by virtue of the exponential decay of $p(x, t, y)$ in y as $|y| \to +\infty$.

PROOF. Let us define on Y the family of probability measures

$$P(\dot{x}, \dot{E}) = \int_{\tilde{Y}} p_0(\dot{x}, 1, \dot{y}) \chi_{\dot{E}}(y) \, dy,$$

i.e.

(3.2.13) $$P(\dot{x}, d\dot{y}) = p_0(\dot{x}, 1, \dot{y}) \, d\dot{y};$$

then (3.1.5), (3.1.6), (3.1.7) are satisfied with

$$p(\dot{x}, \dot{y}) = p_0(\dot{x}, 1, \dot{y}),$$
$$\mu(d\dot{y}) = d\dot{y}.$$

Moreover (3.1.8) is also satisfied, since for instance for any ball U_0 of $\tilde{Y}$, whose closure is contained in Y, then $p(x, 1, y) \geq \delta > 0$, $\delta > 0$, $x \in Y$, and $y \in U_0$, and from (3.2.11)

$$p(\dot{x}, 1, \dot{y}) \geq \delta > 0, \qquad \dot{x} \in \tilde{Y}, \qquad \dot{y} \in U_0.$$

The constant δ depends only on β and on the bounds of a_{ij}, b_i. Now we have

$$Ef(y_x(n)) = P^n f(\dot{x})$$

which follows easily from the Chapman-Kolmogorov property of $p(x, t, y)$, i.e.,

$$p(x, t + s, y) = \int p(x, t, z) p(z, s, y) \, dz.$$

Therefore applying Theorem 3.1, we can state that

(3.2.14) $$\sup_x \left| Ef(y_x(n)) - \int_{\tilde{Y}} f\pi(dy) \right| \leq K\|f\| e^{-\rho n}$$

and there exists one and only one invariant measure on the torus, such that (3.2.14) holds true.

Now for $t \geq n$, we have

$$Ef(y_x(t)) = \int p(x, t - n, z) Ef(y_z(n)) \, dz,$$

therefore

(3.2.15)
$$\left| Ef(y_x(t)) - \int_{\tilde{Y}} f\pi(dy) \right| = \left| \int p(x, t - n, z) \left(Ef(y_z(n)) - \int_{\tilde{Y}} f\pi(dy) \right) dz \right|$$
$$\leq K\|f\| e^{-\rho n}.$$

Applying (3.2.15) with t and $n = \lfloor t \rfloor$ and possibly modifying the constant K, we complete the proof of the desired result.   ∎

REMARK 3.3. The measure $\pi(d\dot{y})$ is invariant in the sense of (3.1.2), i.e.

(3.2.16) $$\int_{\tilde{Y}} (d\dot{x}) P(\dot{x}, \dot{E}) = \pi(\dot{E}).$$

Also

(3.2.17) $$\int_{\tilde{Y}} \int_{\tilde{Y}} \pi(d\dot{x}) p_0(x, 1, y) \chi_{\dot{E}}(\dot{y}) \, dy = \int_{\tilde{Y}} \pi(dy) \chi_{\dot{E}}(\dot{y}).$$

As a consequence of the Chapman Kolmogorov property, stated in the preceding theorem, we have

$$p_0(\dot{x}, t + s, \dot{y}) = \int_{\tilde{Y}} p_0(\dot{x}, t, \dot{z}) p_0(\dot{z}, s, \dot{y}) \, d\dot{z}$$

hence it easily follows from (3.2.17) that

(3.2.18) $$\int_{\tilde{Y}} \int_{\tilde{Y}} \pi(d\dot{x}) p_0(\dot{x}, t, \dot{y}) \chi_{\dot{E}}(\dot{y}) \, d\dot{y} = \int_{\tilde{Y}} \pi(dy) \chi_{\dot{E}}(\dot{y})$$

at least for $t \geq 1$. Actually (3.2.18) is true for any $t > 0$, since one can take in the proof of Theorem 3.2,

$$p(\dot{x}, \dot{y}) = p_0\left(\dot{x}, \frac{1}{k}, \dot{y}\right), \qquad k \text{ fixed},$$

and necessarily obtain the same invariant measure.

THEOREM 3.4. *The assumptions are those of Theorem 3.2. Let $\phi(x)$ be a Borel periodic bounded function such that*

(3.2.19) $$\int_{\tilde{Y}} \phi \pi(d\dot{y}) = 0.$$

Then there exists one and only one (up to a constant) solution of

(3.2.20) $$- a_{ij} \frac{\partial^2 z}{\partial x_i \partial x_j} - b_i \frac{\partial z}{\partial x_i} = \phi,$$
$$z \in W^{2,p,\mu}(\mathbb{R}^n),^9 \qquad p \geq 1, \ p < \infty, \qquad z \text{ periodic}.$$

PROOF. Without assuming (3.2.19), but only that $\phi \in L^\infty(\mathbb{R}^n)$, one can solve for $\alpha > 0$, the equation

(3.2.21) $$- a_{ij} \frac{\partial^2 z_\alpha}{\partial x_i \partial x_j} - b_i \frac{\partial z_\alpha}{\partial x_i} + \alpha z_\alpha = \phi$$
$$z \in W^{2,p,\mu}(\mathbb{R}^n).$$

There exists one and only one solution of (3.2.21). This can be proved by using variational techniques and the iterative scheme

(3.2.22) $$- a_{ij} \frac{\partial^2 z^{n+1}}{\partial x_i \partial x_j} - b_i \frac{\partial^2 z_{\alpha}lpha^{n+1}}{\partial x_i} + \alpha z_\alpha^{n+1} + \lambda z_\alpha^{n+1} = \lambda z_\alpha^n + \phi,$$
$$z_\alpha^{n+1} \in W^{2,p,\mu}(\mathbb{R}^n),$$

as in Bensoussan-Lions [10]. Details are omitted. Furthermore $z_\alpha(x)$ has a probabilistic representation, namely

(3.2.23) $$z_\alpha(x) = E \int_0^\infty e^{-\alpha t} \phi(y_x(t)) \, dt.$$

Now if (3.2.19) holds true, then from (3.2.12) we have

$$|E\phi(y_x(t))| \leq K \|\phi\| e^{-\rho t}$$

---

[9]Sobolev space with weights; $z \in W^{2,p,\mu} \iff z, \frac{\partial z}{\partial x_i} \frac{\partial^2 z}{\partial x_i \partial x_j} \in L^{p,\mu}$ where $L^{p,\mu} = \{\phi \mid \int_{\mathbb{R}^n} |\phi(x)|^p \exp(-p\mu|x|) \, dx < \infty\}$.

which along with (3.2.23) easily leads to $|z_\alpha(x)| \leq C$, as $\alpha \to 0$. This estimate and the equation (3.2.21) insures that z_α remains bounded in $W^{2,p,\mu}(\mathbb{R}^n)$, $p \geq 1$, $p < \infty$. One can then let $\alpha \to 0$ and obtain a solution of (3.2.20). The uniqueness is obtained as follows. Take $\phi = 0$, then by Itô's formula (see (1.4.17)) we have

$$(3.2.24) \qquad z(x) - Ez(y_x(t)), \qquad t \geq 0.$$

Now $z_\alpha(x)$ is periodic, for any α. Hence the limit $z(x)$ is also periodic. Applying Theorem 3.2 again, it follows that as $t \to \infty$,

$$z(x) = \int_{\tilde{Y}} z\pi(d\dot{y})$$

i.e., z is a constant. This completes the proof of the theorem. ∎

3.3. Invariant measure and the Fredholm alternative. We can interpret the results of the preceding paragraph (especially Theorem 3.4) in terms of the Fredholm alternative. This will also allow us to describe explicitly the unique invariant measure.

We rewrite the operator A (given by (3.2.6)) as follows

$$(3.3.1) \qquad \begin{aligned} A &= -\frac{\partial}{\partial x_i}\left(a_{ij}\frac{\partial}{\partial x_j}\right) - \left(b_i - \frac{\partial a_{ij}}{\partial x_j}\right)\frac{\partial}{\partial x_i} \\ &= -\frac{\partial}{\partial x_i}\left(a_{ij}\frac{\partial}{\partial x_j}\right) - a_i\frac{\partial}{\partial x_i}. \end{aligned}$$

The formal adjoint operator is (since a_{ij} are symmetric)

$$(3.3.2) \qquad A^* = -\frac{\partial}{\partial x_i}\left(a_{ij}\frac{\partial}{\partial x_j}\right) + \frac{\partial}{\partial x_i}a_i + a_i\frac{\partial}{\partial x_i}.$$

We recall (cf. Chapter 1, (2.2.10), (2.2.14)), that

$$\begin{aligned} W(Y) &= \{\phi \mid \phi \in H^1(Y), \ \phi \text{ periodic}\}, \\ W^\circ(Y) &= W(Y)/R. \end{aligned}$$

THEOREM 3.5. *The assumptions are those of Theorem 3.2. We consider the homogeneous equations*

$$(3.3.3) \qquad Az = 0, \qquad z \in W^{2,p,\mu}(\mathbb{R}^n) \qquad p \geq 2, \ \mu > 0, \ z \text{ periodic},$$

$$(3.3.4) \qquad A^*m = 0, \qquad m \in W^{2,p,\mu}(\mathbb{R}^n) \qquad p \geq 2, \ \mu > 0, \ m \text{ periodic}.$$

There exists one and only one solution (up to a multiplicative constant) of (3.3.3), (3.3.4) (actually $z = 1$). Let ϕ, ψ be Borel periodic bounded functions such that

$$(3.3.5) \qquad \int_Y \phi(y)m(y)\,dy = 0,$$

$$(3.3.6) \qquad \int_Y \phi(y)\,dy = 0;$$

then there exists one and only one solution of the inhomogeneous equations
(3.3.7)

$$Az_0 = \phi, \qquad z_0 \in W^{2,p,\mu}(\mathbb{R}^n), \qquad p \geq 2, \ \mu > 0, \ z_0 \text{ periodic}, \ \int_Y z_0\,dy = 0,$$

(3.3.8)

$$A^* m_0 = \psi, \qquad m_0 \in W^{2,p,\mu}(\mathbb{R}^n), \qquad p \geq 2, \ \mu > 0, \ m_0 \ periodic, \ \int_Y m_0 \, dy = 1.$$

REMARK 3.6. Obviously 1 is a solution of (3.3.3). The normalization constant of the solution of (3.3.4) is taken such that

$$\int_Y m(y) \, dy = 1.$$

REMARK 3.7. Theorem 3.5 expresses the Fredholm alternative for the operator A, with periodic conditions.

REMARK 3.8. Let f be Borel periodic bounded; we can consider the Cauchy problem

(3.3.9) $$\frac{\partial u}{\partial t} + Au = 0, \qquad u(x,0) = f(x),$$

which has one and only one solution which is bounded periodic, and for instance which is C^2 in x, C^1 in t in $\mathbb{R}^n \times]0, \infty[$. It is given explicitly by

$$u(x,t) = \int p(x,t,y) f(y) \, dy.$$

According to Theorem 3.2 we have

(3.3.10) $$u(x,t) \to C \text{ (a constant) as } t \to \infty, \ \forall x.$$

Now let m be the solution of (3.3.4) satisfying the condition of Remark 3.6; we have from (3.3.9) and (3.3.4)

$$\frac{d}{dt} \int u(x,t) m(x) \, dx = 0,$$

hence

$$\int_Y u(x,t) m(x) \, dx = \int_Y f(y) m(y) \, dy.$$

Let $t \to +\infty$, it follows from Lebesgue's theorem that

$$C = \int_Y f(y) m(y) \, dy.$$

However we know already that

$$C = \int_{\tilde{Y}} f(\dot{y}) \, \pi(d\dot{y}).$$

Therefore

$$m(\dot{y}) \, d\dot{y} = \pi(d\dot{y}).$$

In particular $m(y) \geq 0$. In fact $m(y) > 0$, $\forall y$. Indeed if $f(x)$ is positive and $\neq 0$ on a set of positive Lebesgue measure, then $u(x,t) > 0$, $\forall x$, $\forall t > 0$. Since for t positive $u(x,t)$ is continuous and periodic in x, then we can say for instance that $u(x,1) \geq \delta > 0$, $\forall x$. For $t \geq 1$ we have by the maximum principle

$$u(x,t) \geq \inf_x u(x,1) \geq \delta.$$

Therefore letting $t \to +\infty$, we can assert that $C > 0$. Since this is true for any f as above, it follows that $m(y) > 0$ a.e. Actually $m(y)$ cannot be 0 inside Y, since 0 would be the minimum, contradicting the maximum principle. It cannot be 0 on the boundary because of the periodicity, hence $m(y) > 0$, $\forall y$.

PROOF OF THEOREM 3.5. Let λ be large enough. Then we can solve the problem

$$(3.3.11) \qquad Az_\lambda + \lambda z_\lambda = g, \qquad z_\lambda \in W(Y).$$

For any $g \in L^2(Y)$, there exists one and only one solution of (3.3.11). Thus we define an operator

$$G_\lambda : L^2(Y) \to L^2(Y), \text{ by setting } G_\lambda g = z_\lambda.$$

Since the injection of $W(Y)$ into $L^2(Y)$ is compact, the operator G_λ is compact.

Let $\phi \in L^2(Y)$; the problem

$$(3.3.12) \qquad Az = \phi, \qquad z \in W(Y)$$

is equivalent to the problem

$$(3.3.13) \qquad (I - \lambda G_\lambda)z = G_\lambda \phi, \qquad z \in L^2(Y).$$

Similarly let $\psi \in L^2(Y)$; the problem

$$(3.3.14) \qquad A^*m = \psi, \qquad m \in W(Y),$$

is equivalent to the problem

$$(3.3.15) \qquad (I - \lambda G_\lambda^*)m = G_\lambda^* \psi, \qquad m \in W(HY),$$

where G_λ^* is the adjoint of G_λ (in $L^2(Y)$).

Since G_λ is compact, the Fredholm alternative applies. We must find the number of linearly independent solutions of

$$(I - \lambda G_\lambda)z = 0,$$

i.e.,

$$(3.3.16) \qquad Az = 0, \qquad z \in W(Y).$$

Let us consider a solution of (3.3.16). By regularity, $z \in H^2(Y)$ and (from Green's formula) $\sum_j a_{ij} \frac{\partial z}{\partial x_j}$ is periodic in x_i. This ensures that if one extends z by periodicity then $Az \in L^2_{loc}(\mathbb{R}^n)$ and

$$Az = 0, \text{ a.e. in } \mathbb{R}^n.$$

Furthermore $z \in W^{1,2,\mu}(\mathbb{R}^n)$, $\forall \mu > 0$. But then we can consider that z satisfies the equation

$$(3.3.17) \qquad Az + \lambda_0 z = \lambda_0 z, \qquad z \in W^{1,2,\mu}(\mathbb{R}^n).$$

For λ_0 large enough, and for a right hand side which is given in $L^{2,\mu}(\mathbb{R}^n)$, the equation (3.3.17) has one and only one solution in $W^{1,2,\mu}(\mathbb{R}^n)$. This solution is actually in $W^{2,2,\mu}(\mathbb{R}^n)$. Hence $z \in W^{2,2,\mu}(\mathbb{R}^n)$. Suppose for instance that $n > 4$, then $z \in L^{q,\mu}(\mathbb{R}^n)$, with $\frac{1}{q} = \frac{1}{2} - \frac{2}{n}$, i.e., $q > 2$. The solution of (3.3.17) is then in $W^{2,q,\mu}(\mathbb{R}^n)$. Step by step, by a bootstrap argument which is standard, we see that actually $z \in W^{2,p,\mu}(\mathbb{R}^n)$, $\forall p \geq 2$, $\mu > 0$. Thus (3.3.16) is equivalent to

$$(3.3.18) \qquad Az = 0, \qquad z \in W^{2,p,\mu}(\mathbb{R}^n), \qquad \forall p \geq 2, \ \mu > 0, \ z \text{ periodic.}$$

Suppose that $z \not\equiv$ constant. By the strong maximum principle, z cannot reach its maximum or its minimum inside Y. Suppose that it reaches its maximum on the facet $x_i = 0$ and in a point inside the facet. Then consider the function z on the cube $-\frac{1}{2} < x_i < \frac{1}{2}$, $x_j \in]0,1[$ for $j \neq 1$; we see that it reaches its maximum inside the cube (by the periodicity) contradicting the strong maximum principle. By similar reasonings we see that necessarily the solution of (3.3.18) is a constant.

The Fredholm alternative implies that there exists one and only one solution of

$$A^*m = 0, \qquad m \in W(Y), \qquad \int_Y m \, dy = 1.$$

By the same bootstrap argument as above, we complete the proof of the first part of the theorem (i.e., proving that (3.3.3), (3.3.4) have one and only one solution up to a multiplicative constant).

The second part is a direct consequence of the Fredholm alternative, and the regularity argument. ∎

REMARK 3.9. The fact that (3.3.3) has one solution and only one (up to multiplicative constant) has been proved also in Theorem 3.4 using a probabilistic argument (see (3.2.24)). Since the solution of (3.3.3) is also a constant, the fact that the constant is multiplicative or additive does not matter.

REMARK 3.10. Suppose that

$$(3.3.19) \qquad b_i = \frac{\partial a_{ij}}{\partial x_j}.$$

Then $A = A^*$ (the operator is self-adjoint). In that case $m = 1$, and the invariant measure is the Lebesgue measure.

4. Homogenization with a constant coefficients limit operator

4.1. Orientation. In this paragraph we start the study of homogenization using probabilistic methods. Throughout the paragraph the limit operator will be an operator with constant coefficients. The general case will be considered in the next paragraph.

4.2. Diffusion without drift. We consider functions $a_{ij}(x)$ satisfying the conditions (3.2.1), (3.2.2), (3.2.3), and defining σ by (3.2.4), we introduce the stochastic differential equation (depending on a small parameter ε)

$$(4.2.1) \qquad dy^\varepsilon = \sigma\left(\frac{y^\varepsilon}{\varepsilon}\right) dw(t), \qquad y^\varepsilon(0) = x.$$

The solution of (4.2.1) is constructed on an arbitrary system $(\Omega, \mathcal{A}, P, \mathcal{F}^t, w(t))$.

The solution $y^\varepsilon(t)$ is a continuous process with values in $\mathbb{R}^n$. If one defines

$$(4.2.2) \qquad \psi = C([0, \infty[; \mathbb{R}^n),$$

the ψ is a separable Fréchet space for the topology of uniform convergence on compact sets of $[0, \infty[$ (cf. (2.1.3)). The Borel σ-algebra on ψ, denoted by $\mathcal{B}$ is generated by the sets $\{y(t_j)\} \in B_j$, where $y(\cdot)$ denotes an element of ψ, and $t_j \in \mathbb{R}$, B_j Borel subset of $\mathbb{R}^n$. We denote by μ_x^ε the probability measure on ψ defined by the stochastic process $y_x^\varepsilon(t)$.

We want to study the behavior of μ_x^ε, as $\varepsilon \to 0$. We have

$$E|y^\varepsilon(t) - y^\varepsilon(s)|^4 = E\left|\int_s^t \sigma\left(\frac{y^\varepsilon}{\varepsilon}\right) dw\right|^4 \leq C(t-s)^2$$

by standard properties of stochastic integrals, and since σ is bounded. From Prohorov's compactness theorem (cf. Prohorov [94]) it follows that

(4.2.3) $\forall x$, μ_x^ε remains in a weakly compact set of

$$\mathcal{M}_+^1(\psi) = \text{set of probability measures on } \psi.$$

We introduce the differential operator A defined by

(4.2.4) $$A = -a_{ij}(x)\frac{\partial^2}{\partial x_i \partial x_j}.$$

According to Theorem 3.5, there exists one and only one solution of

(4.2.5) $A^*m = 0$, $m \in W^{2,p,\mu}(\mathbb{R}^n)$, $p \geq 2$, $\mu > 0$, m periodic,

$$\int_Y m(y)\, dy = 1,$$

and we know from Remark 3.8 that

(4.2.6) $$m(y) > 0, \ \forall y.$$

We define

(4.2.7) $$q_{ij} = \int_Y a_{ij}(y)m(y)\, dy.$$

Clearly q_{ij} is a symmetric positive definite matrix. We denote by μ_x the **Gaussian measure** on ψ defined by the process

(4.2.8) $$dy = \sqrt{2}\sqrt{q}\,w(t), \qquad y(0) = x.$$

THEOREM 4.1. *If the a_{ij} satisfy the conditions (3.2.1), (3.2.2), (3.2.3), then we have*

(4.2.9) $$\forall x, \ \mu_x^\varepsilon \to \mu_x \text{ weakly in } \mathcal{M}_+^1(\psi).$$

REMARK 4.2. In Section 4.4 we will see that the result (4.2.9) (and similar results later on) implies homogenization results for elliptic PDE along the lines of the results of Chapter 1.

The proof of Theorem 4.1 will rely on the following.

LEMMA 4.3. *Let ϕ be Borel periodic bounded, such that*

(4.2.10) $$\int_Y \phi(y)m(y)\, dy = 0.$$

Then for any $s \leq t$, we have

(4.2.11) $$E\left[\left(\int_s^t \phi\left(\frac{y^\varepsilon(\lambda)}{\varepsilon}\right) dy\right)^2 \Big| \mathcal{F}^s\right] \to 0.$$

We are going to give two proofs of Lemma 4.3, using different aspects of ergodic theory. One of the proofs will be valid only for more regular ϕ.[10]

---

[10]In general this will be enough for the applications.

First proof of Lemma 4.3. For this proof we require the additional assumption

$$(4.2.12) \qquad \phi \text{ differentiable and } \frac{\partial \phi}{\partial x_i} \in L^p_{loc}(\mathbb{R}^n), \qquad p > \frac{n}{2}.$$

Since (4.2.10) is satisfied, we can assert, using Theorem 3.5, that there exists one and only one function such that

$$(4.2.13) \quad Az = \phi, \qquad z \in W^{2,p,\mu}(\mathbb{R}^n), \qquad p \geq 2, \ \mu > 0, \ z \text{ periodic}, \int_Y z \, dy = 0.$$

By virtue of (4.2.12), z is actually more regular, namely $z \in W^{3,p}_{loc}(\mathbb{R}^n)$, $p > \frac{n}{2}$. In particular $z \in C^2(\mathbb{R}^n)$. Therefore we can apply Itô's formula, which yields

$$z\left(\frac{y^\varepsilon(t)}{\varepsilon}\right) = z\left(\frac{y^\varepsilon(s)}{\varepsilon}\right) + \frac{1}{\varepsilon}\int_s^t \frac{\partial z}{\partial x} \cdot \sigma\left(\frac{y^\varepsilon(\lambda)}{\varepsilon}\right) dw(\lambda) + \frac{1}{\varepsilon^2}\int_s^t Az\left(\frac{y^\varepsilon(\lambda)}{\varepsilon}\right) d\lambda$$

hence

$$\int_s^t \phi\left(\frac{\phi^\varepsilon(\lambda)}{\varepsilon}\right) d\lambda = -\varepsilon^2 z\left(\frac{y^\varepsilon(t)}{\varepsilon}\right) + \varepsilon^2 z\left(\frac{y^\varepsilon(s)}{\varepsilon}\right) + \varepsilon \int_s^t \frac{\partial z}{\partial x} \cdot \sigma\left(\frac{y^\varepsilon(\lambda)}{\varepsilon}\right) dw(\lambda).$$

Therefore

$$E\left[\left(\int_s^t \phi\left(\frac{y^\varepsilon(\lambda)}{\varepsilon}\right) d\lambda\right)^2 \Big| \mathcal{F}^s\right] \leq 2\left[\varepsilon^4 E\left\{\left(z\left(\frac{y^\varepsilon(t)}{\varepsilon}\right) - z\left(\frac{y^\varepsilon(s)}{\varepsilon}\right)\right)^2 \Big| \mathcal{F}^s\right\}\right.$$

$$\left. + \varepsilon^2 E\left(\int_s^t \left|\sigma^* \frac{\partial z}{\partial x}\left(\frac{y^\varepsilon(\lambda)}{\varepsilon}\right)\right|^2 d\lambda \Big| \mathcal{F}^s\right)\right]$$

$$\leq 2\left[4\varepsilon^4 \sup_x |z|^2 + \varepsilon^2(t-s)\sup_x \left|\sigma^*\frac{\partial z}{\partial x}\right|^2\right],$$

hence (4.2.11) follows.   ∎

Second proof of Lemma 4.3. We define the process

$$(4.2.14) \qquad \eta^\varepsilon(t) = \frac{1}{\varepsilon}g^\varepsilon(\varepsilon^2 t)$$

hence from (4.2.1), $\eta^\varepsilon(t)$ is a solution of the Itô equation

$$(4.2.15) \qquad d\eta^\varepsilon(t) = \frac{1}{\varepsilon}\sigma(\eta^\varepsilon(t))\, dw(\varepsilon^2 t), \qquad \eta^\varepsilon(0) = \frac{x}{\varepsilon}.$$

But, if we set

$$w^\varepsilon(t) = \frac{w(\varepsilon^2 t)}{\varepsilon},$$

then $w^\varepsilon(t)$ is also a Wiener process (although it is not a martingale with respect to $\mathcal{F}^t$). We can write

$$(4.2.16) \qquad d\eta^\varepsilon(t) = \sigma(\eta^\varepsilon(t))\, dw^\varepsilon(t), \qquad \eta^\varepsilon(0) = \frac{x}{\varepsilon}.$$

It is easy to check that the conditions of application of Theorem 3.2 are realized, and therefore we can assert that (by virtue of (4.2.10))

$$(4.2.17) \qquad \sup_\xi |E\phi(\eta^\varepsilon_\xi(t))| \leq K\|\phi\|e^{-\rho t}$$

where $\eta^\varepsilon_\xi(t)$ stands for the solution of (4.2.16) with initial condition $\eta^\varepsilon_\xi(0) = \xi$. The constants K, ρ do not depend on ε.

From (4.2.17) it follows in particular that

$$(4.2.18) \qquad \left| E\phi\left(\frac{y_x^\varepsilon(t)}{\varepsilon}\right) \right| \le K\|\phi\|e^{-\rho\frac{t}{\varepsilon^2}}, \qquad x \in \mathbb{R}^n,$$

where $y_x^\varepsilon(t) = y^\varepsilon(t)$ (defined by (4.2.1)).

We now consider

$$
\begin{aligned}
X &= E\left[\left(\int_s^t \phi\left(\frac{y^\varepsilon(\lambda)}{\varepsilon}\right) d\lambda\right)^2 \Bigg| \mathcal{F}^s\right] \\
(4.2.19) \qquad &= E\left[\int_s^t \int_s^t d\lambda\, d\mu\, \phi\left(\frac{y^\varepsilon(\lambda)}{\varepsilon}\right) \phi\left(\frac{y^\varepsilon(\mu)}{\varepsilon}\right) \Bigg| \mathcal{F}^s\right] \\
&= 2E\left[\int_s^t d\lambda \int_\lambda^t d\mu\, \phi\left(\frac{y^\varepsilon(\lambda)}{\varepsilon}\right) \phi\left(\frac{y^\varepsilon(\mu)}{\varepsilon}\right) \Bigg| \mathcal{F}^s\right] \\
&= 2E\left[\int_s^t d\lambda\, \phi\left(\frac{y^\varepsilon}{\varepsilon}\right) \int_\lambda^t E\left(\phi\left(\frac{y^\varepsilon(\mu)}{\varepsilon}\right) \Bigg| \mathcal{F}^\lambda\right) d\mu \Bigg| \mathcal{F}^s\right].
\end{aligned}
$$

If we set

$$H^\varepsilon(x,t) = E\phi\left(\frac{y_x^\varepsilon(t)}{\varepsilon}\right),$$

then it follows from the Markov property of $y_x^\varepsilon(t)$ that

$$E\left(\phi\left(\frac{y^\varepsilon(\mu)}{\varepsilon}\right) \Bigg| \mathcal{F}^\lambda\right) = H^\varepsilon(y^\varepsilon(\lambda), \mu - \lambda) \text{ for } \mu \ge \lambda.$$

Using (4.2.18), we can state

$$
\begin{aligned}
X &\le 2E\left[\int_s^t d\lambda\, \left|\phi\left(\frac{y^\varepsilon(\lambda)}{\varepsilon}\right)\right| \int_\lambda^t K\|\phi\|e^{-\rho\frac{\mu-\lambda}{\varepsilon^2}}\, d\mu \Bigg| \mathcal{F}^s\right] \\
&\le \frac{2K\|\phi\|^2}{\rho}\varepsilon^2(t-s),
\end{aligned}
$$

hence again (4.2.11).   ■

REMARK 4.4. The second proof of Lemma 4.3 is due to Freĭdlin [**40**].

We can now use the results of Lemma 4.3 to prove the following.

LEMMA 4.5. *The finite dimensional distributions of the process $y_x^\varepsilon(t)$ (solution of (4.2.1)) converge toward the finite dimensional distributions of the process $y_x(t)$ (solution of (4.2.8)).*

PROOF. We have to show that $\forall k$, $\forall t_1, \dots, t_k$, $\forall \lambda_1, \dots, \lambda_k$, $(\lambda_i \in \mathbb{R}^n)$ then

$$(4.2.20) \quad E[\exp i(\lambda_1 \cdot y_x^\varepsilon(t_1) + \cdots + \lambda_k \cdots y_x^\varepsilon(t_k))]$$
$$\to \exp i(\lambda_1 + \cdots + \lambda_k) \cdot x \exp\left[-\sum q\lambda_i \cdot \lambda_j \min(t_i, t_j)\right].$$

Since

$$y_x^\varepsilon(t) = x + \int_0^t \sigma\left(\frac{y^\varepsilon}{\varepsilon}(s)\right) dw(s)$$

showing (4.2.20) is equivalent to showing that

$$(4.2.21) \quad E \exp i \left[\int_0^{t_1} \lambda_1 \cdot \sigma \left(\frac{y^\varepsilon}{\varepsilon} \right) dw(s) + \cdots + \int_0^{t_k} \lambda_k \cdot \sigma \left(\frac{y^\varepsilon}{\varepsilon} \right) dw(s) \right]$$

$$\rightarrow \exp \left(- \sum_{i,j} q \lambda_i \cdot \lambda_j \min(t_i, t_j) \right).$$

We can assume without loss of generality that

$$0 \leq t_1 \leq t_2 \leq \cdots \leq t_k.$$

We can prove (4.2.21) by induction on k. Let us take $k = 1$, then we have to prove that

$$(4.2.22) \quad E \exp i \int_0^{t_1} \lambda_1 \cdot \sigma \left(\frac{y^\varepsilon}{\varepsilon} \right) dw(s) \rightarrow \exp - (q \lambda_1 \cdot \lambda_1) t_1.$$

More generally, we will prove the following result. Let us assume that h is given such that

$$(4.2.23) \quad h \text{ is a Borel periodic bounded function from } \mathbb{R}^n \rightarrow \mathbb{R}^n.$$

We write

$$(4.2.24) \quad \bar{h} = \int_Y |h(y)|^2 m(y) \, dy;$$

then we have

$$(4.2.25) \quad E \left[\exp i \int_s^t h \left(\frac{y^\varepsilon}{\varepsilon} \right) dw(\lambda) \,\Big|\, \mathcal{F}^s \right] \rightarrow \exp - \frac{1}{2} \bar{h}(t - s).$$

If we prove (4.2.25), then we can apply this result with $s = 0$, $t = t_1$, $h = \sigma^* \lambda_1$. Then

$$\bar{h} = \int_Y \sigma \sigma^* \lambda_1 \cdot \lambda_1 m(y) \, dy = \int_Y 2a(y) \lambda_1 \cdot \lambda_1 m(y) \, dy = 2q \lambda_1 \cdot \lambda_1,$$

hence (4.2.22) follows.

Let us prove (4.2.25). We start with the identity

$$(4.2.26) \quad E \left[\left(\exp i \int_s^t h \left(\frac{y^\varepsilon}{\varepsilon} \right) \cdot dw(\lambda) \right) \exp \frac{1}{2} \int_s^t |h|^2 \left(\frac{y^\varepsilon}{\varepsilon} \right) d\lambda \,\Big|\, \mathcal{F}^s \right] = 1.$$

Let us set

$$X_\varepsilon = \exp i \int_s^t h \left(\frac{y^\varepsilon}{\varepsilon} \right) dw(\lambda),$$

$$Y_\varepsilon = \exp \frac{1}{2} \int_s^t |h|^2 \left(\frac{y^\varepsilon}{\varepsilon} \right) d\lambda = \exp X_\varepsilon$$

$$C = \exp \frac{1}{2} \bar{h}(t - s).$$

From Lemma 4.3, we can assert that

$$(4.2.27) \quad E \left| \left(\frac{1}{2} \int_s^t |h|^2 \left(\frac{y^\varepsilon}{\varepsilon} \right) d\lambda - \frac{1}{2} \bar{h}(t - s) \right)^2 \,\Big|\, \mathcal{F}^s \right| \rightarrow 0 \text{ a.s.}$$

But

$$(4.2.28) \qquad |Y_\varepsilon - C| \le \left| \frac{1}{2} \int_s^t |h|^2 \left(\frac{y^\varepsilon}{\varepsilon} \right) d\lambda - \frac{1}{2}\bar{h}(t-s) \right| X_\varepsilon$$

where X_ε is bounded, $(0 \le \chi_\varepsilon \le$ constant, since h is bounded). From (4.2.27), (4.2.28) and the Cauchy-Schwarz inequality, it follows that

$$(4.2.29) \qquad E[(Y_\varepsilon - C)^s \mid \mathcal{F}^s] \to 0, \text{ a.s.}$$

We can now complete the proof of (4.2.25), which in our notation, amounts to proving that

$$(4.2.30) \qquad E\left[X_\varepsilon - \frac{1}{C} \,\Big|\, \mathcal{F}^s \right] \to 0, \text{ a.s.}$$

But

$$E\left[X_\varepsilon - \frac{1}{C} \,\Big|\, \mathcal{F}^s \right] = \frac{1}{C} E[X_\varepsilon(-Y_\varepsilon + C) \mid \mathcal{F}^s].$$

Hence

$$\left| E\left[X_\varepsilon - \frac{1}{C} \,\Big|\, \mathcal{F}^s \right] \right| \le \frac{1}{C}[E(|X_\varepsilon|^s \mid \mathcal{F}^s)]^{1/2}[E((-Y_\varepsilon + C)^s \mid \mathcal{F}^s)]^{1/2}$$

$$\le \frac{1}{C}[E(Y_\varepsilon + C)^s \mid \mathcal{F}^s]^{1/2} \to 0^{[11]}$$

from (4.2.29).

We therefore have proved (4.2.22), and we can proceed with the induction. We have

$$E \exp i \left[\int_0^{t_1} \lambda_1 \cdot \sigma\left(\frac{y^\varepsilon}{\varepsilon} \right) dw(s) + \cdots + \int_0^{t_{k+1}} \lambda_k \cdot \sigma\left(\frac{y^\varepsilon}{\varepsilon} \right) dw(s) \right]$$

$$= E\left\{ \left(\exp i \left[\int_0^{t_1} \lambda_1 \cdot \sigma\left(\frac{y^\varepsilon}{\varepsilon} \right) dw(s) + \cdots + \int_0^{t_{k-1}} \lambda_{k-1} \cdot \sigma\left(\frac{y^\varepsilon}{\varepsilon} \right) dw(s) \right] \right) \right.$$

$$\cdot \left(\exp i \int_0^{t_k} (\lambda_k + \lambda_{k+1}) \cdot \sigma\left(\frac{y^\varepsilon}{\varepsilon} \right) dw(s) \right)$$

$$\left. \cdot \left(\exp i \int_{t_k}^{t_{k+1}} \lambda_{k+1} \cdot \sigma\left(\frac{y^\varepsilon}{\varepsilon} \right) dw \,\Big|\, \mathcal{F}^{t_k} \right) \right\} = E z_\varepsilon R_\varepsilon$$

with

$$Z_\varepsilon = \exp i \left[\int_0^{t_1} \lambda_1 \cdot \sigma\left(\frac{y^\varepsilon}{\varepsilon} \right) dw(s) + \cdots + \int_0^{t_k} (\lambda_k + \lambda_{k+1}) \cdot \sigma\left(\frac{y^\varepsilon}{\varepsilon} \right) dw(s) \right]$$

$$T_\varepsilon = E\left(\exp i \int_{t_k}^{t_{k+1}} \lambda_{k+1} \cdot \sigma\left(\frac{y^\varepsilon}{\varepsilon} \right) dw \,\Big|\, \mathcal{F}^{t_k} \right)$$

Applying again (4.2.25), we can assert that

$$R_\varepsilon \to \exp\left(-(q\lambda_{k+1} \cdot \lambda_{k+1})(t_{k+1} - t_k) \right) \text{ a.s.}$$

Since $|Z_\varepsilon| \le 1$, $|R_\varepsilon| \le 1$, it follows that

$$|EZ_\varepsilon R_\varepsilon - (EZ_\varepsilon) \exp\left(-(q\lambda_{k+1} \cdot \lambda_{k+1})(t_{k+1} - t_k) \right)| \to 0.$$

---

[11]$|X_\varepsilon|$ is the modulus of the complex number X_ε.

However from the induction hypothesis, we have

$$
EZ_\varepsilon \to \left(\exp - \sum_{i,j=1}^{k-1} (q\lambda_i \cdot \lambda_j) \min(t_i, t_j) \right)
$$
$$
\cdot (\exp -t_k (q(\lambda_k + \lambda_{k+1}) \cdot (\lambda_k + \lambda_{k+1})))
$$
$$
\cdot \exp -2 \sum_{i,j=1}^{k-1} q\lambda_i \cdot (\lambda_k + \lambda_{k+1}) \min(t_i, t_k).
$$

Hence the result follows. ∎

PROOF OF THEOREM 4.1. This follows immediately from (4.2.3) and Lemma 4.5, since μ_x^ε is the measure of $y_x^\varepsilon(\cdot)$, μ_x is the measure of $y_x(\cdot)$, μ_x^ε is weakly compact, and the finite dimensional distribution of $y_x^\varepsilon(\cdot)$ converge toward the finite dimensional distributions of $y_x(\cdot)$ (cf. Gikhman-Skorokhod [**47**]). ∎

We can now give an extension of Theorem 4.1.

We are going to consider the equation

$$(4.2.31) \qquad dy^\varepsilon = b\left(\frac{y^\varepsilon}{\varepsilon}\right) dt + \sigma\left(\frac{y^\varepsilon}{\varepsilon}\right) dw(t), \qquad y^\varepsilon(0) = x.$$

In (4.2.31), the functions b and σ (derived from a) satisfy the conditions (3.2.1), (3.2.2), (3.2.3).

We denote again by μ_x^ε the probability measure of $y_x^\varepsilon(\cdot)$. Since b is bounded, we again have (4.2.3).

We still consider m as defined by (4.2.5), q_{ij} defined by (4.2.7) and

$$(4.2.32) \qquad \bar{b} = \int_Y b(y)m(y)\, dy.$$

We next introduce the process $y(t) = y_x(t)$ solution of

$$(4.2.33) \qquad dy = \bar{b}\, dt + \sqrt{2}\sqrt{q}\, dw(t), \qquad y(0) = x,$$

and let μ_x be the corresponding Gaussian measure on ψ. Then we have the following:

THEOREM 4.6. *Under the assumptions (3.2.1), (3.2.2), (3.2.3), then we have*

$$(4.2.34) \qquad \forall x,\ \mu_x^\varepsilon \to \mu_x \text{ weakly in } \mathcal{M}_+^1(\psi).$$

PROOF. We first extend Lemma 4.3, whose statement remained unchanged. Let us see what are the modifications in the two proofs which have been given. In the first proof, which carries over easily, we still consider the function defined by (4.2.13). Applying Itô's formula, we obtain

$$
\int_s^t \phi\left(\frac{y^\varepsilon(\lambda)}{\varepsilon}\right) d\lambda = -\varepsilon^2 z\left(\frac{y^\varepsilon(t)}{\varepsilon}\right) + \varepsilon^2 z\left(\frac{y^\varepsilon(s)}{\varepsilon}\right)
$$
$$
+ \varepsilon \int_s^t \frac{\partial z}{\partial x} \cdot \sigma\left(\frac{y^\varepsilon(\lambda)}{\varepsilon}\right) dw(\lambda)
$$
$$
+ \varepsilon \int_s^t \frac{\partial z}{\partial x} \cdot b\left(\frac{y^\varepsilon(\lambda)}{\varepsilon}\right) d\lambda
$$

from which the desired result easily follows.

In the second proof, we still define $\eta^\varepsilon(t)$ by (4.2.14) and now $\eta^\varepsilon(t)$ is a solution of

$$(4.2.35) \qquad d\eta^\varepsilon(t) = \varepsilon b(\eta^\varepsilon(t))\, dt + \sigma(\eta^\varepsilon(t))\, dw^\varepsilon(t), \qquad \eta^\varepsilon(0) = \frac{x}{\varepsilon}.$$

We then introduce the operator

$$
\begin{aligned}
A^\varepsilon &= -a_{ij}\frac{\partial^2}{\partial x_i \partial x_j} - \varepsilon b_i \frac{\partial}{\partial x_i} \\
&= -\frac{\partial}{\partial x_i} a_{ij} \frac{\partial}{\partial x_j} - \left(\varepsilon b_i - \frac{\partial a_{ij}}{\partial x_j} \right) \frac{\partial}{\partial x_i} \\
&= A - \varepsilon b_i \frac{\partial}{\partial x_i},
\end{aligned}
\tag{4.2.36}
$$

$$(4.2.37) \qquad (A^\varepsilon)^* = A^* + \varepsilon \frac{\partial}{\partial x_i}(b_i).$$

Let m^ε be defined analogously to m, i.e.

$$(A^\varepsilon)^* m^\varepsilon = 0, \qquad m^\varepsilon \in W^{2,p,\mu}(\mathbb{R}^n), \qquad p \geq 2,\ \mu > 0,\ m^\varepsilon \text{ periodic},$$

$$(4.2.38) \qquad \int_Y m^\varepsilon(y)\, dy = 1.$$

Denoting by $\eta_\xi^\varepsilon(t)$ the process solution of (4.2.35), with initial data $\eta_\xi^\varepsilon(0) = \xi$, then we can replace (4.2.17) by

$$(4.2.39) \qquad \sup \left| E\phi(\eta_\xi^\varepsilon(t)) - \int_Y \phi m^\varepsilon(y)\, dy \right| \leq K\|\phi\| e^{-\rho t}.$$

Let us prove that

$$(4.2.40) \qquad m^\varepsilon \to m \text{ in } W^{2,p,\mu}(\mathbb{R}^n), \qquad p \geq 2,\ \mu > 0,\ \text{weakly}.$$

We first obtain a priori estimate. We can consider m^ε as the solution of

$$(4.2.41) \qquad A^* m^\varepsilon = -\varepsilon \frac{\partial}{\partial x_i}(b_i m^\varepsilon), \qquad m^\varepsilon \text{ in } W(Y).$$

We make the change of functions

$$m^\varepsilon = m\tilde{m}^\varepsilon \ (\text{note that } m > 0).$$

Then we have since $A^* m = 0$,

$$A^* m^\varepsilon = -\frac{\partial}{\partial x_i}\left(a_{ij} m \frac{\partial \tilde{m}^\varepsilon}{\partial x_j} \right) - \frac{\partial}{\partial x_j}(a_{ij} m)\frac{\partial \tilde{m}^\varepsilon}{\partial x_i}.$$

Therefore

$$
\begin{aligned}
(A^* m^\varepsilon, \tilde{m}^\varepsilon)_Y &= \int_Y a_{ij} m \frac{\partial \tilde{m}^\varepsilon}{\partial x_j}\frac{\partial \tilde{m}^\varepsilon}{\partial x_i}\, dx - \int_Y \frac{\partial}{\partial x_j}(a_{ij} m)\frac{\partial \tilde{m}^\varepsilon}{\partial x_i}\tilde{m}^\varepsilon\, dx \\
&= \int_Y a_{ij} m \frac{\partial \tilde{m}^\varepsilon}{\partial x_j}\frac{\partial \tilde{m}^\varepsilon}{\partial x_i}\, dx + \frac{1}{2}\int_Y \frac{\partial^2}{\partial x_i \partial x_j}(a_{ij} m)(\tilde{m}^\varepsilon)^2\, dx \\
&= \int_Y a_{ij} m \frac{\partial \tilde{m}^\varepsilon}{\partial x_j}\frac{\partial \tilde{m}^\varepsilon}{\partial x_i}\, dx.
\end{aligned}
$$

Hence we obtain the identity

$$(4.2.42) \qquad \int a_{ij}m\frac{\partial \tilde{m}^\varepsilon}{\partial x_j}\frac{\partial \tilde{m}^\varepsilon}{\partial x_i}\,dx = -\varepsilon \int_Y \frac{\partial}{\partial x_i}(b_i m\tilde{m}^\varepsilon)\tilde{m}^\varepsilon\,dx$$

$$= +\varepsilon \int_Y b_i m\tilde{m}^\varepsilon \frac{\partial \tilde{m}^\varepsilon}{\partial x_i}\,dx.$$

From (4.2.42) it follows that

$$\beta \int_Y m|\nabla \tilde{m}^\varepsilon|^2\,dx \le \varepsilon C \left(\int_Y m(\tilde{m}^\varepsilon)^2\,dx\right)^{1/2}\left(\int_Y m|\nabla \tilde{m}^\varepsilon|^2\,dx\right)^{1/2}$$

hence

$$(4.2.43) \qquad \beta \left(\int_Y m|\nabla \tilde{m}^\varepsilon|^2\,dx\right)^{1/2} \le \varepsilon C \left(\int_Y m(\tilde{m}^\varepsilon)^2\,dx\right)^{1/2}.$$

Since m is positive continuous and periodic, it is bounded below by a positive constant, hence (4.2.43) implies

$$(4.2.44) \qquad |\tilde{m}^\varepsilon|_{1,Y} \le \varepsilon C'|\tilde{m}^\varepsilon|_{L^2(Y)}.$$

But

$$\left|\tilde{m}^\varepsilon - \int \tilde{m}^\varepsilon\right|_{L^2(Y)} \le C''|\tilde{m}^\varepsilon|_{1,Y}.$$

Hence

$$|\tilde{m}^\varepsilon|_{L^2(Y)} \le \left|\int \tilde{m}^\varepsilon\right| + C''|\tilde{m}^\varepsilon|_{1,Y}$$

and since $\tilde{m}^\varepsilon \ge 0$, $\int m\tilde{m}^\varepsilon\,dx = 1$ and m is bounded below, it follows that

$$|\tilde{m}^\varepsilon|_{L^2(Y)} \le C''[1 + |\tilde{m}^\varepsilon|_{1,Y}].$$

Therefore from (4.2.44), we see that

$$|\tilde{m}^\varepsilon|_{1,Y} \le K\varepsilon[1 + |\tilde{m}^\varepsilon|_{1,Y}],$$

hence

$$(4.2.45) \qquad |\tilde{m}^\varepsilon|_{1,Y} \to 0, \qquad |\tilde{m}^\varepsilon|_{L^2(Y)} \le C.$$

It follows from (4.2.45) that (at least for a subsequence)

$$(4.2.46) \qquad \tilde{m}^\varepsilon \to \tilde{m} \text{ in } H^1(Y) \text{ weakly, in } L^2(Y) \text{ strongly,}$$

$$\tilde{m} = \text{ constant.}$$

Since

$$1 = \int m^\varepsilon = \int m\tilde{m}^\varepsilon \to \tilde{m},$$

we see that $\tilde{m}^\varepsilon \to 1$ in $H^1(Y)$ weakly. It follows that

$$(4.2.47) \qquad m^\varepsilon \to m \text{ in } H^1(Y) \text{ weakly.}$$

The statement (4.2.40) follows from a bound in $W^{2,p}$, which can be deduced from (4.2.7) and the equation (4.2.41). Actually (4.2.47) is sufficient to assert that

$$(4.2.48) \qquad \int_Y \phi m^\varepsilon(y)\,dy \to \int_Y \phi m(y)\,dy.$$

Therefore from (4.2.39), and (4.2.48) we can deduce that

$$(4.2.49) \qquad \sup_{\xi} \left| E\phi(\eta_\xi^\varepsilon(t)) - \int_Y \phi m(y)\, dy \right| \le h_\phi(\varepsilon) + K\|\phi\|e^{-\rho t},$$

where $h_\phi(\phi) \to 0$ as $\varepsilon \to 0$, and does not depend on t. We can then easily transform the end of the argument of the second proof of Lemma 4.3. We have instead of (4.2.18), using (4.2.10)

$$(4.2.50) \qquad \left| E\phi\left(\frac{y_x^\varepsilon(t)}{\varepsilon}\right) \right| \le h_\phi(\varepsilon) + K\|\phi\|e^{-\rho\frac{t}{\varepsilon^2}}.$$

The end of the argument can be carried over without any difficulty. ∎

We next prove Lemma 4.5, where of course $y_x^\varepsilon(t)$ now denotes the solution of (4.2.31), and $y_x(t)$ the solution of (4.2.33).

We have to show that (analogue of (4.2.21))

$$(4.2.51) \quad E \exp i \left[\int_0^{t_1} \lambda_1 \cdot b\left(\frac{y_\varepsilon}{\varepsilon}\right)\, dx + \cdots + \int_0^{t_k} \lambda_k \cdot b\left(\frac{y^\varepsilon}{\varepsilon}\right)\, ds \right.$$

$$\left. + \int_0^{t_1} \lambda_1 \cdot \sigma\left(\frac{y^\varepsilon}{\varepsilon}\right)\, dw(s) + \cdots + \int_0^{t_k} \lambda_k \cdot \sigma\left(\frac{y^\varepsilon}{\varepsilon}\right)\, dw(s) \right]$$

$$\to \exp\left[i(\lambda_1 \cdot \bar{b}t_1 + \cdots + \lambda_k \cdot \bar{b}t_k) - \sum_{i,j} q\lambda_i \cdot \lambda_j \min(t_i, t_j) \right].$$

If we set

$$X_\varepsilon = \exp i \left[\int_0^{t_1} \lambda_1 \cdot b\left(\frac{y^\varepsilon}{\varepsilon}\right)\, ds + \cdots + \int_0^{t_k} \lambda_k \cdot b\left(\frac{y^\varepsilon}{\varepsilon}\right)\, ds \right],$$

$$Y_\varepsilon = \exp i \left[\int_0^{t_1} \lambda_1 \cdot \sigma\left(\frac{y^\varepsilon}{\varepsilon}\right)\, dw(s) + \cdots + \int_0^{t_k} \lambda_k \cdot \sigma\left(\frac{y^\varepsilon}{\varepsilon}\right)\, dw(s) \right],$$

then since Lemma 4.3 holds, the proof of Lemma 4.5 shows immediately that

$$X_\varepsilon \to \varepsilon i(\lambda_1 \cdot \bar{b}t_1 + \cdots + \lambda_k \cdot \bar{b}t_k) \text{ in } L^2(\Omega, \mathcal{A}, P),$$

$$\varepsilon Y_\varepsilon \to \exp - \sum_{i,j} q\lambda_i \cdot \lambda_j \min(t_i, t_j).$$

Since the limits are constants and since $|X_\varepsilon| = 1$, we can conclude (4.2.51). From the weak compactness and the convergence of the finite dimensional distributions, we obtain (4.2.34).

REMARK 4.7. Theorems 4.6 (and of course Theorem 4.1 which is a particular case) are due to Freĭdlin [**40**].

4.3. Diffusion with unbounded drift. We now consider, instead of (4.2.31), the equation

$$(4.3.1) \qquad dy\varepsilon = \frac{1}{\varepsilon}b\left(\frac{y^\varepsilon}{\varepsilon}\right)\, dt + \sigma\left(\frac{y^\varepsilon}{\varepsilon}\right)\, dw(t), \qquad y^\varepsilon(0) = x.$$

Again we assume that (3.2.1), (3.2.2), (3.2.3) hold true.

We denote by μ_x^ε the probability measure of $y_x^\varepsilon(\cdot)$. This time, contrary to the cases considered above, we are not guaranteed that μ_x^ε remains in a weakly compact subset of $\mathcal{M}_+^1(\psi)$. In general, this will not be true unless we make an additional assumption on b, which we will state below.

We introduce the differential operator A given by

$$(4.3.2) \qquad A = -a_{ij}(x)\frac{\partial^2}{\partial x_i \partial x_j} - b_i(x)\frac{\partial}{\partial x_i},$$

and m will again be the unique solution of
$$(4.3.3)$$
$$A^*m = 0, \qquad m \in W^{2,p,\mu}(\mathbb{R}^n), \qquad p \geq 2,\ \mu > 0,\ m \text{ periodic}, \int m(y)\, dy = 1.$$

Before we state a theorem of convergence concerning μ_x^ε, let us note that Lemma 4.3 is still valid mutatis mutandis. We repeat it for convenience (we will only emphasize the differences with the proof already given).

LEMMA 4.8. *Let ϕ be Borel periodic bounded, such that*

$$(4.3.4) \qquad \int_Y \phi(y)m(y)\, dy = 0.$$

Then for any $s \leq t$, we have

$$(4.3.5) \qquad E\left[\left(\int_s^t \phi\left(\frac{y^\varepsilon(\lambda)}{\varepsilon}\right)\, d\lambda\right)^2 \,\middle|\, \mathcal{F}^s\right] \to 0.^{[12]}$$

FIRST PROOF. We assume (4.2.12) and apply Itô's formula to a function z solution of

$$(4.3.6) \qquad Az = \phi, \qquad z \text{ periodic}, \qquad \int_Y z\, dy = 0,$$

and we know that there exists a solution which is C^2.

We apply Itô's formula and we follow exactly the same calculation as in Lemma 4.3 (we have changed the definition of A in the adequate manner). Hence (4.3.5) follows. ∎

SECOND PROOF. We again define $\eta^\varepsilon(t)$ by (4.2.14). This time we get that $\eta^\varepsilon(t)$ is a solution of

$$(4.3.7) \qquad d\eta^\varepsilon(t) = b(\eta^\varepsilon(t))\, dt + \sigma(\eta^\varepsilon(t))\, dw^\varepsilon(t), \qquad \eta^\varepsilon(0) = \frac{x}{\varepsilon}.$$

We see that m has been defined in the right way to make the calculations of the second proof of Lemma 4.3 valid, without any change. ∎

We now make the assumption

$$(4.3.8) \qquad \int_Y b(y)m(y)\, dy = 0.$$

Applying Theorem 3.5, we can then assert that there exist functions $\chi^\ell(x)$ such that
$$(4.3.9)$$
$$A\chi^\ell = -b_\ell, \qquad \chi^\ell \in W^{2,p,\mu}(\mathbb{R}^n), \qquad p \geq 2,\ \mu > 0,\ \chi^\ell \text{ periodic}, \int_Y \chi^\ell(y)\, dy = 0.$$

---

[12]The property (4.3.5) does not depend on the fact that the measure μ_x^ε remains or not in a weakly compact set of $\mathcal{M}_+^1(\psi)$.

We define now the matrix q by (compare with (4.2.7))

$$(4.3.10) \qquad q = \int_Y \left(I - \frac{\partial \chi}{\partial y} \right) a \left(I - \frac{\partial \chi}{\partial y} \right)^* m(y) \, dy.$$

Let then μ_x be the Gaussian measure on ψ defined by the process

$$(4.3.11) \qquad dy = \sqrt{2}\sqrt{q} \, dw(t), \qquad y(0) = x.$$

Then we have the

THEOREM 4.9. *Under the assumptions (3.2.1), (3.2.2), (3.2.3) and (4.3.8), we have*

$$(4.3.12) \qquad \forall x, \ \mu_x^\varepsilon \to \mu_x \text{ weakly in } \mathcal{M}_+^1(\psi).$$

PROOF. We first prove that μ_x^ε remains in a weakly compact subset of $\mathcal{M}_+^1(\psi)$. For this we note that $\chi^\ell(y)$ are C^2 functions, hence we can apply Itô's formula, which gives using (4.3.9)

$$(4.3.13) \quad \chi\left(\frac{y^\varepsilon(t)}{\varepsilon}\right) = \chi\left(\frac{x}{\varepsilon}\right) + \frac{1}{\varepsilon^2} \int_0^t b\left(\frac{y^\varepsilon}{\varepsilon}\right) \, ds + \frac{1}{\varepsilon} \int_0^t \frac{\partial \chi}{\partial y} \sigma \left(\frac{y^\varepsilon}{\varepsilon}\right) \, dw(s).$$

Combining (4.3.13) and (4.3.1) to eliminate the b term, we obtain

$$(4.3.14) \quad y^\varepsilon(t) = x + \varepsilon \chi\left(\frac{y^\varepsilon(t)}{\varepsilon}\right) - \varepsilon \chi\left(\frac{x}{\varepsilon}\right) + \int_0^t \left(I - \frac{\partial \chi}{\partial y}\right) \sigma\left(\frac{y^\varepsilon}{\varepsilon}\right) \, dw(s).$$

To show that μ_x^ε remains in a weakly compact subset of $\mathcal{M}_+^1$, we can apply the following criterion (cf. Parthasarathy [91]):

$$(4.3.15) \quad \lim_{\delta \downarrow 0} \overline{\lim_{\varepsilon \downarrow 0}} P\left[\sup_{\substack{0 \le t_1 < t_2 \le T \\ |t_1 - t_2| \le \delta}} |y_x^\varepsilon(t_2) - y_x(t_1)| \ge \eta \right] = 0, \qquad \eta > 0, \ T < \infty,$$

$$(4.3.16) \qquad \lim_{M \uparrow \infty} \overline{\lim_{\varepsilon \downarrow 0}} P[|y_x^\varepsilon(0)| \ge M] = 0.$$

The second condition is trivially realized since $y_x^\varepsilon(0) = x$. We have next

$$\sup_{|t_1 - t_2| < \delta} y^\varepsilon(t_2) - y^\varepsilon(t_1)| \le C_1 \varepsilon + \sup_{|t_1 - t_2| \le \delta} \left| \int_{t_1}^{t_2} \left(I - \frac{\partial \chi}{\partial y}\right) \sigma\left(\frac{y^\varepsilon}{\varepsilon}\right) \, dw(s) \right|.$$

Hence

$$P\left[\sup_{|t_2 - t_1| \le \delta} |y^\varepsilon(t_2) - y^\varepsilon(t_1)| \ge \eta \right]$$

$$\le P\left[\sup_{\substack{|t_2 - t_1| \le \delta \\ 0 \le t_1 < t_2 \le T}} \left| \int_{t_1}^{t_2} \left(I - \frac{\partial \chi}{\partial y}\right) \sigma\left(\frac{y^\varepsilon}{\varepsilon}\right) \, dw(s) \right| \ge \eta - c_1 \varepsilon \right]$$

$$= P[X \ge \eta - C_1 \varepsilon].$$

But

$$\chi \le 2 \sup_{0 \le j \le T/\delta} \sup_{j\delta \le t \le (j+1)\delta} \left| \int_{j\delta}^t \left(I - \frac{\partial \chi}{\partial y}\right) \sigma\left(\frac{y^\varepsilon}{\varepsilon}\right) \, dw(s) \right|$$

so that

$$P[X \geq \eta - c_1\varepsilon] \leq \sum_{j \leq T/\delta} P\left[\sup_{j\delta \leq t \leq (j+1)\delta} \left|\int_{j\delta} \left(I - \frac{\partial\chi}{\partial y}\right)\sigma\, dw\right| \geq \frac{\eta - c_1\varepsilon}{2}\right]$$

$$\leq \sum_{j \leq T/\delta} \frac{2^4}{(\eta - c_1\varepsilon)^4} E \sup_{j\delta \leq t \leq (j+1)\delta} \left|\int_{j\delta} t\left(I - \frac{\partial\chi}{\partial y}\right)\sigma\, dw\right|^4$$

$$\leq \sum_{j \leq T/\delta} \frac{C\delta^2}{(\eta - c_1\varepsilon)^4} \leq \frac{C'T\delta}{(\eta - c_1\varepsilon)^2},$$

and (4.3.15) follows immediately.

We turn now to the convergence of the finite dimensional distributions. We have to prove (4.2.20), where $y_x^\varepsilon(t)$ is the process given by (4.3.14) and where q is defined by (4.3.10). Since χ is bounded, it is clear that it is enough to prove that

(4.3.17)

$$E \exp i \left|\int_0^{t_1} \lambda_1 \cdot \left(I - \frac{\partial\chi}{\partial y}\right)\sigma\left(\frac{y^\varepsilon}{\varepsilon}\right) dw(s) + \cdots + \int_0^{t_k} \lambda_k \cdot \left(I - \frac{\partial\chi}{\partial y}\right)\sigma\left(\frac{y^\varepsilon}{\varepsilon}\right) dw(s)\right|$$

$$\rightarrow \exp - \sum_{i,j} q\lambda_i \cdot \lambda_j \min(t_i, t_j).$$

But from 4.8, from the proof of Lemma 4.5, and by virtue of the choice of q, it follows that (4.3.17) holds true, which completes the proof of the theorem. ∎

We can give an extension of Theorem 4.9 similar to the one of Theorem 4.6 with respect to Theorem 4.1. We only state it, the proof being left to the reader.

We consider the equation

$$(4.3.18) \qquad dy^\varepsilon = \frac{1}{\varepsilon}b\left(\frac{y^\varepsilon}{\varepsilon}\right) dt + c\left(\frac{y^\varepsilon}{\varepsilon}\right) dt + \sigma\left(\frac{y^\varepsilon}{\varepsilon}\right) dw(t), \qquad y^\varepsilon(0) = x.$$

We assume that (3.2.1), (3.2.2), (3.2.3) are satisfied and that c satisfies the same conditions as b. We denote by m the solution of (4.3.3). We again assume (4.3.8). Define the χ^ℓ functions by (4.3.9), q by (4.3.10) and

$$(4.3.19) \qquad \bar{c} = \int_Y \left(I - \frac{\partial\chi}{\partial y}\right) c(y)m(y)\, dy.$$

We consider the Gaussian process $y_x(t)$ solution of

$$(4.3.20) \qquad dy = \bar{c}\, dt + \sqrt{2}\sqrt{q}\, dw(t), \qquad y(0) = x,$$

and we denote by μ_x^ε and μ_x the measures of $y_x^\varepsilon(\cdot)$ and $y_x(\cdot)$. We have

THEOREM 4.10. *Under the assumptions*[13] *(3.2.1), (3.2.2), (3.2.3), c satisfying the same assumptions as b, and (4.3.8) then*

$$(4.3.21) \qquad \forall x, \ \mu_x^\varepsilon \rightarrow \mu_x \ weakly \ in \ \mathcal{M}_+^1(\psi).$$

REMARK 4.11. In all the above Theorems 4.1 through 4.10, one cannot hope for a stronger convergence such as

$$(4.3.22) \qquad \forall t, \ y_x^\varepsilon(t) \rightarrow y_x(t) \ in \ L^2(\Omega, \mathcal{A}, P; \mathbb{R}^n).$$

---

[13]In particular $a_{ij} \in W^{2,\infty}(\mathbb{R}^n)$.

Indeed, let us consider for example y^ε defined by (4.2.1). If (4.3.22) was correct then we would have

$$\left| \int_0^t \left(\sigma\left(\frac{y^\varepsilon}{\varepsilon}\right) - \sqrt{2}\sqrt{q} \right) dw(t) \right|^2 \to 0.$$

But

$$E\left| \int_0^t \left(\sigma\left(\frac{y^\varepsilon}{\varepsilon}\right) - \sqrt{2}\sqrt{q} \right) dw(t) \right|^2 = E \int_0^t \mathrm{tr}(\sigma - \sqrt{2}q)(\sigma^* - \sqrt{2}q)\left(\frac{y^\varepsilon}{\varepsilon}\right) ds$$

$$\to \int_Y \mathrm{tr}(\sigma - \sqrt{2}q)(\sigma^* - \sqrt{2}q)(y)m(y)\, dy,$$

which is different from 0, since $\sqrt{\int_Y a\, dy} \neq \int \sqrt{a}\, dy$.

4.4. Convergence of functionals and probabilistic proof of homogenization. Let $\mathcal{O}$ be an open bounded subset of $\mathbb{R}^n$, whose boundary is denoted by Γ; $\mathcal{O}$ will be supposed regular in the following sense:

(4.4.1)

$$\mathcal{O} = \{x \mid \phi(x) < 0\}, \text{ where } \phi \in C^2(\mathbb{R}^n),\ \phi \text{ bounded, } \frac{\partial \phi}{\partial x_i}, \frac{\partial^2 \phi}{\partial x_i \partial x_j} \text{ bounded,}$$

$$\left\| \frac{\partial \phi}{\partial x} \right\| \geq \delta > 0,\ \forall x \in \Gamma.$$

Defining a_{ij}, b_i, c_i as in Theorem 4.10, and

(4.4.2) a_0 continuous bounded, periodic, $a_0 \geq \alpha_0 > 0$.

We may consider the solution $u_\varepsilon(x)$ of the Dirichlet problem
(4.4.3)

$$-a_{ij}\left(\frac{x}{\varepsilon}\right)\frac{\partial^2 u_\varepsilon}{\partial x_i \partial x_j} - \frac{1}{\varepsilon}b_i\left(\frac{x}{\varepsilon}\right)\frac{\partial u_\varepsilon}{\partial x_i} - c_i\left(\frac{x}{\varepsilon}\right)\frac{\partial u_\varepsilon}{\partial x_i} + a_0\left(\frac{x}{\varepsilon}\right)u_\varepsilon = f, \qquad u_\varepsilon \mid_\Gamma = 0,$$

where

(4.4.4) $f \in C^0(\overline{\mathcal{O}}).$

We know (cf. Section 1.4) that there exists one and only one solution of (4.4.3) in $W^{2,p}(\mathcal{O}), \forall p \geq 2$.

The homogenized operator is defined by the formula

(4.4.5) $$\mathcal{A} = -q_{ij}\frac{\partial^2}{\partial x_i \partial x_j} - \bar{c}_i\frac{\partial}{\partial x_i} + \bar{a}_0$$

where

(4.4.6) $$\bar{a}_0 = \int_Y a(y)m(y)\, dy.$$

We also consider the solution u of the homogenized equation

(4.4.7) $$\mathcal{A}u = f, \qquad u \mid_\Gamma = 0.$$

We can state the following

THEOREM 4.12. *Under the assumptions of Theorem 4.10, and (4.4.1), (4.4.2), (4.4.4), we have*

(4.4.8) $$u_\varepsilon(x) \to u(x),\ \forall x \in \overline{\mathcal{O}}.$$

REMARK 4.13. We can relate the result of Theorem 4.12, with the homogenization results obtained in Chapter 1. Let us take

$$(4.4.9) \qquad\qquad c_i = 0, \qquad b_i = \frac{\partial a_{ij}}{\partial x_j}.$$

Then (4.4.3) can be rewritten as

$$-\frac{\partial}{\partial x_i} a_{ij}\left(\frac{x}{\varepsilon}\right) \frac{\partial u_\varepsilon}{\partial x_j} + a_0\left(\frac{x}{\varepsilon}\right) u_\varepsilon = f, \qquad u_\varepsilon \mid_\Gamma = 0,$$

which is the problem considered in Chapter 1 (of course here we require much more regularity on the data than in Chapter 1). If we turn to the operator A defined by (4.3.2), we see that

$$A = -\frac{\partial}{\partial x_i} a_{ij}(x) \frac{\partial}{\partial x_j} = A^*$$

so that $m(x) \equiv 1$. The equations for the χ^ℓ functions (see (4.3.9)) reduce to the ones obtained in Chapter 1 (cf. (2.3.5)), and we recover the usual homogenization formulas. One advantage of the probabilistic method is to obtain a pointwise convergence result. A very serious drawback is the need of stringent regularity assumptions.

REMARK 4.14. The assumption of symmetry $a_{ij} = a_{ji}$ is not really a loss of generality, since (4.4.3) can always be rewritten with symmetric coefficients. This is not true for the divergence form.

PROOF OF THEOREM 4.12. We consider the process $y_x^\varepsilon(t)$, defined by (4.3.18) and the process

$$(4.4.10) \qquad\qquad \xi^\varepsilon(t) = \int_0^t a_0\left(\frac{y^\varepsilon(s)}{\varepsilon}\right) ds.$$

From Lemma (4.8) we can assert that

$$(4.4.11) \qquad\qquad E[\xi^\varepsilon(t) - \bar{a}_0 t]^s \to 0, \quad \forall t.$$

We denote by τ_x^ε the exit time of $y_x^\varepsilon(t)$ from $\mathcal{O}$. Then we ca write an explicit formula for $u_\varepsilon(x)$ (see (1.4.8)), namely

$$(4.4.12) \qquad\qquad u_\varepsilon(x) = E \int_0^{\tau_x^\varepsilon} f(y_x^\varepsilon(t))(\exp -\xi^\varepsilon(t)) \, dt.$$

It will be convenient to give another formulation of the right hand side of (4.4.12). We introduce

$$\tilde{\Psi} = C([0,\infty[; \mathbb{R}^n) \times C([0,\infty[; \mathbb{R}).$$

An element of $\tilde{\Psi}$, namely $\tilde{\psi}$, will be a pair

$$\tilde{\psi} = (y(\cdot), \xi(\cdot)).$$

We provide $\tilde{\Psi}$ with the Borel σ-algebra $\tilde{\mathcal{B}}$. It is generated by the sets $\{y(t_j), \xi(t_j)\} \in B_j$, $t_j \in \mathbb{R}$, B_j Borel sets of $\mathbb{R}^{n+1}$. The process $y_x^\varepsilon(\cdot)$, $\xi^\varepsilon(\cdot)$ induces on $\tilde{\Psi}$ a probability measure denoted by $\tilde{\mu}_x^\varepsilon$.

Similarly, we denote by $\tilde{\mu}_x$ the measure induced by $y_x(\cdot)$, $\xi(\cdot)$ where $y_x(t)$ is defined by (4.3.20) and

$$(4.4.13) \qquad\qquad \xi(t) = \bar{a}_0 t.$$

The family μ_x^ε (for fixed x) remains in a weakly compact subset of $\mathcal{M}_+^1(\psi)$. To prove it, it is enough to modify slightly the proof of Theorem 4.9. One has then to prove that

$$\limsup_{\delta\downarrow 0}\limsup_{\varepsilon\downarrow 0} P\left[\sup_{\substack{0\leq t_1<t_2\leq T\\ t_2-t_1\leq\delta}}(|y_x^\varepsilon(t_2)-y_x^\varepsilon(t_1)|+|\xi^\varepsilon(t_2)-\xi^\varepsilon(t_1)|)\geq\eta\right]=0,$$

for any $\eta>0$, $T<\infty$. Since

$$|\xi^\varepsilon(t_2)-\xi^\varepsilon(t_1)|\leq C|t_2-t_1|$$

this can be done easily by extending the argument of Theorem 4.9. The next thing to prove is that the finite dimensional distributions of $(y_x^\varepsilon(\cdot),\xi^\varepsilon(\cdot))$ converge toward the corresponding finite dimensional distributions of $(y_x(\cdot),\xi(\cdot))$. This is also easy to obtain as in Theorem 4.9, taking into account the property (4.4.11).

We now define on (y,ξ) a functional $F(\tilde\psi)$ by

$$(4.4.14)\qquad F(\tilde\psi)=\begin{cases}\int_0^\tau(\exp-\xi(t))f(y(t))\,dt, & \text{if }\xi(t)\geq\alpha_0 t,\ \forall t,\\ +\infty & \text{otherwise,}\end{cases}$$

where we have set if $\tilde\psi=(y(\cdot),\xi(\cdot))$

$$(4.4.15)\qquad \tau(y)=\tau=\inf\{t\mid t\geq 0,\ y(t)\notin\mathcal{O}\}.$$

Then we can assert the following properties

$$(4.4.16)\qquad F\text{ is }\tilde{\mathcal{B}}\text{ measurable, it is bounded a.s. with respect to }\tilde\mu_x^\varepsilon\text{ and }\tilde\mu_x;$$

$$(4.4.17)\qquad F\text{ is continuous a.s. with respect to }\tilde\mu_x^\varepsilon\text{ and }\tilde\mu_x.$$

The property (4.4.16) is clear, since $a_0\geq\alpha_0>0$. The property (4.4.17) is a standard result for diffusions, and it is where the specific assumptions on the domain (4.4.1) are used (cf. Lemma 4.15 below). We can then rewrite $u_\varepsilon(x)$ as follows

$$(4.4.18)\qquad u_\varepsilon(x)=\int_{\tilde Y}F(\tilde\psi)\,d\tilde\mu_x^\varepsilon$$

and similarly

$$(4.4.19)\qquad u(x)=\int_{\tilde\Psi}F(\tilde\psi)\,d\tilde\mu_x.$$

We can now rely on a theorem of Gikhman-Skorokhod [47], to assert that since $\tilde\mu_x^\varepsilon$ converges weakly toward $\tilde\mu_x$, and since F is $\tilde\mu_x$ a.s. continuous and bounded, then

$$\int_{\tilde\Psi}F(\tilde\psi)\,d\tilde\mu_x^\varepsilon\to\int_{\tilde\Psi}F(\tilde\psi)\,d\tilde\mu_x$$

i.e., (4.4.8).   ∎

LEMMA 4.15. *The functional F is $\tilde\mu_x$ a.s. continuous.*

PROOF. Let Σ be the complement of $\mathcal{O}$ and $\overset{\circ}{\Sigma}$ be the interior of Σ. Let us set

$$(4.4.20)\qquad \Lambda(y)=\inf\{t\geq\tau(y)\mid y(t)\in\overset{\circ}{\Sigma}\}.$$

We shall prove that

$$(4.4.21)\qquad \Lambda(y)=\tau(y),\qquad \tilde\mu_x\text{ a.s.}$$

If (4.4.21) holds true, then $y \to \tau(y)$ is continuous. Indeed if

(4.4.22) $\qquad\qquad y_n \to y$ in $C(0[0, \infty[\; \mathbb{R}^n)$, then $\tau(y_n) \to \tau(y)$.

To prove (4.4.22), let us consider three cases.

(1) $\tau(y) = +\infty$. Then $y(t) \in \mathcal{O}, \forall t \geq 0$. Let T be an arbitrary number; the set

$$K_T = \{y(t), t \in [0, T]\},$$

is compact and contained in $\mathcal{O}$. Hence there exists $\alpha_T > 0$ such that if $\xi \in K_T$ and $|\bar{\xi} - \xi| < d_T$, then $\bar{\xi} \in \mathcal{O}$. For $n \geq N_T$, one has $|y_n(t) - y(t)| < d_T, \forall t \in [0, T]$, hence $y_n(t) \in \mathcal{O}, \forall t \in [0, T]$, which implies

$$n \geq N_T \to \tau(y_n) \geq T.$$

Since T is arbitrary, we have proved that $\tau(y_n) \to +\infty$.

(2) $\tau(y) = 0$. By (4.4.21) we also have $\Lambda(y) = 0$. Therefore $\forall \delta > 0, \exists t_\delta$ such that $0 < t_\delta \leq \delta$ and $y(t_\delta) \in \overset{\circ}{\Sigma}$. But then $\exists N_\delta$ such that

$$n \geq N_\delta \to y_n(t_\delta) \in \overset{\circ}{\Sigma} \to \tau(y_n) \leq \Lambda(y_n) \leq \delta.$$

Therefore $\tau(y_n) \to 0$.

(3) $0 < \tau(y) < +\infty$. Let $0 < \delta < \tau(y)$ and $K_\delta = \{y(t), t \in [0, \tau(y) - \delta]\}$ which is a compact subset of $\mathcal{O}$. Let d_δ be such that

$$\xi \in K_\delta, \qquad |\xi - \bar{\xi}| < d_\delta \to \bar{\xi} \in \mathcal{O}.$$

There exists N_δ such that for $n \geq N_\delta$, then

$$y_n(t) - y(t)| \leq d_\delta, \; \forall t \in [0, \tau(y)].$$

Hence for $t \in [0, \tau(y) - \delta]$ one has $y_n(t) \in \mathcal{O}$, which implies

(4.4.23) $\qquad\qquad n \geq N_\delta \to \tau(y_n) \geq \tau(y) - \delta.$

On the other hand since

$$\delta + \tau(y) = \delta + \Lambda(y),$$

there exists $t_\delta \in [\tau(y), \delta + \tau(y)]$ such that $y(t_\delta) \in \overset{\circ}{\Sigma}$.

Also, there exists N_δ^1 such that $n \geq N_\delta^1 \to y_n(t_\delta) \in \overset{\circ}{\Sigma}$, hence

(4.4.24) $\qquad\qquad n \geq N_\delta^1 \to \tau(y_n) \leq \delta + \tau(y)$

and (4.4.23), (4.4.24) prove that $\tau(y_n) \to \tau(y)$.

It remains to prove (4.4.21). This will follows from the fact that

$$\tilde{\mu}_x \text{ a.s.,} \qquad y(t) = x + \sqrt{2}\sqrt{q}w(t)$$

and properties of Brownian motion. Here the special assumptions on $\mathcal{O}$ will be explicitly used.

We will follow ideas of Stroock-Varadhan [**114**]. Actually, we shall prove that

(4.4.25) $\qquad\qquad T \wedge \Lambda(y) = T \wedge \tau(y), \; \forall T, \text{ a.s}$

Let $\delta_0 < \delta$ (where δ is defined in (4.4.1)). Let us set

(4.4.26) $\qquad\qquad \tau_{\delta_0} = \inf \left\{ t \mid t \geq \tau, \left\| \frac{\partial \phi}{\partial x}(y(t)) \right\| \leq \delta_0 \right\}$

which makes sense since $\frac{\partial \phi}{\partial x}(y(t))$ is continuous and $\left\| \frac{\partial \phi}{\partial x}(y(\tau)) \right\| \geq \delta$ (since $y(\tau) \in \partial \mathcal{O}$). Let us set

(4.4.27) $$\tau' = \tau \wedge T, \qquad \Lambda' = \Lambda \wedge \tau_{\delta_0} \wedge T,$$

which are stopping times with respect to $\mathcal{F}^{t+0}$. Let us set for $\lambda > 0$

$$\beta(t) = \lambda \chi_{\Lambda'}(t)(1 - \chi_{\tau'}(t))\sqrt{2}\sqrt{q}\frac{\partial \phi}{\partial x}(y(t))$$

where

$$\chi_s(t) = \begin{cases} 1 & \text{if } t \leq s, \\ 0 & \text{otherwise.} \end{cases}$$

If $\beta \neq 0$, then $\tau' < \Lambda'$ and $\tau' = \tau$, hence $[\tau', \Lambda'] \subset [\tau, \Lambda]$, which implies that $y(t) \in \overline{\mathcal{O}}$, $t \in [\tau', \Lambda']$, and therefore $|\beta(t)| \leq C$. Since $w(t)$ is an $\mathcal{F}^{t+0}$ martingale, we have

(4.4.28) $$E \exp\left[\int_0^T \beta(t) \cdot w(t) - \frac{1}{2} \int_0^T |\beta(t)|^2 \, dt \right] = 1.$$

Therefore

(4.4.29) $$E \exp\left[\lambda \int_{\tau'}^{\Lambda'} \sqrt{2}\frac{\partial \phi}{\partial x} \cdot q^{1/2} \, dw(t) - \lambda^2 \int_{\tau'}^{\Lambda'} q\frac{\partial \phi}{\partial x} \cdot \frac{\partial \phi}{\partial x} \, dt \right] = 1.$$

From Itô's formula, one has

$$\lambda \phi(y(\Lambda')) = \lambda \phi(y(\tau')) + \lambda \int_{\tau'}^{\Lambda'} \sqrt{2}\frac{\partial \phi}{\partial x} \cdot q^{1/2} \, dw(t) + \lambda \int_{\tau'}^{\Lambda'} \operatorname{tr} \frac{\partial^2 \phi}{\partial x^2} \cdot q(y(t)) \, dt.$$

If $\Lambda' > \tau'$, one has since $y(t) \in \overline{\mathcal{O}}$, for $t \in [\tau', \Lambda']$

(4.4.30) $$\lambda \int_{\tau'}^{\Lambda'} \sqrt{2}\frac{\partial \phi}{\partial x} \cdot q^{1/2} \, dw(t) \leq \lambda M(\Lambda' - \tau')$$

taking into account that $\tau' = \tau$, and $\phi(y(\tau')) = 0$, and $\phi(y(\Lambda')) \leq 0$. Again if $\Lambda' > \tau'$, one has

$$t \in [\tau', \Lambda'] \rightarrow t \in [\tau, \tau_{\delta_0}]$$

and therefore

$$\left\| \frac{\partial \phi}{\partial x}(y(t)) \right\| > \delta_0.$$

Therefore we can assert that

(4.4.31)
$$\lambda \int_{\tau'}^{\Lambda'} \sqrt{2}\frac{\partial \phi}{\partial x} \cdot q^{1/2} \, dw(t) - \lambda^2 \int_{\tau'}^{\Lambda'} q\frac{\partial \phi}{\partial x} \cdot \frac{\partial \phi}{\partial x} \, dt \leq (\lambda M - \lambda^2 M_1 \delta_0)(\Lambda' - \tau'), \text{ if } \Lambda' > \tau'.$$

Clearly (4.4.31) is also true if $\Lambda' = \tau'$.

It follows from (4.4.29) and (4.4.31) that

(4.4.32) $$E \exp(\lambda M - \lambda^2 M_1 \delta_0)(\Lambda' - \tau') \geq 1.$$

If $\Omega_0 = \{\omega \mid \Lambda' > \tau'\}$, then (4.4.32) can be rewritten as

$$P(\Omega - \Omega_0) + \int_{\Omega_0} \exp(\Lambda' - \tau')(\lambda M - \lambda^2 M_1 \delta_0) \, dP(\omega) \geq 1.$$

Letting $\lambda \to +\infty$, we obtain

$$P(\Omega - \Omega_0) = 1$$

which proves that $\tau' = \Lambda'$, $\tilde{\mu}_x$ a.s.

But $\tau_{\delta_0} > \tau \geq \tau_\Lambda T$, hence also

$$T \wedge \tau = T \wedge \Lambda,$$

i.e. (4.4.25) and the proof of the Lemma is complete. ∎

5. Analytic approach to the problem (4.4.3)

The problem (4.4.3) has not be considered in Chapter 1 (aside form the case when $b_i = \partial a_{ij}/\partial x_j$). We will show in this paragraph that the method of asymptotic expansions as well as the energy method can be carried over to this case, with some adequate changes.

5.1. The method of asymptotic expansions. We write (4.4.3) as

$$A^\varepsilon u_\varepsilon = f, \qquad u_\varepsilon\mid_\Gamma = 0.$$

We set (as in (2.2.2), (2.2.3) of Chapter 1)

(5.1.1) $$A^\varepsilon = \varepsilon^{-2} A_1 + \varepsilon^{-1} A_2 + \varepsilon^0 A_3,$$

where

(5.1.2)
$$A_1 = -a_{ij}(y)\frac{\partial^2}{\partial y_i \partial y_j} - b_i(y)\frac{\partial}{\partial y_i},$$

$$A_2 = -a_{ij}(y)\frac{\partial^2}{\partial y_i \partial x_j} - a_{ij}(y)\frac{\partial^2}{\partial x_i \partial y_j} - b_i(y)\frac{\partial}{\partial x_i} - c_i(y)\frac{\partial}{\partial y_i},$$

$$A_3 = -a_{ij}(y)\frac{\partial^2}{\partial x_i \partial x_j} - c_i(y)\frac{\partial}{\partial x_i} + a_0.$$

We define

(5.1.3) $$\tilde{u}_\varepsilon(x) = u_0\left(x, \frac{x}{\varepsilon}\right) + \varepsilon u_1\left(x, \frac{x}{\varepsilon}\right) + \varepsilon^2 u_2\left(x, \frac{x}{\varepsilon}\right).$$

The functions u_0, u_1, u_2 are chosen to satisfy if possible the conditions

(5.1.4) $$A_1 u_0 = 0,$$

(5.1.5) $$A_1 u_1 + A_2 u_0 = 0,$$

(5.1.6) $$A_1 u_2 + A_2 u_1 + A_3 u_0 = f.$$

From Theorem 3.5[14] we can deduce that

(5.1.7) $$u_0(x, y) \equiv u(x).$$

Then (5.1.5) becomes

(5.1.8) $$A_1 u_1 - b_i(y)\frac{\partial u}{\partial x_i} = 0.$$

Since $b_i(y)$ satisfies the solvability condition $\left(\int_Y b_i(y)m(y)\,dy = 0\right)$, and recalling the definition of the functions χ^ℓ (cf. (4.3.9)), we can assert that

(5.1.9) $$u_1(x, y) = -\frac{\partial u}{\partial x_i}\chi^i(y) + \tilde{u}_1(x).$$

---

[14] A_1 is the operator denoted by A in Theorem 3.5.

We now turn to (5.1.6), which yields

(5.1.10)
$$A_1 u_2 + a_{ij}\frac{\partial^2 u}{\partial x_\ell \partial x_j}\frac{\partial \chi^\ell}{\partial y_i} + a_{ij}\frac{\partial^2 u}{\partial x_\ell \partial x_i}\frac{\partial \chi^\ell}{\partial y_j} + b_i \chi^\ell \frac{\partial^2 u}{\partial x_\ell \partial x_i} + c_i \frac{\partial u}{\partial x_\ell}\frac{\partial \chi^\ell}{\partial y_i} - b_i \frac{\partial \tilde{u}_1}{\partial x_i}$$
$$- a_{ij}\frac{\partial^2 u}{\partial x_i \partial x_j} - c_i \frac{\partial u x_i}{\partial +} a_0 u = f.$$

Before we write the solvability condition for (5.1.10), we notice that we can rewrite (5.1.10) as follows

(5.1.11) $A_1 u_2 - \dfrac{\partial^2 u}{\partial \partial x_i \partial x_j}\left\{ a_{ij} - a_{kj}\dfrac{\partial \chi^i}{\partial y_k} - a_{ki}\dfrac{\partial \chi^j}{\partial y_k} - \dfrac{1}{2}(b_i \chi^j + b_j \chi^i)\right\}$
$$- \frac{\partial u}{\partial x_i}\left(c_i - c_k \frac{\partial \chi^i}{\partial y_k}\right) - b_i \frac{\partial \tilde{u}_1}{\partial x_i} + a_0 u = f.$$

Writing the solvability condition (3.3.5) we obtain

(5.1.12)
$$-q_{ij}\frac{\partial^2 u}{\partial x_i \partial x_j} - \bar{c}_i \frac{\partial u}{\partial x_i} + \bar{a}_0 u = f,$$

where we have set

(5.1.13)
$$q_{ij} = \int_Y m(y)\left[a_{ij} - a_{kj}\frac{\partial \chi^i}{\partial y_k} - a_{ki}\frac{\partial \chi^j}{\partial y_k} - \frac{1}{2}(b_i \chi^j + b_j \chi^i)\right]\,dy,$$

(5.1.14)
$$\bar{c}_i = \int_Y m(y)\left(c_i - c_k \frac{\partial \chi^i}{\partial y_k}\right)\,dy.$$

Clearly (5.1.14) and (4.3.19) are identical. We need to show that (5.1.13) and (4.3.10) are identical, as it should be (although it is not completely obvious).

We use the equation for χ^j which is

$$-a_{k\ell}\frac{\partial^2 \chi^j}{\partial y_k \partial y_\ell} - b_k \frac{\partial \chi^j}{\partial y_k} = -b_j.$$

Multiplying by $m\chi^i$ and integrating over Y, we obtain

$$\int_Y b_j \chi^i m\,dy = \int_Y \left(a_{k\ell}\frac{\partial^2 \chi^j}{\partial y_k \partial y_\ell} + b_k \frac{\partial \chi^j}{\partial y_k}\right)\chi^i m\,dy$$
$$= \int_Y \chi^j \frac{\partial^2}{\partial y_k \partial y_\ell}(a_{k\ell}\chi^i m)\,dy - \int_Y \chi^j \frac{\partial}{\partial y_k}(b_k \chi^i m)\,dy$$

and by virtue of the equation satisfied by m
(5.1.15)
$$\int_Y b_j \chi^i m\,dy = \int_Y \chi^j a_{k\ell}m\frac{\partial^2 \chi^i}{\partial y_k \partial y_\ell}\,dy + 2\int_Y \chi^j \frac{\partial \chi^i}{\partial y_\ell}\frac{\partial}{\partial y_k}(a_{k\ell}m)\,dy - \int_Y \chi^j \frac{\partial \chi^i}{\partial y_k}b_k m\,dy.$$

But

(5.1.16)
$$\int_Y b_i \chi^j m\,dy = \int_Y \left(a_{k\ell}\frac{\partial^2 \chi}{\partial y_k \partial y_\ell} + b_k \frac{\partial \chi^i}{\partial y_k}\right)\chi^j m\,dy.$$

By addition of (5.1.15) and (5.1.16) we obtain

(5.1.17)
$$\frac{1}{2}\int_Y b_j\chi^i m\,dy + \frac{1}{2}\int_Y b_i\chi^j m\,dy = \int_Y \chi^k a_{k\ell}m\frac{\partial^2\chi^i}{\partial y_k\partial y_\ell}\,dy + \int_Y \chi^i\frac{\partial\chi^i}{\partial y_\ell}\frac{\partial}{\partial y_k}(a_{k\ell}m)\,dy$$
$$= \int_Y \chi^i\frac{\partial}{\partial y_k}\left(a_{k\ell}m\frac{\partial\chi}{\partial y_\ell}\right)\,dy$$
$$= -\int_Y \frac{\partial\chi^j}{\partial y_k}\chi\frac{\partial\chi^i}{\partial y_\ell}a_{k\ell}m\,dy.$$

Using the expression in (5.1.13) we obtain

$$q_{ij} = \int_Y m\left[a_{ij} + a_{k\ell}\frac{\partial\chi^j}{\partial y_k}\frac{\partial\chi^i}{\partial y_\ell} - a_{kj}\frac{\partial\chi^i}{\partial y_k} - a_{ki}\frac{\partial\chi^j}{\partial y_k}\right]\,dy,$$

and it is easy then to check that this last expression is the same as (4.3.19).

We turn back to the solution of (5.1.11). We take $\tilde{u}_1 = 0$, and we can take

(5.1.18)
$$u_2(x,y) = \frac{\partial^2 u}{\partial x_i\partial x_j}\chi^{ij}(y) - \frac{\partial u}{\partial x_i}\psi^i(y) + u\chi^0(y)$$

where $\chi^{ij}, \psi^i, \chi^0$ are defined by

(5.1.19) $A_1\chi^{ij} = a_{ij} - a_{kj}\dfrac{\partial\chi^i}{\partial y_k} - a_{ki}\dfrac{\partial\chi^j}{\partial y_k} - \dfrac{1}{2}(b_i\chi^j + b_j\chi^i) - q_{ij}$, χ^{ij} periodic,

(5.1.20) $A_1\psi^i = -\left(c_i - c_k\dfrac{\partial\chi^i}{\partial y_k}\right) + \bar{c}_i$, ψ^i periodic,

(5.1.21) $A_1\chi^0 = -(a_0 - \bar{a}_0)$, χ^0 periodic.

With the preceding choice, we can check that

(5.1.22) $A^\varepsilon\tilde{u}_\varepsilon = f + \varepsilon A_2 u_2 + \varepsilon A_3 u_1 + \varepsilon^2 A_3 u_2 = f + \varepsilon g_\varepsilon,$

where

$$(5.1.23) \qquad g_\varepsilon = -a_{ij}\frac{\partial}{\partial y_i}\chi^{k\ell}\frac{\partial^3 u}{\partial x_j \partial x_j \partial x_\ell} - a_{ij}\frac{\partial}{\partial y_j}\chi^{k\ell}\frac{\partial^3 u}{\partial x_i \partial x_j \partial x_\ell}$$

$$- b_i\chi^{k\ell}\frac{\partial^3 u}{\partial x_i \partial x_k \partial x_\ell} - c_i\frac{\partial}{\partial y_i}\chi^{k\ell}\frac{\partial^2 u}{\partial x_k \partial x_\ell} + a_{ij}\frac{\partial \psi^\ell}{\partial y_k}\frac{\partial^2 u}{\partial x_\ell \partial x_j}$$

$$+ a_{ij}\frac{\partial \psi^\ell}{\partial y_j}\frac{\partial^2 u}{\partial x_i \partial x_\ell} + b_i\psi^\ell\frac{\partial^2 u}{\partial x_i \partial x_\ell} + c_i\frac{\partial \psi^\ell}{\partial y_i}\frac{\partial u}{\partial x_\ell}$$

$$- a_{ij}\frac{\partial u}{\partial x_j}\frac{\partial \chi^0}{\partial y_i} - a_{ij}\frac{\partial u}{\partial x_i}\frac{\partial \chi^0}{\partial y_j} - b_i\frac{\partial u}{\partial x_i}\chi^0 - c_i u\frac{\partial \chi^0}{\partial y_i}$$

$$+ a_{ij}\chi^\ell\frac{\partial^3 u}{\partial x_i \partial x_j \partial x_\ell} + c_i\frac{\partial^2 u}{\partial x_i \partial x_\ell}\chi^\ell - a_0\frac{\partial u}{\partial x_\ell}\chi^\ell$$

$$- \varepsilon a_{ij}\chi^{k\ell}\frac{\partial^4 u}{\partial x_i \partial x_j \partial x_k \partial x_\ell} - \varepsilon c_i\chi^{k\ell}\frac{\partial^3 u}{\partial x_i \partial x_k \partial x_\ell}$$

$$+ a_0\chi^{k\ell}\frac{\partial^2 u}{\partial x_k \partial x_\ell} + \varepsilon a_{ij}\psi^\ell\frac{\partial^3 u}{\partial x_i \partial x_j \partial x_\ell} + \varepsilon c_i\psi^\ell\frac{\partial^2 u}{\partial x_i \partial x_\ell}$$

$$- \varepsilon a_0\frac{\partial u}{\partial x_\ell}\psi^\ell - \varepsilon a_{ij}\chi^0\frac{\partial^2 u}{\partial x_i \partial x_j} - \varepsilon c_i\frac{\partial u}{\partial x_i}\chi^0 + \varepsilon a_0 u\chi^0.$$

If f is in $W^{3,p}(\mathcal{O})$, $p > n$, then the derivatives of u (solution of (5.1.12) with $u\mid_\Gamma = 0$) up to the order 4 will be bounded, and thus we may assert that

$$(5.1.24) \qquad \|g_\varepsilon\|_{L^\infty(\mathcal{O})} \leq C.$$

Therefore if we set

$$z_\varepsilon = u_\varepsilon - \tilde{u}_\varepsilon$$

we see that

$$A^\varepsilon z_\varepsilon = \varepsilon g_\varepsilon, \qquad z_\varepsilon\mid_\Gamma = -\varepsilon u_1 - \varepsilon^2 u_2\mid_\Gamma .$$

Therefore

$$\|z_\varepsilon\|_{L^\infty} \leq C\varepsilon.$$

We can then state the

THEOREM 5.1. *Under the assumptions of Theorem 4.12, and if $f \in W^{3,p}(\mathcal{O})$, $p > n$, we have*

$$(5.1.25) \qquad \|u_\varepsilon - u\|_{L^\infty(\mathcal{O})} \leq C\varepsilon.$$

5.2. The method of energy. It is based upon obtaining a priori estimates and "adjoint" expansion (cf. Chapter 1, Section 3.3). We note that since we have assumed that $a_0 \geq \alpha_0 > 0$, the maximum principle gives an estimate, namely that

$$\|u_\varepsilon\|_{L^\infty} \leq \frac{1}{\alpha_0}\|f\|_{L^\infty}.$$

However we are not going to use here this estimate, since we assume that

$$(5.2.1) \qquad a_0 \geq 0, \qquad a_0 \text{ bounded and periodic.}$$

Also we will weaken the assumptions on f to

$$(5.2.2) \qquad f \in L^2(\mathcal{O}).$$

On the coefficients a_{ij}, b_i, c_i we will keep the same assuptions as in Theorem 4.10. This guarantees the existence of functions $m^\varepsilon(y)$, $m(y)$ solutions of

(5.2.3)
$$-\frac{\partial^2}{\partial y_i \partial y_j}(a_{ij}(y)m^\varepsilon) + \frac{\partial}{\partial y_i}(b_i(y)m^\varepsilon) + \varepsilon\frac{\partial}{\partial y_i}(c_i(y)m^\varepsilon) = 0,$$

$$m^\varepsilon \in W^{2,p,\mu}(\mathbb{R}^n), \qquad p \geq 2, \ \mu > 0,$$

$$m^\varepsilon \text{ periodic}, \qquad \int_Y m^\varepsilon(y)\, dy = 1.$$

(5.2.4)
$$-\frac{\partial^2}{\partial y_i \partial y_j}(a_{ij}(y)m) + \frac{\partial}{\partial y_i}(b_i(y)m) = 0,$$

$$m \in W^{2,p,\mu}(\mathbb{R}^n), \qquad p \geq 2, \ \mu > 0,$$

$$m \text{ periodic}, \qquad \int_Y m(y)\, dy = 1.$$

Moreover we know that

(5.2.5)
$$m^\varepsilon(y) > 0, \qquad m(y) > 0,$$

$$m^\varepsilon(y) \to m(y) \text{ in } C^0(\bar{Y}).$$

This follows from Theorem 3.5, Remark 3.8, and the proof of Theorem 4.6. In particular we can assert that

(5.2.6)
$$M \geq m^\varepsilon(y) \geq \gamma > 0, \ \forall y.$$

We then multiply the equation $(4.4.3)$[15] by $m^\varepsilon(y)$, and we set

(5.2.7)
$$\tilde{a}_{ij}^\varepsilon = a_{ij}m^\varepsilon, \qquad \tilde{b}_i^\varepsilon = b_i m^\varepsilon, \qquad \tilde{c}_i^\varepsilon = c_i m^\varepsilon,$$

$$\tilde{a}_9^\varepsilon = a_0 m^\varepsilon, \qquad \tilde{f}^\varepsilon = f m^\varepsilon.$$

We obtain

$$-\tilde{a}_{ij}^\varepsilon \frac{\partial^2 u_\varepsilon}{\partial x_i \partial x_j} - \frac{1}{\varepsilon}\tilde{b}_i^\varepsilon \frac{\partial u_\varepsilon}{\partial x_i} - c_i^\varepsilon \frac{\partial u_\varepsilon}{\partial x_i} + \tilde{a}_0^\varepsilon u_\varepsilon = \tilde{f}^\varepsilon.$$

We now set

(5.2.8)
$$\tilde{\beta}_i^\varepsilon(y) = \tilde{b}_i^\varepsilon(y) + \varepsilon\tilde{c}_i^\varepsilon(y) - \frac{\partial}{\partial y_j}\tilde{a}_{ij}^\varepsilon(y),$$

so that u_ε is a solution of

(5.2.9)
$$-\frac{\partial}{\partial x_i}\tilde{a}_{ij}^\varepsilon \frac{\partial u_\varepsilon}{\partial x_j} - \frac{1}{\varepsilon}\tilde{\beta}_i^\varepsilon\left(\frac{x}{\varepsilon}\right)\frac{\partial u_\varepsilon}{\partial x_i} + \tilde{a}_0^\varepsilon u_\varepsilon = \tilde{f}^\varepsilon, \qquad u_\varepsilon \mid_\Gamma = 0.$$

We note that from (5.2.3) we have

(5.2.10)
$$\sum_i \frac{\partial}{\partial y_i}\tilde{\beta}_i^\varepsilon = 0.$$

Thus multiplying (5.2.9) by u_ε and integrating over $\mathcal{O}$, we obtain, by virtue of (5.2.10)

$$\int_{\mathcal{O}} \tilde{a}_{ij}^\varepsilon \frac{\partial u_\varepsilon}{\partial x_j}\frac{\partial u_\varepsilon}{\partial x_i}\, dx + \int_{\mathcal{O}} \tilde{a}_0^\varepsilon(u_\varepsilon)^2\, dx = \int_{\mathcal{O}} \tilde{f}^\varepsilon u_\varepsilon\, dx.$$

From (5.2.6) it follows that

$$|u_\varepsilon|_{1,\mathcal{O}} \leq C,$$

---

[15]The solution of (4.4.3) is in $H^2(\mathcal{O}) \cap H_0^1(\mathcal{O})$ if $\mathcal{O}$ is regular.

and since $u_\varepsilon \in H^1(\mathcal{O})$, it follows from Poincaré's inequality, that

(5.2.11)
$$\|u_\varepsilon\|_{H_0^1(\mathcal{O})} \leq C.$$

We now set instead of (5.2.7)

(5.2.12)
$$\tilde{a}_{ij} = a_{ij}m, \qquad \tilde{b}_i = b_im, \qquad \tilde{c}_i = c_im,$$
$$\tilde{a}_0 = a_0m, \qquad f^\varepsilon = fm\left(\frac{x}{\varepsilon}\right).$$

We next set

(5.2.13)
$$\tilde{\beta}_i(y) = \tilde{b}_i(y) - \frac{\partial}{\partial y_j}\tilde{a}_{ij}(y).$$

With these definitions, and by doing a calculation similar to the one for (5.2.9), we obtain that u_ε is a solution of

(5.2.14)
$$-\frac{\partial}{\partial x_i}\tilde{a}_{ij}\left(\frac{x}{\varepsilon}\right)\frac{\partial u_\varepsilon}{\partial x_j} - \frac{1}{\varepsilon}\tilde{\beta}_i\left(\frac{x}{\varepsilon}\right)\frac{\partial u_\varepsilon}{\partial x_i} - \tilde{c}_i\left(\frac{x}{\varepsilon}\right)\frac{\partial u_\varepsilon}{\partial x_i} + \tilde{a}_0\left(\frac{x}{\varepsilon}\right)u_\varepsilon = f^\varepsilon, \qquad u_\varepsilon\mid_\Gamma = 0.$$

The variational form of (5.2.14) is then

(5.2.15)
$$\int \tilde{a}_{ij}\left(\frac{x}{\varepsilon}\right)\frac{\partial u_\varepsilon}{\partial x_j}\frac{\partial v}{\partial x_i}\,dx - \int \left(\frac{1}{\varepsilon}\tilde{\beta}_i\left(\frac{x}{\varepsilon}\right) + \tilde{c}_i\left(\frac{x}{\varepsilon}\right)\right)\frac{\partial u_\varepsilon}{\partial x_i}v\,dx$$
$$+ \int \tilde{a}_0\left(\frac{x}{\varepsilon}\right)u_\varepsilon v\,dx = \int f^\varepsilon v\,dx, \qquad \forall v \in H_0^1(\mathcal{O}).$$

We introduce next the function $\chi(y)$ solution of

(5.2.16)
$$-\frac{\partial}{\partial y_i}\tilde{a}_{ij}(y)\frac{\partial \chi}{\partial y_i} + \tilde{\beta}_i(y)\frac{\partial \chi}{\partial y_i} = -\frac{\partial \tilde{c}_i}{\partial y_i},$$
$$\chi \in W^{2,p,\mu}(\mathbb{R}^n), \qquad p \geq 2, \ \mu > 0, \ \chi \text{ periodic},$$
$$\int_Y \chi\,dy = 0.$$

The problem (5.2.16) has one and only one solution, by virtue of Theorem 3.5, applied to the operator

$$A = -\frac{\partial}{\partial y_i}\tilde{a}_{ij}\frac{\partial}{\partial y_j} - \tilde{\beta}_i\frac{\partial}{\partial y_i},$$

and noting that the corresponding $m = 1$, since

$$A^\varepsilon = -ddxy_i\tilde{a}_{ij}\frac{\partial}{\partial y_j} + \tilde{\beta}_i\frac{\partial}{\partial y_i}$$

by virtue of the fact that $\frac{\partial \tilde{\beta}_i}{\partial y_i} = 0$.

For ε small enough, the problem (5.2.15) is equivalent to the one with v changed into $v\left(1 + \varepsilon\chi\left(\frac{x}{\varepsilon}\right)\right)$. Hence we have

(5.2.17)
$$\int \tilde{a}_{ij}\left(\frac{x}{\varepsilon}\right)\frac{\partial u_\varepsilon}{\partial x_j}\frac{\partial v}{\partial x_i}(1 + \varepsilon\chi)\,dx + \int \tilde{a}_{ij}\frac{\partial u_\varepsilon}{\partial x_j}v\frac{\partial \chi}{\partial y_i}\,dx$$
$$+ \int u_\varepsilon\left\{\left(\frac{1}{\varepsilon}\tilde{\beta}_i + \tilde{c}_i\right)(1 + \varepsilon\chi)\frac{\partial v}{\partial x_i} + v\left(\frac{1}{\varepsilon}\tilde{\beta}_i + \tilde{c}_i\right)\frac{\partial \chi}{\partial y_i} + v(1 + \varepsilon\chi)\frac{1}{\varepsilon}\frac{\tilde{c}_i}{y_i}\right\}$$
$$+ \int \tilde{a}_0 u_\varepsilon(1 + \varepsilon\chi)v = \int f^\varepsilon(1 + \varepsilon\chi)v.^{16}$$

Next writing (5.2.16) at point $\frac{x}{\varepsilon}$ and recalling that $\frac{\partial}{\partial x_i} = \frac{1}{\varepsilon}\frac{\partial}{\partial y_i}$, we have

$$-\varepsilon\frac{\partial}{\partial x_i}\left(\tilde{a}_{ij}\left(\frac{x}{\varepsilon}\right)\frac{\partial\chi}{\partial y_j}\left(\frac{x}{\varepsilon}\right)\right) + \tilde{\beta}_i\left(\frac{x}{\varepsilon}\right)\frac{\partial\chi}{\partial y_i}\left(\frac{x}{\varepsilon}\right) = -\frac{\partial\tilde{c}_i}{\partial y_i}\left(\frac{x}{\varepsilon}\right).$$

We multiply by $u_\varepsilon v$ and integrate over $\mathcal{O}$. We obtain, using integrating by parts

(5.2.18)
$$\frac{1}{\varepsilon}\int_{\mathcal{O}} u_\varepsilon v\frac{\partial\tilde{c}_i}{\partial y_i}\,dx + \frac{1}{\varepsilon}\int_{\mathcal{O}} u_\varepsilon v\tilde{\beta}_i\frac{\partial\chi}{\partial y_i}\,dx + \int_{\mathcal{O}}\tilde{a}_{ij}\frac{\partial\chi}{\partial y_j}\frac{\partial u_\varepsilon}{\partial x_i}v\,dx = -\int_{\mathcal{O}}\tilde{a}_{ij}\frac{\partial\chi}{\partial y_j}u_\varepsilon\frac{\partial v}{\partial x_i}\,dx.$$

Combining (5.2.17) and (5.2.18), we obtain

(5.2.19) $$\int_{\mathcal{O}}\tilde{a}_{ij}\left(\frac{x}{\varepsilon}\right)\frac{\partial u_\varepsilon}{\partial x_j}\frac{\partial v}{\partial x_i}(1+\varepsilon\chi)\,dx$$

$$+ \int_{\mathcal{O}} u_\varepsilon\frac{\partial v}{\partial x_i}\left[\left(\frac{1}{\varepsilon}\tilde{\beta}_i + \tilde{c}_i\right)(1+\varepsilon\chi) - \tilde{a}_{ij}\frac{\partial\chi}{\partial y_j}\right]dx$$

$$+ \int_{\mathcal{O}} u_\varepsilon v\left(\frac{\partial(\tilde{c}_i\chi)}{\partial y_i} + \tilde{a}_0(1+\varepsilon\chi)\right)dx = \int_{\mathcal{O}} f^\varepsilon(1+\varepsilon\chi)v\,dx, \qquad \forall v \in H_0^1(\mathcal{O}).$$

Let now $\phi \in C_0^\infty(\mathcal{O})$ and let us take in (5.2.19)

(5.2.20) $$v \equiv v^\varepsilon \equiv \phi(x) + \varepsilon\phi_1\left(x, \frac{x}{\varepsilon}\right).$$

The function $\phi_1(x, y)$ will be chosen below. Let us denote by

$$A^{\varepsilon*} = -\frac{\partial}{\partial x_i}\tilde{a}_{ij}\left(\frac{x}{\varepsilon}\right)\frac{\partial}{\partial x_j}$$

$$+ \left[\frac{1}{\varepsilon}\tilde{\beta}_i\left(\frac{x}{\varepsilon}\right) + \tilde{c}_i\left(\frac{x}{\varepsilon}\right) + \tilde{\beta}_i\chi - \tilde{a}_{ij}\left(\frac{x}{\varepsilon}\right)\frac{\partial\chi}{\partial y_j}\left(\frac{x}{\varepsilon}\right)\right]\frac{\partial}{\partial x_i}$$

$$= (\varepsilon^{-2}\tilde{A}_1^* + \varepsilon^{-1}\tilde{A}_2^\varepsilon + \varepsilon^0\tilde{A}_3^\varepsilon)|_{y=x/\varepsilon},$$

where

(5.2.21)
$$A_1^* = -\frac{\partial}{\partial y_i}\tilde{a}_{ij}(y)\frac{\partial}{\partial y_j} + \tilde{\beta}_i(y)\frac{\partial}{\partial y_i},$$

$$A_2^* = -\frac{\partial}{\partial x_i}\tilde{a}_{ij}(y)\frac{\partial}{\partial y_j} - \frac{\partial}{\partial y_i}\tilde{a}_{ij}(y)\frac{\partial}{\partial x_j} + \tilde{\beta}_i(y)\frac{\partial}{\partial x_i}$$

$$+ \left(\tilde{c}_i(y) - \tilde{a}_{ij}(y)\frac{\partial\chi}{\partial y_j}(y)\right)\frac{\partial}{\partial y_i} + \tilde{\beta}_i\chi\frac{\partial}{\partial y_i},$$

$$A_3^* = -\tilde{a}_{ij}(y)\frac{\partial}{\partial x_i\partial x_j} + \left(\tilde{c}_i(y) - \tilde{a}_{ij}(y)\frac{\partial\chi}{\partial y_j}(y)\right)\frac{\partial}{\partial x_i} + \tilde{\beta}_i\chi\frac{\partial}{\partial x_i}.$$

The function ϕ_1 will be chosen to satisfy

$$A_1^*\phi_1 + \left(-\frac{\partial}{\partial y_j}\tilde{a}_{ji}(y) + \tilde{\beta}_i(y)\right)\frac{\partial\phi}{\partial x_i} = 0,$$

i.e.,

(5.2.22) $$\phi_1(x, y) = -\frac{\partial\phi}{\partial x_i}(x)\tilde{\chi}^i(y),$$

---

[16]We recall that in (5.2.17) the argument of periodic functions is x/ε.

where the $\tilde{\chi}^\ell$ functions are defined by

$$(5.2.23) \qquad \tilde{A}_1^* \tilde{\chi}^\ell = -\frac{\partial}{\partial y_j} \tilde{a}_{j\ell}(y) + \tilde{\beta}_\ell(y), \qquad \tilde{\chi}^\ell \text{ periodic},$$

$$\tilde{\chi}^\ell \in W^{2,p,\mu}(\mathbb{R}^n), \qquad p \geq 2, \ \mu > 0.$$

Such functions exist, since by the assumptions on b_i, the integral of the right hand side of (5.2.23) over Y is 0. We have

$$(5.2.24) \qquad \tilde{A}^{\varepsilon *} v^\varepsilon = +\frac{\partial^2 \phi}{\partial x_i \partial x_\ell} \tilde{a}_{ij} \frac{\partial \tilde{\chi}^\ell}{\partial y_j} + \frac{\partial^2 \phi}{\partial x_j \partial x_\ell} \frac{\partial}{\partial y_i} \tilde{a}_{ij} \tilde{\chi}^\ell - \frac{\partial^2 \phi}{\partial x_i \partial x_\ell} \tilde{\beta}_i \tilde{\chi}^\ell$$
$$- \left(\tilde{c}_i - \tilde{a}_{ij} \frac{\partial \chi}{\partial y_j} \right) \frac{\partial \phi}{\partial x_i} + \varepsilon \tilde{a}_{ij} \tilde{\chi}^\ell \frac{\partial^\ell \phi}{\partial x_i \partial x_j \partial x_\ell} + \tilde{\beta}_i \chi \frac{\partial \phi}{\partial x_i}$$
$$- \varepsilon \left(\tilde{c}_i - \tilde{a}_{ij} \frac{\partial \chi}{\partial y_j} \right) \tilde{\chi}^\ell \frac{\partial^2 \phi}{\partial x_i \partial x_\ell} - \varepsilon \tilde{\beta}_i \chi \frac{\partial^2 \phi}{\partial x_i \partial x_\ell} \tilde{\chi}^\ell.$$

We also note that (5.2.19) can be rewritten as

$$(5.2.25) \qquad \int_{\mathcal{O}} u_\varepsilon \tilde{A}^{\varepsilon *} v^\varepsilon \, dx + \varepsilon \int_{\mathcal{O}} \tilde{a}_{ij} \frac{\partial u_\varepsilon}{\partial x_j} \frac{\partial v^\varepsilon}{\partial x_i} \chi \, dx + \int_{\mathcal{O}} u_\varepsilon \frac{\partial v_\varepsilon}{\partial x_i} (\varepsilon \tilde{c}_i) \chi \, dx$$
$$+ \int_{\mathcal{O}} u_\varepsilon v^\varepsilon \left(\frac{\partial (\tilde{c}_i \chi)}{\partial y_i} + a_0 (1 + \varepsilon \chi) \right) dx = \int_{\mathcal{O}} f^\varepsilon (1 + \varepsilon \chi) v^\varepsilon.$$

Since $\|u_\varepsilon\|_{H_0^1} \leq C$, we can extract a subsequence converging weakly toward u satisfying

$$(5.2.26)$$

$$\int_{\mathcal{O}} u \left[+\frac{\partial^2 \phi}{\partial x_i \partial x_\ell} \overline{\tilde{a}_{ij} \frac{\partial \tilde{\chi}^\ell}{\partial y_j}} - \frac{\partial^2 \phi}{\partial x_i \partial x_\ell} \overline{\tilde{\beta}_i \tilde{\chi}^\ell} \right.$$
$$- \overline{\left(\tilde{c}_i - \tilde{a}_{ij} \frac{\partial \chi}{\partial y_j} \right) \frac{\partial \tilde{\chi}^\ell}{\partial y_i}} \frac{\partial \phi}{\partial x_\ell} - \overline{\tilde{a}_{ij}} \frac{\partial^2 \phi}{\partial x_i \partial x_j} + \overline{\left(\tilde{c}_i - \tilde{a}_{ij} \frac{\partial \chi}{\partial y_j} \right)} \frac{\partial \phi}{\partial x_i} \right] dx$$
$$+ \int_{\mathcal{O}} u \frac{\partial \phi}{\partial x_i} \overline{\tilde{\beta}_i \chi} \, dx - \int_{\mathcal{O}} \frac{\partial \phi}{\partial x_\ell} \overline{\tilde{\beta}_i \chi \frac{\partial \tilde{\chi}^\ell}{\partial y_i}} \, dx + \int_{\mathcal{O}} u \overline{\phi} \tilde{a}_0 \, dx = \int_{\mathcal{O}} f \phi \, dx. {}^{[17]}$$

We set

$$(5.2.27) \qquad q_{ij} = \int_Y m(y) \left[a_{ij}(y) - \frac{1}{2} a_{ik} \frac{\partial \tilde{\chi}^j}{\partial y_k} + \frac{1}{2} \tilde{\chi}^j \left(b_i - \frac{1}{m} \frac{\partial}{\partial y_m} (a_{ikm}) \right) \right] dy$$
$$+ \int_Y m(y) \left[-\frac{1}{2} a_{jk} \frac{\partial \tilde{\chi}^i}{\partial y_k} + \frac{1}{2} \tilde{\chi}^i \left(b_j - \frac{1}{m} \frac{\partial}{\partial y_k} (a_{jkm}) \right) \right] dy$$

which is in (5.2.26) the coefficient (in symmetric form) of $-\int_{\mathcal{O}} u \frac{\partial^2 \phi}{\partial x_j \partial x_j} \, dx$. Let us identify the expression (5.2.27) with the expression obtained in (5.1.13). First of all, we remark that (5.2.27) simplifies as follow

$$(5.2.28) \qquad q_{ij} = \int_Y m(y) \left[a_{ij}(y) + \frac{1}{2} \tilde{\chi}^j b_i + \frac{1}{2} \tilde{\chi}^i b_j \right] dy.$$

---

[17] In (5.2.26), $\tilde{h} = \int_Y h(y) \, dy$.

We note that $\tilde{\chi}^\ell$ is a solution of (cf. (5.2.23))

$$(5.2.29) \qquad -\frac{\partial}{\partial y_i \partial y_j}(a_{ij} m \tilde{\chi}^\ell) + \frac{\partial}{\partial y_i}(b_i m \tilde{\chi}^\ell) = -2\frac{\partial}{\partial y_j}(a_{j\ell} m) + b_\ell m,$$

where χ^k is a solution of

$$(5.2.30) \qquad -a_{ij}\frac{\partial^2 \chi^k}{\partial y_i \partial y_j} - b_i \frac{\partial \chi^k}{\partial y_i} = -b_k.$$

From (5.2.29) and (5.2.30), it easily follows that

$$-\int_Y b_k m \tilde{\chi}^\ell \, dy = \int_Y \chi^k \left[b_\ell m - 2\frac{\partial}{\partial y_j}(a_{j\ell} m) \right] dy.$$

Hence

$$q_{ij} = \int_Y m(y) \left[a_{ij}(y) - \frac{1}{2}\chi^i b_j - \frac{1}{2}\chi^j b_i + \frac{\chi^i}{m}\frac{\partial}{\partial y_k}(a_{kj} m) + \frac{\chi^j}{m}\frac{\partial}{\partial y_k}(a_{ki} m) \right] dy$$

and integrating by parts we recover the expression (5.1.13).

We next set

$$(5.2.31)$$

$$\bar{c}_\ell = \int_Y \left[\tilde{\beta}_\ell \chi - \tilde{a}_{\ell j}\frac{\partial \chi}{\partial y_j} + c_\ell \right] dy - \int_Y \left[\left(\tilde{c}_i - \tilde{a}_{ij}\frac{\partial \chi}{\partial y_j} \right)\frac{\partial \tilde{\chi}^\ell}{\partial y_i} + \tilde{\beta}_i \chi \frac{\partial \tilde{\chi}^\ell}{\partial y_i} \right] dy$$

which is the coefficient of $\int_{\mathcal{O}} u\frac{\partial \phi}{\partial x_\ell} \, dx$ in (5.2.26).

We must identify the expression (5.2.31) with (5.1.14). We first notice that

$$\tilde{\beta}_\ell \chi - \tilde{a}_{\ell j}\frac{\partial \chi}{\partial y_j} = \tilde{b}_\ell \chi - \frac{\partial}{\partial y_j}(\tilde{a}_{\ell j}\chi),$$

hence

$$(5.2.32)$$

$$\bar{c}_\ell = \int_Y (\tilde{c}_\ell + b_\ell \chi) \, dy - \int_Y \frac{\partial \tilde{\chi}^\ell}{\partial y_i} \left(\tilde{c}_i + \tilde{b}_i \chi - \frac{\partial}{\partial y_j}(\tilde{a}_{ij}\chi) \right) dy$$

$$= \int_Y \left(\tilde{c}_\ell + \tilde{b}_\ell \chi - \tilde{c}_i \frac{\partial \tilde{\chi}^\ell}{\partial y_i} \right) dy + \int_Y \tilde{\chi}^\ell \left(\tilde{b}_i \frac{\partial \chi}{\partial y_i} - 2\frac{\partial \tilde{a}_{ij}}{\partial y_i}\frac{\partial \chi}{\partial y_i} - \tilde{a}_{ij}\frac{\partial^2 \chi}{\partial y_i \partial y_j} \right) dy.$$

But from (5.2.16) we deduce that

$$-\tilde{a}_{ij}\frac{\partial^2 \chi}{\partial y_i \partial y_j} + \tilde{b}_i \frac{\partial \chi}{\partial y_i} - 2\frac{\partial \tilde{a}_{ij}}{\partial y_j}\frac{\partial \chi}{\partial y_i} = -\frac{\partial \tilde{c}_i}{\partial y_i}$$

hence from (5.2.32) we see that

$$(5.2.33) \qquad \bar{c}_\ell = \int_Y (\tilde{c}_\ell + \tilde{b}_\ell \chi) \, dy.$$

However an easy calculation shows that

$$(5.2.34) \qquad -\frac{\partial^2}{\partial y_i \partial y_j}(\tilde{a}_{ij}\chi) + \frac{\partial}{\partial y_i}(\tilde{b}_i \chi) = -\frac{\partial \tilde{c}_i}{\partial y_i}.$$

Since χ^ℓ is a solution of

$$-\tilde{a}_{ij}\frac{\partial^2 \chi^\ell}{\partial y_i \partial y_j} - \tilde{b}_i \frac{\partial \chi^\ell}{\partial y_i} = -\tilde{b}^\ell,$$

we see by integration by parts that

$$-\int_Y \tilde{b}^\ell \chi \, dy = -\int_Y \tilde{\chi}^\ell \frac{\partial \tilde{c}_i}{\partial y_i} \, dy = \int_Y \frac{\partial \chi^\ell}{\partial y_i} \tilde{c}_i \, dy$$

hence finally

$$\bar{c}_\ell = \int_Y \left(\tilde{c}_\ell - \frac{\partial \chi^\ell}{\partial y_i} \tilde{c}_i \right) \, dy,$$

which is nothing other than (5.1.14), since $\bar{c}_\ell = mc_\ell$. Therefore we can write (5.2.26) as follows

$$\int_\mathcal{O} \left[-q_{ij} \frac{\partial^2 \phi}{\partial x_i \partial x_j} + \bar{c}_i \frac{\partial \phi}{\partial x_i} + \bar{a}_0 \phi \right] \, dx = \int_\mathcal{O} f \phi \, dx,$$

and since ϕ is arbitrary in $C_0^\infty(\mathcal{O})$, we see that u is a solution of the homogenized equation.

We can summarize the results which have been obtained in the following

THEOREM 5.2. *We assume (5.2.1), (5.2.2) that $\mathcal{O}$ is regular and that the coefficients $a_{ij}(y)$, $b_i(y)$, $c_i(y)$ satisfy the assumptions of Theorem 4.10. Then the solution u of (4.4.3) converges in $H_0^1(\mathcal{O})$ weakly toward the solution of the homogenized equation (4.4.7).*

REMARK 5.3. The method of energy requires no regularity assumptions on f, and does not require that a_0 be bounded below. Like in the self adjoint case, it has therefore a considerable advantage with respect to the probabilistic method or with respect to the method of asymptotic expansions. However the type of convergence is different in the three methods.

REMARK 5.4. For operators considered here, none of the three methods seem to extend to boundary conditions other than Dirichlet. This shows the importance of having the problem in the framework of Chapters 1 and 2 (divergence form of the operator).

6. Operators with locally periodic[19] coefficients

6.1. Setting of the problem. We consider the analogue of problem (4.4.3) with coefficients depending on $x, \frac{x}{\varepsilon}$. We will consider functions $a_{ij}(x,y)$, $b_i(x,y)$, $c_i(x,y)$, $a_0(x,y)$. We assume that

(6.1.1)   for x fixed, a_{ij}, b_i, c_i satisfy as functions of y the assumptions
           (3.2.1), (3.2.2), (3.2.3) (c_i satisfying the same conditions as b_i).
           The a_{ij} are uniformly strictly elliptic;

(6.1.2) a_{ij}, b_i, c_i are continuous, bounded in x, y and for y fixed are
           twice continuously differentiable in x. All the partial derivatives
           with respect to x or y or mixed are globally bounded and continuous;

(6.1.3) a_0 is continuous in x, y, bounded and periodic in y, $a_0 \geq \alpha_0 > 0$.

Let $\mathcal{O}$ be a domain of $\mathbb{R}^n$ and f a function on $\mathcal{O}$, such that

(6.1.4) $\mathcal{O}$ satisfies (4.4.1); $f \in C^0(\overline{\mathcal{O}})$.

---

[19]Also referred to as nonuniformly oscillating coefficients

We are going to study the behavior as $\varepsilon \to 0$, of the solution u_ε of the Dirichlet problem

(6.1.5)
$$-a_{ij}\left(x,\frac{x}{\varepsilon}\right)\frac{\partial^2 u_\varepsilon}{\partial x_i \partial x_j} - \frac{1}{\varepsilon}b_i\left(x,\frac{x}{\varepsilon}\right)\frac{\partial u_\varepsilon}{\partial x_i} - c_i\left(x,\frac{x}{\varepsilon}\right)\frac{\partial u_\varepsilon}{\partial x_i} + a_0\left(x,\frac{x}{\varepsilon}\right)u = f(x),$$

$$u_\varepsilon\mid_\Gamma = 0, \qquad u_\varepsilon \in W^{2,p}(\mathcal{O}), \qquad p \geq 2.$$

We denote by A the differential operator (in y)

(6.1.6)
$$A = -a_{ij}(x,y)\frac{\partial^2}{\partial y_i \partial y_j} - b_i(x,y)\frac{\partial}{\partial y_i},$$

and by A^* its formal adjoint

(6.1.7)
$$A^* = -\frac{\partial}{\partial y_i}\left(a_{ij}(x,y)\frac{\partial}{\partial y_j}\right) + \frac{\partial}{\partial y_i}(b_i\cdot) - \frac{\partial}{\partial y_i}\left(\frac{\partial}{\partial y_j}a_{ij}\cdot\right)$$

$$= -\frac{\partial}{\partial y_i \partial y_j}(a_{ij}(x,y)\cdot) + \frac{\partial}{\partial y_i}(b_i\cdot).$$

Applying Theorem 3.5, there exists for any x fixed, one and only one function $m(x,y)$ such that

(6.1.8)
$$A^*m = 0, \qquad y \to m(x,y) \in W^{2,p,\mu}(\mathbb{R}^n), \qquad p \geq 2,\ \mu > 0,$$

$$\int_Y m(y)\,dy = 1, \qquad y \to m(x,y) \text{ periodic}, \qquad m(x,y) > 0.$$

The function $m(x,y)$ is globally continuous in x, y. Actually only the values for $x \in \overline{\mathcal{O}}$ will matter. Since $\mathcal{O}$ is bounded, we can without loss of generality assume that

(6.1.9)
$$M \geq m(x,y) \geq m_0 > 0.$$

As in the preceding paragraphs the vector b will satisfy

(6.1.10)
$$\int_Y m(x,y)b(y)\,dy = 0, \ \forall x.$$

6.2. Probabilistic approach. We define $\sigma(x,y)$ such that

(6.2.1)
$$\frac{1}{2}\sigma(x,y)\sigma^*(x,y) = a(x,y).$$

We construct on an arbitrary probability system $(\Omega, \mathcal{A}, P, \mathcal{F}^t, w(t))$ the Itô process $y^\varepsilon(t) = y_x^\varepsilon(t)$ solution of

(6.2.2)
$$dy^\varepsilon = \frac{1}{\varepsilon}b\left(y^\varepsilon,\frac{y^\varepsilon}{\varepsilon}\right)dt + c\left(y^\varepsilon,\frac{y^\varepsilon}{\varepsilon}\right)dt + \sigma\left(y^\varepsilon,\frac{y^\varepsilon}{\varepsilon}\right)dw(t),$$

$$y^\varepsilon(0) = x.$$

We denote by μ_x^ε the probability measure induced by $y_x^\varepsilon(\cdot)$ on $\Psi = C[0,\infty;\mathbb{R}^n]$.

LEMMA 6.1. *The family μ_x^ε as ε varies, remains in a weakly compact subset of $\mathcal{M}_+^1(\Psi)$.*

PROOF. We consider the functions $\chi^\ell(x, y)$ solutions of

$$(6.2.3) \qquad A\chi^\ell = -b_\ell(x, y), \qquad \chi^\ell \text{ periodic in } y, \qquad \int_Y \chi^\ell(x, y) \, dy = 0.$$

By the regularity on the coefficients we can assert that there exists one and only one solution of (6.2.3) which is $C^2(\mathbb{R}^n \times \mathbb{R}^n)$ and bounded. Applying Itô's formula, we obtain, (noting $\chi = (\chi^1, \dots, \chi^n)$),

$$\chi\left(y^\varepsilon(t), \frac{y^\varepsilon(t)}{\varepsilon}\right) = \chi\left(x, \frac{x}{\varepsilon}\right) + \int_0^t \left(\frac{\partial\chi}{\partial x} + \frac{1}{\varepsilon}\frac{\partial\chi}{\partial y}\right)\left(\frac{1}{\varepsilon}b + c\right)\left(y^\varepsilon, \frac{y^\varepsilon}{\varepsilon}\right) dx$$
$$+ \int_0^t \left(\frac{\partial\chi}{\partial x} + \frac{1}{\varepsilon}\frac{\partial\chi}{\partial y}\right)\sigma\left(y^\varepsilon, \frac{y^\varepsilon}{\varepsilon}\right) dw(s)$$
$$+ \int_0^t \operatorname{tr} a\left(\frac{\partial\chi}{\partial x^2} + \frac{1}{\varepsilon}\frac{\partial^2\chi}{\partial y\partial x} + \frac{1}{\varepsilon}\frac{\partial^2\chi}{\partial x\partial y} + \frac{1}{\varepsilon}\frac{\partial^2\chi}{\partial y^2}\right)\left(y^\varepsilon, \frac{y^\varepsilon}{\varepsilon}\right) ds.$$

But

$$\operatorname{tr}\left[a\frac{\partial^2\chi}{\partial y^2} + \frac{\partial\chi}{\partial y}b\right] = b,$$

hence

$$(6.2.4) \qquad \varepsilon\chi\left(y^\varepsilon(t), \frac{y^\varepsilon(t)}{\varepsilon}\right) = \varepsilon\chi\left(x, \frac{x}{\varepsilon}\right) + \int_0^t \frac{1}{\varepsilon}b\left(y^\varepsilon, \frac{y^\varepsilon}{\varepsilon}\right) ds$$
$$+ \int_0^t \left(\frac{\partial\chi}{\partial y}c + \frac{\partial\chi}{\partial x}b\right)\left(y^\varepsilon, \frac{y^\varepsilon}{\varepsilon}\right) ds$$
$$+ \int_0^t \left(\operatorname{tr} a\frac{\partial^2\chi}{\partial y\partial x} + \operatorname{tr} a\frac{\partial^2\chi}{\partial x\partial y}\right)\left(y^\varepsilon, \frac{y^\varepsilon}{\varepsilon}\right) ds$$
$$+ \varepsilon\int_0^t \left(\frac{\partial\chi}{\partial x}c + \operatorname{tr} a\frac{\partial^2\chi}{\partial x^2}\right)\left(y^\varepsilon, \frac{y^\varepsilon}{\varepsilon}\right) ds$$
$$+ \int_0^t \left(\varepsilon\frac{\partial\chi}{\partial x} + \frac{\partial\chi}{\partial y}\right)\sigma\left(y^\varepsilon, \frac{y^\varepsilon}{\varepsilon}\right) dw(s),$$

we eliminate the term $\int_0^t \frac{1}{\varepsilon}b$ between (6.2.4) and (6.2.2). One can then use a reasoning similar to the one in Theorem 4.9, to complete the proof. Actually one has

$$\sup_{|t_1-t_2|\leq\delta}|y^\varepsilon(t_2)-y^\varepsilon(t_1)| \leq c_1\varepsilon+c_2\delta+ \sup_{|t_1-t_2|\leq\delta}\left|\int_{t_1}^{t_2}\left(I - \varepsilon\frac{\partial\chi}{\partial x} - \frac{\partial\chi}{\partial y}\right)\sigma\left(y^\varepsilon, \frac{y^\varepsilon}{\varepsilon}\right) dw\right|,$$

$$P\left|\sup_{\substack{|t_1-t_2|\leq\delta \\ 0\leq t_1<t_2\leq T}}|y^\varepsilon(t_2)-y^\varepsilon(t_1)| \geq \eta\right| \leq \frac{c'T\delta}{(\eta - c_1\varepsilon - c_2\delta)}$$

hence the compactness property follows.   ■

LEMMA 6.2. *Let $\phi(x, y)$ be a function which is C^2 in x, C^1 in y, with all partial derivatives continuous and bounded globally in x, y. Suppose also that ϕ is periodic in y for fixed x, and satisfies*

$$(6.2.5) \qquad \int_Y \phi(x, y)m(x, y) \, dy = 0.$$

Then we have

$$(6.2.6) \qquad E\left[\left(\int_s^t \phi\left(y^\varepsilon(\lambda), \frac{y^\varepsilon(\lambda)}{\varepsilon}\right) d\lambda\right)^2 \,\Big|\, \mathcal{F}^s\right] \to 0.$$

PROOF. We can solve the equation

$$(6.2.7) \qquad A\psi = \phi, \qquad \psi \text{ periodic.}$$

From the regularity assumptions on the coefficients and ϕ, we can assert that ψ is $C^2(\mathbb{R}^n \times \mathbb{R}^n)$ and ψ bounded as well as its partial derivatives.

From Itô's formula we have

$$(6.2.8) \quad \psi\left(y^\varepsilon(t), \frac{y^\varepsilon(t)}{\varepsilon}\right) = \psi\left(y^\varepsilon(s), \frac{y^\varepsilon(s)}{\varepsilon}\right)$$
$$+ \int_s^t \left(\frac{\partial \psi}{\partial x} + \frac{1}{\varepsilon}\frac{\partial \psi}{\partial y}\right)\left(\frac{1}{\varepsilon}b + c\right)\left(y^\varepsilon, \frac{y^\varepsilon}{\varepsilon}\right) d\lambda + \int_s^t \left(\frac{\partial \psi}{\partial x} + \frac{1}{\varepsilon}\frac{\partial \psi}{\partial y}\right)\sigma\left(y^\varepsilon, \frac{y^\varepsilon}{\varepsilon}\right) d\lambda$$
$$+ \int_s^t \operatorname{tr} a\left(\frac{\partial^2 \psi}{\partial x^2} + \frac{2}{\varepsilon}\frac{\partial^2 \psi}{\partial x \partial y} + \frac{1}{\varepsilon^2}\frac{\partial^2 \psi}{\partial y^2}\right)\left(y^\varepsilon, \frac{y^\varepsilon}{\varepsilon}\right) d\lambda.$$

From (6.2.7) it follows that

$$\frac{\partial \psi}{\partial y} \cdot b + \operatorname{tr} a \frac{\partial^2 \psi}{\partial y^2} = -\phi.$$

Multiplying (6.2.8) by ε^2, and reasoning as in the proof of Lemma 4.3 (first proof), one easily obtains the desired result. ∎

REMARK 6.3. The second proof of Lemma 4.3 does not seem to extend easily to the present context. This explains why we assume more regularity on the function ϕ.

We now define the functions

$$(6.2.9)$$
$$q(x) = \int_Y \left(I - \frac{\partial \chi}{\partial y}\right)(x,y)a(x,y)\left(I - \frac{\partial \chi}{\partial y}\right)^*(x,y)m(x,y)\,dy,$$

$$(6.2.10)$$
$$r(x) = \int_Y \left[\left(I - \frac{\partial \chi}{\partial y}\right)(x,y)c(x,y) - \frac{\partial \chi}{\partial x}(x,y)b(x,y) - 2\operatorname{tr} a \frac{\partial^2 \chi}{\partial y \partial x}(x,y)\right]m(x,y)\,dy.$$

We consider the Itô process defined by

$$(6.2.11) \qquad dy = r(y)\,dt + \sqrt{2}\sqrt{q(y)}\,dw(t), \qquad y(0) = x,$$

and let μ_x be the probability measure on $C([0,\infty[;\mathbb{R}^n)$ defined by $y_x(\cdot)$. Then we have the

THEOREM 6.4. *Under assumptions (6.1.1), (6.1.2), (6.1.10) we have*

$$(6.2.12) \qquad \forall x, \ \mu_x^\varepsilon \to \mu_x \ \text{in } \mathcal{M}_+^1(\Psi) \text{ weakly.}$$

REMARK 6.5. The situation is more complex than in the situation of Section 4, since the limit μ_x is the measure of a diffusion process which is no longer Gaussian. Hence we cannot express easily its finite dimensional distributions. Here is where the martingale formulation of diffusions is quite useful (cf. Section 2).

PROOF OF THEOREM 6.4. We know that μ_x^ε remains in a weakly compact subset of $\mathcal{M}_+^1(\Psi)$. Let us set

$$(6.2.13) \qquad z^\varepsilon(t) = y^\varepsilon(t) - \varepsilon\chi\left(y^\varepsilon(t), \frac{y^\varepsilon(t)}{\varepsilon}\right).$$

It follows from (6.2.4) and (6.2.2) that we have

$$(6.2.14) \quad z^\varepsilon(t) = x - \varepsilon\chi\left(x, \frac{x}{\varepsilon}\right) + \int_0^t \left[\left(I - \frac{\partial\chi}{\partial y}\right)c - \frac{\partial\chi}{\partial z}b - \operatorname{tr} a\frac{\partial^2\chi}{\partial y\partial x}\right.$$
$$\left. - \operatorname{tr} a\frac{\partial^2\chi}{\partial x\partial y} - \varepsilon\frac{\partial\chi}{\partial x}c - \varepsilon\operatorname{tr} a\frac{\partial^2\chi}{\partial x^2}\right]\left(y^\varepsilon(s), \frac{\partial y^\varepsilon(s)}{\partial\varepsilon}\right) ds$$
$$+ \int_0^t \left(I - \frac{\partial\chi}{\partial y} - \varepsilon\frac{\partial\chi}{\partial x}\right)\sigma\left(y^\varepsilon(s), \frac{y^\varepsilon(s)}{\varepsilon}\right) dw(s).$$

We are going now to use the martingale formulation of diffusions. For this we consider the space $\Psi = C[0, \infty; \mathbb{R}^n]$ and the canonical process $x(t; \psi) = \psi(t)$. We set $\mathcal{G}^t = \sigma(x(s), 0 \le s \le t)$. The σ-algebra $\mathcal{G}^\infty = \mathcal{B} = \sigma$-algebra of Borel sets on Ψ.

From the theory of weak solutions of stochastic differential equations (see Section 2), we can construct a process $w^\varepsilon(t)$ adapted to $\mathcal{G}^t$, such that when $\Psi, \mathcal{B}$ is equipped with the probability μ_x^ε, then $w^\varepsilon(t)$ is a Wiener process with respect to $\mathcal{G}^t$, and

$$(6.2.15) \qquad dx(t) = \frac{1}{\varepsilon}b\left(x, \frac{x}{\varepsilon}\right) dt + \sigma\left(x, \frac{x}{\varepsilon}\right) dw^\varepsilon(t), \qquad \mu^\varepsilon(x(0) = x) = 1.$$

In that context, we set (analog of (6.2.13))

$$(6.2.16) \qquad \zeta^\varepsilon(t) = x(t) - \varepsilon\chi\left(x(t), \frac{x(t)}{\varepsilon}\right),$$

and it is clear that $\zeta^\varepsilon(t)$ will satisfy the relation (analog to (6.2.14)),

$$(6.2.17) \qquad \zeta^\varepsilon(t) = x - \varepsilon\chi\left(x, \frac{x}{\varepsilon}\right)$$
$$+ \int_0^t \left[\left(I - \frac{\partial\chi}{\partial y}\right)c - \frac{\partial\chi}{\partial x}b - \operatorname{tr} a\frac{\partial^2\chi}{\partial y\partial x} - \operatorname{tr} a\frac{\partial^2\chi}{\partial x\partial y}\right.$$
$$\left. - \varepsilon\frac{\partial\chi}{\partial x}c - \varepsilon\operatorname{tr} a\frac{\partial^2\chi}{\partial x^2}\right]\left(x(s), \frac{x(s)}{\varepsilon}\right) ds$$
$$+ \int_0^t \left(I - \frac{\partial\chi}{\partial y} - \varepsilon\frac{\partial\chi}{\partial x}\right)\sigma\left(x(s), \frac{x(s)}{\varepsilon}\right) dw^\varepsilon(s).$$

Let $\phi \in C_0^\infty(\mathbb{R}^n)$. By Itô's formula, we have

(6.2.18)

$$
\phi(\zeta^\varepsilon(t)) = \phi(\zeta^\varepsilon(s))
$$
$$
+ \int_s^t \frac{\partial\phi}{\partial x}(\zeta^\varepsilon(\lambda)) \cdot \left[\left(I - \frac{\partial\chi}{\partial y} \right) c - \frac{\partial\chi}{\partial x} b - \operatorname{tr} a \frac{\partial^2\chi}{\partial y \partial x} \right.
$$
$$
\left. - \operatorname{tr} a \frac{\partial^2\chi}{\partial x \partial y} - \varepsilon \frac{\partial\chi}{\partial x} c - \varepsilon \operatorname{tr} a \frac{\partial^2\chi}{\partial x^2} \right] \left(x(\lambda), \frac{x(\lambda)}{\varepsilon} \right) \, d\lambda
$$
$$
+ \int_s^t \operatorname{tr} \frac{\partial^2\phi}{\partial x^2}(\zeta^\varepsilon(\lambda)) \cdot \left(I - \frac{\partial\chi}{\partial y} - \varepsilon \frac{\partial\chi}{\partial x} \right) a \left(I - \frac{\partial\chi}{\partial y} - \varepsilon \frac{\partial\chi}{\partial x} \right)^* \left(x(\lambda), \frac{x(\lambda)}{\varepsilon} \right) \, d\lambda
$$
$$
+ \int_s^t \frac{\partial\phi}{\partial x}(\zeta^\varepsilon(\lambda)) \left(I - \frac{\partial\chi}{\partial y} - \varepsilon \frac{\partial\chi}{\partial x} \right) \sigma \left(x(\lambda), \frac{x(\lambda)}{\varepsilon} \right) \, dw^\varepsilon(\lambda).
$$

We define

(6.2.19) $\qquad \tilde{q}(x, y) = \left(I - \dfrac{\partial\chi}{\partial y} \right) a \left(I - \dfrac{\partial\chi}{\partial y} \right)^* - q(x),$

(6.2.20) $\qquad r(x, y) = \left(I - \dfrac{\partial\chi}{\partial y} \right) c - \dfrac{\partial\chi}{\partial x} b - \operatorname{tr} a \dfrac{\partial^2\chi}{\partial y \partial x} - \operatorname{tr} a \dfrac{\partial^2\chi}{\partial x \partial y} - r(x).$

We then obtain from (6.2.18)

(6.2.21)
$$
E_x^\varepsilon \left[\phi(x(t)) - \int_s^t \operatorname{tr} q(x(\lambda)) \frac{\partial^2\phi}{\partial x^2}(x(\lambda)) \, d\lambda - \int_s^t r(x(\lambda)) \cdot \frac{\partial\phi}{\partial x}(x(\lambda)) \, d\lambda \;\middle|\; \mathcal{G}^s \right]
$$
$$
= \phi(x(s)) + E_x^\varepsilon [\phi(x(t)) - \phi(\zeta_\varepsilon(t)) \mid \mathcal{G}^s] + \phi(\zeta_\varepsilon(s)) - \phi(x(s))
$$
$$
+ E_x^\varepsilon \left[\int_s^t \operatorname{tr} \left(\frac{\partial^2\phi}{\partial x^2}(\zeta^\varepsilon(\lambda)) - \frac{\partial^2\phi}{\partial x^2}(x(\lambda)) \right) \left(I - \frac{\partial\chi}{\partial y} - \varepsilon \frac{\partial\chi}{\partial x} \right) a \left(I - \frac{\partial\chi}{\partial y} - \varepsilon \frac{\partial\chi}{\partial x} \right)^* \right.
$$
$$
\left. \cdot \left(x(\lambda), \frac{x(\lambda)}{\varepsilon} \right) \, d\lambda \;\middle|\; \mathcal{G}^s \right]
$$
$$
+ E_x^\varepsilon \left[\int_s^t \left(\frac{\partial\phi}{\partial x}(\zeta^\varepsilon(\lambda)) - \frac{\partial\phi}{\partial x}(x(\lambda)) \right) \cdot \left(I - \frac{\partial\chi}{\partial y} \right) c - \frac{\partial\chi}{\partial x} b \right.
$$
$$
\left. - \operatorname{tr} a \frac{\partial^2\chi}{\partial y \partial x} - \operatorname{tr} a \frac{\partial^2\chi}{\partial x \partial y} - \varepsilon \frac{\partial\chi}{\partial x} c - \operatorname{tr} a \frac{\partial^2\chi}{\partial x^2} \left(x(\lambda), \frac{x(\lambda)}{\varepsilon} \right) \, d\lambda \;\middle|\; \mathcal{G}^s \right]
$$
$$
+ E_x^\varepsilon \left[\int_s^t \frac{\partial\phi}{\partial x}(x(\lambda)) \cdot \tilde{r} \left(x(\lambda), \frac{x(\lambda)}{\varepsilon} \right) \;\middle|\; \mathcal{G}^s \right] + O(\varepsilon),\; [20]
$$

where $|O(\varepsilon)| \le C\varepsilon$, a.s., C being a deterministic constant. We notice that

$$(6.2.22) \quad E_x^\varepsilon \left| E_x^\varepsilon \left[\int_s^t \operatorname{tr} \frac{\partial^2 \phi}{\partial x^2}(x(\lambda))\tilde{q}\left(x(\lambda), \frac{x(\lambda)}{\varepsilon}\right) \,\bigg|\, \mathcal{G}^s \right] \right|$$

$$\le E_x^\varepsilon \left| \int_s^t \operatorname{tr} \frac{\partial^2 \phi}{\partial x^2}(x(\lambda))\tilde{q}\left(x(\lambda), \frac{x(\lambda)}{\varepsilon}\right) \, d\lambda \right|$$

$$= E \left| \int_s^t \operatorname{tr} \frac{\partial^2 \phi}{\partial x^2}(y^\varepsilon(\lambda))\tilde{q}\left(y^\varepsilon(\lambda), \frac{y^\varepsilon(\lambda)}{\varepsilon}\right) \, d\lambda \right| \to 0$$

from Lemma 6.2.

Similarly we have

$$(6.2.23) \qquad E_x^\varepsilon \left| E_x^\varepsilon \left[\int_s^t \frac{\partial \phi}{\partial x}(x(\lambda)) \cdot \tilde{r}\left(x(\lambda), \frac{x(\lambda)}{\varepsilon}\right) \, d\lambda \,\bigg|\, \mathcal{G}^s \right] \right| \to 0.$$

Using (6.2.22), (6.2.23) and (6.2.16), it easily follows from (6.2.21) that

$$(6.2.24) \quad E_x^\varepsilon \left| E_x^\varepsilon \left[\phi(x(t)) - \int_s^t \operatorname{tr} q(x(\lambda)) \frac{\partial^2 \phi}{\partial x^2}(x(\lambda)) \, d\lambda \right. \right.$$

$$\left. \left. - \int_s^t r(x(\lambda)) \cdot \frac{\partial \phi}{\partial x}(x(\lambda)) d\lambda - \phi(x(s)) \,\bigg|\, \mathcal{G}^s \right] \right| \to 0, \text{ as } \varepsilon \to 0.$$

Since at least for a subsequence $\mu_x^\varepsilon \to \mu_x$ weakly in $\mathcal{M}_+^1$, (6.2.23) implies

$$(6.2.25) \quad E_x \left[\phi(x(t)) - \int_s^t \operatorname{tr} q(x(\lambda)) \frac{\partial^2 \phi}{\partial x^2}(x(\lambda)) \, d\lambda \right.$$

$$\left. \int_s^t r(s(\lambda)) \cdot \frac{\partial \phi}{\partial x}(x(\lambda)) \, d\lambda - \phi(x(s)) \,\bigg|\, \mathcal{G}^s \right] = 0.$$

Hence μ_x solves the martingale problem relatively to the pair $q(x)$, $r(x)$ and by the uniqueness theorem of Stroock-Varadhan [112] and [113], (see Section 2), it follows that the whole sequence μ_x^ε converges toward μ_x, which is the probability measure on Ψ induced by $y_x(\cdot)$ defined by (6.2.11). This completes the proof of Theorem 6.4. ∎

We can now obtain the limit of $u_\varepsilon(x)$, solution of (6.1.5). We introduce the homogenized equation

$$(6.2.26) \qquad -q_{ij}(x)\frac{\partial^2 u}{\partial x_i \partial x_j} - r_i(x)\frac{\partial u}{\partial x_i} + \bar{a}_0(x)u = f(x),$$

$$u\,|_\Gamma = 0, \qquad u \in W^{2,p}(\mathcal{O}), \qquad p \ge 2,$$

where

$$(6.2.27) \qquad \bar{a}_0(x) = \int_Y a(x,y)m(x,y) \, dy.$$

We then have the following

THEOREM 6.6. *Under the assumptions of Theorem 6.4 and (6.1.3), (6.1.4), we have*

$$(6.2.28) \qquad u_\varepsilon(x) \to u(x), \qquad \forall x \in \mathcal{O}.$$

---

[20]E_x^ε (respectively, E_x) denotes the mathematical expectation with respect to μ_x^ε (respectively, μ_x).

PROOF. We have the explicit formula

$$(6.2.29) \qquad u_\varepsilon(x) = E \int_0^{\tau_x^\varepsilon} f(y_x^\varepsilon(t)) \left(\exp - \int_0^t a_0 \left(y_x^\varepsilon(s), \frac{y_x^\varepsilon(s)}{\varepsilon} \right) ds \right) dt.$$

Let us first prove that

$$(6.2.30) \quad X^\varepsilon = \left| E \int_0^{\tau_x^\varepsilon} f(y_x^\varepsilon(t)) \left[\exp - \int_0^t a_0 \left(y_x^\varepsilon(s), \frac{y_x^\varepsilon(s)}{\varepsilon} \right) ds \right. \right.$$
$$\left. \left. - \exp - \int_0^t \bar{a}_0(y_x(s)) \, ds \right] dt \right| \to 0, \text{ as } \varepsilon \to 0.$$

We note that

$$(6.2.31) \qquad\qquad X^\varepsilon \le CE \int_0^\infty Y^\varepsilon(t) \, dt,$$

where

$$Y^\varepsilon(t) = \left| \exp - \int_0^t a_0 \left(y_x^\varepsilon(s), \frac{y_x^\varepsilon(s)}{\varepsilon} \right) ds - \exp - \int_0^t \bar{a}_0(y_x^\varepsilon(s)) \, ds \right|.$$

But

$$EY^\varepsilon(t) \le E \left| \int_0^t a_0 \left(y^\varepsilon(s), \frac{y^\varepsilon(s)}{\varepsilon} \right) ds - \int_0^t \bar{a}_0(y^\varepsilon(s)) \, ds \right| \to 0$$

from Lemma 6.2. Moreover,

$$EY^\varepsilon(t) \le 2 \exp - \alpha_0 t.$$

From Lebesgue's theorem, we obtain (6.2.30).

It is thus enough to study the limit of

$$(6.2.32) \qquad v^\varepsilon(x) = E \int_0^{\tau_x^\varepsilon} f(y_x^\varepsilon(t)) \left(\exp - \int_0^t \bar{a}_0(y_x^\varepsilon(s)) \, ds \right) dt.$$

But we can rewrite $v^\varepsilon(x)$ as

$$v^\varepsilon(x) = \int_\Psi F(\psi) \, d\mu_x^\varepsilon$$

where

$$F(\psi) = \int_0^\tau f(y(t)) \exp \left(- \int_0^t \bar{a}_0(y(s)) \, ds \right) dt.$$

As in the proof of Theorem 4.12, we can assert that F is μ_x a.s. continuous and bounded. This and the result of Theorem 6.4, imply (6.2.28). ∎

6.3. Remarks on the martingale approach and the adjoint expansion method. The method of "adjoint expansion" used in the proof of Theorem 5.2 can also be used within the context of the martingale approach. Indeed the martingale formulation of (6.2.2), leads to

$$(6.3.1)$$
$$E_x^\varepsilon \left[\phi(x(t)) - \int_s^t \frac{\partial \phi}{\partial x}(x(\lambda)) \cdot \left(\frac{1}{\varepsilon} b \left(x(\lambda), \frac{x(\lambda)}{\varepsilon} \right) d\lambda + c \left(x(\lambda), \frac{x(\lambda)}{\varepsilon} \right) \right) d\lambda \right.$$
$$\left. - \int_s^t \operatorname{tr} \frac{\partial^2 \phi}{\partial x^2}(x(\lambda)) a \left(x(\lambda), \frac{x(\lambda)}{\varepsilon} \right) d\lambda \,\middle|\, \mathcal{G}^s \right] = \phi(x(s)),$$

which we rewrite as

$$(6.3.2) \qquad E_x^\varepsilon \left[\phi(x(t)) + \int_s^t A^\varepsilon \phi(x(\lambda)) \, d\lambda \; \middle| \; \mathcal{G}^s \right] = \phi(x(s)),$$

and (6.3.2) is valid for any $\phi \in C_0^\infty(\mathbb{R}^n)$.

Clearly by approximation, (6.3.2) is valid for any $\phi \in C_b^2(\mathbb{R}^n)$.[21] We are going to take then ϕ as follows:

$$\phi = \phi^\varepsilon(x) = \phi_0(x) + \varepsilon \phi_1\left(x, \frac{x}{\varepsilon}\right),$$

where $\phi_0 \in C_0^\infty(\mathbb{R}^n)$, and $\phi_1(x, y)$ periodic in y determined from an asymptotic expansion, namely

$$(6.3.3) \qquad -a_{ij}(x,y) \frac{\partial^2 \phi_1}{\partial y_i \partial y_j} - b_i(x,y) \frac{\partial \phi_1}{\partial y_i} = b_i \frac{\partial \phi_0}{\partial x_i}.$$

Therefore using the χ^ℓ functions defined in (6.2.3), we choose

$$\phi_1(x,y) = -\frac{\partial \phi_0}{\partial x_\ell} \chi^\ell(x,y).$$

Hence we have

$$\begin{aligned}
A^\varepsilon \phi^\varepsilon = &-a_{ij} \frac{\partial^2 \phi_0}{\partial x_i \partial x_j} - c_i \frac{\partial \phi_0}{\partial x_i} + a_{ij} \frac{\partial^2 \phi_0}{\partial x_i \partial x_\ell} \frac{\partial \chi^\ell}{\partial y_j} \\
&+ a_{ij} \frac{\partial \phi_0}{\partial x_\ell} \frac{\partial^2 \chi^\ell}{\partial x_i \partial y_j} + a_{ij} \frac{\partial^2 \phi_0}{\partial x_j \partial x_\ell} \frac{\partial \chi^\ell}{\partial y_i} \\
&+ a_{ij} \frac{\partial \phi_0}{\partial x} \frac{\partial^2 \chi_\ell}{\partial x_j \partial y_i} + b_i \frac{\partial^2 \phi_0}{\partial x_i \partial x_\ell} \chi^\ell + b_i \frac{\partial \phi_0}{\partial x_\ell} \frac{\partial \chi_\ell}{\partial x_i} \\
&+ c_i \frac{\partial \phi_0}{\partial x_\ell} \frac{\partial \chi^\ell}{\partial y_i} + O(\varepsilon)
\end{aligned}$$

with $|O(\varepsilon)| \leq C\varepsilon$ where C is a constant depending on ϕ_0. Hence (6.3.2), we obtain

(6.3.4)

$$\begin{aligned}
E_x^\varepsilon \Bigg[\phi_0(x(t)) - \int_s^t \frac{\partial^2 \phi_0}{\partial x_i \partial x_j} \left(a_{ij} - a_{ik} \frac{\partial \chi^j}{\partial y_k} - a_{jk} \frac{\partial \chi^i}{\partial y_k} - \frac{1}{2} b_i \chi^j - \frac{1}{2} b_j \chi^i \right) d\lambda \\
- \int_s^t \frac{\partial \phi_0}{\partial x_i} (r_i + r_i) \, d\lambda \; \Bigg| \; \mathcal{G}^s \Bigg] = \phi_0(x(s)) + O_1(\varepsilon)
\end{aligned}$$

where

$$|O_1(\varepsilon)| \leq C_1 \varepsilon$$

where C_1 is a deterministic constant. Using Lemma 6.1, and 6.2 as in the proof of Theorem 6.4, it follows that if for a subsequence $\mu_x^\varepsilon \to \mu_x$, then

(6.3.5)

$$E_x \left[\phi_0(x(t)) - \int_s^t \frac{\partial^2 \phi_0}{\partial x_i \partial x_j} q_{ij}(x(\lambda)) - \int_s^t \frac{\partial \phi_0}{\partial x_i} r_i(x(\lambda)) \, d\lambda \; \middle| \; \mathcal{G}^s \right] = \phi_0(x(s)),$$

---

[21]Functions which are twice continuously differentiable, bounded as well as its derivative up to the order 2.

where we have set

$$(6.3.6) \qquad q_{ij}(x) = \int_Y m(x,y) \left[a_{ij}(x,y) - a_{ik}(x,y) \frac{\partial \chi^j}{\partial y_k}(x,y) - a_{jk} \frac{\partial \chi^i}{\partial y_k}(x,y) \right.$$
$$\left. - \frac{1}{2} b_i(x,y) \chi^j(x,y) - \frac{1}{2} b_j(x,y) \chi^i(x,y) \right] dy.$$

This formula for q_{ij} is the same as (6.2.9), as it has been shown in (5.1.13). Here x is merely a parameter which plays no role.

The above reasoning thus leads to an alternative proof of Theorem 6.4 (but Lemmas 6.1 and 6.2 are used in both proofs).

6.4. Analytic approach to problem (6.1.5). We do not repeat here the method of asymptotic expansions, which is an easy extension of what has been done in Section 5.1. We will concentrate on the energy method for which some differences with Section 5.2 will be encountered.

We will assume that

$$(6.4.1) \qquad a_0(x,y) \geq 0, \qquad a_0 \text{ bounded, periodic in } y,$$

$$(6.4.2) \qquad f \in L^2(\mathcal{O}), \qquad \mathcal{O} \text{ regular and bounded.}$$

By virtue of the regularity assumptions on the coefficients and (6.4.1), (6.4.2) there exists one and only one solution u_ε of

$$(6.4.3)$$
$$- a_{ij}\left(x, \frac{x}{\varepsilon}\right) \frac{\partial^2 u_\varepsilon}{\partial x_i \partial x_j} - \frac{1}{\varepsilon} b_i\left(x, \frac{x}{\varepsilon}\right) \frac{\partial u_\varepsilon}{\partial x_i} - c_i\left(x, \frac{x}{\varepsilon}\right) \frac{\partial u_\varepsilon}{\partial x_i} + a_0\left(x, \frac{x}{\varepsilon}\right) u_\varepsilon = f,$$
$$u_\varepsilon \mid_\Gamma = 0, \qquad u_\varepsilon \in H^2(\mathcal{O}) \cap H_0^1(\mathcal{O}).$$

Similarly there exists one and only one solution u of the homogenized equation

$$(6.4.4) \qquad - q_{ij}(x) \frac{\partial^2 u}{\partial x_i \partial x_j} - r_i(x) \frac{\partial u}{\partial x_i} + \bar{a}_0(x) u = f,$$
$$u \mid_\Gamma = 0, \qquad u \in H^2(\mathcal{O}) \cap H_0^1(\mathcal{O}).$$

We are going to prove the following.

THEOREM 6.7. *Under the assumptions (6.1.1), (6.1.2), (6.1.10) and (6.4.1), (6.4.2), then $u_\varepsilon \to u$ in $H_0^1(\mathcal{O})$ weakly.*

PROOF. Our first objective is to obtain a priori estimates in $H_0^1(\mathcal{O})$. Let $m^\varepsilon(x,y)$ be a function which will be made precise below and such that

$$(6.4.5) \qquad M \geq m^\varepsilon(x,y) \geq m_0 > 0.$$

We set

$$(6.4.6) \qquad \tilde{a}_{ij}^\varepsilon = a_{ij} m^\varepsilon, \qquad \tilde{b}_i^\varepsilon = b_i m^\varepsilon, \qquad \tilde{c}_i^\varepsilon = c_i m^\varepsilon,$$
$$\tilde{a}_0^\varepsilon = a_0 m^\varepsilon, \qquad \tilde{f} = f m^\varepsilon.$$

We multiply (6.4.3) by m^ε and rewrite the result in variational form. We obtain

$$(6.4.7) \qquad - \frac{\partial}{\partial x_i} \tilde{a}_{ij}^\varepsilon \frac{\partial u_\varepsilon}{\partial x_j} - \frac{1}{\varepsilon} \tilde{\beta}_i^\varepsilon\left(x, \frac{x}{\varepsilon}\right) \frac{\partial u_\varepsilon}{\partial x_i} + \tilde{a}_0^\varepsilon u^\varepsilon = \tilde{f}^\varepsilon, \qquad u_\varepsilon \mid_\Gamma = 0,$$

where we have set

(6.4.8)
$$\tilde{\beta}_i^\varepsilon(x,y) = \tilde{b}_i^\varepsilon + \varepsilon \tilde{c}_i^\varepsilon - \frac{\partial}{\partial y_j}\tilde{a}_{ij}^\varepsilon - \varepsilon\frac{\partial}{\partial x_j}\tilde{a}_{ij}^\varepsilon.$$

We are going to choose m^ε in order that

$$\frac{1}{\varepsilon^2}\frac{\partial\tilde{\beta}_i^\varepsilon}{\partial y_i} + \frac{1}{\varepsilon}\frac{\partial\tilde{\beta}_i^\varepsilon}{\partial x_i} = O(\varepsilon).$$

Expliciting, we obtain the desired relationship

(6.4.9)
$$-\frac{1}{\varepsilon^2}\frac{\partial^2}{\partial y_i \partial y_j}(a_{ij}m^\varepsilon) + \frac{1}{\varepsilon^2}\frac{\partial}{\partial y_i}(b_i m^\varepsilon) + \frac{1}{\varepsilon}\frac{\partial}{\partial y_i}(c_i m^\varepsilon)$$
$$-\frac{1}{\varepsilon}\frac{\partial^2}{\partial x_i \partial y_i}(a_{ij}m^\varepsilon) + \frac{1}{\varepsilon}\frac{\partial}{\partial x_i}(b_i m^\varepsilon) - \frac{1}{\varepsilon}\frac{\partial^2}{\partial x_i \partial y_j}(a_{ij}m^\varepsilon)$$
$$+\frac{\partial}{\partial x_i}(c_i m) - \frac{\partial^2}{\partial x_i \partial x_j}(a_{ij}m^\varepsilon) = O(\varepsilon).$$

We are going to choose m^ε as follows

(6.4.10)
$$m^\varepsilon(x,y) = m_0(x,y) + \varepsilon m_1(x,y) + \varepsilon^2 m_2(x,y),$$

where

(6.4.11)
$$-\frac{\partial^2}{\partial y_i \partial y_j}(a_{ij}m_0) + \frac{\partial}{\partial y_i}(b_i m_0) = 0, \qquad m_0 \text{ periodic in } y,$$

(6.4.12)
$$-\frac{\partial^2}{\partial y_i \partial y_j}(a_{ij}m_1) + \frac{\partial}{\partial y_i}(b_i m_1) + \frac{\partial}{\partial y_i}(c_i m_0)$$
$$-\frac{\partial^2}{\partial x_j \partial y_i}(a_{ij}m_0) + \frac{\partial}{\partial x_i}(b_i m_0) - \frac{\partial^2}{\partial x_i \partial y_j}(a_{ij}m_0) = 0,$$
$$m_1 \text{ periodic in } y,$$

(6.4.13)
$$-\frac{\partial^2}{\partial y_i \partial y_j}(a_{ij}m_2) + \frac{\partial}{\partial y_i}(b_i m_2) + \frac{\partial}{\partial y_i}(c_i m_1) - \frac{\partial^2}{\partial x_j \partial y_i}(a_{ij}m_1)$$
$$+\frac{\partial}{\partial x_i}(b_i m_1) - \frac{\partial^2}{\partial x_i \partial y_j}(a_{ij}m_1) + \frac{\partial}{\partial x_i}(c_i m_0)$$
$$-\frac{\partial^2}{\partial x_i \partial x_j}(a_{ij}m_0) = 0, \qquad m_2 \text{ periodic in } y.$$

We take

$$m_0(x,y) = m(x,y)\lambda(x).$$

From the fact that $\int b_i m \, dy = 0 \; \forall x$, it follows that the solvability condition for (6.4.12) is satisfied, hence (6.4.12) has a solution which we write as

$$m_1(x,y) = -\frac{\partial\lambda}{\partial x_j}\tilde{m}^j + \lambda\tilde{n}$$

where $\tilde{m}^i(x, y)$, $n(x, y)$ are solutions of

$$(6.4.14) \qquad -\frac{\partial}{\partial y_i \partial y_j}(a_{ij}\tilde{m}^\ell) + \frac{\partial}{\partial y_i}(b_i \tilde{m}^\ell) = +b_\ell m - 2\frac{\partial}{\partial y_j}(a_{\ell j}m),$$

$$\tilde{m}^\ell \text{ periodic}, \qquad \int_Y \tilde{m}^\ell(x, y)\, dy = 1,$$

$$(6.4.15)$$
$$-\frac{\partial^2}{\partial y_i \partial y_j}(a_{ij}\tilde{n}) + \frac{\partial}{\partial y_i}(b_i \tilde{n}) = \frac{\partial^2}{\partial x_j \partial y_i}(a_{ij}m) + \frac{\partial^2}{\partial x_i \partial y_j}(a_{ij}m) - \frac{\partial}{\partial x_i}(b_i m) - \frac{\partial}{\partial y_i}(c_i m)$$

$$\tilde{n} \text{ periodic in } y, \qquad \int_Y \tilde{n}(x, y)\, dy = 1.$$

We turn next to (6.4.13), which is rewritten as follows

$$(6.4.16) \qquad -\frac{\partial^2}{\partial y_i \partial y_j}(a_{ij}m_2) + \frac{\partial}{\partial y_i}(b_i m_2) + \frac{\partial}{\partial y_i}(c_i m_1)$$

$$-2\frac{\partial^2}{\partial x_j \partial y_i}(a_{ij}m_1) - \frac{\partial}{\partial x_i}\left(b_i \frac{\partial \lambda}{\partial x_j}\tilde{m}^j\right) + \frac{\partial}{\partial x_i}(\lambda b_i \tilde{n})$$

$$+\frac{\partial}{\partial x_i}(c_i \lambda m) - \frac{\partial^2}{\partial x_i \partial x_j}(a_{ij}\lambda m) = 0, \qquad m \text{ periodic in } y.$$

We obtain the following solvability condition
$$(6.4.17)$$
$$-\frac{\partial}{\partial x_i}\int_Y b_i \frac{\partial \lambda}{\partial x_j}\tilde{m}^j\, dy + \frac{\partial}{\partial x_i}\int_Y \lambda b_i \tilde{n}\, dy + \frac{\partial}{\partial x_i}\int_Y c_i \lambda m\, dy - \frac{\partial^2}{\partial x_i \partial x_j}\int_Y a_{ij}\lambda m\, dy = 0.$$

Comparing (6.4.14) with (5.2.23), we see that $\tilde{m}^\ell$ and $m\tilde{\chi}^\ell$ are identical when the coefficients do not depend on x. From the identification made to obtain (5.2.28), we can assert since x does not play any specific role in such an identification, that

$$(6.4.18) \qquad q_{ij}(x) = \int_Y a_{ij}m\, dy + \frac{1}{2}\int_Y b_i \tilde{m}^j\, dy + \frac{1}{2}\int_Y b_j \tilde{m}^i\, dy.$$

We next consider the quantity

$$s_i(x) = \int_Y b_i \tilde{n}\, dy + \int_Y c_i m\, dy + \frac{\partial}{\partial x_j}\int_Y b_i \tilde{m}^j\, dy.$$

Recalling the equations for the χ^ℓ functions (6.2.3), the equation for $\tilde{n}$, (6.4.15), it follows from an integration by parts calculation that
$$(6.4.19)$$
$$\int_Y b_i \tilde{n}\, dy = 2\int_Y \frac{\partial \chi^i}{\partial y_\ell}\frac{\partial}{\partial x_k}(a_{k\ell}m)\, dy - \int_Y \frac{\partial \chi^i}{\partial y_\ell}c_\ell m\, dy + \int_Y \chi^i \frac{\partial}{\partial x_k}(b_k m)\, dy.$$

Similarly, we can check that

$$(6.4.20) \qquad \int_Y b_i \tilde{m}^j\, dy = -\int_Y \chi^i\left(b_j m - 2\frac{\partial}{\partial y_k}(a_{jk}m)\right)\, dy$$

$$= -\int_Y \chi^i b_j m\, dy - 2\int_Y \frac{\partial \chi^i}{\partial y_\ell}a_{j\ell}m\, dy.$$

We easily deduce from (6.4.19), (6.4.20) that

$$s_i(x) = \int_Y \left(c_i - \frac{\partial \chi^i}{\partial y_\ell} c_\ell \right) m \, dy - 2 \int_Y \frac{\partial^2 \chi^i}{\partial x_k \partial y_\ell} a_{k\ell} m \, dy - \int_Y \frac{\partial \chi^i}{\partial x_k} b_k m \, dy.$$

Comparing with (6.2.10), one checks that $s_i(x) = r_i(x)$. Taking this and (6.4.18) into account, we can rewrite (6.4.17) as follows:

$$(6.4.21) \qquad -\frac{\partial^2}{\partial x_i \partial x_j}(\lambda(x) q_{ij}(x)) + \frac{\partial}{\partial x_i}(\lambda(x) r_i(x)) = 0.$$

Our problem is to find a function $\lambda(x)$ such that (6.4.10) is guaranteed. We notice that (6.4.21) is an equation without boundary conditions. But of course the values of x which matter are those in $\overline{\mathcal{O}}$. Therefore we must find a function $\lambda(x)$ which satisfies (6.4.21) in $\mathcal{O}$, bounded below and above by positive constants in $\overline{\mathcal{O}}$. To do that we consider a regular domain $\widetilde{\mathcal{O}}$ such that $\overline{\mathcal{O}} \subset \widetilde{\mathcal{O}}$, a function $\tilde{g} \in L^\infty(\widetilde{\mathcal{O}})$, $\tilde{g} \geq 0$, $\tilde{g} \neq 0$ in a set of positive Lebesgue measure and $\tilde{g} = 0$ on $\mathcal{O}$. We then solve the problem

$$(6.4.22) \qquad -\frac{\partial^2}{\partial x_i \partial x_j}(\tilde{\lambda}(x) q_{ij}(x)) + \frac{\partial}{\partial x_i}(\tilde{\lambda}(x) r_i(x)) = \tilde{g}(x) \text{ in } \mathcal{O},$$

$$\tilde{\lambda}\,|_{\partial \widetilde{\mathcal{O}}} = 0, \qquad \tilde{\lambda} \in W^{2,p}(\widetilde{\mathcal{O}}), \qquad p \geq 2.^{22}$$

The existence and uniqueness of $\tilde{\lambda}$ follows by duality from the existence and uniqueness of

$$(6.4.23) \qquad - q_{ij}(x) \frac{\partial^2 \tilde{\mu}}{\partial x_i \partial x_j} - r_i(x) \frac{\partial \tilde{\mu}}{\partial x_i} = h(x) \text{ in } \widetilde{\mathcal{O}},$$

$$\tilde{\mu}\,|_{\partial \widetilde{\mathcal{O}}} = 0, \qquad \tilde{\mu} \in W^{2,p}(\widetilde{\mathcal{O}}), \qquad p \geq 2,$$

where $\tilde{h} \in L^\infty(\widetilde{\mathcal{O}})$. The Fredholm alternative must be applied in a routine way. Moreover we have the relation

$$\int_{\widetilde{\mathcal{O}}} \tilde{g} \tilde{\mu} \, dx = \int_{\widetilde{\mathcal{O}}} \tilde{h} \tilde{\lambda} \, dx.$$

Since for $\tilde{h} \geq 0$, then $\tilde{\mu} \geq 0$, it follows that if $\tilde{g} \geq 0$, then $\int \tilde{h} \tilde{\lambda} \, dx \geq 0$, for any $\tilde{h} \geq 0$, hence $\tilde{\lambda} \geq 0$ if $\tilde{g} \geq 0$.

We can always assume that $\tilde{g}$ is sufficiently smooth (since $\mathcal{O}$ is regular), and therefore we may assume that $\tilde{\lambda} \in C^4(\overline{\widetilde{\mathcal{O}}})$. From the strong maximum principle it follows that $\tilde{\lambda} > 0$ in $\widetilde{\mathcal{O}}$. Taking $\lambda = \tilde{\lambda}\,|_{\overline{\mathcal{O}}}$, we see that we can construct a function λ such that

$$\lambda \in C^4(\overline{\mathcal{O}}), \qquad \lambda > 0 \text{ in } \overline{\mathcal{O}}, \text{ and } \lambda \text{ satisfies (6.4.21) in } \mathcal{O}.$$

We can then exhibit a choice for m_2 as usual in the method of asymptotic expansions. We can also define what is the function $O(\varepsilon)$ involved in (6.4.9). By virtue of the regularity of $\lambda(x)$ we have $|O(\varepsilon)| \leq C\varepsilon$. We thus have completely defined the function $m^\varepsilon(x, y)$.

We can now turn back to (6.4.7). Multiplying by u_ε and integrating we obtain, by virtue of the choice of m^ε,

$$(6.4.24) \qquad \int_{\mathcal{O}} \tilde{a}_{ij}^\varepsilon \frac{\partial u_\varepsilon}{\partial x_j} \frac{\partial u_\varepsilon}{\partial x_i} \, dx + \int_{\mathcal{O}} O(\varepsilon)(u_\varepsilon)^2 + \int_{\mathcal{O}} \tilde{a}_0^\varepsilon(u_\varepsilon)^2 \, dx = \int_{\mathcal{O}} f^\varepsilon u_\varepsilon \, dx.$$

---

[22]Note that $q_{ij}(x)$, $r_i(x)$ have been defined in $\mathbb{R}^n$.

For ε small enough, we deduce from (6.4.24) and Poincaré's inequality that

$$(6.4.25) \qquad \|u_\varepsilon\|_{H_0^1(\mathcal{O})} \le C.$$

We next proceed as in the proof of Theorem 5.2. We define

$$(6.4.26) \qquad \tilde{a}_{ij} = a_{ij}m, \qquad \tilde{b}_i = b_i m, \qquad \tilde{c}_i = c_i m,$$

$$\tilde{a}_0 = a_0 m, \qquad f^\varepsilon = f(x)m\left(x, \frac{x}{\varepsilon}\right).$$

Thus u_ε is a solution of

$$(6.4.27) \quad -\tilde{a}_{ij}\left(x, \frac{x}{\varepsilon}\right)\frac{\partial^2 u_\varepsilon}{\partial x_i \partial x_j} - \frac{1}{\varepsilon}\tilde{b}_i\left(x, \frac{x}{\varepsilon}\right)\frac{\partial u_\varepsilon}{\partial x_i}$$

$$-\tilde{c}_i\left(x, \frac{x}{\varepsilon}\right)\frac{\partial u_\varepsilon}{\partial x_i} + a_0\left(x, \frac{x}{\varepsilon}\right)u_\varepsilon = f^\varepsilon, \qquad u_\varepsilon \mid_\Gamma = 0.$$

Let $v \in H_0^1(\mathcal{O}) \cap H^2(\mathcal{O})$. If A^ε denotes the operator in (6.4.27), then we have, since u_ε, v vanish on Γ,

$$(6.4.28) \qquad \int u_\varepsilon(\tilde{A}^\varepsilon)^* v \, dx = \int f^\varepsilon v \, dx,$$

where

$$(\tilde{A}^\varepsilon)^* = \varepsilon^{-2}\tilde{A}_1^* + \varepsilon^{-1}\tilde{A}_2^* + \varepsilon^0 \tilde{A}_3^*$$

with

$$\tilde{A}_1^* = -\frac{\partial^2}{\partial y_i \partial y_j}(\tilde{a}_{ij} \cdot) + \frac{\partial}{\partial y_i}(b_i \cdot),$$

$$\tilde{A}_2^* = -\frac{\partial^2}{\partial x_i \partial y_j}(\tilde{a}_{ij} \cdot) - \frac{\partial^2}{\partial y_i \partial x_j}(\tilde{a}_{ij} \cdot) + \frac{\partial}{\partial x_i}(\tilde{b}_i) + \frac{\partial}{\partial y_i}(\tilde{c}_i \cdot),$$

$$\tilde{A}_3^* = -\frac{\partial^2}{\partial x_i \partial x_j} + \frac{\partial}{\partial x_i}(\tilde{c}_i \cdot) + \tilde{a}_0.$$

The unique solution of

$$\tilde{A}_1^* \psi = 0, \qquad \psi \text{ periodic}$$

is $\psi = 1$ (up to a multiplicative constant) by virtue of the uniqueness of the solution of (6.1.8). The equation

$$\tilde{A}_1 \phi = 0, \qquad \phi \text{ periodic},$$

has also $\phi = 1$ as a solution (thus unique up to a multiplicative constant). We look for a function v in (6.4.28) of the form

$$v\left(x, \frac{x}{\varepsilon}\right) = \phi(x) + \varepsilon\phi_1\left(x, \frac{x}{\varepsilon}\right)$$

where $\phi(x) \in C_0^\infty(\mathcal{O})$ and $\phi_1(x, y)$ are defined such that

$$\tilde{A}_1^* \phi_1 + A_2^* \phi = 0, \qquad \phi_1 \text{ periodic in } y.$$

We have

$$\tilde{A}_1^* \phi_1 - 2\frac{\partial\phi}{\partial x_i}\frac{\partial}{\partial y_j}\tilde{a}_{ij} - 2\phi\frac{\partial^2 a_{ij}}{\partial x_i \partial y_j} + \frac{\partial}{\partial x_i}\tilde{b}_i + \phi\frac{\partial}{\partial x_i}\tilde{b}_i + \phi\frac{\partial\tilde{c}_i}{\partial y_i} = 0,$$

hence

$$\phi_1(x, y) = -\frac{\partial\phi}{\partial x_\ell}\tilde{\chi}^\ell + \phi\chi,$$

where

$$\tilde{A}_1^* \tilde{\chi}^\ell = \tilde{b}_\ell - 2\frac{\partial}{\partial y_j}\tilde{a}_{\ell j},$$

$$\tilde{A}_1^* \chi = -\frac{\partial \tilde{b}_i}{\partial x_i} - \frac{\partial \tilde{c}_i}{\partial y_i} + 2\frac{\partial^2 \tilde{a}_{ij}}{\partial x_i \partial y_j}.$$

We have next

$$(\tilde{A}^\varepsilon)^* v = \tilde{A}_3^* \phi + \tilde{A}_2^* \phi_1 + \varepsilon A_3^* \phi_1,$$

and

$$\tilde{A}_3^* \phi + \tilde{A}_2^* \phi_1 = -\frac{\partial^2}{\partial x_i \partial x_j}(\phi \tilde{a}_{ij}) + \frac{\partial}{\partial x_i}(\tilde{c}_i \phi) + \tilde{a}_0 \phi$$

$$+ \frac{\partial^2}{\partial x_i \partial x_j}\left(\tilde{a}_{ij}\frac{\partial \phi}{\partial x_\ell}\tilde{\chi}^\ell\right) + \frac{\partial^2}{\partial y_i \partial x_j}\left(\tilde{a}_{ij}\frac{\partial \phi}{\partial x_\ell}\tilde{\chi}^\ell\right)$$

$$- \frac{\partial}{\partial x_i}\left(\tilde{b}_i\frac{\partial \phi}{\partial x_\ell}\tilde{\chi}^\ell\right) - \frac{\partial}{\partial y_i}\left(\tilde{c}_i\frac{\partial \phi}{\partial x_\ell}\tilde{\chi}^\ell\right)$$

$$- \frac{\partial^2}{\partial x_i \partial y_j}(\tilde{a}_{ij}\phi\chi) - \frac{\partial^2}{\partial y_i \partial x_j}(\tilde{a}_{ij}\phi\chi)$$

$$+ \frac{\partial}{\partial x_i}(\tilde{b}_i\phi\chi) + \frac{\partial}{\partial y_i}(\tilde{c}_i\phi\chi).$$

The mean value of $\tilde{A}_3^* \phi + \tilde{A}_2^* \phi_1$ in y, is equal to

$$(6.4.29) \quad \overline{A_3^* \phi + A_2^* \phi_1} = -\frac{\partial^2}{\partial x_i \partial x_j}(\phi \overline{\tilde{a}_{ij}}) + \frac{\partial}{\partial x_i}(\phi \overline{\tilde{c}_i}) + \overline{\tilde{a}_0}\phi$$

$$- \frac{\partial^2}{\partial x_i \partial x_\ell}(\phi \overline{\tilde{b}_i\tilde{\chi}^\ell}) + \frac{\partial}{\partial x_i}\left(\phi \frac{\partial}{\partial x_\ell}(\overline{\tilde{b}_i\tilde{\chi}^\ell})\right) + \frac{\partial}{\partial x_i}(\phi \overline{\tilde{b}_i\chi}).$$

We notice that (cf. (5.2.28))

$$(6.4.30) \qquad\qquad q_{ij} = \tilde{a}_{ij} + \frac{1}{2}\overline{\tilde{b}_i\tilde{\chi}^j} + \frac{1}{2}\overline{\tilde{b}_j\tilde{\chi}^i}.$$

Let us set

$$(6.4.31) \qquad\qquad s_i(x) = \overline{\tilde{c}_i} + \frac{\partial}{\partial x_\ell}(\overline{\tilde{b}_i\tilde{\chi}^\ell}) + \overline{\tilde{b}_i\chi}.$$

We recall that (cf. (6.2.3))

$$\tilde{A}_1\chi^i = -\tilde{b}_i.$$

Combining this with the χ equation and integrating by parts, yields

$$\overline{\tilde{b}_i\chi} = \int_Y \chi^i \left(\frac{\partial \tilde{b}_\ell}{\partial x_\ell} + \frac{\partial \tilde{c}_\ell}{\partial y_\ell} - 2\frac{\partial^2 \tilde{a}_{\ell j}}{\partial x_\ell \partial y_j}\right) dy$$

$$= -\int_Y \tilde{c}_\ell \frac{\partial \chi^i}{\partial y_\ell} dy + 2\int_Y \frac{\partial \chi^i}{\partial y_\ell}\frac{\partial \tilde{a}_{\ell j}}{\partial x_\ell} dy + \int_Y \chi^i \frac{\partial \tilde{b}_\ell}{\partial x_\ell} dy.$$

Hence

$$(6.4.32)$$

$$s_i = \int_Y \left(\tilde{c}_i - \tilde{c}_\ell\frac{\partial \chi^i}{\partial y_\ell}\right) dy + \frac{\partial}{\partial x_\ell}\int_Y \tilde{b}_i\tilde{\chi}^\ell dy + 2\int_Y \frac{\partial \chi^i}{\partial x_\ell}\frac{\partial \tilde{a}_{\ell j}}{\partial x_\ell} dy + \int_Y \chi\frac{\partial \tilde{b}_\ell}{\partial x_\ell} dy.$$

Using now the equations of χ^i and $\tilde{\chi}^\ell$, we obtain

$$\int_Y \tilde{b}_i \tilde{\chi}^\ell \, dy = \int_Y \chi^i \left(-\tilde{b}_\ell + 2\frac{\partial}{\partial y_j}\tilde{a}_{\ell j} \right) dy,$$

hence

$$\frac{\partial}{\partial x_\ell} \int_Y \tilde{b}_i \tilde{\chi}^\ell \, dy = \int_Y \frac{\partial \chi^i}{\partial x_\ell} \left(-\tilde{b}_\ell + 2\frac{\partial}{\partial y_j}\tilde{a}_{\ell j} \right) dy + \int_Y \chi^i \left(\frac{\partial \tilde{b}_\ell}{\partial x_\ell} + 2\frac{\partial^2}{\partial x_\ell \partial y_j}\tilde{a}_{ij} \right) dy.$$

Using this in (6.4.32), we obtain

$$s_i = \int_Y \left(\tilde{c}_i - \tilde{c}_\ell \frac{\partial \chi^i}{\partial y_\ell} \right) dy - \int_Y \frac{\partial \chi^i}{\partial x_\ell} \tilde{b}_\ell \, dy + 2 \int_Y \frac{\partial^2 \chi^i}{\partial x_\ell \partial y_j} \tilde{a}_{\ell j} \, dy$$

and thus $s_i(x) = r_i(x)$ (cf. (6.2.10)).

Collecting results we can assert that

$$\overline{\tilde{A}_3^* \phi + \tilde{A}_2^* \phi_1} = -\frac{\partial^2}{\partial x_i \partial x_j}(\phi q_{ij}) + \frac{\partial}{\partial x_i}(\phi r_i) + \bar{a}_0 \phi.$$

Therefore from (6.4.28) and for a convergent subsequence $u_\varepsilon \to u$, we have

$$\int_{\mathcal{O}} u \mathcal{A}^* \phi \, dx = \int_{\mathcal{O}} f\phi \, dx,$$

hence

$$\mathcal{A}u = f, \qquad u\,|_\Gamma = 0.$$

From the uniqueness of the limit we conclude the proof of the theorem. ∎

REMARK 6.8. We have used in the proof of Theorem 6.7 a method slightly different from the one used in Theorem 5.2. The functions $\tilde{\chi}^\ell$ and χ have been used in both proofs, but in a different manner. But both proofs can be used in each theorem.

7. Reiterated homogenization

We consider in this paragraph, the probabilistic counterpart of Section 8, Chapter 1.

7.1. Setting of the problem. Let us consider functions $b^1(x, y, z)$, $b^2(x, y, z)$, $c(x, y, z)$ from $\mathbb{R}^n \to \mathbb{R}^n$, and $\sigma(x, y, z) : \mathbb{R}^n \to \mathcal{L}(\mathbb{R}^n; \mathbb{R}^n)$, satisfying the following conditions

(7.1.1)
b^1, b^2, c, σ are "sufficiently" smooth in x, y, z with bounded partial derivatives,[23]

(7.1.2)
b^1, b^2, c, σ are periodic in y, and periodic in z with period 1 in each component,[24]

(7.1.3) $$a = \frac{1}{2}\sigma\sigma^* \geq \alpha I, \qquad \alpha > 0, \text{ constant.}$$

---

[23]This of course can be made more precise, considering what will be needed in the following.
[24]Again the choice of 1 as the period is a matter of convenience.

We construct on an arbitrary probability space $(\Omega, \mathcal{A}, P, \mathcal{F}^t, w(t))$, the Itô process $y^\varepsilon(t) = y_x^\varepsilon(t)$ solution of

$$(7.1.4) \qquad dy^\varepsilon = \frac{1}{\varepsilon} b^2 \left(y^\varepsilon, \frac{y^\varepsilon}{\varepsilon}, \frac{y^\varepsilon}{\varepsilon^2} \right) dt + \frac{1}{\varepsilon^2} b^1 \left(y^\varepsilon, \frac{y^\varepsilon}{\varepsilon}, \frac{y^\varepsilon}{\varepsilon^2} \right) dt$$

$$+ c \left(y^\varepsilon, \frac{y^\varepsilon}{\varepsilon}, \frac{y^\varepsilon}{\varepsilon^2} \right) dt + c \left(y^\varepsilon, \frac{y^\varepsilon}{\varepsilon}, \frac{y^\varepsilon}{\varepsilon^2} \right) dw(t),$$

$$y^\varepsilon(0) = x.$$

We denote by μ_x^ε the probability measure induced by $y_x^\varepsilon(\cdot)$ on $\Psi = C([0, \infty; \mathbb{R}^n])$.

We want to study the behavior of μ_x^ε for fixed x, as ε goes to 0. As we have seen in the previous paragraphs, the weak convergence of μ_x^ε toward some μ_x in $\mathcal{M}_+^1(\Psi)$, is related to the convergence of the solution of Dirichlet problems, with rapidly varying coefficients.

We introduce the differential operator connected with (7.1.4), namely

$$A^\varepsilon = -\left(\frac{b^1}{\varepsilon^2} + \frac{b^2}{\varepsilon} + c \right) \cdot \nabla - \operatorname{div} a\nabla$$

and operating on functions of $x, \frac{x}{\varepsilon}, \frac{x}{\varepsilon^2}$, we write with standard notation

$$A = -\left(\frac{b_i^1}{\varepsilon^2} + \frac{b_i^2}{\varepsilon} + c_i \right) \left(\frac{\partial}{\partial x_i} + \frac{1}{\varepsilon} \frac{\partial}{\partial y_i} + \frac{1}{\varepsilon^2} \frac{\partial}{\partial z_i} \right)$$

$$- a_{ij} \left(\frac{\partial}{\partial x_i} + \frac{1}{\varepsilon} \frac{\partial}{\partial y_i} + \frac{1}{\varepsilon^2} \frac{\partial}{\partial z_i} \right) \left(\frac{\partial}{\partial x_j} + \frac{1}{\varepsilon} \frac{\partial}{\partial y_j} + \frac{1}{\varepsilon^2} \frac{\partial}{\partial z_j} \right)$$

$$= \frac{1}{\varepsilon^4} A_1 + \frac{1}{\varepsilon^3} A_2 + \frac{1}{\varepsilon^2} A_3 + \frac{1}{\varepsilon} A_4 + A_5$$

with

$$A_1 = -b_i^1 \frac{\partial}{\partial z_i} - a_{ij} \frac{\partial^2}{\partial z_i \partial z_j},$$

$$A_2 = -b_i^1 \frac{\partial}{\partial y_i} - b_i^2 \frac{\partial}{\partial z_i} - 2a_{ij} \frac{\partial^2}{\partial y_i \partial z_j},$$

$$(7.1.5) \qquad A_3 = -c_i \frac{\partial}{\partial z_i} - b_i^1 \frac{\partial}{\partial x_i} - b_i^2 \frac{\partial}{\partial y_i} - a_{ij} \frac{\partial^2}{\partial y_i \partial y_j} - 2a_{ij} \frac{\partial^2}{\partial z_i \partial x_j},$$

$$A_4 = -c_i \frac{\partial}{\partial y_i} - b_i^2 \frac{\partial}{\partial x_i} - a_{ij} \frac{\partial^2}{\partial y_i \partial x_j},$$

$$A_5 = -c_i \frac{\partial}{\partial x_i} - a_{ij} \frac{\partial^2}{\partial x_i \partial x_j}.$$

We first define $m(x, y, z)$ such that

$$(7.1.6) \qquad A_1^* m = 0, \qquad m \text{ periodic in } z, \qquad \int_Z m(x, y, z)\, dz = 1.^{25}$$

From Theorem 3.5, we know that there exists one and only one function m of z $(x, y$ being merely parameters), which is "sufficiently" smooth and $m > 0$. The

———

[25] $Z = Y =]0, 1[^n$.

smoothness of m is easily determined by the smoothness of the coefficients. We first assume that

(7.1.7) $$\int_Z b^1(x, y, z)m(x, y, z)\, dz = 0.$$

The assumption (7.1.7) guarantees the existence of functions $\chi^\ell(x, y, z)$ such that

(7.1.8) $$A_1\chi^\ell = -b^1_\ell, \qquad \chi^\ell \text{ periodic in } z, \qquad \int_Z \chi^\ell(x, y, z)\, dz = 0.$$

To proceed in a reasonably intuitive manner, we will recall the result obtained in the self-adjoint case (cf. Section 8, Chapter 1). The homogenized coefficients were obtained in a two stage procedure: first homogenization in z, letting y be a parameter, next homogenization in y of the operator obtained after the first stage.

Within the present context, we can rely on the results of the previous paragraphs to assert that the homogenized coefficients after the first stage are give by (cf. (6.3.6) and (6.2.10))

(7.1.9) $$q^1_{ij}(x, y) = \int_Z m(x, y, z)\left[a_{ij}(x, y, z) - a_{ik}\frac{\partial \chi^j}{\partial z_k} - a_{jk}\frac{\partial \chi^i}{\partial z_k}\right.$$
$$\left. - \frac{1}{2}b^1_i(x, y, z)\chi^j - \frac{1}{2}b^1_j(x, y, z)\chi^i\right] dy,$$

(7.1.10) $$r^1_i(x, y) = \int_Z m(x, y, z)\left[b^2_i - \frac{\partial \chi^i}{\partial z_j}b^2_j - \frac{\partial \chi^i}{\partial y_j}b^1_j - 2a_{jk}\frac{\partial^2 \chi^i}{\partial y_j \partial z_k}\right] dz.$$

Formulas (7.1.9), (7.1.10) correspond to the homogenization problem where y is merely a parameter, the diffusion term is $a\left(x, y, \frac{x}{\varepsilon}\right)$, the drift term is $\frac{1}{\varepsilon}b^1\left(x, y, \frac{x}{\varepsilon}\right) + b^2\left(x, y, \frac{x}{\varepsilon}\right)$.

We next define

(7.1.11) $$\mathcal{A}^\varepsilon_1 = -q^1_{ij}\left(x, \frac{x}{\varepsilon}\right)\frac{\partial^2}{\partial x_i \partial x_j} - \frac{1}{\varepsilon}r^1_i\left(x, \frac{x}{\varepsilon}\right)\frac{\partial}{\partial x_i}.$$

We write with standard notation

(7.1.12) $$\mathcal{A}^\varepsilon_1 = \frac{1}{\varepsilon^2}\mathcal{A}_1 + \frac{1}{\varepsilon}\mathcal{A}_2 + \mathcal{A}_3,$$

where

(7.1.13) $$\mathcal{A}_1 = -q^1_{ij}\frac{\partial^2}{\partial y_i \partial y_j} + r^1_i\frac{\partial}{\partial y_i},$$
$$\mathcal{A}_2 = -2q^1_{ij}\frac{\partial^2}{\partial x_i \partial y_j} - r^1_i\frac{\partial}{\partial x_i},$$
$$\mathcal{A}_3 = -q^1_{ij}\frac{\partial^2}{\partial x_i \partial x_j}.$$

We then define the function $m^1(x, y)$ satisfying the analogue of (7.1.6), i.e.

(7.1.14) $$\mathcal{A}^*_1 m^1 = 0, \qquad m^1 \text{ periodic in } y, \qquad \int_Y m^1(x, y)\, dy = 1.$$

We have then to assume that

(7.1.15) $$\int_Y r^1(x, y)m^1(x, y)\, dy = 0.$$

Then we can define the functions $K^\ell(x, y)$ solutions of

$$(7.1.16) \qquad \mathcal{A}_1 K^\ell = -r_\ell^1, \qquad K^\ell \text{ periodic in } y, \qquad \int_Y K^\ell(x, y)\, dy = 0.$$

The final homogenization diffusion term will be defined as follows
$$(7.1.17)$$
$$q_{ij}(x) = \int_Y m^1(x, y) \left[q_{ij}^1(x, y) - q_{ik}^1 \frac{\partial K^j}{\partial y_k} - q_{jk}^1 \frac{\partial K^i}{\partial y_k} - \frac{1}{2} r_i^1 K^j - \frac{1}{2} r_j^1 K^i \right] dy.$$

To make precise the final homogenization drift term, we need some other functions. We introduce the functions ζ^ℓ solutions of

$$(7.1.18) \qquad A_1 \zeta^\ell = -\frac{\partial^2 K^\ell}{\partial y_i \partial y_j} \left(a_{ij} - a_{ik} \frac{\partial \chi^j}{\partial z_k} - a_{jk} \frac{\partial \chi^i}{\partial z_k} - \frac{1}{2} b_i^1 \chi^j - \frac{1}{2} b_j^1 \chi^i \right)$$
$$\frac{\partial K^\ell}{\partial y_i} \left(b_i^2 - b_j^2 \frac{\partial \chi^i}{\partial z_j} - \frac{\partial \chi^i}{\partial y_j} b_j^1 - 2 a_{jk} \frac{\partial^2 \chi^i}{\partial y_j \partial z_k} \right)$$
$$+ b^2 - \frac{\partial \chi^\ell}{\partial z_j} b_j^2 - \frac{\partial \chi^\ell}{\partial y_j} b_j^1 - 2 a_{jk} \frac{\partial^2 \chi^\ell}{\partial y_j \partial z_k},$$

$$\zeta^\ell \text{ periodic in } z, \qquad \int_Z \zeta^\ell \, dy = 0.$$

The functions ζ^ℓ are well defined since the mean value in z of the right hand side of (7.1.18), in z using the weight $m(x, y, z)$, is nothing other than $\mathcal{A}_1 K^\ell + r_\ell^1$, which is 0 according to the choice of functions K^ℓ.

With this definition we set

$$(7.1.19) \qquad r_i(x) = \int_Y dy\, m^1(x, y) \int_Z dz\, m(x, y, z) \left[b_j^2 \frac{\partial \zeta^i}{\partial z_j} + b_j^1 \frac{\partial \zeta^i}{\partial y_j} \right.$$
$$+ 2 a_{jk} \frac{\partial^2 \zeta^i}{\partial y_k \partial z_j} + b_j^2 \frac{\partial}{\partial y_j} \left(-\chi^i + \chi^\ell \frac{\partial K^i}{\partial y_\ell} \right)$$
$$+ a_{jk} \frac{\partial^2}{\partial y_j \partial y_k} \left(\chi^\ell \frac{\partial K^i}{\partial y_\ell} \right) - a_{jk} \frac{\partial^2 \chi^i}{\partial y_k \partial y_j}$$
$$+ 2 a_{jk} \frac{\partial^2}{\partial x_j \partial z_k} \left(-\chi^i + \chi^\ell \frac{\partial K^i}{\partial y_\ell} \right)$$
$$+ b_j^1 \frac{\partial}{\partial x_j} \left(-\chi^i + \chi^\ell \frac{\partial K^i}{\partial y_\ell} \right) + c_j \frac{\partial}{\partial z_j} \left(-\chi^i + \chi^\ell \frac{\partial K^i}{\partial y_\ell} \right)$$
$$\left. - 2 a_{jk} \frac{\partial^2 K^i}{\partial x_j \partial y_k} - b_j^2 \frac{\partial K^i}{\partial x_j} - c_j \frac{\partial K^i}{\partial y_j} + c_i \right].$$

With the coefficients q_{ij}, r_i, one can define the Itô process $y(t) = y_x(t)$ give by

$$(7.1.20) \qquad dy = r(y)\, dt + \sqrt{2} \sqrt{q(t)}\, dw(t), \qquad y(0) = x.$$

Let μ_x be the corresponding probability measure on Ψ. One can state the following

THEOREM 7.1. *Under the assumptions (7.1.1), (7.1.2), (7.1.3) and (7.1.7), (7.1.15), we have*

$$(7.1.21) \qquad \mu_x^\varepsilon \to \mu_x \text{ in } \mathcal{M}_+^1(\Psi) \text{ weakly.}$$

REMARK 7.2. To obtain the q_{ij} one uses the reiterated homogenization rule. We first consider the coefficients $a_{ij}\left(x, y, \frac{x}{\varepsilon}\right)$, $\frac{1}{\varepsilon}b_i^1\left(x, y, \frac{x}{\varepsilon}\right) + b_i^2\left(x, y, \frac{x}{\varepsilon}\right)$ where y is a parameter, then we define the corresponding homogenization coefficients $q_{ij}^1(x, y), r_i^1(x, y)$.

We next consider the homogenization problem for the diffusion with coefficients $q_{ij}^1\left(x, \frac{x}{\varepsilon}\right), \frac{1}{\varepsilon}r i^1\left(x, \frac{x}{\varepsilon}\right)$. We apply the standard homogenization rule to obtain $q_{ij}(x)$. To obtain $r_i(x)$ a more complicated procedure is required.

REMARK 7.3. Let us consider the case when

$$b_i^1(x, y, z) = \frac{\partial}{\partial z_j} a_{ij}(x, y, z),$$

$$b_i^2(x, y, z) = \frac{\partial}{\partial y_j} a_{ij}(x, y, z),$$

$$c_i(x, y, z) = \frac{\partial}{\partial x_j} a_{ij}(x, y, z).$$

This is the situation when

$$A = -\frac{\partial}{\partial x_i}\left(a_{ij}\left(x, \frac{x}{\varepsilon}, \frac{x}{\varepsilon^2}\right)\frac{\partial}{\partial x_j}\right).$$

We clearly have $m = 1$, and (7.1.7) is satisfied. We have

$$q_{ij}^1(x, y) = \int_Z \left[a_{ij} - \frac{1}{2}a_{ik}\frac{\partial \chi^j}{\partial z_k} - \frac{1}{2}a_{jk}\frac{\partial \chi^i}{\partial z_k}\right] dz$$

$$= \int_Z \left[a_{ij} - a_{jk}\frac{\partial \chi^i}{\partial z_k}\right] dz$$

since

$$\int_Z a_{jk}\frac{\partial \chi^i}{\partial z_k} dy = \int_Z a_{ik}\frac{\partial \chi^j}{\partial z_k} dz = \int_Z a_{kj}\frac{\partial \chi^j}{\partial z_\ell}\frac{\partial \chi^i}{\partial z_k} dz.$$

Next we have

$$r_i^1(x, y) = \frac{\partial}{\partial y_j} q_{ij}^1.$$

It follows that $\mathcal{A}_1 = \mathcal{A}_1^*$ and $m^1 = 1$. The assumption (7.1.15) is satisfied. We then have

$$q_{ij}(x) = \int_Y \left[q_{ij}^1(x, y) - q_{jk}^1\frac{\partial K^i}{\partial y_k}\right] dy$$

i.e.,

(7.1.22) $$q = \int_Y dy \int_Z dz \left(I - \frac{\partial K}{\partial y}\right)\left(I - \frac{\partial \chi}{\partial z}\right) a.$$

Let us compute the drift term $r_i(x)$, using (7.1.19). We first have

$$b_j^2\frac{\partial \zeta}{\partial z_j} + b_j^1\frac{\partial \zeta^i}{\partial y_j} + 2a_{jk}\frac{\partial^2 \zeta^i}{\partial y_k \partial z_j}$$

$$= \frac{\partial a_{jk}}{\partial y_k}\frac{\partial \zeta^i}{\partial z_j} + \frac{\partial a_{jk}}{\partial z_k}\frac{\partial \zeta^i}{\partial y_j} + 2a_{jk}\frac{\partial^2 \zeta^i}{\partial y_k \partial z_j}$$

$$= \frac{\partial}{\partial y_k}\left(a_{jk}\frac{\partial \zeta^i}{\partial z_j}\right) + \frac{\partial}{\partial z_j}\left(a_{jk}\frac{\partial \zeta^i}{\partial y_k}\right),$$

and the mean $\iint_{Y \times Z} dy\, dz$ vanishes.

Then we have

$$
b_j^2 \frac{\partial}{\partial y_j} \left(-\chi^i + \chi^\ell \frac{\partial K^i}{\partial y_\ell} \right) + a_{jk} \frac{\partial^2}{\partial y_j \partial y_k} \left(\chi^\ell \frac{\partial \chi^i}{\partial y_\ell} \right) - a_{jk} \frac{\partial^2 \chi^i}{\partial y_k \partial y_j}
$$

$$
= \frac{\partial a_{kj}}{\partial y_k} \frac{\partial}{\partial y_j} \left(-\chi^i + \chi^\ell \frac{\partial K^i}{\partial y_\ell} \right) + a_{jk} \frac{\partial^2}{\partial y_j \partial y_k} \left(\chi^\ell \frac{\partial K^i}{\partial y_\ell} \right) - a_{jk} \frac{\partial^2 \chi^i}{\partial y_k \partial y_j}
$$

$$
= -\frac{\partial}{\partial y_k} \left(a_{kj} \frac{\partial \chi^i}{\partial y_j} \right) + \frac{\partial}{\partial y_k} \left(a_{kj} \frac{\partial}{\partial y_j} \left(\chi^\ell \frac{\partial K^i}{\partial y_\ell} \right) \right)
$$

and the corresponding mean is again 0. From (7.1.19)

$$
r_i(x) = \int_Y dy \int_Z dz \left[2a_{jk} \frac{\partial^2}{\partial x_j \partial z_k} \left(-\chi^i + \chi^\ell \frac{\partial K^i}{\partial y_\ell} \right) \right.
$$

$$
+ \frac{\partial a_{jk}}{\partial z_k} \frac{\partial}{\partial x_j} \left(-\chi^i + \chi^\ell \frac{\partial K^i}{\partial y_\ell} \right) + \frac{\partial a_{jk}}{\partial x_j} \frac{\partial}{\partial z_k} \left(-\chi^i + \chi^\ell \frac{\partial K^i}{\partial y_\ell} \right)
$$

$$
\left. -2a_{jk} \frac{\partial^2 K^i}{\partial x_j \partial y_k} - \frac{\partial a_{jk}}{\partial y_k} \frac{\partial K^i}{\partial x_j} - \frac{\partial a_{jk}}{\partial x_k} \frac{\partial K^i}{\partial y_j} + \frac{\partial a_{ij}}{\partial x_j} \right]
$$

$$
= \int_Y dy \int_Z dz \left[\frac{\partial}{\partial x_j} a_{jk} \frac{\partial}{\partial z_k} \left(-\chi^i + \chi^\ell \frac{\partial K^i}{\partial y_\ell} \right) \right.
$$

$$
\left. -\frac{\partial}{\partial x_j} \left(a_{jk} \frac{\partial K^i}{\partial y_k} \right) + \frac{\partial a_{ij}}{\partial x_j} \right]
$$

$$
= \frac{\partial}{\partial x_j} q_{ij}.
$$

Thus the homogenized operator is

$$
\mathcal{A} = -\frac{\partial}{\partial x_i} q_{ij}(x) \frac{\partial}{\partial x_j},
$$

where q_{ij} is given by the reiterated homogenization procedure. We recover the results obtained in Chapter 1, Section 8.

7.2. Proof of Theorem 7.1. We write $K(x, y)$ for the vector $(K^1, \ldots, K^\ell, \ldots, K^n)$ and $\chi(x, y, z)$ for $(\chi^1, \ldots, \chi^n)$. We set

$$
(7.2.1) \qquad\qquad L(x, y, z) = + \left(I - \frac{\partial K}{\partial y} \right) \chi.
$$

Then by construction L is a solution of

$$
(7.2.2) \qquad\qquad A_1 L + A_2 K = -b^1.
$$

We next define $M(x, y, z)$ as being the solution of

$$
(7.2.3) \qquad\qquad A_1 M + A_2 L + A_3 K = -b^2, \qquad M \text{ periodic in } z.
$$

To show that M is well defined, we note that

$$b_\ell^2 - A_2 L^\ell - A_3 K^\ell = b_\ell^2 - b_i^1 \frac{\partial}{\partial y_i} \left(\chi^\ell - \frac{\partial K^\ell}{\partial y_k} \chi^k \right)$$
$$- b_i^2 \frac{\partial}{\partial z_i} \left(\chi^\ell - \frac{\partial K^\ell}{\partial y_k} \chi^k \right)$$
$$- 2 a_{ij} \frac{\partial}{\partial y_i \partial z_j} \left(\chi^\ell - \frac{\partial K^\ell}{\partial y_k} \chi^k \right)$$
$$- b_i^1 \frac{\partial}{\partial x_i} K^\ell - b_i^2 \frac{\partial K^\ell}{\partial y_i} - a_{ij} \frac{\partial^2 K^\ell}{\partial y_i \partial y_j}.$$

Taking the mean in z (with the weight m), we obtain

$$\int_Z m(b_\ell^2 - A_2 L^\ell - A_3 K^\ell) \, dz = r_\ell^1(x,y) + \mathcal{A}_1 K^\ell = 0,$$

and thus the solvability condition for (7.2.3) is satisfied.

Let us define next the vector function

(7.2.4) $$\theta^\varepsilon \left(x, \frac{x}{\varepsilon}, \frac{x}{\varepsilon^2} \right) = \varepsilon K \left(x, \frac{x}{\varepsilon} \right) + \varepsilon^2 L \left(x, \frac{x}{\varepsilon}, \frac{x}{\varepsilon^2} \right) + \varepsilon^3 M \left(x, \frac{x}{\varepsilon}, \frac{x}{\varepsilon^2} \right).$$

We have from the expression of A^ε, and (7.2.2), (7.2.3)

$$A^\varepsilon \theta^\varepsilon = -\frac{b^1}{\varepsilon^2} - \frac{b^2}{\varepsilon} + O^\varepsilon(1)$$

where $|O^\varepsilon(1)| \leq C$.

We then apply Itô's formula to compute $\theta^\varepsilon \left(y^\varepsilon(t), \frac{y^\varepsilon(t)}{\varepsilon}, \frac{y^\varepsilon(t)}{\varepsilon^2} \right)$ and we obtain

$$\theta^\varepsilon \left(y^\varepsilon(t), \frac{y\varepsilon(t)}{\varepsilon}, \frac{y^\varepsilon(t)}{\varepsilon^2} \right) = \theta^\varepsilon \left(x, \frac{x}{\varepsilon}, \frac{x}{\varepsilon^2} \right) + \int_0^t D\theta^\varepsilon \cdot \sigma \, dw(s)$$
$$- \int_0^t A^\varepsilon \theta^\varepsilon \left(y^\varepsilon(s), \frac{y^\varepsilon(s)}{\varepsilon}, \frac{y^\varepsilon(s)}{\varepsilon^2} \right) \, ds.$$

Therefore

(7.2.5) $$y^\varepsilon(t) - \theta^\varepsilon \left(y^\varepsilon(t), \frac{y^\varepsilon(t)}{\varepsilon}, \frac{y^\varepsilon(t)}{\varepsilon^2} \right) = x - \theta^\varepsilon \left(x, \frac{x}{\varepsilon}, \frac{x}{\varepsilon^2} \right)$$
$$+ \int_0^t \left(c \left(y^\varepsilon(s), \frac{y^\varepsilon(s)}{\varepsilon}, \frac{y^\varepsilon(s)}{\varepsilon^2} \right) + O^\varepsilon(1) \right) \, ds$$

() $$+ \int_0^t (I - D\theta^\varepsilon) \cdot \sigma \, dw(s).*$$

Since

$$D\theta^\varepsilon = \frac{\partial K}{\partial y} + \frac{\partial L}{\partial z} + O(\varepsilon),$$

it follows from (7.2.5), reasoning as after (4.3.14), that μ_x^ε remains in a compact subset of $\mathcal{M}_+^1(\Psi)$.

To prove that μ_x^ε has a limit and to identify it, we proceed as in Section 6.3. From the martingale formulation we have

(7.2.6) $$E_x^\varepsilon \left[\phi(x(t)) + \int_s^t A^\varepsilon \phi(x(\lambda)) \, d\lambda \; \middle| \; \mathcal{G}^s \right] = \phi(x(s))$$

for any $\phi \in C_b^2(\mathbb{R}^n)$. We choose in (7.2.6)

$$\phi = \phi^\varepsilon(x) = \phi_0(x) + \varepsilon\phi_1\left(x, \frac{x}{\varepsilon}\right) + \varepsilon^2\phi_2\left(x, \frac{x}{\varepsilon}, \frac{x}{\varepsilon^2}\right)$$
$$+ \varepsilon^3\phi_3\left(x, \frac{x}{\varepsilon}, \frac{x}{\varepsilon^2}\right) + \varepsilon^4\phi_4\left(x, \frac{x}{\varepsilon}, \frac{x}{\varepsilon^2}\right),$$

where $\phi_0 \in C_0^\infty(\mathbb{R}^n)$, and $\phi_1, \phi_2, \phi_3, \phi_4$ are determined by an asymptotic expansion. We write the conditions

(7.2.7) $$A_1\phi_2 + A_2\phi_1 + A_3\phi = 0,$$

(7.2.8) $$A_1\phi_3 + A_2\phi_2 + A_3\phi_1 + A_4\phi = 0,$$

(7.2.9) $$A_1\phi_4 + A_2\phi_3 + A_3\phi_2 + A_4\phi_1 + A_5\phi = \text{ function of } x \text{ only.}$$

Easy but tedious calculations show that

$$\phi_1(x, y) = -\frac{\partial\phi}{\partial x_k}K^k(x, y),$$

$$\phi_2(x, y, z) = -\chi^\ell(x, y, z)\left(\frac{\partial\phi}{\partial x_\ell} - \frac{\partial\phi}{\partial x_k}\frac{\partial K^k}{\partial y_\ell}\right) + \tilde{\phi}_2(x, y),$$

$$\phi_3(x, y, z) = -\frac{\partial\tilde{\phi}_2}{\partial y_k}\chi^k + \frac{\partial^2\phi}{\partial x_k\partial x_\ell}K^\ell\chi^k + \frac{\partial\phi}{\partial x_\ell}\zeta^\ell.$$

Writing the solvability condition for (7.2.9), we get an equation for $\tilde{\phi}_2(x, y)$, for which a solvability condition is again needed. Expressing it, one checks that one must choose in (7.2.9) the function of x only, to be equal to $\mathcal{A}\phi_0$, where of course

$$\mathcal{A} = -r_i\frac{\partial}{\partial x_i} - q_{ij}\frac{\partial^2}{\partial x_i\partial x_j}$$

with the definitions (7.1.19) and (7.1.17). Therefore (7.2.6) yields

$$E_x^\varepsilon\left[\phi^\varepsilon(x(t)) + \int_s^t (\mathcal{A}\phi_0 + O(\varepsilon))\, d\lambda \,\middle|\, \mathcal{G}^s\right] = \phi^\varepsilon(x(s)).$$

Taking a subsequence of μ_x^ε, converging weakly in $\mathcal{M}_+^1(\Psi)$ toward μ_x, we obtain

$$E_x\left[\phi_0(x(t)) + \int_0^t \mathcal{A}\phi_0(x(\lambda))\, d\lambda \,\middle|\, \mathcal{G}^s\right] = \phi_0(x(s))$$

and from the uniqueness of the solution of the martingale problem, we obtain that μ_x is the diffusion connected with (7.1.20), which completes the proof of the theorem. ∎

REMARK 7.4. We leave to the reader to express the convergence result for the corresponding Dirichlet problem.

REMARK 7.5. We can consider in the same manner drift terms of the form

$$\frac{1}{\varepsilon^3}b_1 + \frac{1}{\varepsilon^2}b_2 + \frac{1}{\varepsilon}b_3 + c,$$

with adequate assumptions.

8. Problems with potentials

We consider here the situation of potentials a_0 which are not always positive and we consider also some generalizations to include bounded potentials.

8.1. A variant of Theorem 6.7. We will first notice that Theorem 6.7 holds true with the assumption (6.4.1) replaced by

(8.1.1) $a_0(x, y)$ measurable and bounded, periodic in y,

twice continuously differentiable in x, once differentiable

in y, with all the derivatives being globally bounded,

(8.1.2) $$\int_Y a_0(x, y)m(x, y) \, dy \geq 0.$$

We can state the

THEOREM 8.1. *Under the assumptions of Theorem 6.7 with (6.4.1) replaced by (8.1.1), (8.1.2), then the same conclusions hold true.*

REMARK 8.2. Theorem 8.1 is not exactly a generalization of Theorem 6.7, since some smoothness is required for a_0, unlike Theorem 6.7. However, as far as the positivity of a_0 is concerned, it states that only the positivity in the average is important.

PROOF OF THEOREM 8.1. We have denoted by $\bar{a}_0(x)$ the quantity (cf. (6.2.27))

$$\bar{a}_0(x) = \int_Y a_0(x, y)m(x, y) \, dy.$$

We consider the equation (6.4.3). Since a_0 is no more positive, the existence of u_ε is not a consequence of standard results. However we will first prove that at least for ε small enough, there exists one and only one solution of (6.4.3). This will follow from a change of unknown functions. First of all we introduce the function $\theta(x, y)$ solution of

(8.1.3) $$- a_{ij}\frac{\partial^2 \theta}{\partial y_i \partial y_j} - b_i\frac{\partial \theta}{\partial y_i} = -a_0 + \bar{a}_0,$$

$$\theta \text{ periodic in } y, \qquad \int_Y \theta \, dy = 0.$$

By virtue of the regularity assumptions on a_0, we can assert that

(8.1.4) $$\frac{\partial \theta}{\partial x_i}, \frac{\partial \theta}{\partial y_i}, \frac{\partial^2 \theta}{\partial y_i \partial x_j}, \frac{\partial^2 \theta}{\partial x_i \partial x_j} \text{ are bounded globally in } x, y.$$

We look for u_ε in the following form:

(8.1.5) $$u_\varepsilon(x) = \left(1 + \varepsilon^2 \theta\left(x, \frac{x}{\varepsilon}\right)\right) \tilde{u}_\varepsilon(x).$$

For ε small enough the knowledge of u_ε is equivalent to the knowledge of $\tilde{u}_\varepsilon$. But from (6.4.3) we easily check that $\tilde{u}_\varepsilon$ must be a solution of (raking (8.1.3) into account)

$$(8.1.6) \quad -a_{ij}(1+\varepsilon^2\theta)\frac{\partial^2\tilde{u}_\varepsilon}{\partial x_i\partial x_j} - \frac{1}{\varepsilon}b_i\frac{\partial\tilde{u}_\varepsilon}{\partial x_i} - c_i\frac{\partial\tilde{u}_\varepsilon}{\partial x_i}$$

$$-\varepsilon\frac{\partial\tilde{u}_\varepsilon}{\partial x_i}\left[2a_{ij}\frac{\partial\theta}{\partial y_j} + \varepsilon\frac{\partial\theta}{\partial x_j}a_{ij} + b_i + \varepsilon\theta c_i\right]$$

$$+\bar{a}_0 u_\varepsilon + \varepsilon\tilde{u}_\varepsilon\left[-2a_{ij}\frac{\partial^2\theta}{\partial x_i\partial y_j} - \varepsilon a_{ij}\frac{\partial^2\theta}{\partial x_i\partial x_j} - b_i\frac{\partial\theta}{\partial x_i}\right.$$

$$\left. -c_i\frac{\partial\theta}{\partial y_i} - \varepsilon c_i\frac{\partial\theta}{\partial x_i} + \varepsilon a_0\theta\right] = f.$$

Since $\bar{a}_0 \geq 0$, it follows that at least for ε small enough the problem (8.1.6) with the conditions

$$(8.1.7) \qquad \tilde{u}_\varepsilon\mid_\Gamma = 0, \qquad \tilde{u}_\varepsilon \in H^2(\mathcal{O})\cap H_0^1(\mathcal{O})$$

is well posed. Therefore $\tilde{u}_\varepsilon$ (and also u_ε) are well defined, in a unique way. It is also clear that u_ε and $\tilde{u}_\varepsilon$ have the same limit in H_0^1, since

$$\|u_\varepsilon - \tilde{u}_\varepsilon\|_{H_0^1(\mathcal{O})} \to 0 \text{ as } \varepsilon \to 0,$$

provided that we prove that $\tilde{u}_\varepsilon$ remains in a bounded subset of $H_0^1(\mathcal{O})$. Therefore it is enough to study the limit of $\tilde{u}_\varepsilon^1$ in $H_0^1(\mathcal{O})$. This can be done in an easy variant of the proof of Theorem 6.7. We will just briefly sketch the main simple changes. Using the notation $\tilde{a}_{ij}^\varepsilon$, $\tilde{b}_i^\varepsilon$, $\tilde{c}_i^\varepsilon$, $\tilde{f}^\varepsilon$ with the same meaning as in the proof of Theorem 6.7, we easily check that $\tilde{u}_\varepsilon$ is a solution of

$$(8.1.8) \qquad -\frac{\partial}{\partial x_i}(\tilde{a}_{ij}^\varepsilon + \tilde{b}_{ij}^\varepsilon)\frac{\partial\tilde{u}_\varepsilon}{\partial x_j} - \frac{1}{\varepsilon}\beta_i^\varepsilon\frac{\partial\tilde{u}_\varepsilon}{\partial x_i} - \varepsilon\tilde{\gamma}_i^\varepsilon\frac{\partial\tilde{u}_\varepsilon}{\partial x_i}$$

$$+\bar{a}_0 m^\varepsilon\tilde{u}_\varepsilon + \tilde{b}_0^\varepsilon\tilde{u}_\varepsilon = \tilde{f}^\varepsilon, \qquad \tilde{u}_\varepsilon\mid_\Gamma = 0,$$

where we have set

$$(8.1.9) \qquad \tilde{b}_{ij}^\varepsilon(x,y) = \varepsilon a_{ij}(x,y)\theta(x,y)m^\varepsilon(x,y),$$

$$(8.1.10) \qquad \tilde{\gamma}_i^\varepsilon(x,y) = m^\varepsilon(x,y)\left[2a_{ij}(x,y)\frac{\partial\theta}{\partial y_j} + \varepsilon\frac{\partial\theta}{\partial x_j}a_{ij} + \theta b_i + \varepsilon\theta c_i\right.$$

$$\left. -\frac{\partial}{\partial y_j}(a_{ij}m^\varepsilon\theta) - \varepsilon\frac{\partial}{\partial x_j}(a_{ij}m^\varepsilon\theta)\right],$$

$$(8.1.11) \qquad b_0(x,y) = -2a_{ij}\frac{\partial^2\theta}{\partial x_i\partial y_j} - a_{ij}\frac{\partial^2\theta}{\partial x_i\partial x_j} - b_i\frac{\partial\theta}{\partial x_i}$$

$$-c_i\frac{\partial\theta}{\partial y_i} - c_i\frac{\partial\theta}{\partial x_i} + \varepsilon a_0\theta.$$

Multiplying (8.1.8) by $\tilde{u}_\varepsilon$ and integrating, we get by virtue of the properties of $\bar{\beta}_i^\varepsilon$, that

$$(8.1.12) \quad \int\tilde{a}_{ij}^\varepsilon\frac{\partial\tilde{u}_\varepsilon}{\partial x_j}\frac{\partial\tilde{u}_\varepsilon}{\partial x_i}\,dx + \varepsilon\int\tilde{b}_{ij}^\varepsilon\frac{\partial\tilde{u}_\varepsilon}{\partial x_j}\frac{\partial\tilde{u}_\varepsilon}{\partial x_i}\,dx + \int O(\varepsilon)(\tilde{u}_\varepsilon)^2\,dx$$

$$-\varepsilon\int\tilde{\gamma}^\varepsilon\frac{\partial\tilde{u}_\varepsilon}{\partial x_i}\tilde{u}_\varepsilon\,dx + \int\bar{a}_0 m^\varepsilon(\tilde{u}_\varepsilon)^2\,dx + \varepsilon\int\tilde{b}^\varepsilon(\tilde{u}_\varepsilon)^2\,dx = \int\tilde{f}^\varepsilon\tilde{u}_\varepsilon\,dx.$$

Since $\tilde{b}_{ij}^\varepsilon$, $\tilde{\gamma}_i^\varepsilon$, $\tilde{b}_0^\varepsilon$ are bounded, we deduce from (8.1.12) and from Poincaré's inequality, that at least for ε small enough, we have

$$(8.1.13) \qquad \qquad \|\tilde{u}_\varepsilon\|_{H_0^1(\mathcal{O})} \le C.$$

Therefore also $\|u_\varepsilon\|_{H_0^1(\mathcal{O})} \le C$. To identify the limit of u_ε, we proceed with the equation of u_ε (and not with the equation of $\tilde{u}_\varepsilon$), as in the proof of Theorem 6.7. The proof is actually identical, hence the result follows. ∎

8.2. A general problem with potentials. We consider here the following problem

$$(8.2.1) \qquad -a_{ij}\left(x,\frac{x}{\varepsilon}\right)\frac{\partial^2 u_\varepsilon}{\partial x_i \partial x_j} - \left(\frac{1}{\varepsilon}b_i\left(x,\frac{x}{\varepsilon}\right) + c_i\left(x,\frac{x}{\varepsilon}\right)\right)\frac{\partial u_\varepsilon}{\partial x_i}$$

$$+ \left(\frac{1}{\varepsilon^2}w_1\left(x,\frac{x}{\varepsilon}\right) + \frac{1}{\varepsilon}w_2\left(x,\frac{x}{\varepsilon}\right) + a_0\left(x,\frac{x}{\varepsilon}\right)\right)u_\varepsilon = f,$$

$$u_\varepsilon\mid_\Gamma = 0.$$

We will show mainly that (under some adequate assumptions) one can reduce problem (8.2.1) to the problem considered in Theorem 8.1, although equation (6.4.3) seems to be only a very particular case of (8.2.1), (when w_1, w_2 are equal to 0).

We assume that

$$(8.2.2) \qquad a_{ij}, b_i, c_i \text{ satisfy conditions } (6.1.1), (6.1.2);$$

$$(8.2.3) \qquad \text{there exists a solution } \psi(x,y) \text{ of}$$

$$-a_{ij}\frac{\partial^2\psi}{\partial y_i \partial y_j} - b_i\frac{\partial\psi}{\partial y_i} + w_1\psi = 0,$$

periodic in y, $\psi(x,y) \ge \delta > 0$, and ψ is sufficiently

regular in x, y, so that the quantities

$$a_{ij}\psi, \qquad 2a_{ij}\frac{\partial\psi}{\partial y_j} + \psi b_i, \qquad 2a_{ij}\frac{\partial\psi}{\partial x_j} + \psi c_i,$$

satisfy the same regularity assumptions as a_{ij}, b_i, c_i,

respectively;

$$(8.2.4) \qquad \text{the quantity } 2a_{ij}\frac{\partial^2\psi}{\partial x_i \partial y_j} + b_i\frac{\partial\psi}{\partial x_i} + c_i\frac{\partial\psi}{\partial y_i} + 2_w\psi \text{ is}$$

periodic in y, is twice continuously differentiable

in x, once in y, all derivatives being bounded.

We denote by A^{w_1} (or $A_1^{w_1}$) the differential operator

$$(8.2.5) \qquad A^{w_1} = -a_{ij}\frac{\partial^2}{\partial y_i \partial y_j} - b_i\frac{\partial}{\partial y_i} + w_1,$$

and by $A = A_1$ the differential operator

$$(8.2.6) \qquad A = A_1 = -a_{ij}\psi\frac{\partial^2}{\partial y_i \partial y_j} - \left(2a_{ij}\frac{\partial\psi}{\partial y_j} + \psi b_i\right)\frac{\partial}{\partial y_i}.$$

When $w_1 = 0$, we have $\psi = 1$ and $A^0 = A = A_1$, the usual operator which has been considered up to now. Since $a_{ij}, 2a_{ij}\frac{\partial\psi}{\partial y_j} + \psi b_i$, satisfy the same regularity assumptions as a_{ij}, b_i respectively and since $\psi \ge \delta > 0$, the Fredholm alternative applies to the operator A, and therefore there exists one and only one function m such that

$$(8.2.7) \qquad -\frac{\partial}{\partial y_i \partial y_j}(a_{ij}\psi m) + \frac{\partial}{\partial y_i}\left(\left(2a_{ij}\frac{\partial \psi}{\partial y_j} + \psi b_i\right)m\right) = 0,$$

$$\int_Y m(x,y)\,dy = 1, \qquad m \text{ periodic in } y, \qquad m > 0,$$

with the standard regularity assumptions on m.

A simple calculation shows that m is a solution of

$$(8.2.8) \qquad -\frac{\partial^2}{\partial y_i \partial y_j}(a_{ij}m) + \frac{\partial}{\partial y_i}(b_i m) + W_1 m = 0.$$

Since conversely a solution of (8.2.8) satisfies (8.2.7), it follows from the uniqueness of the solution of (8.2.7), that (8.2.8) has a unique solution satisfying the additional conditions of (8.2.7).

Applying again the Fredholm alternative to (8.2.8) which is the formal adjoint of A^{W_1}, we can assert that the function ψ satisfying the conditions (8.2.3) is necessarily unique (up to a multiplicative constant in y).

We then assume that

$$(8.2.9) \qquad \int_Y m\left(b_i\psi + 2a_{ij}\frac{\partial \psi}{\partial y_j}\right)dy = 0,$$

$$(8.2.10) \qquad \int_Y m\left(2a_{ij}\frac{\partial^2 \psi}{\partial x_i \partial y_j} + b_i\frac{\partial \psi}{\partial x_i} + c_i\frac{\partial \psi}{\partial y_i} + W_2\psi\right)dy = 0.$$

From the assumption (8.2.10) it follows that there exists one and only one solution $\chi^0(x,y)$ of the problem

$$(8.2.11) \qquad A^{W_1}\chi^0 = 2a_{ij}\frac{\partial^2 \psi}{\partial x_i \partial y_j} + b_i\frac{\partial \psi}{\partial x_i} + c_i\frac{\partial \psi}{\partial y_i} + W_2\psi,$$

$$\chi^0 \text{ periodic in } y, \qquad \int_Y \chi^0\,dy = 0.$$

We will finally assume that

$$(8.2.12) \qquad \int_Y m\left[a_0\psi - c_i\frac{\partial \psi}{\partial x_i} - a_{ij}\frac{\partial^2 \psi}{\partial x_i \partial x_j} + W_2\chi^0 - c_i\frac{\partial \chi^0}{\partial y_i}\right.$$
$$\left. -b_i\frac{\partial \chi^0}{\partial x_i} - 2a_{ij}\frac{\partial^2 \chi^0}{\partial x_i \partial y_j}\right]dy \geq 0.$$

We also define the functions χ^ℓ solutions of

$$(8.2.13) \qquad A^{W_1}\chi^\ell = -2a_{\ell j}\frac{\partial \psi}{\partial y_j} - b_\ell\psi,$$

$$\chi^\ell \text{ periodic in } y, \qquad \int_Y \chi^\ell\,dy = 0.$$

We define the homogenized coefficients as follows

$$(8.2.14) \qquad q_{ij}(x) = \int_Y m \left[a_{ij}\psi - a_{ik}\frac{\partial \chi^j}{\partial y_k} - a_{jk}\frac{\partial \chi^i}{\partial y_k} \right.$$
$$\left. -\frac{1}{2}b_i\chi^j - \frac{1}{2}b_j\chi^i \right] dy,$$

$$(8.2.15) \qquad r_i(x) = \int_Y m \left[c_i\psi - c_k\frac{\partial \chi^i}{\partial y_k} - 2a_{jk}\frac{\partial^2 \chi^i}{\partial x_k \partial y_k} - 2a_{jk}\frac{\partial^2 \chi^i}{\partial x_k \partial y_j} - b_k\frac{\partial \chi^i}{\partial x_k} \right.$$
$$\left. +2a_{ij}\frac{\partial \chi}{\partial x_j} + \chi^i W_2 + b_i\chi^0 + 2a_{ij}\frac{\partial \chi^0}{\partial y_j} \right] dy,$$

$$(8.2.16) \qquad \bar{a}_0(x) = \int_Y m \left[a_0\psi - c_i\frac{\partial \psi}{\partial x_i} - a_{ij}\frac{\partial^2 \psi}{\partial x_i \partial x_j} + W_2\chi^0 - c_i\frac{\partial \chi^0}{\partial y_i} \right.$$
$$\left. -b_i\frac{\partial \chi^0}{\partial x_i} - 2a_{ij}\frac{\partial^2 \chi^0}{\partial x_i \partial x_j} \right] dy.$$

Let u be the solution of

$$(8.2.17) \qquad -q_{ij}\frac{\partial^2 u}{\partial x_i \partial x_j} - r_i\frac{\partial u}{\partial x_i} + \bar{a}_0 u = f, \qquad u\mid_\Gamma = 0.$$

Then we can state the following

THEOREM 8.3. *Under the assumptions (8.2.2), (8.2.3), (8.2.4), (8.2.9), (8.2.10), (8.2.12) then we have*

$$(8.2.18) \qquad \frac{u_\varepsilon(x)}{\psi\left(x, \frac{x}{\varepsilon}\right)} \to u(x) \text{ in } H_0^1(\mathcal{O}) \text{ weakly.}$$

PROOF. We will reduce the proof of Theorem 8.3 to the one of theorem 8.1, by successive transformations.

We look for u_ε in the following form

$$(8.2.19) \qquad u_\varepsilon(x) = \left(\psi\left(x, \frac{x}{\varepsilon}\right) + \varepsilon\chi^0\left(x, \frac{x}{\varepsilon}\right) \right) \tilde{u}_\varepsilon(x).$$

Again for ε small enough, (8.2.19) defines $\tilde{u}_\varepsilon$ in a unique way. Taking into account the equations for ψ and χ^0, it follows from an easy calculation that $\tilde{u}_\varepsilon$ satisfies

$$(8.2.20) \qquad -a_{ij}(\psi + \varepsilon\chi^0)\frac{\partial^2 \tilde{u}_\varepsilon}{\partial x_i \partial x_j} - \frac{1}{\varepsilon}\frac{\partial \tilde{u}_\varepsilon}{\partial x_i}\left(2a_{ij}\frac{\partial \psi}{\partial y_j} + \psi b_i \right)$$
$$-\frac{\partial \tilde{u}_\varepsilon}{\partial x_i}\left(2a_{ij}\frac{\partial \psi}{\partial x_j} + 2a_{ij}\frac{\partial \chi^0}{\partial y_j} + \psi c_i + \chi^0 b_i \right)$$
$$-\frac{\partial \tilde{u}_\varepsilon}{\partial x_i}\left(2a_{ij}\frac{\partial \chi}{\partial x_j} + \chi^0 c_i \right) + \tilde{u}_\varepsilon\left(a_0\psi - c_i\frac{\partial \psi}{\partial x_i} - c_i\frac{\partial \chi^0}{\partial y_i} \right.$$
$$\left. -b_i\frac{\partial \chi^0}{\partial x_i} - a_{ij}\frac{\partial^2 \psi}{\partial x_i \partial x_j} - 2a_{ij}\frac{\partial^2 \chi}{\partial x_i \partial y_j} + W_2\chi^0 \right)$$
$$+\varepsilon\tilde{u}_\varepsilon\left(a_0\chi^0 - c_i\frac{\partial \chi}{\partial x_i} - a_{ij}\frac{\partial^2 \chi^0}{\partial x_i \partial x_j} \right) = f, \qquad \tilde{u}_\varepsilon\mid_\Gamma = 0.$$

Therefore $\tilde{u}_\varepsilon$ satisfies a problem similar to the one considered in Theorem 8.1. The assumptions made, especially (8.2.9) and (8.2.12) guarantee that $\tilde{u}_\varepsilon \to u$ in $H_0^1(\mathcal{O})$ weakly.

From (8.2.19) we see that $\frac{u_\varepsilon(x)}{\psi(x,\frac{x}{\varepsilon})}$ remains in a bounded subset of $H_0^1(\mathcal{O})$ and has the same limit as $\tilde{u}_\varepsilon$ in $L^2(\mathcal{O})$ strongly. Hence the desired result follows. ∎

9. Homogenization of reflected diffusion processes

9.1. Homogenization of the reflected diffusion processes. Let $\mathcal{O}$ be an open bounded subset of $\mathbb{R}^n$, satisfying the conditions of (4.4.1). We consider functions $a_{ij}(x,y)$, $b_i(x,y)$ satisfying the assumptions (6.1.1), (6.1.2).

Let next $\gamma(x)$ be a function such that

$$(9.1.1) \qquad \gamma : \partial\mathcal{O} \to \mathbb{R}^n, \; text continuous bounded, and such that$$
$$\gamma(x) \cdot \nabla\phi(x) \geq \beta > 0, \qquad x \in \partial\mathcal{O}, [26]$$

Defining $\sigma(x,y)$ such that

$$(9.1.2) \qquad\qquad \frac{1}{2}\sigma\sigma^* = a,$$

we can construct on an arbitrary system $(\Omega, \mathcal{A}, P, \mathcal{F}^t, w(t))$ the diffusion process reflected at the boundary of $\mathcal{O}$.

More precisely, for $\varepsilon > 0$ fixed, there exists a unique pair $y^\varepsilon(t)$, $\xi^\varepsilon(t)$ of processes satisfying the conditions

$$(9.1.3) \qquad dy^\varepsilon = b\left(y^\varepsilon, \frac{y^\varepsilon}{\varepsilon}\right) dt + \sigma\left(y^\varepsilon, \frac{y^\varepsilon}{\varepsilon}\right) dw(t) - \gamma(y^\varepsilon) d\xi^\varepsilon(t),$$
$$y^\varepsilon 0) = x \in \overline{\mathcal{O}}, \qquad \xi^\varepsilon(0) = 0,$$
$$y^\varepsilon(t), \; \xi^\varepsilon(t) \text{ are adapted continuous processes}$$
$$\text{with values in } \mathbb{R}^n \text{ and } \mathbb{R} \text{ respectively; } \xi^\varepsilon(t) \text{ is an}$$
$$\text{increasing process;}$$
$$y^\varepsilon(t) \in \overline{\mathcal{O}}, \; \forall t,$$
$$\chi_\mathcal{O}(y^\varepsilon(t)) d\xi^\varepsilon(t) = 0.$$

The last condition (9.1.3) means that $\int_{t_1}^{t_2} \chi_\mathcal{O}(y^\varepsilon(t)) d\xi^\varepsilon(t) = 0,[27]$ a.s., $\forall t_1, t_2$ with $t_1 \leq t_2$. For details see Watanabe [123] and [124].

We will denote by μ_x^ε the probability measure defined by $y^\varepsilon(\cdot)$ on $C^0[0; \infty; \overline{\mathcal{O}}]$.

Setting $\Psi = C^0[0; \infty; \overline{\mathcal{O}}]$, and $\psi \equiv x(\cdot) \in \Psi$, $x(t) = x(t; \psi)$, one can characterize μ_x^ε as the unique probability measure on Ψ, such that for any $\phi(x,t) \in C^{2,1}(\mathbb{R}^n \times [0, \infty[)$ with compact support, satisfying

$$(9.1.4) \qquad\qquad \frac{\partial\phi}{\partial x} \cdot \gamma(x) \geq 0 \text{ on } [0, \infty[\times\partial\mathcal{O},$$

---

[26] $\frac{\nabla\phi}{|\nabla\phi|} = n =$ unit outward normal.

[27] This is a Stieltjes integral.

then one has

$$(9.1.5) \quad E_x \left[\phi(x(t),t) - \phi(x(s),s) - \int_s^t \frac{\partial \phi}{\partial t}(x(\lambda),\lambda)\, d\lambda \right.$$

$$- \int_s^t \frac{\partial \phi}{\partial x}(x(\lambda),\lambda) \cdot b\left(x(\lambda),\frac{x(\lambda)}{\varepsilon}\right)\, d\lambda$$

$$\left. - \int_s^t \operatorname{tr} \frac{\partial^2 \phi}{\partial x^2}(x(\lambda),\lambda) a\left(x(\lambda),\frac{x(\lambda)}{\varepsilon}\right)\, d\lambda \,\Big|\, \mathcal{G}^s \right] \le 0,$$

$$x(0) = x, \ \text{a.s.} \ \mu_x^\varepsilon.$$

We recall that $\mathcal{G}^s = \sigma(x(\lambda), \lambda \le s)$.

This is the submartingale formulation of the reflected diffusion process (Stroock-Varadhan [114]).

We want to study the limit of μ_x^ε as ε goes to 0. We introduce the operator

$$(9.1.6) \qquad A = A_1 = -a_{ij}(x,y)\frac{\partial^2}{\partial y_i \partial y_j}$$

and let $m(x,y)$ be the unique solution of

$$(9.1.7) \qquad A_1^* m = 0, \qquad m \text{ periodic in } y, \qquad \int_Y m(x,y)\, dy = 1,$$

(cf. (6.1.8)). Let us consider the homogenized coefficients defined as follows

$$(9.1.8) \qquad q_{ij}(x) = \int_Y a_{ij}(x,y)m(x,y)\, dy$$

$$r_i(x) = \int_Y b_i(x,y)m(x,y)\, dy.$$

Let us consider now the reflected diffusion process corresponding to the coefficients $q_{ij}(x)$, $r_i(x)$, i.e., the pair of processes $\gamma(t)$, $\xi(t)$ satisfying the conditions

$$(9.1.9) \qquad dy = r(y)\, dt + \sqrt{2}\sqrt{q}(y)\, dw(t) - \gamma(y)\, d\xi(t),$$

$$y(0) = x \in \overline{\mathcal{O}}, \qquad \xi(0) = 0,$$

$$y(t), \ \xi(t) \text{ are adapted continuous processes with values}$$

$$\text{in } \mathbb{R}^n \text{ and } \mathbb{R}, \text{ respectively; } \xi(t) \text{ is an increasing process;}$$

$$y(t) \in \overline{\mathcal{O}}, \ \forall t,$$

$$\chi_{\mathcal{O}}(y(t))\, d\xi(t) = 0.$$

Let μ_x be the probability measure defined by $y(t)$ on Ψ. Then we can state the following.

THEOREM 9.1. *We assume that $\mathcal{O}$ satisfies (4.4.1) and a_{ij}, b_i satisfy (6.1.1), (6.1.2). We also assume (9.1.1). Then we have*

$$(9.1.10) \qquad \mu_x^\varepsilon \to \mu_x \ \text{in } \mathcal{M}_+^1(\Psi) \text{ weakly.}$$

9.2. Proof of convergence.

LEMMA 9.2. *The family μ_x^ε remains in a weakly compact subset of $\mathcal{M}_+^1(\Psi)$.*

PROOF. Let us apply Itô's formula to the functions ϕ and ϕ^2. We have

$$(9.2.1) \qquad \phi(y^\varepsilon(t_2)) - \phi(y^\varepsilon(t_1)) = \int_{t_1}^{t_2} \frac{\partial\phi}{\partial x}(y^\varepsilon(s)) \cdot b\left(y^\varepsilon, \frac{y^\varepsilon}{\varepsilon}\right) ds$$

$$+ \int_{t_1}^{t_2} \operatorname{tr} \frac{\partial^2\phi}{\partial x^2}(y^\varepsilon(s)) a\left(y^\varepsilon, \frac{y^\varepsilon}{\varepsilon}\right) ds$$

$$+ \int_{t_1}^{t_2} \frac{\partial\phi}{\partial x}(y^\varepsilon(s)) \cdot \sigma\left(y^\varepsilon, \frac{y^\varepsilon}{\varepsilon}\right) dw(s)$$

$$+ \int_{t_1}^{t_2} \gamma(y^\varepsilon(s)) \frac{\partial\phi}{\partial x}(y^\varepsilon(s)) \, d\xi^\varepsilon(s)$$

and since $\phi \mid_\Gamma = 0$

$$(9.2.2) \qquad \phi^2(y^\varepsilon(t_2)) - \phi^2(y^\varepsilon(t_1)) = \int_{t_1}^{t_2} 2\phi \frac{\partial\phi}{\partial x} \cdot \left(b\left(y^\varepsilon, \frac{y^\varepsilon}{\varepsilon}\right) ds\right.$$

$$\left. + \sigma\left(y^\varepsilon, \frac{y^\varepsilon}{\varepsilon}\right) dw(s)\right)$$

$$+ 2 \int_{t_1}^{t_2} \left(a \frac{\partial\phi}{\partial x} \cdot \frac{\partial\phi}{\partial x} + \phi \operatorname{tr} a \frac{\partial^2\phi}{\partial x^2}\right) ds.$$

We note the identity

$$(9.2.3) \quad (\phi(y^\varepsilon(t_2)) - \phi(y^\varepsilon(t_1)))^2 = \phi^2(y^\varepsilon(t_2)) - \phi^2(y(t_1))$$

$$- 2\phi(y^\varepsilon(t_1)) \cdot (\phi(y^\varepsilon(t_2)) - \phi(y^\varepsilon(t_1))).$$

Since $-\phi(y^\varepsilon(t_1)) \geq 0$ and $\gamma \cdot \frac{\partial\phi}{\partial x} \mid_\Gamma \geq 0$, it follows from (9.2.1), (9.2.2) and (9.2.3) that

$$(9.2.4) \quad (\phi(y^\varepsilon(t_2)) - \phi(y^\varepsilon(t_1)))^2$$

$$\leq \int_{t_1}^{t_2} 2\phi \frac{\partial\phi}{\partial x} \cdot (b \, ds + \sigma \, dw(s)) + 2 \int_{t_1}^{t_2} \left(a \frac{\partial\phi}{\partial x} \cdot \frac{\partial\phi}{\partial x} + \phi \operatorname{tr} a \frac{\partial^2\phi}{\partial x^2}\right) ds$$

$$- 2\phi(y^\varepsilon(t_1)) \left|\int_{t_1}^{t_2} \frac{\partial\phi}{\partial x} b \, ds + \int_{t_1}^{t_2} \operatorname{tr} \frac{\partial^2\phi}{\partial x^2} a \, ds + \int_{t_1}^{t_2} \frac{\partial\phi}{\partial x} \cdot \sigma \, dw\right|.$$

From (9.2.4) it follows that, using standard estimates on stochastic integrals, we have

$$(9.2.5) \qquad E(\phi(y^\varepsilon(t_2)) - \phi(y^\varepsilon(t_1)))^6 \leq C(t_2 - t_1)^{3/2}.$$

Going back to (9.2.1) we deduce that

$$(9.2.6) \qquad E\left(\int_{t_1}^{t_2} \gamma(y^\varepsilon(s)) \cdot \frac{\partial\phi}{\partial x}(y^\varepsilon(s)) \, d\xi^\varepsilon(s)\right)^6 \leq C(t_2 - t_1)^{3/2}.$$

In (9.2.5) and (9.2.6) C denotes a constant independent of ε, whose value need not be made precise. But

$$\int_{t_1}^{t_2} \gamma(y^\varepsilon(s)) \cdot \frac{\partial\phi}{\partial x}(y^\varepsilon(s)) \, d\xi^\varepsilon(s) \geq \beta(\xi^\varepsilon(t_2) - \xi^\varepsilon(t_1))$$

hence we can assert that

$$(9.2.7) \qquad E(\xi^\varepsilon(t_2) - \xi^\varepsilon(t_1))^6 \leq C(t_2 - t_1)^{3/2}.$$

If we go back to the equation (9.1.3), we deduce from (9.2.7) that

$$(9.2.8) \qquad E|y^\varepsilon(t_2) - y^\varepsilon(t_1)|^6 \leq C(t_2 - t_1)^{3/2},$$

which insures that the family μ_x^ε remains in a weakly compact subset of $\mathcal{M}_+^1(\Psi)$.[28] ∎

LEMMA 9.3. *Let $\phi(x,y)$ be a function which is C^2 in x, C^1 in y, with all partial derivatives continuous and bounded globally in x, y. If ϕ is periodic in y and satisfies*

$$(9.2.9) \qquad \int_Y \phi(x,y)m(x,y)\,dy = 0,$$

then we have

$$(9.2.10) \qquad E\left[\left(\int_0^t \phi\left(y^\varepsilon(\lambda), \frac{y^\varepsilon(\lambda)}{\varepsilon}\right) d\lambda\right)^2\right] \to 0.$$

PROOF. We proceed as in the proof of Lemma 6.2. We first solve the equation

$$A\psi = \phi, \qquad \psi \text{ periodic},$$

and use Itô's formula. We obtain

$$(9.2.11) \qquad \psi\left(y^\varepsilon(t), \frac{y^\varepsilon(t)}{\varepsilon}\right) = \psi\left(y^\varepsilon(s), \frac{y^\varepsilon(s)}{\varepsilon}\right) + \int_s^t \left(\frac{\partial\psi}{\partial x} + \frac{1}{\varepsilon}\frac{\partial\psi}{\partial y}\right) b\,d\lambda$$
$$+ \int_s^t \left(\frac{\partial\psi}{\partial x} + \frac{1}{\varepsilon}\frac{\partial\psi}{\partial y}\right) \cdot \sigma\,dw$$
$$+ \int_s^t \operatorname{tr} a\left(\frac{\partial^2\psi}{\partial x^2} + \frac{2}{\varepsilon}\frac{\partial^2\psi}{\partial x\partial y} + \frac{1}{\varepsilon^2}\frac{\partial^2\psi}{\partial y^2}\right) d\lambda$$
$$- \int_s^t \left(\frac{\partial\psi}{\partial x} + \frac{1}{\varepsilon}\frac{\partial\psi}{\partial y}\right) \cdot \gamma\,d\xi^\varepsilon.$$

By construction we have

$$\operatorname{tr} a\frac{\partial^2\psi}{\partial y^2} = -\phi$$

hence it follows from (9.2.11) that

$$(9.2.12) \qquad \int_s^t \phi\left(y^\varepsilon(\lambda), \frac{y^\varepsilon(\lambda)}{\varepsilon}\right) d\lambda = \varepsilon^2\psi\left(y^\varepsilon(s), \frac{y^\varepsilon(s)}{\varepsilon}\right) - \varepsilon^2\psi\left(y^\varepsilon(t), \frac{y^\varepsilon(t)}{\varepsilon}\right)$$
$$+ \int_s^t \left(\varepsilon^2\frac{\partial\psi}{\partial x} + \varepsilon\frac{\partial\psi}{\partial y}\right) \cdot b\,d\lambda$$
$$+ \int_s^t \left(\varepsilon^2\frac{\partial\psi}{\partial x} + \varepsilon\frac{\partial\psi}{\partial y}\right) \cdot \sigma\,dw$$
$$+ \int_s^t \varepsilon^2 \operatorname{tr} a\frac{\partial^2\psi}{\partial x^2}\,d\lambda + 2\varepsilon\int_s^t \operatorname{tr} a\frac{\partial^2\psi}{\partial x\partial y}\,d\lambda$$
$$- \int_s^t \left(\varepsilon^2\frac{\partial\psi}{\partial x} + \varepsilon\frac{\partial\psi}{\partial y}\right) \cdot \gamma\,d\xi^\varepsilon.$$

---

[28]We have used in the proof some ideas of Stroock-Varadhan [**114**] in a different context.

But from (9.2.7) it follows that

$$E(\xi^\varepsilon(t) - \xi^\varepsilon(s))^2 \le C(t-s)^{1/2}.$$

Using this estimate in (9.2.12), we easily obtain that

$$E \left(\int_s^t \phi \left(y^\varepsilon(\lambda), \frac{y^\varepsilon(\lambda)}{\varepsilon} \right) d\lambda \right)^2 \le C\varepsilon,$$

hence (9.2.10) follows.   ∎

PROOF OF THEOREM 9.1. We can now proceed with the proof of Theorem 9.1. We will rely on the submartingale formulation of the diffusion with reflection. Let $\phi(x,t) \in C_0^\infty(\mathbb{R}^n \times [0,\infty[)$ satisfying (9.1.4). From (9.1.5) we have

(9.2.13)

$$\begin{aligned}
X^\varepsilon = E_x^\varepsilon \Bigg[& \phi(x(t),t) - \phi(x(s),s) - \int_s^t \frac{\partial\phi}{\partial t}(x(\lambda),\lambda)\, d\lambda \\
& - \int_s^t \frac{\partial\phi}{\partial x}(x(\lambda),\lambda) \cdot r(x(\lambda))\, d\lambda - \int_s^t \mathrm{tr}\, \frac{\partial^2\phi}{\partial x^2}(x(\lambda),\lambda) \cdot q(x(\lambda))\, d\lambda \, \Bigg| \, \mathcal{G}^s \Bigg] \\
\le & E_x^\varepsilon \Bigg[\int_s^t \frac{\partial\phi}{\partial x}(x(\lambda),\lambda) \cdot b \left(x(\lambda), \frac{x(\lambda)}{\varepsilon} \right) d\lambda \, \Bigg| \, \mathcal{G}^s \Bigg] \\
& + E_x^\varepsilon \Bigg[\int_s^t \mathrm{tr}\, \frac{\partial^2\phi}{\partial x^2} \tilde{a} \left(x(\lambda), \frac{x(\lambda)}{\varepsilon} \right) d\lambda \, \Bigg| \, \mathcal{G}^s \Bigg],
\end{aligned}$$

where we have set

$$\tilde{b}(x,y) = b(x,y) - r(x), \qquad \tilde{a}(x,y) = a(x,y) - q(x).$$

From Lemma 9.3, it follows that

$$E_x^\varepsilon \left| E^\varepsilon \left[\int_s^t \frac{\partial\phi}{\partial x}(x(\lambda),\lambda) b \left(x(\lambda), \frac{x(\lambda)}{\varepsilon} \right) d\lambda \, \Bigg| \, \mathcal{G}^s \right] \right| \to 0,$$

$$E_x^\varepsilon \left| E_x^\varepsilon \left[\int_s^t \mathrm{tr}\, \frac{\partial^2\phi}{\partial x^2} \tilde{a} \left(x(\lambda), \frac{x(\lambda)}{\varepsilon} \right) d\lambda \, \Bigg| \, \mathcal{G}^s \right] \right| \to 0.$$

This and (9.2.13) imply at least for a converging subsequence of μ_x^ε that

$$\begin{aligned}
E_x \Bigg[& \phi(x(t),t) - \phi(x(s),s) - \int_s^t \frac{\partial\phi}{\partial t}(x(\lambda),\lambda)\, d\lambda \\
& - \int_s^t \frac{\partial\phi}{\partial x}(x(\lambda),\lambda) \cdot r(x(\lambda))\, d\lambda \\
& - \int_s^t \mathrm{tr}\, \frac{\partial^2\phi}{\partial x^2}(x(\lambda),\lambda) q(x(\lambda))\, d\lambda \, \Bigg| \, \mathcal{G}^s \Bigg] \le 0,
\end{aligned}$$

and from the uniqueness of the solution of the submartingale problem we complete the proof of the theorem.   ∎

9.3. Applications to partial differential equations.

9.3.1. *Homogenization of a Neumann problem.* With the notation of the previous paragraphs, we take

$$(9.3.1) \qquad\qquad a(x,y) = a(x),$$

$$(9.3.2) \qquad\qquad \gamma_j(x) = a_{ij}(x)n_i, \qquad (n_i = \cos(n, x_i)).$$

Let us consider

$$(9.3.3) \qquad f \in C^0(\overline{\mathcal{O}}), \qquad a_0(x,y) \geq \alpha_0 > 0 \text{ large enough,}$$

$$a_0 \text{ continuous and bounded,} \qquad a_0 \text{ periodic in } y.$$

We consider the following Neumann problem

$$(9.3.4) \qquad - a_{ij}(x) \frac{\partial^2 u_\varepsilon}{\partial x_i \partial x_j} - b_i\left(x, \frac{x}{\varepsilon}\right) \frac{\partial u_\varepsilon}{\partial x_i} + a_0\left(x, \frac{x}{\varepsilon}\right) u_\varepsilon = f,$$

$$a_{ij}(x) \left.\frac{\partial u_\varepsilon}{\partial x_j}\right|_\Gamma = 0,$$

$$u_\varepsilon \in W^{2,p}(\mathcal{O}), \qquad p \geq 2.$$

The structure of this operator is very special and actually the analytic treatment is easy (see Remark 9.5).

We can rewrite (9.3.4) in divergence form, namely

$$-\frac{\partial}{\partial x_i} a_{ij}(x) \frac{\partial u_\varepsilon}{\partial x_j} + \frac{\partial u_\varepsilon}{\partial x_i}\left(\frac{\partial a_{ij}}{\partial x_j} - b_i\left(x, \frac{x}{\varepsilon}\right)\right) + a_0\left(x, \frac{x}{\varepsilon}\right) u_\varepsilon = f.$$

Setting

$$(9.3.5) \qquad a_i\left(x, \frac{x}{\varepsilon}\right) = \frac{\partial a_{ij}}{\partial x_j} - b_i\left(x, \frac{x}{\varepsilon}\right),$$

we can give the variational formulation of (9.3.4), namely

$$(9.3.6) \quad \int_{\mathcal{O}} a_{ij}(x) \frac{\partial u_\varepsilon}{\partial x_j} \frac{\partial v}{\partial x_i} \, dx + \int_{\mathcal{O}} a_i\left(x, \frac{x}{\varepsilon}\right) \frac{\partial u_\varepsilon}{\partial x_i} v \, dx + \int_{\mathcal{O}} a_0\left(x, \frac{x}{\varepsilon}\right) u^\varepsilon v \, dx$$

$$= \int_{\mathcal{O}} fv \, dx, \qquad \forall v \in H^1(\mathcal{O}).$$

For α_0 large enough, the bilinear form on the left hand side of (9.3.6) is coercive and therefore u_ε is defined in a unique way in $H^1(\mathcal{O})$. The regularity follows from standard results (Agmon-Douglas-Nirenberg [**2**]).

Considering the process $y^\varepsilon(t)$ solution of (9.1.3), the function $u_\varepsilon(x)$ is given explicitly by the following formula

$$(9.3.7) \qquad u_\varepsilon(x) = E \int_0^{+\infty} \left(\exp - \int_0^t a_0\left(y_x(s), \frac{y^\varepsilon(s)}{\varepsilon}\right) ds\right) f(y_x(t)) \, dt.$$

Clearly $m = 1$ is the unique solution of

$$(9.3.8) \qquad -\frac{\partial^2}{\partial y_i \partial y_j}(a_{ij}(x)m) = 0, \qquad m \text{ periodic,} \qquad \int_Y m(x,y) \, dy = 1.$$

We set

$$(9.3.9) \qquad r_i(x) = \int_Y b_i(x,y) \, dy, \qquad \bar{a}_0 = \int_Y a_0(x,y) \, dy.$$

We consider the homogenized Neumann problem

$$(9.3.10) \qquad -a_{ij}(x)\frac{\partial^2 u}{\partial x_i \partial x_j} - r_i(x)\frac{\partial u}{\partial x_i} + \bar{a}_0(x)u = f,$$

$$a_{ij}(x)n_i\frac{\partial u}{\partial x_j}\bigg|_\Gamma = 0,$$

$$u \in W^{2,p}(\mathcal{O}), \qquad p \geq 2.$$

We then have the following

THEOREM 9.4. *Under the assumptions of Theorem 9.1 and (9.3.1), (9.3.2), we have*

$$(9.3.11) \qquad u_\varepsilon(x) \to u(x), \qquad x \notin \mathcal{O}.$$

PROOF. The proof is identical to the proof of Theorem 6.6, as a consequence of Theorem 9.1. ∎

REMARK 9.5. It is not surprising that we can give a convergence result which is valid pointwise up to the boundary of $\mathcal{O}$. Indeed if we look at (9.3.4) from the point of view of a priori estimates, we can see that u_ε remains in a bounded subset of $W^{2,p}(\mathcal{O})$, $p \geq 2$. Therefore we can pass to the limit in (9.3.6) without any difficulty, obtaining the convergence of u_ε toward u in $W^{2,p}(\mathcal{O})$ weakly for any $p \geq 2$.

9.3.2. *The Neumann problem in a half space.* We take here $\mathcal{O} = \{x_n > 0\}$. We note that (4.4.1) is satisfied if for instance we set

$$\phi(x) = -\frac{x_n}{1|+x+^2}.$$

The fact that $\mathcal{O}$ was supposed bounded in the previous paragraph was actually not used. Only (4.4.1) has played a role. We take $\gamma = (0, \ldots, 0, -1)$, and consider the Neumann problem

$$(9.3.12) \qquad -a_{ij}\left(x, \frac{x}{\varepsilon}\right)\frac{\partial^2 u_\varepsilon}{\partial x_i \partial x_j} - b_i\left(x, \frac{x}{\varepsilon}\right)\frac{\partial u_\varepsilon}{\partial x_i} + a_0\left(x, \frac{x}{\varepsilon}\right)u_\varepsilon = f,$$

$$\frac{\partial u_\varepsilon}{\partial x_n}\bigg|_{x_n} = 0.$$

We assume that

$$a_{nj}(x, y) = 0, \qquad j \neq n,$$
$$(9.3.13) \qquad f \in C^0(\overline{\mathcal{O}}), \qquad f \in L^2(\mathcal{O}) \cap L^\infty(\mathcal{O}),$$
$$a_0(x, y) \geq \alpha_0 > 0 \text{ large enough, continuous and bounded.}$$

Then (9.3.12) has one and only one solution in

$$(9.3.14) \qquad u_\varepsilon \in W^{2,p}(\mathcal{O}), \qquad \forall p \geq 2.$$

Indeed the problem (9.3.12) admits a variational formulation as follows

$$(9.3.15) \quad \int a_{ij}\left(x, \frac{x}{\varepsilon}\right)\frac{\partial u_\varepsilon}{\partial x_j}\frac{\partial v}{\partial x_i}\, dx + \int \frac{\partial u_\varepsilon}{\partial x_i}\left(\frac{\partial a_{ij}}{\partial x_j} + \frac{1}{\varepsilon}\frac{\partial a_{ij}}{\partial y_j} - b_i\right)v\, dx$$

$$+ \int a_0 u_\varepsilon v\, dx = \int fv\, dx, \qquad \forall v \in H^1(\mathcal{O}).$$

Indeed the conormal derivative corresponding to the matrix a_{ij} and the domain $\mathcal{O}$ is $a_{nn}\frac{\partial}{\partial x_j}$, and therefore applying Green's formula, we see that the boundary condition corresponding to (9.3.15) is

$$a_{nn}\left(x, \frac{x}{\varepsilon}\right) \frac{\partial u_\varepsilon}{\partial x_n}\bigg|_\Gamma = 0,$$

and since $a_{nn} > 0$, this is equivalent to $\frac{\partial u_\varepsilon}{\partial x_n}\big|_\Gamma = 0$. The regularity follows from standard arguments. We again have the explicit formula (9.3.7).

The function m solution of (9.1.7) is no more identical to 1. We set

$$(9.3.16) \qquad \bar{a}_0(x) = \int_Y a_0(x,y)m(x,y)\,dy,$$

and consider the homogenized Neumann problem

$$(9.3.17) \qquad -q_{ij}(x)\frac{\partial^2 u}{\partial x_i \partial x_j} - r_i(x)\frac{\partial u}{\partial x_i} + \bar{a}_0(x)u = f, \qquad \frac{\partial u}{\partial x_n}\bigg|_\Gamma = 0$$

(note that $q_{nj} = 0$ if $j \neq n$). Then again $u \in W^{2,p}(\mathcal{O})$, $p \geq 2$. We can state the following.

THEOREM 9.6. *Under the assumptions of Theorem 9.1 and (9.3.13), we have*

$$(9.3.18) \qquad u_\varepsilon(x) \to u(x), \qquad x \in \mathcal{O}.$$

10. Evolution problems

10.1. Orientation. We will consider in this paragraph some of the problems studied in Chapter 2, from the probabilistic viewpoint. We will not give a treatment as detailed as in the stationary case, and we shall mainly emphasize the differences between the stationary and the evolutionary case.

10.2. Notation and setting of problems. We consider here functions $a_{ij}(y,\tau)$, $b_i(y,\tau)$, $c_i(y,\tau)$ satisfying the following conditions

(10.2.1) $\qquad a_{ij}, b_i, c_i$ are continuously differentiable in y, τ

$\qquad\qquad a_{ij}$ are twice continuously differentiable in y,

$\qquad\qquad\qquad a_{ij}b_i$ are twice continuously differentiable in τ.

(10.2.2) $\qquad a_{ij}\xi_i\xi_j \geq \alpha|\xi|^2, \qquad \xi \in \mathbb{R}^n, \qquad a_{ij} = a_{ji},$

(10.2.3) $\qquad a_{ij}, b_i, c_i$ are periodic in y with period 1 in all

$\qquad\qquad$ components, and periodic in τ with period τ_0.

Setting as usual

$$(10.2.4) \qquad a = \frac{\sigma\sigma^*}{2},$$

we can construct on an arbitrary probability space $(\Omega, \mathcal{A}, P, \mathcal{F}_t^s, w(s))$ the solution of the stochastic differential equation

$$(10.2.5) \qquad dy^\varepsilon = \frac{1}{\varepsilon}b\left(\frac{y^\varepsilon}{\varepsilon}, \frac{s}{\varepsilon^k}\right)ds + c\left(\frac{y^\varepsilon}{\varepsilon}, \frac{s}{\varepsilon^k}\right)ds + \sigma\left(\frac{y^\varepsilon}{\varepsilon}, \frac{s}{\varepsilon^k}\right)dw(s),$$

$$y^\varepsilon(t) = x.$$

The process $y_{xt}(s) \equiv y^\varepsilon(s)$ defines a probability measure μ^ε_{xt} on the space $\Psi = C^0[t, \infty; \mathbb{R}^n]$. We will denote as usual by $x(\cdot)$ the canonical element of Ψ, and by $\mathcal{G}^s_t$ the σ-algebra generated by $x(\lambda)$, $t \leq \lambda \leq s$. We are mainly concerned in studying the weak limit of μ^ε_{xt}, as $\varepsilon \to 0$. As in the previous paragraphs, this will imply convergence results for PDE (here parabolic PDE).

10.3. Fredholm alternative for an evolution operator. We will be concerned in this paragraph with the Fredholm alternative for the operator

$$(10.3.1) \qquad R = R_1 = -\frac{\partial}{\partial \tau} - b_i \frac{\partial}{\partial y_i} - a_{ij} \frac{\partial^2}{\partial y_i \partial y_j}, {}^{29}$$

with periodic boundary conditions in τ, y. We denote by R^* the formal adjoint

$$(10.3.2) \qquad R^* = +\frac{\partial}{\partial \tau} + \frac{\partial}{\partial y_i}(b_i \cdot) - \frac{\partial^2}{\partial y_i \partial y_j}(a_{ij} \cdot).$$

THEOREM 10.1. *We assume that a_{ij}, b_i satisfy (10.2.1), (10.2.2). We consider the homogeneous equations*

$$(10.3.3) \qquad Rz = 0, \qquad z \in C^{2,1}, \qquad z \text{ periodic in } y, \tau,$$

$$(10.3.4) \qquad R^*p = 0, \qquad p \in C^{2,1}, \qquad m \text{ periodic in } y, \tau.$$

There exists one and only one solution (up to a multiplicative constant) of (10.3.3), (10.3.4).[30] *Let ϕ, ψ be continuously differentiable functions such that*

$$(10.3.5) \qquad \int_0^{\tau_0} \int_Y \phi(y, \tau) p(y, \tau) \, dy \, d\tau = 0,$$

$$(10.3.6) \qquad \int_0^{\tau_0} \int_Y \psi(y, \tau) \, dy \, d\tau = 0,$$

then there exists one and only one solution of the inhomogeneous equations

$$(10.3.7) \qquad R\zeta = \phi, \qquad \zeta \in C^{2,1}, \qquad \zeta \text{ periodic}, \qquad \int_0^{\tau_0} \int_Y \zeta \, dy \, d\tau = 0,$$

$$(10.3.8) \qquad R^*\pi = \psi, \qquad \pi \in C^{2,1}, \qquad \pi \text{ periodic}, \qquad \frac{1}{\tau_0} \int_0^{\tau_0} \int_Y \pi \, dy \, d\tau = 1.$$

PROOF. Let λ be large enough; we consider the problem

$$(10.3.9) \qquad Rz_\lambda + \lambda z_\lambda = g,$$

for $g \in L^2(0, \tau_0; L^2(Y))$. To give a meaning to the equation (10.3.9), we first rewrite R under divergence form, namely

$$\begin{aligned}
R &= -\frac{\partial}{\partial \tau} - b_i \frac{\partial}{\partial y_i} + \frac{\partial}{\partial y_j} a_{ij} \frac{\partial}{\partial y_i} - \frac{\partial}{\partial y_i}\left(a_{ij} \frac{\partial}{\partial y_j}\right) \\
&= -\frac{\partial}{\partial \tau} + a_i \frac{\partial}{\partial y_i} - \left(a_{ij} \frac{\partial}{\partial y_j}\right) \\
&= -\frac{\partial}{\partial \tau} + A(\tau).
\end{aligned}$$

---

[29]The notation R_i has been chosen in connection with Chapter 2, (1.3.7).
[30]z is obviously a constant.

We look for a solution $z_\lambda \in L^2(0, \tau_0; W(Y))$, $\frac{\partial z_\lambda}{\partial \tau} \in L^2(0, \tau_0; W(Y))$ with periodic conditions $z_\lambda(0) = z_\lambda(\tau_0)$. If $g = 0$, then $z_\lambda = 0$ since then

$$0 = \int_0^{\tau_0} \int_Y a_i \frac{\partial z_\lambda}{\partial y_i} z_\lambda \, dy \, d\tau + \int_0^{\tau_0} \int_Y a_{ij} \frac{\partial z_\lambda}{\partial y_j} \frac{\partial z_\lambda}{\partial y_i} \, dy \, d\tau + \lambda \int_0^{\tau_0} \int_Y z_\lambda^2 \, dy \, d\tau$$

and for λ large enough, this implies $z_\lambda = 0$.

To prove the existence, we notice that the problem

(10.3.10)
$$-\frac{\partial z}{\partial \tau} + A(\tau)z + \lambda z = h \in L^2(0, \infty; L^2(Y)),$$
$$z \in L^2(0, \infty; W(Y)),$$

has one and only one solution.[31] However $g \in L^2(0, \tau_0; L^2(Y))$ implies $g e^{-\gamma \tau} \in L^2(0, \infty; L^2(Y))$ assuming that g has been extended by periodicity. Taking $h = g e^{-\gamma \tau}$ and setting $z = \zeta e^{-\gamma \tau}$ we obtain the existence of ζ such that $\zeta e^{-\gamma \tau} \in L^2(0, \infty; W(Y))$ and

$$-\frac{\partial \zeta}{\partial \tau} + A(\tau)\zeta + (\lambda + \gamma)\zeta = g.$$

Changing λ into $\lambda - \gamma$ (γ is > 0 arbitrary), we see that there exists one and only one solution of

$$-\frac{\partial z}{\partial \tau} + A(\tau)z + \lambda z = g, \qquad z e^{-\gamma \tau} \in L^2(0, \infty; W(Y)).$$

The function z is clearly periodic (since $z(\tau + \tau_0)$ is also a solution). Hence we can define a mapping

$$T_\lambda : L^2(0, \tau_0; L^2(Y)) \to L^2(0, \tau_0; L^2(Y)) \text{ by setting } T_\lambda g = z_\lambda.$$

The operator T_λ is compact. For $\phi, \psi \in L^2(0, \tau_0; L^2(Y))$, the problem

(10.3.11)
$$Rz = \phi,\,[32]$$
$$R^* p = \psi,$$

are equivalent to the problem

(10.3.12)
$$(I - \lambda T_\lambda)z = T_\lambda \phi$$
$$(I - \lambda T_\lambda^*)p = T_\lambda^* \psi.$$

The Fredholm alternative applies, hence we must find the number of linearly independent solutions of

(10.3.13)
$$Rz = 0, \qquad z \in L^2(0, \tau_0; W(Y)),$$
$$z' \in L^2(0, \tau_0; W'(Y)), \qquad z(0) = z(\tau_0).$$

Since the solution of (10.3.13) satisfies

$$-\frac{\partial z}{\partial \tau} + A(\tau)z + \lambda z = \lambda z, \qquad z e^{-\gamma \tau} \in L^2(0, \infty; W(Y)),$$

---

[31] Use a truncation argument; consider $-\frac{\partial z^T}{\partial \tau} + A(\tau)z^T + \lambda z^T = h$, $z^T(T) = 0$ and extend $z^T(\tau)$ by 0 for $\tau > T$; then as $T \to +\infty$, $z^T \to z$ solution of (10.3.10).

[32] Solutions are of course in the space of functions $z \in L^2(0, \tau_0; W(Y))$ with $\frac{\partial z}{\partial \tau} \in L^2(0, \tau_0; W'(Y))$ and $z(0) = z(\tau_0)$.

$a_{ij} = a_{ji}$ and the coefficients are differentiable in τ, we see that $\frac{\partial z}{\partial \tau} \in L^2(0, \tau_0; L^2(Y))$. Since the coefficients are differentiable in y, $z \in L^2(0, \tau_0; H^2(Y))$. Using the Sobolev spaces with weights we see that we can consider that

$$-\frac{\partial z}{\partial \tau} + A(\tau)z + \lambda z = \lambda z \quad \text{a.e. in } \mathbb{R}^n$$

and $z \in \mathcal{W}^{2,1,2,\mu}(\mathbb{R}^n \times (0, \infty))$.[33] By a bootstrap argument like in the proof of Theorem 3.5, we obtain that $z \in \mathcal{W}^{2,1,p,\mu}$, $p \geq 2$, $\mu > 0$. Since the a_{ij}'s are twice continuously differentiable in y, the b_i's are continuously differentiable in y, and both coefficients are twice continuously differentiable in τ, we can also show that $\frac{\partial z}{\partial \tau}, \frac{\partial z}{\partial y_i} \in \mathcal{W}^{2,1,p,\mu}$. Hence $z \in C^{2,1}(\mathbb{R}^n \times (0, \infty))$. The strong maximum principle implies that z is a constant. The remainder of the results are a consequence of the Fredholm alternative and regularity arguments. ∎

We shall need in some of the results to follow (Theorem 10.8) the assumption,

(10.3.14)   the solution p of (10.3.4) is such that

$$\frac{1}{\tau_0} \int_0^{\tau_0} \int_Y p(y, \tau)\, dy\, d\tau = 1, \qquad \text{is} \geq 0.$$

Let us notice that if (10.3.14) is satisfied then from the strong maximum principle it follows that

(10.3.15) $p > 0.$

Let us notice that if $a_0 = 0$ ($b_i = \frac{\partial a_{ij}}{\partial y_j}$) then $R = R^*$, hence $p = 1$. It is possible to give a sufficient condition for (10.3.14) to hold true, which is less restrictive than $a_i = 0$. Indeed we have

PROPOSITION 10.2. *If* $\sum_i |a_i|^2 \leq k^2$, *and* $\left|\frac{\partial a_i}{\partial y_i}\right| \leq k$ *where* k *is small enough then* (10.3.14) *holds true.*

PROOF. We proceed using elliptic regularization. We consider p^ε solution of

$$\frac{\partial p^\varepsilon}{\partial \tau} - \varepsilon \frac{\partial^2 p^\varepsilon}{\partial \tau^2} - \frac{\partial}{\partial y_i}(a_i p^\varepsilon) - \frac{\partial}{\partial y_i}\left(a_{ij} \frac{\partial p^\varepsilon}{\partial y_j}\right) = 0,$$

$$p^\varepsilon \text{ periodic in } y, \tau, \qquad \frac{1}{\tau_0} \int_0^{\tau_0} \int_Y p^\varepsilon\, dy\, d\tau = 1.$$

From the elliptic theory, we know that $p^\varepsilon \geq 0$. But multiplying by p^ε and integrating we obtain

$$0 = \varepsilon \int_Y \int_0^{\tau_0} \left|\frac{\partial p^\varepsilon}{\partial \tau}\right|^2 dy\, d\tau + \int_Y \int_0^{\tau_0} a_i p^\varepsilon \frac{\partial p^\varepsilon}{\partial y_i}\, dy\, d\tau$$

$$+ \int_Y \int_0^{\tau_0} a_{ij} \frac{\partial p^\varepsilon}{\partial y_j} \frac{\partial p^\varepsilon}{\partial y_i}\, dy\, d\tau,$$

---

[33]

$$\mathcal{W}^{2,1,p,\mu}(\mathbb{R}^n \times (0, \infty)) = \left\{ z, \frac{\partial z}{\partial x_i}, \frac{\partial^2 z}{\partial x_i \partial x_j}, \frac{\partial z}{\partial \tau} \in L^{p,\mu}(\mathbb{R}^n \times (0, \infty)) \right\}.$$

$$L^{p,\mu}(\mathbb{R}^n \times (0, \infty)) = \left\{ z \ \bigg| \ \int_0^{+\infty} \int_{\mathbb{R}^n} |z|^p (\exp -\mu|x| \exp -\gamma\tau)\, dx\, d\tau < \infty \right\}.$$

hence

$$(10.3.16) \qquad \alpha|\nabla p^\varepsilon|_{L^2(0,\tau_0;L^2(Y))} \le k|p^\varepsilon|_{L^2(0,\tau_0;L^2(Y))}.$$

Multiplying the equation by $\frac{\partial p^\varepsilon}{\partial \tau}$ and integrating in y only we obtain

$$\int_Y \left(\frac{\partial p^\varepsilon}{\partial \tau}\right)^2 dy - \frac{\varepsilon}{2}\frac{d}{d\tau}\int_Y \left(\frac{\partial p^\varepsilon}{\partial \tau}\right)^2 dy - \int_Y \frac{\partial}{\partial y_i}(a_i p^\varepsilon)\frac{\partial p^\varepsilon}{\partial \tau} dy$$
$$+ \frac{1}{2}\frac{d}{d\tau}\int_Y a_{ij}\frac{\partial p^\varepsilon}{\partial y_j}\frac{\partial p^\varepsilon}{\partial y_i} dy - \frac{1}{2}\int_Y \frac{\partial a_{ij}}{\partial \tau}\frac{\partial p^\varepsilon}{\partial y_j}\frac{\partial p^\varepsilon}{\partial y_i} dy = 0.$$

Integrating in τ yields

$$(10.3.17) \qquad 0 = \int_0^{\tau_0}\int_Y \left(\frac{\partial p^\varepsilon}{\partial \tau}\right)^2 dy\, d\tau - \int_0^{\tau_0}\int_Y \frac{\partial}{\partial y_i}(a_i p^\varepsilon)\frac{\partial p^\varepsilon}{\partial \tau} dy\, d\tau$$
$$- \frac{1}{2}\int_0^{\tau_0}\int_Y \frac{\partial a_{ij}}{\partial \tau}\frac{\partial p^\varepsilon}{\partial y_j}\frac{\partial p^\varepsilon}{\partial y_i} dy\, d\tau.$$

Hence

$$(10.3.18) \qquad \left|\frac{\partial p^\varepsilon}{\partial \tau}\right|^2_{L^2(0,\tau_0;L^2(Y))} \le C|\nabla p^\varepsilon|^2_{L^2(0,\tau_0;L^2(Y))}$$
$$+ C'k^2|p^\varepsilon|^2_{L^2(0,\tau_0;L^2(Y))}.$$

From (10.3.16), (10.3.18) it follows that

$$|\nabla p^\varepsilon|^2_{L^2(0,\tau_0;L^2(Y))} + \left|\frac{\partial p^\varepsilon}{\partial \tau}\right|^2_{L^2(0,\tau_0;L^2(Y))} \le C''k^2|p^\varepsilon|^2_{L^2(0,\tau_0;L^2(Y))}.$$

But from Poincaré's inequality

$$|p^\varepsilon|^2_{L^2(0,\tau_0;L^2(Y))} \le P\left(|\nabla p^\varepsilon|^2 + \left|\frac{\partial p^\varepsilon}{\partial \tau}\right|^2 + 1\right)$$

and therefore if k is small enough $(1 > PC''k^2)$, we obtain

$$|p^\varepsilon|_{L^2(0,\tau_0;L^2(Y))}, \ |\nabla p^\varepsilon|_{L^2(0,\tau_0;L^2(Y))}, \ \left|\frac{\partial p^\varepsilon}{\partial \tau}\right|_{L^2(0,\tau_0;L^2(Y))} \le C.$$

We can then extract a subsequence, which clearly converges weakly toward p. Hence $p \ge 0$. ∎

10.4. The case $k < 2$. We begin now the study of (10.2.5), assuming here that $k < 2$. We set

$$A_1(\tau) = -b_i(y,\tau)\frac{\partial}{\partial y_i} - a_{ij}(y,\tau)\frac{\partial^2}{\partial y_i \partial y_j}.$$

We consider τ as a parameter. From the elliptic theory there exists $m(y,\tau)$ solution of

$$(10.4.1) \qquad A_1^* m = 0, \qquad m \text{ periodic in } y, \qquad \int_Y m\, dy = 1.$$

We have $m(y,\tau) > 0$, $\forall y,\tau$, and m is also periodic in τ. We make the assumption

$$(10.4.2) \qquad \int_Y b(y,\tau)m(y,\tau)\, dy = 0.$$

We can then define the functions $\chi^j(y, \tau)$ as the solutions of

$$(10.4.3) \qquad A_1\chi^j = -b_j, \qquad \chi^j \text{ periodic in } y, \qquad \int_Y m\chi^j \, dy = 0.$$

From the regularity of the coefficients, we can ensure that $\chi^j \in C^{2,1}$, and is periodic in y, τ.

LEMMA 10.3. *The family of measures μ_{xt}^ε remains as $\varepsilon \to 0$, in a weakly compact subset of $\mathcal{M}_+^1(\Psi)$.*

PROOF. We first assume that $k \leq 1$. From Itô's formula, we have

$$(10.4.4) \quad \chi^j\left(\frac{y^\varepsilon(s)}{\varepsilon}, \frac{s}{\varepsilon^k}\right) = \chi^i\left(\frac{x}{\varepsilon}t\varepsilon^k\right) + \int_t^s \left(\frac{\partial\chi^j}{\partial\tau}\frac{1}{\varepsilon^k} + \frac{1}{\varepsilon^2}\frac{\partial\chi^j}{\partial y_k}b_k + \frac{1}{\varepsilon}\frac{\partial\chi^j}{\partial y_k}c_k\right.$$
$$\left. + \frac{1}{\varepsilon^2}\frac{\partial^2\chi^j}{\partial y_k \partial y_\ell}a_{k\ell}\right) d\lambda + \int_t^s \frac{1}{\varepsilon}\frac{\partial\chi^j}{\partial y} \cdot \sigma \, dw.$$

Using (10.4.3) and setting $\chi = (\chi^1, \ldots, \chi^n)$, it follows from (10.4.4) that

$$(10.4.5) \qquad \chi\left(\frac{y^\varepsilon(s)}{\varepsilon}, \frac{s}{\varepsilon^k}\right) = \chi\left(\frac{x}{\varepsilon}, \frac{t}{\varepsilon^k}\right) + \int_t^s \left(\frac{\partial\chi}{\partial\tau}\frac{1}{\varepsilon^k} + \frac{\chi}{y} \cdot c\frac{1}{\varepsilon}\right) d\lambda$$
$$+ \int_t^s \frac{1}{\varepsilon^2}b \, d\lambda + \int_t^s \frac{1}{\varepsilon}\frac{\partial\chi}{\partial y} \cdot \sigma \, dw.$$

From (10.2.5) and (10.4.5) it follows, eliminating the b term, that

$$(10.4.6) \qquad y^\varepsilon(s) - \varepsilon\chi\left(\frac{y^\varepsilon(s)}{\varepsilon}, \frac{s}{\varepsilon^k}\right) = x - \varepsilon\chi\left(\frac{x}{\varepsilon}, \frac{t}{\varepsilon^k}\right)$$
$$+ \int_t^s \left[\left(I - \frac{\partial\chi}{\partial y}\right) \cdot c - \frac{\partial\chi}{\partial\tau}\varepsilon^{1-k}\right] d\lambda$$
$$+ \int_t^s \left(I - \frac{\partial\chi}{\partial y}\right)\sigma \, dw.$$

Since $1 - k \geq 0$, it easily follows from (10.4.6) that μ_{xt}^ε remains in a compact subset of $\mathcal{M}_+^1(\Psi)$.

We next consider the case when $1 \leq k < \frac{3}{2}$. We set $r = 2 - k$, and define $\chi_1(y, \tau)$ by

$$(10.4.7) \qquad A_1\chi_1 = \frac{\partial\chi}{\partial\tau}, \qquad \chi_1 \text{ periodic}, \qquad \int_Y \chi_1 m \, dy = 0.$$

Since we have chosen χ in order that $\int_Y \chi m \, dy = 0$, then the solvability condition is satisfied for (10.4.7), hence χ_1 is well defined. We apply Itô's formula to $\chi + \varepsilon^{2-k}\chi_1$, to obtain

$$(10.4.8)$$
$$\varepsilon\chi\left(\frac{y^\varepsilon(s)}{\varepsilon}, \frac{s}{\varepsilon^k}\right) + \varepsilon^{3-k}\chi_1\left(\frac{y^\varepsilon(s)}{\varepsilon}, \frac{s}{\varepsilon^k}\right)$$
$$= \varepsilon\chi\left(\frac{x}{\varepsilon}, \frac{t}{\varepsilon^k}\right) + \varepsilon^{3-k}\chi_1\left(\frac{x}{\varepsilon}, \frac{t}{\varepsilon^k}\right) + \int_t^s \left(\frac{\partial\chi}{\partial y} \cdot c + \varepsilon^{2-k}\frac{\partial\chi_1}{\partial y}c\right) d\lambda$$
$$+ \int_t^s \frac{1}{\varepsilon}b \, d\lambda + \int_t^s \varepsilon^{3-2k}\frac{\partial\chi_1}{\partial\tau} \, d\lambda + \int_t^s \frac{\partial\chi}{\partial y} \cdot \sigma \, dw + \int_t^s \varepsilon^{2-k}\frac{\partial\chi_1}{\partial y} \cdot \sigma \, dw.$$

From (10.2.5) and (10.4.8), we can again eliminate the term in $\frac{1}{\varepsilon}b$, and obtain the compactness.

From here on, we need to assume one more degree of differentiability in τ than in (10.2.1). We consider the case $\frac{3}{2} \leq k < \frac{5}{3}$. We define χ_2 by

$$(10.4.9) \qquad A_1\chi_2 = \frac{\partial \chi_1}{\partial \tau}, \qquad \int_Y \chi_2 m \, dy = 0,$$

and we apply Itô's formula to $\chi + \varepsilon^{2-k}\chi_1 + \varepsilon^{4-2k}\chi_2$.

Reasoning as before we obtain again the compactness. We next consider the case when $\frac{5}{3} \leq k < \frac{7}{4}$ and so on. Provided that there is enough differentiability in τ, we can consider any $k < 2$. Hence the desired result follows. ∎

LEMMA 10.4. *Let $\phi(x, y, \tau, s)$ be a function which is C^1 in y, τ, s, C^2 in x, periodic in y, τ and such that*

$$(10.4.10) \qquad \int_Y \phi(y, \tau)m(y, \tau) \, dy = 0;$$

then we have

$$(10.4.11) \qquad E\left[\left(\int_{s_1}^{s_2} \phi\left(y^\varepsilon(\lambda), \frac{y^\varepsilon(\lambda)}{\varepsilon}, \frac{\lambda}{\varepsilon^k}, \lambda\right) d\lambda\right)^2 \middle| \mathcal{F}_t^{s_1}\right] \to 0 \ a.s.$$

PROOF. We define ψ such that

$$(10.4.12) \qquad A_1\psi = \phi, \qquad \psi \text{ periodic.}$$

From the regularity on ϕ and on the coefficients it follows that $\psi \in C^{2,1}$. From Itô's formula, we have (for convenience we just consider the case when ϕ is independent of x, s)

$$\psi\left(\frac{y^\varepsilon(s_2)}{\varepsilon}, \frac{s_2}{\varepsilon^k}\right) = \psi\left(\frac{y^\varepsilon(s_1)}{\varepsilon}, \frac{s_1}{\varepsilon^k}\right) + \int_{s_1}^{s_2}\left(\frac{\partial\psi}{\partial\tau}\frac{1}{\varepsilon^k} + \frac{\partial\psi}{\partial y}\cdot c\frac{1}{\varepsilon}\right) d\lambda$$
$$+ \int_{s_1}^{s_2}\frac{1}{\varepsilon^2}\left(\frac{\partial\psi}{\partial y}\cdot b + \operatorname{tr} a\frac{\partial^2\psi}{\partial y^s}\right) d\lambda + \int_{s_1}^{s_2}\frac{1}{\varepsilon}\frac{\partial\psi}{\partial y}\cdot\sigma \, dw$$

and from (10.4.12) it follows that

$$\int_{s_1}^{s_2}\phi\left(\frac{y^\varepsilon(\lambda)}{\varepsilon}, \frac{\lambda}{\varepsilon^k}\right) d\lambda = \varepsilon^2\psi\left(\frac{y^\varepsilon(s_1)}{\varepsilon}, \frac{s_1}{\varepsilon^k}\right)$$
$$+ \int_{s_1}^{s_2}\left(\frac{\partial\psi}{\partial\tau}\varepsilon^{2-k} + \varepsilon\frac{\partial\psi}{\partial y}\cdot c\right) d\lambda$$
$$+ \int_{s_1}^{s_2}\varepsilon\frac{\partial\psi}{\partial y}\cdot\sigma \, dw.$$

Taking the square and the conditional expectation we easily obtain the property (10.4.11). ∎

We now define the homogenized coefficients

$$(10.4.13) \qquad q = \frac{1}{\tau_0}\int_0^{\tau_0}\int_Y\left(I - \frac{\partial\chi}{\partial y}\right)a\left(I - \frac{\partial\chi}{\partial y}\right)^* m \, dy \, d\tau,$$

$$(10.4.14) \qquad r = \frac{1}{\tau_0}\int_0^{\tau_0}\int_Y m\left(I - \frac{\partial\chi}{\partial y}\right)c \, dy \, d\tau;$$

then we can define the measure μ_{xt} on Ψ corresponding to the diffusion with constant coefficients q, r, namely

$$(10.4.15) \qquad dy = r\,ds + \sqrt{2}\sqrt{q}\,dw(s), \qquad y(t) = x.$$

We can now state the

THEOREM 10.5. *We assume* (10.2.1), (10.2.2), (10.2.3), (10.4.2) *and* $k < 2$. *If* $\frac{2p-1}{p} \leq k < \frac{2p+1}{p+1}$ ($p \geq 2$), *then we assume that* a_{ij}, b_i *are* $p+1$ *times continuously differentiable in* τ. *Then we have*

$$(10.4.16) \qquad \forall x, t, \qquad \mu_{xt}^{\varepsilon} \to \mu_{xt} \text{ in } \mathcal{M}_+^1(\Psi) \text{ weakly}.$$

PROOF. We use the martingale formulation of diffusions as in the proof of Theorem 6.4. We consider the canonical process $x(s; \psi) = \psi(s)$ and $\mathcal{G}_t^s = \sigma(x(\lambda), t \leq \lambda \leq s)$. From the theory of weak solutions of stochastic differential equations, we can construct a process w which is a Wiener process when Ψ is equipped with the probability measure μ_{xt}^{ε}, and such that

$$(10.4.17) \qquad dx = \frac{1}{\varepsilon}b\left(\frac{x}{\varepsilon}, \frac{s}{\varepsilon^k}\right)ds + \sigma\left(\frac{x}{\varepsilon}, \frac{s}{\varepsilon^k}\right)dw^{\varepsilon}(s) + c\left(\frac{x}{\varepsilon}, \frac{s}{\varepsilon^k}\right)ds,$$
$$\mu_{xt}^{\varepsilon}(x(t) = x) = 1.$$

To fix the ideas we assume that $\frac{3}{2} \leq k < \frac{5}{3}$ (but the reasoning is completely general). We consider the function $\chi(y, \tau) + \varepsilon^{2-k}\chi_1(y, \tau) + \varepsilon^{4-2k}\chi_2(y, \tau)$, and define

$$(10.4.18) \qquad \zeta^{\varepsilon}(s) = x(s) - \varepsilon\chi\left(\frac{x(s)}{\varepsilon}, \frac{s}{\varepsilon^k}\right) - \varepsilon^{3-k}\chi_1\left(\frac{x(s)}{\varepsilon}, \frac{s}{\varepsilon^k}\right)$$
$$- \varepsilon^{5-2k}\chi_2\left(\frac{x(s)}{\varepsilon}, \frac{s}{\varepsilon^k}\right).$$

By Itô's formula we have (taking into account the equations of χ, χ_1, χ_2)

$$(10.4.19) \qquad \zeta^{\varepsilon}(s) = x - \varepsilon\chi\left(\frac{x}{\varepsilon}, \frac{t}{\varepsilon^k}\right) - \varepsilon^{3-k}\chi_1\left(\frac{x}{\varepsilon}, \frac{t}{\varepsilon^k}\right) - \varepsilon^{5-2k}\chi_1\left(\frac{x}{\varepsilon}, \frac{t}{\varepsilon^k}\right)$$
$$+ \int_t^s \left(I - \frac{\partial\chi}{\partial y} - \varepsilon^{2-k}\frac{\partial\chi_1}{\partial y} - \varepsilon^{4-2k}\frac{\partial\chi_2}{\partial y}\right)\sigma\,dw^{\varepsilon}$$
$$+ \int_t^s \left(I - \frac{\partial\chi}{\partial y} - \varepsilon^{2-k}\frac{\partial\chi_1}{\partial y} - \varepsilon^{4-2k}\frac{\partial\chi_2}{\partial y}\right)c\,d\lambda$$
$$- \int_t^s \varepsilon^{5-3k}\frac{\partial\chi_2}{\partial \tau}\,d\lambda.$$

Let now $\phi \in C_0^{\infty}(\mathbb{R}^n \times [0, \infty[)$. From Itô's formula we have using (10.4.19)

$$(10.4.20) \quad \phi(\zeta^{\varepsilon}(s_2), s_2) = \phi(\zeta^{\varepsilon}(s_1), s_1) + \int_{s_1}^{s_2}\left\{\frac{\partial\phi}{\partial s} + \frac{\partial\phi}{\partial x}\cdot\left[\left(I - \frac{\partial\chi}{\partial y}\right)c + O(\varepsilon)\right]\right.$$
$$\left. + \operatorname{tr}\frac{\partial^2\phi}{\partial x^2}\left(I - \frac{\partial\chi}{\partial y}\right)a\left(I - \frac{\partial\chi}{\partial y}\right)^* + O(\varepsilon)\right\}ds$$
$$+ \int_{s_1}^{s_2}\frac{\partial\phi}{\partial x}\left(I - \frac{\partial\chi}{\partial y}\right)\sigma\,dw^{\varepsilon}$$
$$+ \int_{s_1}^{s_2}\frac{\partial\phi}{\partial x}\left(-\varepsilon^{2-k}\frac{\partial\chi_1}{\partial y} - \varepsilon^{4-2k}\frac{\partial\chi_2}{\partial y}\right)\sigma\,dw^{\varepsilon}.$$

We set

$$q(\tau) = \int_Y \left(I - \frac{\partial \chi}{\partial y}\right) a \left(I - \frac{\partial \chi}{\partial y}\right)^* m(y, \tau) \, dy,$$

$$r(\tau) = \int_Y \left(I - \frac{\partial \chi}{\partial y}\right) cm \, dy.$$

We obtain using (10.4.18) and (10.4.20)

$$(10.4.21) \quad E_{xt}^\varepsilon \left[\phi(x(s_2), s_2) - \phi(x(s_1), s_1) - \int_{s_1}^{s_2} \operatorname{tr} q \left(\frac{s}{\varepsilon^k}\right) \frac{\partial^2 \phi}{\partial x^2}(x(s), s) \, ds \right.$$

$$\left. - \int_{s_1}^{s_2} r \left(\frac{s}{\varepsilon^k}\right) \cdot \frac{\partial \phi}{\partial x}(x(s), s) \, ds - \int_{s_1}^{s_2} \frac{\partial \phi}{\partial s} \, \middle| \, \mathcal{F}_t^{s_1}\right]$$

$$= E_{xt} \left[\int_{s_1}^{s_2} \operatorname{tr} \frac{\partial^2 \phi}{\partial x^2}(x(s), s) q \left(\frac{x(s)}{\varepsilon}, \frac{s}{\varepsilon^k}\right) \, ds \, \middle| \, \mathcal{F}_t^{s_1}\right]$$

$$+ E_{xt}^\varepsilon \left[\int_{s_1}^{s_2} \frac{\partial \phi}{\partial x}(x(s), s) r \left(\frac{x(s)}{\varepsilon}, \frac{s}{\varepsilon^k}\right) \, ds \, \middle| \, \mathcal{F}_t^{s_1}\right] + O(\varepsilon),$$

where

$$\tilde{q}(y, \tau) = q(y, \tau) - q(\tau),$$

$$\tilde{r}(y, \tau) = r(y, \tau) - r(\tau).$$

From Lemma 10.4, we have

$$E_{xt}^\varepsilon \left| E_{xt}^\varepsilon \left[\int_{s_1}^{s_2} \operatorname{tr} \frac{\partial^2 \phi}{\partial x^2}(x(s), s) q \left(\frac{x(s)}{\varepsilon}, \frac{s}{\varepsilon^k}\right) \, ds \, \middle| \, \mathcal{F}_t^{s_1}\right]\right| \to 0,$$

$$E_{xt}^\varepsilon \left| E_{xt}^\varepsilon \left[\int_{s_1}^{s_2} \frac{\partial \phi}{\partial x}(x(s), s) \tilde{r} \left(\frac{x(s)}{\varepsilon}, \frac{s}{\varepsilon^k}\right) \, ds \, \middle| \, \mathcal{F}_t^{s_1}\right]\right| \to 0,$$

hence

$$(10.4.22) \quad E_{xt}^\varepsilon \left| E_{xt}^\varepsilon \left[\phi(x(s_2), s_2) - \phi(x(s_1), s_1) - \int_{s_1}^{s_2} \operatorname{tr} q \left(\frac{s}{\varepsilon^k}\right) \frac{\partial^2 \phi}{\partial x^2}(x(s), s) \, ds \right.\right.$$

$$\left.\left. - \int_{s_1}^{s_2} r \left(\frac{s}{\varepsilon^k}\right) \cdot \frac{\partial \phi}{\partial x}(x(s), s) \, ds - \int_{s_1}^{s_2} \frac{\partial \phi}{\partial s} \, ds \, \middle| \, \mathcal{F}_t^{s_1}\right]\right| \to 0.$$

If $H(\psi) = H(x(\cdot))$ is a continuous bounded functional which is $\mathcal{F}_t^{s_1}$ measurable, it follows from (10.4.22) that

$$(10.4.23) \quad E_{xt} \left\{\left[\phi(x(s_2), s_2) - \phi(x(s_1), s_1) - \int_{s_1}^{s_2} \operatorname{tr} q \left(\frac{s}{\varepsilon^k}\right) \frac{\partial^2 \phi}{\partial x^2}(x(s), s) \, ds \right.\right.$$

$$\left.\left. - \int_{s_1}^{s_2} r \left(\frac{s}{\varepsilon^k}\right) \cdot \frac{\partial \phi}{\partial x}(x(s), s) - \int_{s_1}^{s_2} \frac{\partial \phi}{\partial s} \, ds\right] H\right\} \to 0.$$

Now at least for a converging subsequence of $\mu_{xt}^\varepsilon \to \mu_{xt}$,

$$(10.4.24) \quad E_{xt}^\varepsilon \left\{\left[\phi(x(s_2), s_2) - \phi(x(s_1), s_1) - \int_{s_1}^{s_2} \frac{\partial \phi}{\partial s} \, ds\right]\right\}$$

$$\to E_{xt} \left\{\left[\phi(x(s_2), s_2) - \phi(x(s_1), s_1) - \int_{s_1}^{s_2} \frac{\partial \phi}{\partial s} \, ds\right] H\right\}.$$

Next we have

$$(10.4.25) \quad E_{xt}^\varepsilon \left\{ \left[\int_{s_1}^{s_2} \operatorname{tr} q\left(\frac{s}{\varepsilon^k}\right) \frac{\partial^2 \phi}{\partial x^2}(x(s), s)\, ds \right] H \right\}$$

$$= \int_{s_1}^{s_2} \operatorname{tr} q\left(\frac{s}{\varepsilon^k}\right) E_{xt}^\varepsilon \left(\frac{\partial^2 \phi}{\partial x^2}(x(s), s) H \right)\, ds.$$

However setting

$$\phi^\varepsilon(s) = E_{xt}^\varepsilon \frac{\partial^2 \phi}{\partial x^2}(x(s), s) H,$$

we have

$$\rho^\varepsilon(s) \to \rho(s) = E_{xt} \frac{\partial^2 \phi}{\partial x^2}(x(s), s) H,$$

and $\rho^\varepsilon(s)$ is bounded, hence $\rho^\varepsilon(s) \to \rho(s)$ in $L^2(s_1, s_2; \mathcal{L}(\mathbb{R}^n; \mathbb{R}^n))$. But

$$q\left(\frac{s}{\varepsilon^k}\right) \to q \text{ in } L^2(s_1, s_2; \mathcal{L}(\mathbb{R}^n; \mathbb{R}^n)) \text{ weakly.}$$

Therefore we can assert that

$$(10.4.26) \quad E_{xt}^\varepsilon \left(\int_{s_1}^{s_2} \operatorname{tr} q\left(\frac{s}{\varepsilon^k}\right) \frac{\partial^2 \phi}{\partial x^2}(x(s), s)\, ds \right) H$$

$$\to E_{xt} \left(\int_{s_1}^{s_2} \operatorname{tr} q \frac{\partial^2 \phi}{\partial x^2}(x(s), s)\, ds \right) H,$$

and a similar result holds true for the term containing r in (10.4.23).

Gathering results, we obtain

$$E_{xt} \left[\phi(x(s_2), s_2) - \phi(x(s_1), s_1) - \int_{s_1}^{s_2} \operatorname{tr} q \frac{\partial^2 \phi}{\partial x^2}(x(s), s)\, ds \right.$$

$$\left. - \int_{s_1}^{s_2} r \cdot \frac{\partial \phi}{\partial x}(x(s), s)\, ds - \int_{s_1}^{s_2} \frac{\partial \phi}{\partial s}\, ds \right] H = 0.$$

This implies

$$E_{xt} \left[\phi(x(s_2), s_2) - \phi(x(s_1), s_1) - \int_{s_1}^{s_2} \operatorname{tr} q \frac{\partial^2 \phi}{\partial x^2}(x(s), s)\, ds \right.$$

$$\left. - \int_{s_1}^{s_2} r \cdot \frac{\partial \phi}{\partial x}(x(s), s) - \int_{s_1}^{s_2} \frac{\partial \phi}{\partial s}\, ds \,\middle|\, \mathcal{F}_t^{s_1} \right] = 0.$$

Therefore μ_{xt} is the diffusion corresponding to (10.4.15). From the uniqueness, we obtain (10.4.16), which completes the proof.   ∎

10.5. The case $k = 2$. Here we use the results of Section 10.3. We assume that instead of (10.4.2), we have

$$(10.5.1) \qquad\qquad \int_0^{\tau_0} \int_Y b(y, \tau) p(y, \tau)\, dy\, d\tau = 0.$$

Therefore we can define the solutions $\theta^j(y, \tau)$ of

$$(10.5.2) \qquad\qquad -\frac{\partial}{\partial \tau}\theta^j + A_1 \theta^j = -b_j, \qquad \theta^j \text{ periodic in } y, \tau.$$

From the regularity of the coefficients we can ensure that $\theta^j \in C^{2,1}$.

LEMMA 10.6. *The family of measures μ_{xt}^ε remains as $\varepsilon \to 0$, in a weakly compact subset of $\mathcal{M}_+^1(\Psi)$.*

PROOF. We use Itô's formula on the θ^j, hence $(\theta = (\theta^1, \ldots, \theta^n))$

$$(10.5.3) \qquad \theta\left(\frac{y^\varepsilon(s)}{\varepsilon}, \frac{s}{\varepsilon^2}\right) = \theta\left(\frac{x}{\varepsilon}, \frac{t}{\varepsilon^2}\right) + \int_t^s \left(\frac{1}{\varepsilon^2}\frac{\partial\theta}{\partial\tau} + \frac{1}{\varepsilon^2}\frac{\partial\theta}{\partial y}b + \frac{1}{\varepsilon}\frac{\partial\theta}{\partial y}c\right.$$
$$\left. + \frac{1}{\varepsilon^2}\operatorname{tr}\frac{\partial^2\theta}{\partial y}a\right)\, d\lambda + \int_t^s \frac{1}{\varepsilon}\frac{\partial\theta}{\partial y}\cdot\sigma\, dw$$

and from (10.5.2), it follows that

$$(10.5.4) \quad \theta\left(\frac{y^\varepsilon(s)}{\varepsilon}, \frac{s}{\varepsilon^2}\right) = \theta\left(\frac{x}{\varepsilon}, \frac{t}{\varepsilon^2}\right) + \int_t^s \frac{b}{\varepsilon^2}\, d\lambda + \int_t^s \frac{1}{\varepsilon}\frac{\partial\theta}{\partial y}c\, d\lambda + \int_t^s \frac{1}{\varepsilon}\frac{\partial\theta}{\partial y}\cdot\sigma\, dw.$$

Combining with (10.2.5), we obtain
$$(10.5.5)$$
$$y^\varepsilon(s) - \varepsilon\theta\left(\frac{y^\varepsilon(s)}{\varepsilon}, \frac{s}{\varepsilon^2}\right) = x - \varepsilon\theta\left(\frac{x}{\varepsilon}, \frac{t}{\varepsilon^2}\right) + \int_t^s \left(I - \frac{\partial\theta}{\partial y}\right)c\, d\lambda + \int_t^s \left(I - \frac{\partial\theta}{\partial y}\right)\sigma\, dw$$

and this relation implies by standard arguments that μ_{xt}^ε remains in a compact subset of $\mathcal{M}_+^1(\Psi)$. ∎

LEMMA 10.7. *Let $\phi(x, y, \tau, s)$ be a function which is C^1 in y, τ, s, C^2 in x, periodic in y, τ and such that*

$$(10.5.6) \qquad \int_0^{\tau_0}\int_Y \phi(x, y, \tau, s)p(y, \tau)\, dy\, d\tau = 0;$$

then we have

$$(10.5.7) \qquad E\left|\left(\int_{s_1}^{s_2}\phi\left(y^\varepsilon(\lambda), \frac{y^\varepsilon(\lambda)}{\varepsilon}, \frac{\lambda}{\varepsilon^2}, \lambda\right)\, d\lambda\right)^2 \,\middle|\, \mathcal{F}_t^{s_1}\right| \to 0.$$

PROOF. We define ψ such that

$$(10.5.8) \qquad -\frac{\partial\psi}{\partial\tau} + A_1\psi = \phi, \qquad \psi \text{ periodic in } y, \tau.$$

Then ψ is in C^2 in x, y, C^1 in τ, s. Using Itô's formula in a routine way, the result follows as in Lemma 10.4. ∎

We define the homogenized coefficients as follows:

$$(10.5.9) \qquad q = \frac{1}{\tau_0}\int_0^{\tau_0}\int_Y \left(I - \frac{\partial\theta}{\partial y}\right)a\left(I - \frac{\partial\theta}{\partial y}\right)^*p\, dy\, d\tau,$$

$$(10.5.10) \qquad r = \frac{1}{\tau_0}\int_0^{\tau_0}\int_Y \left(I - \frac{\partial\theta}{\partial y}\right)cp\, dy\, d\tau.$$

We define the measure μ_{xt} on Ψ corresponding to the diffusion with constant coefficients q, r, i.e., (10.4.15) and we state the

THEOREM 10.8. *We assume (10.2.1), (10.2.2), (10.2.3), (10.3.14) and (10.5.1). Then we have*

$$(10.5.11) \qquad \forall x, t, \qquad \mu_{xt}^\varepsilon \to \mu_{xt} \text{ in } \mathcal{M}_+^1(\Psi) \text{ weakly.}$$

PROOF. The argument is now standard and we will only sketch it briefly. We use the canonical process, hence

$$(10.5.12) \qquad dx = \frac{1}{\varepsilon} b\left(\frac{x}{\varepsilon}, \frac{s}{\varepsilon^2}\right) ds + \sigma\left(\frac{x}{\varepsilon}, \frac{s}{\varepsilon^2}\right) dw^\varepsilon(s) + c\left(\frac{x}{\varepsilon}, \frac{s}{\varepsilon^2}\right) ds,$$
$$\mu_{xt}(x(t) = x) = 1.$$

We define

$$\zeta^\varepsilon(s) = x(s) - \varepsilon\theta\left(\frac{x(s)}{\varepsilon}, \frac{s}{\varepsilon^2}\right)$$

hence

$$(10.5.13) \quad \zeta^\varepsilon(s) = x - \varepsilon\mathcal{O}\left(\frac{x}{\varepsilon}, \frac{t}{\varepsilon^2}\right) + \int_t^s \left(I - \frac{\partial\theta}{\partial y}\right) c \, d\lambda + \int_t^s \left(I - \frac{\partial\theta}{\partial y}\right) \sigma \, dw.$$

Let $\phi \in C_0^\infty(\mathbb{R}^n \times [0, \infty[)$. Applying again Itô's formula we have

$$\phi(\zeta^\varepsilon(s_2), s_2) = \phi(\zeta^\varepsilon(s_1), s_1) + \int_{s_1}^{s_2} \left\{ \frac{\partial\phi}{\partial s} + \frac{\partial\phi}{\partial x} \cdot \left(I - \frac{\partial\theta}{\partial y}\right) c \right.$$
$$+ \operatorname{tr} \frac{\partial^2\phi}{\partial x^2} \left(I - \frac{\partial\theta}{\partial y}\right) a \left(I - \frac{\partial\theta}{\partial y}\right)^* \right\} ds$$
$$+ \int_{s_1}^{s_2} \frac{\partial\phi}{\partial x} \left(I - \frac{\partial\theta}{\partial y}\right) \sigma \, dw,$$

hence

$$E_{xt}^\varepsilon \left[\phi(x(s_2), s_2) - \phi(x(s_1), s_1) - \int_{s_1}^{s_2} \operatorname{tr} q \frac{\partial^2\phi}{\partial x^2}(x(s), s) \, ds \right.$$
$$\left. - \int_{s_1}^{s_2} r \frac{\partial\phi}{\partial x}(x(s), s) \, ds - \int_{s_1}^{s_2} \frac{\partial\phi}{\partial s} \, ds \,\middle|\, \mathcal{G}_t^{s_1} \right]$$
$$= E_{xt}^\varepsilon \left[\int_{s_1}^{s_2} \operatorname{tr} \tilde{q}\left(\frac{x(s)}{\varepsilon}, \frac{s}{\varepsilon^2}\right) \frac{\partial^2\phi}{\partial x^2}(x(s), s) \, ds \,\middle|\, \mathcal{G}_t^{s_1} \right]$$
$$+ E_{xt}^\varepsilon \left[\int_{s_1}^{s_2} \frac{\partial\phi}{\partial x}(x(s), s) \cdot r\left(\frac{x(s)}{\varepsilon}, \frac{s}{\varepsilon^2}\right) ds \,\middle|\, \mathcal{G}_t^{s_1} \right] + O(\varepsilon),$$

where

$$\tilde{q}(y, \tau) = q(y, \tau) - q,$$
$$\tilde{r}(y, \tau) = r(y, t) - r.$$

Using Lemma 10.7 and the compactness of μ^ε, we easily obtain (10.5.11). ∎

10.6. The case $k > 2$. A situation similar to the case when $k < 2$, will be encountered. We set now $r = k - 2$, and consider an expansion of the form (vector expansion)

$$\phi\left(\frac{y^\varepsilon(s)}{\varepsilon}, \frac{s}{\varepsilon^k}\right) + \varepsilon^r \phi_1\left(\frac{y^\varepsilon(s)}{\varepsilon}, \frac{s}{\varepsilon^k}\right) + \varepsilon^{2r} \phi_2\left(\frac{y^\varepsilon(s)}{\varepsilon}, \frac{s}{\varepsilon^k}\right) + \cdots.$$

Applying Itô's formula, we obtain

(10.6.1) $\varepsilon\phi\left(\dfrac{y^\varepsilon(s)}{\varepsilon},\dfrac{s}{\varepsilon^k}\right) + \varepsilon^{r+1}\phi_1\left(\dfrac{y^\varepsilon(s)}{\varepsilon},\dfrac{s}{\varepsilon^k}\right) + \varepsilon^{2r+1}\phi_2\left(\dfrac{y^\varepsilon(s)}{\varepsilon},\dfrac{s}{\varepsilon^k}\right) + \cdots$

() $* = \varepsilon\phi\left(\dfrac{x}{\varepsilon},\dfrac{t}{\varepsilon^k}\right) + \varepsilon^{r+1}\phi_1\left(\dfrac{x}{\varepsilon},\dfrac{t}{\varepsilon^k}\right) + \varepsilon^{2r+1}\phi_2\left(\dfrac{x}{\varepsilon},\dfrac{t}{\varepsilon^k}\right) + \cdots$

$$+ \int_t^s \left(\frac{\partial\phi}{\partial\tau}\frac{\varepsilon}{\varepsilon^k} + \frac{\varepsilon^{r+1}}{\varepsilon^k}\frac{\partial\phi_1}{\partial\tau} + \frac{\varepsilon^{2r+1}}{\varepsilon^k}\frac{\partial\phi_2}{\partial\tau} + \cdots\right) d\lambda$$

$$+ \int_t^s \left(-\frac{A_1\phi}{\varepsilon} + \frac{\varepsilon^r}{\varepsilon}A_1\phi_1 - \frac{\varepsilon^{2r}}{\varepsilon}A_1\phi_2 + \cdots\right) d\lambda$$

$$+ \int_t^s \left(\frac{\partial\phi}{\partial y}\cdot c + \varepsilon^r\frac{\partial\phi_1}{\partial y}\cdot c + \varepsilon^{2r}\frac{\partial\phi_2}{\partial y}c\right) d\lambda$$

$$+ \int_s^t \left(\frac{\partial\phi}{\partial y}\cdot\sigma\,dw + \varepsilon^r\frac{\partial\phi_1}{\partial y}\cdot\sigma\,dw + \varepsilon^{2r}\frac{\partial\phi_2}{\partial y}\cdot\sigma\,dw + \cdots\right).$$

Our objective is to adjust the expansion (10.6.1) in order that the only term which has a positive power of ε equal to 1 in the denominator be equal to b. Hence we try to satisfy the conditions

(10.6.2) $$\frac{\partial\phi}{\partial\tau} = 0,$$

(10.6.3) $$\frac{\partial\phi_1}{\partial\tau} - A_1\phi = b.$$

At this level a discussion is necessary. If $k \geq 3$, then $r \geq 1$, $2r + 1 - k = r - 1 \geq 0$, and the objective is satisfied. If $3 > k \geq \frac{5}{2}$, then we need to add the condition

(10.6.4) $$\frac{\partial\phi_2}{\partial\tau} - A_1\phi_1 = 0.$$

If $\frac{5}{2} > k \geq \frac{7}{3}$, we need the additional condition

(10.6.5) $$\frac{\partial\phi_3}{\partial\tau} - A_1\phi_2 = 0,$$

and so on, if $\frac{2p+1}{p} > k \geq \frac{2p+3}{p+1}$, we need to go to

(10.6.6) $$\frac{\partial\phi_{p+1}}{\partial\tau} - A_1\phi_p = 0.$$

We must ensure that conditions (10.6.2), (10.6.3), (10.6.6) can be realized. We have $\phi = \phi(y)$ and to satisfy (10.6.3) we must have

$$\int_0^{\tau_0} A_1\phi\,d\tau = -\int_0^{\tau_0} b\,d\tau,$$

or

(10.6.7) $$\bar{A}_1\phi = -\bar{b},$$

where

$$\bar{A}_1 = \frac{1}{\tau_0}\int_0^{\tau_0} A_1(\tau)\,d\tau.$$

We introduce the function $\pi(y)$ solution of

(10.6.8) $\bar{A}_1^*\pi = 0,$ π periodic, $\displaystyle\int_Y \pi\,dy = 1,$ $\pi > 0,$

and the solvability condition for (10.6.7) is

$$(10.6.9) \qquad \int_Y \pi(y)\bar{b}(y) \, dy = \int_Y \int_0^{\tau_0} \pi(y)b(y,\tau) \, dy \, d\tau = 0.$$

Then we have

$$(10.6.10) \qquad \phi_1(y,\tau) = \int_0^\tau (A_1\phi + b) \, ds + \tilde{\phi}_1(y).$$

If we need to satisfy (10.6.4) then we must guarantee that

$$\bar{A}_1\tilde{\phi}_1 + \frac{1}{\tau_0}\int_0^{\tau_0} d\tau \, A_1 \left[\int_0^\tau (A_1\phi + b) \, ds\right] = 0,$$

or

$$(10.6.11) \qquad \bar{A}_1\tilde{\phi}_1 = -\frac{1}{\tau_0}\int_0^{\tau_0} A_1(\tau)\left(\int_0^\tau (A_1(s)\phi + b(s)) \, ds\right) d\tau.$$

To solve (10.6.11) we need the following solvability condition

$$(10.6.12) \qquad \int_0^{\tau_0}\int_Y \pi(y)A_1(\tau)\left(\int_0^\tau (A_1(s)\phi(y) + b(y,s)) \, ds\right) d\tau \, dy = 0.$$

We then have

$$(10.6.13) \qquad \phi_2(y,\tau) = \int_0^\tau A_1\phi_1 \, ds + \tilde{\phi}_2(y),$$

and if we need to solve (10.6.5) we must have the analogue for (10.6.12), i.e.

$$(10.6.14) \qquad \int_0^{\tau_0}\int_Y \pi(y)A_1(\tau)\left(\int_0^\tau A_1(s)\phi_1(y,s) \, ds\right) d\tau \, dy = 0.$$

In general if $\frac{2p+1}{p} > k \geq \frac{2p+3}{p+1}$ we will need the condition

$$(10.6.15) \qquad \int_0^{\tau_0}\int_Y \pi(y)A_1(\tau)\left(\int_0^\tau A_1(s)\phi_{p-1}(y,s) \, ds\right) d\tau = 0, \qquad p \geq 2.$$

The conditions (10.6.9), (10.6.12) and (10.6.15) are conditions on b.

We can now state the convergence result. The homogenized coefficients are defined by

$$(10.6.16) \qquad q = \frac{1}{\tau_0}\int_0^{\tau_0}\int_Y \pi\left(I - \frac{\partial\phi}{\partial y}\right)a\left(I - \frac{\partial\phi}{\partial y}\right)^* dy \, d\tau,$$

$$r = \frac{1}{\tau_0}\int_0^{\tau_0}\int_Y \pi\left(I - \frac{\partial\phi}{\partial y}\right)c \, dy \, d\tau.$$

THEOREM 10.9. *We assume (10.2.1), (10.2.2), (10.2.3), (10.6.9). If $\frac{2p+1}{p} > k \geq \frac{2p+3}{p+1}$, we assume in addition that a_{ij} are $3p+2$ times continuously differentiable in y, b_i are $ep+1$ times continuously differentiable in y and (10.6.12), (10.6.15) hold true. Then we have*

$$(10.6.17) \qquad \forall x,t, \qquad \mu_{xt}^\varepsilon \to \mu_{xt} \text{ in } \mathcal{M}_+^1(\Psi) \text{ weakly.}$$

PROOF. The family μ_{xt}^ε remains in a compact subset of $\mathcal{M}_+^1(\Psi)$, by the construction explained above. Now if $\phi(x,y,\tau,s)$ is C^1 in y,τ,s, C^2 in x and satisfies

$$(10.6.18) \qquad \int_0^{\tau_0}\int_Y \pi(y)\phi(y,\tau) \, dy \, d\tau = 0,$$

then we can state the analogue of Lemma 10.3 or 10.4, i.e.,

$$(10.6.19) \qquad E\left[\left(\int_{s_1}^{s_2} \phi\left(y^\varepsilon(\lambda)\frac{y^\varepsilon(\lambda)}{\varepsilon}, \frac{\lambda}{\varepsilon^k}, \lambda\right) d\lambda\right)^2 \bigg| \mathcal{F}_t^{s_1}\right] \to 0.$$

Indeed let $\psi(y)$ be independent of τ such that

$$\bar{A}_1\psi = \frac{1}{\tau_0}\int_0^{\tau_0} \phi(y, \tau) \, d\tau,$$

and let $\psi(y, \tau)$ be defined by

$$\psi_1(y, \tau) = \int_0^\tau (A_1(s)\psi(y) - \phi(y, s)) \, ds;$$

then ψ_1 is periodic in y, τ and

$$-\frac{\partial\psi_1}{\partial\tau} + A_1\psi - \phi = 0.$$

Therefore, from Itô's formula we have (omitting x, s to simplify)

$$\varepsilon^2\psi\left(\frac{y^\varepsilon(s_2)}{\varepsilon}\right) - \varepsilon^2\psi\left(\frac{y^\varepsilon(s_1)}{\varepsilon}\right) + \varepsilon^{2+r}\psi_1\left(\frac{y^\varepsilon(s_2)}{\varepsilon}, \frac{s_2}{\varepsilon^k}\right)$$
$$- \varepsilon^{2+r}\psi_1\left(\frac{y^\varepsilon(s_1)}{\varepsilon}, \frac{s_1}{\varepsilon^k}\right) = -\int_{s_1}^{s_2} \phi\left(\frac{y^\varepsilon(\lambda)}{\varepsilon}, \frac{\lambda}{\varepsilon^k}\right) d\lambda$$
$$+ \int_{s_1}^{s_2} \left(\varepsilon\frac{\partial\psi}{\partial y} + \varepsilon^{r+1}\frac{\partial\psi_1}{\partial y}\right) \cdot \sigma \, dw$$

and (10.6.19) follows by standard arguments.

We next proceed as in the proof of Theorem 10.5. We define

$$\zeta^\varepsilon(s) = x(s) - \varepsilon\phi\left(\frac{x(s)}{\varepsilon}\right) - \varepsilon^{r+1}\phi_1\left(\frac{y(s)}{\varepsilon}, \frac{s}{\varepsilon^k}\right)$$

and by Itô's formula

$$\zeta^\varepsilon(s) = x - \varepsilon\phi\left(\frac{x}{\varepsilon}\right) - \varepsilon^{r+1}\phi_1\left(\frac{x}{\varepsilon}, \frac{t}{\varepsilon^k}\right)$$
$$+ \int_t^s \left(I - \frac{\partial\phi}{\partial y} - \varepsilon^r\frac{\partial\phi_1}{\partial y} - \cdots\right)\sigma \, dw^\varepsilon$$
$$+ \int_t^s \left(I - \frac{\partial\phi}{\partial y} - \varepsilon^r\frac{\partial\phi_1}{\partial y} - \cdots\right) c \, d\lambda.$$

Let $\phi \in C_0^\infty(\mathbb{R}^n \times [0, \infty[)$. From Itô's formula again we obtain

$$\phi(\zeta^\varepsilon(s_2), s_2) = \phi(\zeta^\varepsilon(s_1), s_1) + \int_{s_1}^{s_2}\left\{\frac{\partial\phi}{\partial s} + \frac{\partial\phi}{\partial x}\cdot\left[\left(I - \frac{\partial\phi}{\partial y}\right)c + O(\varepsilon)\right]\right.$$
$$+ \operatorname{tr}\frac{\partial^2\phi}{\partial x^2}\left(I - \frac{\partial\phi}{\partial y}\right)a\left(I - \frac{\partial\phi}{\partial y}\right)^* + O(\varepsilon)\bigg\} \, ds$$
$$+ \int_{s_1}^{s_2}\frac{\partial\phi}{\partial x}\left(I - \frac{\partial\phi}{\partial y}\right) dw^\varepsilon$$
$$+ \int_{s_1}^{s_2}\frac{\partial\phi}{\partial x}\left(-\varepsilon^r\frac{\partial\phi_1}{\partial y} - \varepsilon^{2r}\frac{\partial\phi_2}{\partial y} - \cdots\right) ds.$$

Hence

$$
E_{xt}^\varepsilon \left[\phi(x(s_2), s_2) - \phi(x(s_1), s_1) - \int_{s_1}^{s_2} \operatorname{tr} q \frac{\partial^2 \phi}{\partial x^2}(x(s), s) \, ds \right.
$$
$$
\left. - \int_{s_1}^{s_2} r \frac{\partial \phi}{\partial x}(x(s), s) \, ds - \int_{s_1}^{s_2} \frac{\partial \phi}{\partial s} \, \Big| \, \mathcal{G}_t^{s_1} \right]
$$
$$
= E_{xt}^\varepsilon \left[\int_{s_1}^{s_2} \operatorname{tr} \tilde{q} \left(\frac{x(s)}{\varepsilon}, \frac{s}{\varepsilon^2} \right) \frac{\partial^2 \phi}{\partial x^2}(x(s), s) \, ds \, \Big| \, \mathcal{G}_t^{s_1} \right]
$$
$$
+ E_{xt}^\varepsilon \left[\int_{s_1}^{s_2} \frac{\partial \phi}{\partial x}(x(s), s) \tilde{r} \left(\frac{x(s)}{\varepsilon}, \frac{s}{\varepsilon^2} \right) \, ds \, \Big| \, \mathcal{G}_t^{s_1} \right] + O(\varepsilon).
$$

Using the property (10.6.19) we easily complete the proof of the desired result. ∎

10.7. Applications to parabolic equations. We can apply the preceding results to the homogenization of the parabolic equation

$$
(10.7.1) \quad -\frac{\partial u_\varepsilon}{\partial t} - a_{ij} \left(\frac{x}{\varepsilon}, \frac{t}{\varepsilon^k} \right) \frac{\partial^2 u_\varepsilon}{\partial x_i \partial x_j} - \left(\frac{1}{\varepsilon} b_i \left(\frac{x}{\varepsilon}, \frac{t}{\varepsilon^k} \right) + c_i \left(\frac{x}{\varepsilon}, \frac{t}{\varepsilon^k} \right) \right) \frac{\partial u_\varepsilon}{\partial x_i} = f,
$$
$$
u_\varepsilon \mid_\Sigma = 0, \qquad u_\varepsilon(x, T) = 0,
$$

where $\Sigma = \Gamma \times]0, T[$, $\underline{Q} = \mathcal{O} \times]0, T[$.

The homogenized problem will be

$$
(10.7.2) \quad -\frac{\partial u}{\partial t} - q_{ij} \frac{\partial^2 u}{\partial x_i \partial x_j} - r_i \frac{\partial u}{\partial x_i} = f, \qquad u \mid_\Sigma = 0, \qquad u(x, T) = 0.
$$

We can state the

THEOREM 10.10. *We assume that $\mathcal{O}$ satisfies (4.4.1), $f \in C^0(\underline{Q})$, and that the assumptions of Theorem 10.5 (if $k < 2$) 10.8 (if $k = 2$), 10.9 (if $k > 2$) are satisfied, then we have*

$$
(10.7.3) \qquad\qquad u_\varepsilon(x, t) \to u(x, t), \qquad x, t \in \overline{\underline{Q}}.
$$

The proof of Theorem 10.10 is now standard and is left to the reader.

REMARK 10.11. The formulas for q, r depend on k in the following way. There is one formula for $k < 2$, (independent of k), one formula for $k = 2$, one formula for $k > 2$ (independent of k).

REMARK 10.12. Let us consider the self-adjoint case, i.e., $c = 0$, $b_i = \frac{\partial a_{ij}}{\partial y_j}$. Then we have $m = 1$, $p = 1$, $\pi = 1$. Then (10.4.2) is satisfied, also (10.5.1) and (10.6.9) are satisfied. Let us check that (10.6.12) and all the conditions (10.6.15) for any prime p are also satisfied. Indeed $A_1(\tau) = -\frac{\partial}{\partial y_i} a_{ij} \frac{\partial}{\partial y_j}$, and $\int_Y A_1(\tau) \psi(Y, \tau) \, dy = 0$ for any periodic function ψ. Therefore when the a_{ij} are sufficiently regular, the results of Theorem 10.10 apply for the problem

$$
\frac{\partial u_\varepsilon}{\partial t} - \frac{\partial}{\partial x_i} a_{ij} \left(\frac{x}{\varepsilon}, \frac{t}{\varepsilon^k} \right) \frac{\partial u_\varepsilon}{\partial x_j} = f, \qquad u_\varepsilon \mid_\Sigma = 0, \qquad u_\varepsilon(x, T) = 0.
$$

We recover the results of Chapter 2.

REMARK 10.13. We can also give some results in the case of Neumann boundary conditions by probabilistic methods. This can be done as in the elliptic case (cf. Section 9) with extensions to the parabolic case similar to those of the present paragraph.

11. Averaging

11.1. Setting of the problem. We consider the following problem

$$(11.1.1) \qquad -\frac{\partial u_\varepsilon}{\partial t} - a_{ij}\left(x, \frac{t}{\varepsilon}\right)\frac{\partial^2 u_\varepsilon}{\partial x_i \partial x_j} - b_i\left(x, \frac{t}{\varepsilon}\right)\frac{\partial u_\varepsilon}{\partial x_i} + a_0\left(x, \frac{t}{\varepsilon}\right)u_\varepsilon = f,$$

$$u_\varepsilon\mid_\Gamma = 0, \qquad u_\varepsilon(x, T) = 0.$$

We will assume that $\mathcal{O}$ satisfies (4.4.1) and

$(11.1.2) \qquad a_{ij}, b_i, a_0$ are measurable and bounded,

$$a_{ij}\xi_i\xi_j \geq \alpha|\xi|^2, \qquad \alpha > 0, \ \forall \xi \in \mathbb{R}^n,$$

$(11.1.3) \qquad$ if $h(x, t)$ denotes one of the functions $a_{ij}(x, t)$,

$b_i(x, t), a_0(x, t)$ then we assume that

$$\sup_{\substack{s \geq 0 \\ }} \sup_{\substack{|x-x'| \leq \delta \\ x, x' \in \overline{\mathcal{O}}}} |h(x, s) - h(x', s)| \leq \rho(\delta),$$

where $\rho(\delta) \to 0$ as $\delta \to 0$,

$(11.1.4) \qquad h(x, t)$ denoting the same functions as in (11.1.3), then

$$\int_t^{t+\theta} (h(x, s) - \bar{h}(x))\, ds \text{ is bounded as } \theta \to +\infty,$$

for x, t fixed,

$(11.1.5) \qquad a_{ij}, b_i, \bar{a}_{ij}, \bar{b}_i$ are continuous in x, t, and bounded,

$(11.1.6) \qquad f \in C^0(\overline{Q}).$

Since the problem is parabolic we can without loss of generality, assume that

$$(11.1.7) \qquad\qquad\qquad a_0 \geq 0.$$

The averaged problem is defined as follows:

$$(11.1.8) \qquad -\frac{\partial u}{\partial t} - \bar{a}_{ij}(x)\frac{\partial^2 u}{\partial x_i \partial x_j} - \bar{b}_i(x)\frac{\partial u}{\partial x_i} + \bar{a}_0(x)u = f,$$

$$u\mid_\Sigma = 0, \qquad u(x, T) = 0.$$

The functions u_ε and u belong to $\mathcal{W}^{2,1,p}(Q)$, $\forall p \geq 2$, hence in particular are continuous in $\overline{Q}$.

We can state the following

THEOREM 11.1. *Under the assumptions (11.1.2), (11.1.3), (11.1.4), (11.1.5), and (11.1.6) we have*

$$(11.1.9) \qquad\qquad u_\varepsilon(x, t) \to u(x, t), \qquad x, t \in \overline{Q}.$$

11.2. Proof of Theorem 11.1. We consider $\sigma(x, t)$ such that

$$(11.2.1) \qquad\qquad\qquad \frac{\sigma\sigma^*}{2} = a,$$

and we extend the functions a_{ij}, b_i, a_0 outside $\overline{\mathcal{O}}$, in such a way that the conditions (11.1.2), (11.1.3), (11.1.4), (11.1.5) are satisfied on $\mathbb{R}^n \times [0, T]$. We set $\Psi = C^0[t, T; \mathbb{R}^n]$ and consider the canonical process $\psi = x(\cdot)$, $x(s) = \psi(s) = x(s, \psi)$ and $\mathcal{G}_t^s = (x(\lambda), t \leq \lambda \leq s)$. There exists one and only one probability measure on Ψ, $\mathcal{G}_t^\infty$, denoted by μ_{xt}^ε, such that there exists a standard Wiener process with respect to $\mathcal{G}_t^s$, denoted by $w^\varepsilon(s)$, for which one has

$$(11.2.2) \qquad dx(s) = b\left(x(s), \frac{s}{\varepsilon}\right) ds + \sigma\left(x(s), \frac{s}{\varepsilon}\right) dw^\varepsilon(s),$$
$$\mu_{xt}^\varepsilon(x(t) = x) = 1.$$

Similarly, we denote by μ_{xt} the probability measure associated with

$$(11.2.3) \qquad dx(s) = \bar{b}(x(s)) ds + \sqrt{s}\bar{\sigma}(x(s)) dw(s), [34]$$
$$\mu_{xt}(x(t) = x) = 1.$$

We are going to prove that

$$(11.2.4) \qquad \mu_{xt}^\varepsilon \to \mu_{xt} \text{ in } \mathcal{M}_+^1(\Psi) \text{ weakly}, \forall x, t.$$

It is clear that μ_{xt} remains in a weakly compact subset of $\mathcal{M}_+^1(\Psi)$. Let $\phi \in C_0^\infty(\mathbb{R}^n)$. We have

$$(11.2.5) \quad E_{xt}^\varepsilon\left[\phi(x(s_2)) - \phi(x(s_1)) - \int_{s_1}^{s_2} \frac{\partial \phi}{\partial x}(x(\lambda)) \cdot b\left(x(\lambda), \frac{\lambda}{\varepsilon}\right) d\lambda\right.$$
$$\left. - \int_{s_1}^{s_2} \text{tr} \frac{\partial^2 \phi}{\partial x^2}(x(\lambda)) a\left(x(\lambda), \frac{\lambda}{\varepsilon}\right) d\lambda \,\middle|\, \mathcal{G}_t^{s_1}\right] = 0.$$

Hence

$$(11.2.6) \quad E_{xt}^\varepsilon\left[\phi(x(s_2)) - \phi(x(s_1)) - \int_{s_1}^{s_2} \frac{\partial \phi}{\partial x}(x(\lambda)) \cdot b(x(\lambda)) d\lambda\right.$$
$$\left. - \int_{s_1}^{s_2} \text{tr} \frac{\partial^2 \phi}{\partial x^2}(x(\lambda)) \bar{a}(x(\lambda)) d\lambda \,\middle|\, \mathcal{G}_t^{s_1}\right]$$
$$= E_{xt}^\varepsilon\left[\int_{s_1}^{s_2} \frac{\partial \phi}{\partial x}(x(\lambda)) \cdot \tilde{b}\left(x(\lambda), \frac{\lambda}{\varepsilon}\right) d\lambda \,\middle|\, \mathcal{G}_t^{s_1}\right]$$
$$+ E_{xt}^\varepsilon\left[\int_{s_1}^{s_2} \text{tr} \frac{\partial^2 \phi}{\partial x^2}(x(\lambda)) \tilde{a}\left(x(\lambda), \frac{\lambda}{\varepsilon}\right) d\lambda \,\middle|\, \mathcal{G}_t^{s_1}\right]$$

where we have set

$$(11.2.7) \qquad \tilde{b}(x, t) = b(x, t) - \bar{b}(x),$$
$$\tilde{a}(x, t) = a(x, t) - \bar{a}(x).$$

We are going to prove that

$$(11.2.8) \qquad E_{xt}^\varepsilon\left|E_{xt}\left[\int_{s_1}^{s_2} \frac{\partial \phi}{\partial x}(x(\lambda)) \cdot \tilde{b}\left(x(\lambda), \frac{\lambda}{\varepsilon}\right) d\lambda \,\middle|\, \mathcal{G}_{t_1}^s\right]\right| \to 0,$$
$$E_{xt}^\varepsilon\left|E_{xt}^\varepsilon\left[\int_{s_1}^{s_2} \text{tr} \frac{\partial^2 \phi}{\partial x^2}(x(\lambda)) \tilde{a}\left(x(\lambda), \frac{\lambda}{\varepsilon}\right) d\lambda \,\middle|\, \mathcal{G}_t^{s_1}\right]\right| \to 0,$$

---

[34]Since the coefficients are not Lipschitz, only the weak sense of stochastic differential equations (11.2.2), (11.2.3) can be considered.

from which, and the weak compactness, the desired result will follow. It is of course enough to prove

$$(11.2.9) \qquad E_{xt}^{\varepsilon} \left(\int_{s_1}^{s_2} \frac{\partial \phi}{\partial x}(x(\lambda)) \cdot b \left(x(\lambda), \frac{\lambda}{\varepsilon} \right) d\lambda \right)^2 \to 0,$$

$$E_{xt}^{\varepsilon} \left(\int_{s_1}^{s_2} \operatorname{tr} \frac{\partial^2 \phi}{\partial x^2}(x(\lambda)) a \left(x(\lambda), \frac{\lambda}{\varepsilon} \right) d\lambda \right)^2 \to 0.$$

This will follow from the

LEMMA 11.2. *Let* $\phi(x, \tau)$ *be a measurable bounded function such that*

$$(11.2.10) \qquad \sup_{s} \sup_{|x-x'| \leq \delta} |\phi(x, s) - \phi(x', s)| \leq \rho(\delta) \to 0, \text{ as } \delta \to 0,$$

$$(11.2.11) \qquad \int_0^{t+\tau} \phi(x, s) \, ds \text{ is bounded as } \tau \to +\infty, \text{ for } x, t \text{ fixed.}$$

Then we have

$$(11.2.12) \qquad E_{xt}^{\varepsilon} \left(\int_{s_1}^{s_2} \phi \left(x(\lambda), \frac{\lambda}{\varepsilon} \right) d\lambda \right)^2 \to 0.$$

REMARK 11.3. Lemma 11.2 is the analogue of Lemma 4.3, and again two proofs will be given. One easy proof, requiring more regularity on ϕ and a more technical one using just the assumptions of the Lemma.

FIRST PROOF OF LEMMA 11.2. We assume here that $\phi(x, \tau)$ is C^2 in x, and that

$$\psi(x, \tau) = \int_0^{\tau} \phi(x, s) \, ds,$$

and $\frac{\partial \psi}{\partial x_i}, \frac{\partial^2 \psi}{\partial x_i \partial x_j}$ are bounded in x, τ.

From Itô's formula, we have

$$\psi \left(x(s_2), \frac{s_2}{\varepsilon} \right) = \psi \left(x(s_1), \frac{s_1}{\varepsilon} \right) + \int_{s_1}^{s_2} \frac{1}{\varepsilon} \phi \left(x(\lambda), \frac{\lambda}{\varepsilon} \right) d\lambda$$

$$+ \int_{s_1}^{s_2} \left(\frac{\partial \psi}{\partial x_i} \left(x(\lambda), \frac{\lambda}{\varepsilon} \right) b_i \left(x(\lambda), \frac{\lambda}{\varepsilon} \right) \right.$$

$$\left. + \frac{\partial^2 \psi}{\partial x_i \partial x_j} \left(x(\lambda), \frac{\lambda}{\varepsilon} \right) a_{ij} \left(x(\lambda), \frac{\lambda}{\varepsilon} \right) \right) d\lambda$$

$$+ \int_{s_1}^{s_2} \frac{\partial \psi}{\partial x} \left(x(\lambda), \frac{\lambda}{\varepsilon} \right) \sigma \left(x(\lambda), \frac{\lambda}{\varepsilon} \right) dw^{\varepsilon}(\lambda).$$

Multiplying by ε, taking the square and the mathematical expectation, we obtain (11.2.12) (and also an estimate of the rate of convergence). ∎

SECOND PROOF OF LEMMA 11.2. For simplicity of notation we take $s_1 = 0$, $s_2 = T$, and we denote by $\mu_x^{\varepsilon} = \mu_{x0}^{\varepsilon}$. We rewrite (11.2.12) as

$$(11.2.13) \qquad X^{\varepsilon} = \int_{\Psi} d\mu_x^{\varepsilon}(\psi) \left(\int_0^T \phi \left(x(s, \psi), \frac{s}{\varepsilon} \right) ds \right)^2 \to 0.$$

The measures μ_x^ε are weakly compact and therefore, there exists, for given $\delta > 0$ a compact set $K_\delta \subset \Psi$ such that

$$\mu_x^\varepsilon(\psi \notin K_\delta) < \delta, \text{ for all } \varepsilon > 0,$$

(x is fixed, so we do not indicate the dependence on x). Therefore, we have if $|\phi| \leq M$

$$(11.2.14) \qquad X^\varepsilon \leq (MT)^2 \delta + \int_{K_\delta} d\mu_x^\varepsilon(\psi) \left(\int_0^T \phi\left(x(s,\psi), \frac{s}{\varepsilon} \right) ds \right)^2.$$

Since K_δ is a compact subset of a complete metric space, we can write for any given η

$$K_\delta = \bigcup_{j=1}^m K_j,$$

where the K_j are disjoint and K_j is contained in a sphere of center ψ_j and diameter η. Therefore

$(11.2.15)$

$$\int_{K_\delta} d\mu_x^\varepsilon(\psi) \left(\int_0^T \phi\left(x(s,\psi), \frac{s}{\varepsilon} \right) ds \right)^2 = \sum_j \int_{K_j} d\mu_x^\varepsilon(\psi) \left(\int_0^T \phi\left(x(s(\psi), \frac{s}{\varepsilon} \right) ds \right)$$

$$\leq 2 \sum_j \int_{K_j} d\mu_x^\varepsilon(\psi) \left| \left(\int_0^T \left(\phi\left(x(s,\psi), \frac{s}{\varepsilon} \right) - \phi\left(x(s,\psi_j), \frac{s}{\varepsilon} \right) \right) ds \right)^2 \right.$$

$$\left. + \left(\int_0^T \phi\left(x(s,\psi)j), \frac{s}{\varepsilon} \right) ds \right)^2 \right|$$

$$\leq 2(\rho(\eta)T)^2 + 2 \sum_{j=1}^m \mu_x^\varepsilon(K_j) \left(\int_0^T \phi\left(x(s,\psi_j), \frac{s}{\varepsilon} \right) ds \right)^2.$$

But $x(s,\psi_j) \in C(0,T;\mathbb{R}^n)$. We consider a splitting of $(0,T)$, $0, k, 2k, \ldots, Nk = T$, and

$$|x(s,\psi_j) - x(nk,\psi_j)| \leq \zeta_j(k) \text{ if } s \in [nk, (n+1)k]$$

with $\zeta_j(k) \to 0$ as $k \to 0$. therefore we have

$$(11.2.16) \quad \left(\int_0^T \phi\left(x(s, \psi_j), \frac{s}{\varepsilon} \right) ds \right)^2$$

$$\leq 2 \left(\sum_{n=0}^{N-1} \int_{nk}^{(n+1)k} \left(\phi\left(x(s, \psi_j), \frac{s}{\varepsilon} \right) - \phi\left(x(nk), \frac{s}{\varepsilon} \right) \right) ds \right)^2$$

$$+ 2 \left(\sum_{n=0}^{N-1} \int_{nk}^{(n+1)k} \phi\left(x(nk, \psi_j), \frac{s}{\varepsilon} \right) ds \right)^2$$

$$\leq 2 \left(\sum_{n=0}^{N-1} k\rho(\zeta_j(k)) \right)^2 + 2 \left(\sum_{n=0}^{N-1} \int_{nk}^{(n+1)k} \phi\left(x(nk, \psi_j), \frac{s}{\varepsilon} \right) ds \right)^2$$

$$\leq 2(T\rho(\zeta_j(k)))^2 + 2 \left(\sum_{n=0}^{N-1} \int_{nk}^{(n+1)k} \phi\left(x(nk, \psi_j), \frac{s}{\varepsilon} \right) ds \right)^2.$$

But

$$\int_{nk}^{(n+1)k} \phi\left(x(nk, \psi_j), \frac{s}{\varepsilon} \right) ds = \int_0^{(n+1)k} - \int_0^{nk}$$

$$= \varepsilon \int_0^{(n+1)k/\varepsilon} \phi(x(nk, \tau)) \, d\tau$$

$$- \varepsilon \int_0^{nk/\varepsilon} \phi(x(nk, \tau)) \, d\tau,$$

hence

$$(11.2.17) \quad \left| \int_{nk}^{(n+1)k} \phi\left(x(nk, \psi_j), \frac{s}{\varepsilon} \right) ds \right| \quad e\varepsilon\theta_n^k.$$

Collecting results we may write

$$X^\varepsilon \leq (MT)^2\delta + 2(\rho(\eta)T)^2 + 2T^2 \sup_{j=1,\ldots,m} \rho^2(\zeta_j(k)) + 2\varepsilon^2 \left(\sum_{n=0}^{N-1} \theta_n^k \right)^2.$$

Letting first ε go to 0, then k, then η, then δ, we obtain $X^\varepsilon \to 0$, which completes the proof of the lemma. ∎

We thus have completely proven (11.2.4). To complete the proof of the theorem, we write

$$u_\varepsilon(x, t) = \int_\Psi d\mu_{xt}^\varepsilon \left(\int_t^{T \wedge \tau} f(x(s), s) \left(\exp - \int_t^s a_0\left(x(\lambda), \frac{\lambda}{\varepsilon} \right) d\lambda \right) ds \right)$$

and we know that $a_0 \geq 0$. We first assert that

$$X^\varepsilon = \left| \int_\Psi d\mu_{xt}^\varepsilon \left(\int_t^{T \wedge \tau} f(x(s), s) \right) \right.$$

$$\cdot \left| \exp - \int_t^s a_0\left(x(\lambda), \frac{\lambda}{\varepsilon} \right) d\lambda - \exp - \int_t^s \bar{a}_0(x(\lambda)) \, d\lambda \right| \left. ds \right| \to 0, \qquad \varepsilon \to 0.$$

But

$$X^\varepsilon \le C \int_t^T ds \left\| \int_\Psi d\mu_{xt}^\varepsilon \left| \int_t^s \left(a_0 \left(x(\lambda), \frac{\lambda}{\varepsilon} \right) - \bar{a}_0(x(\lambda)) \right) d\lambda \right| \right\| \to 0$$

using again Lemma 11.2. The remainder of the proof is an immediate consequence of (11.2.4). The proof of Theorem 11.1 is complete. ∎

11.3. Remarks on generalized averaging. We can put homogenization and averaging in a common framework, which could also include other examples, provided that one can check some assumptions which will be stated below. We will remain a little bit formal in the presentation, since we are more interested in giving ideas than precise statements (there will actually be no new results). We consider a pair of diffusion processes as follows:

$$(11.3.1) \qquad dx^\varepsilon = \left(\frac{1}{\varepsilon} b_1(x^\varepsilon, y^\varepsilon) + c_1(x^\varepsilon, y^\varepsilon) \right) dt + \sigma_1(x^\varepsilon, y^\varepsilon) \, dw_1,$$

$$dy^\varepsilon = \left(\frac{1}{\varepsilon^2} b^2(x^\varepsilon, y^\varepsilon) + \frac{1}{\varepsilon} c_2(x^\varepsilon, y^\varepsilon) + d_2(x^\varepsilon, y^\varepsilon) \right) dt$$
$$+ \frac{\sigma_2}{\varepsilon}(x^\varepsilon, y^\varepsilon) \, dw_2,$$

$$(11.3.2) \qquad x^\varepsilon(0) = x, \qquad y^\varepsilon(0) = y_0 + \frac{y_1}{\varepsilon}.$$

The process x^ε will be called the slow varying process, and y^ε the fast varying process. The Wiener processes w_1, w_2 will in general be correlated, as follows:

$$(11.3.3) \qquad Ew_1(t)w_2^*(s) = Q_{12} \min(t, s).$$

Let us give some examples. First we take

$$b_2 = b_1 = b, \qquad c_2 = c_1 = c, \qquad d_2 = 0,$$
$$(11.3.4) \qquad \sigma_2 = \sigma_1 = \sigma, \qquad w_1 = w_2 = w \to \underline{Q}_{12} = I,$$
$$y_0 = 0, \qquad y_1 = x,$$

then

$$dx^\varepsilon = \frac{1}{\varepsilon} b \left(x^\varepsilon, \frac{x^\varepsilon}{\varepsilon} \right) dt + c \left(x^\varepsilon, \frac{x^\varepsilon}{\varepsilon} \right) dt + \sigma \left(x^\varepsilon, \frac{x^\varepsilon}{\varepsilon} \right) dw,$$
$$x^\varepsilon(0) = x,$$
$$y^\varepsilon(t) = \frac{x^\varepsilon(t)}{\varepsilon},$$

and we recover the homogenization problem.

A second example is the following

$$(11.3.5) \qquad b_1 = 0, \qquad b_2 = 1, \qquad c_2 = 0, \qquad d_2 = 0,$$
$$\sigma_1 = 0\sigma, \qquad w_1 = w, \qquad y_0 = y_1 = 0,$$

then we have $y^\varepsilon(t) = \frac{t}{\varepsilon^2}$ and

$$dx^\varepsilon = c \left(x^\varepsilon, \frac{t}{\varepsilon^2} \right) dt + \sigma \left(x^\varepsilon, \frac{t}{\varepsilon^2} \right) dw,$$
$$x^\varepsilon(0) = x,$$

and we recover the averaging example.

We can even consider a more general model than (11.3.1), which will include for instance reiterated homogenization. Indeed we consider diffusion processes $y_0^\varepsilon(t), y_1^\varepsilon(t), \ldots, y_N^\varepsilon(t)$ solution of the system

$$(11.3.6) \qquad dy_j^\varepsilon(t) = \sum_{k=0}^{N+j} \frac{c_{jk}(y_0^\varepsilon, y_1^\varepsilon, \ldots, y_N^\varepsilon)}{\varepsilon^k} + \frac{\sigma_j(y_0^\varepsilon, y_1^\varepsilon, \ldots, y_N^\varepsilon)\, dw_j}{\varepsilon^j},$$

$$y_j^\varepsilon(0) = y_0 + \frac{y_1}{\varepsilon} + \cdots + \frac{y_j}{\varepsilon^j}, \qquad j = 0, \ldots, N.$$

If we take $N = 1$, and

$$
\begin{aligned}
&c_{00} = c_1, && c_{01} = b_1, \\
&c_{10} = d_2, && c_{11} = c_2, && c_{12} = b_2, \\
&\sigma_0 = \sigma_1, && \sigma_1 = \sigma_2, && w_0 = w_1, && w_1 = w_2, \\
&x^\varepsilon(t) = y_0^\varepsilon(t), && y^\varepsilon(t) = y_1^\varepsilon(t),
\end{aligned}
$$

then we reobtain (11.3.1). If we take $N = 2$ and

$$dx^\varepsilon = \frac{1}{\varepsilon^2}b_1\, dt + \frac{1}{\varepsilon}b_2\, dt + c\, dt + \sigma\, dw,$$

$$y_0^\varepsilon(t) = x^\varepsilon(t), \qquad y_1^\varepsilon(t) = \frac{x^\varepsilon(t)}{\varepsilon}, \qquad y_2^\varepsilon(t) = \frac{x^\varepsilon(t)}{\varepsilon^2}, {}^{35}$$

then we recover the reiterated homogenization.

Therefore the model (11.3.6) is the more general one in the sense that it contains all the problems considered in this chapter.[36] We still call $y_0^\varepsilon(t)$ the slow process, and the problem of interest is to study the limit of μ_x^ε as $\varepsilon \to 0$, where μ_x^ε is the probability measure of $y_0^\varepsilon(\cdot)$, defined on $\Psi = C^0[0, \infty; \mathbb{R}^n]$.

The first objective is to check under what conditions μ_x^ε can remain in a weakly compact subset of $\mathcal{M}_+^1(\Psi)$. For this, we introduce a vector function $\chi^\varepsilon(y_0, y_1, \ldots, y_N)$ and apply Itô's formula, to yield

$$\chi^\varepsilon(y_0^\varepsilon(t), y_1^\varepsilon(t), \ldots, y_N^\varepsilon(t)) = \chi^\varepsilon(y_0^\varepsilon(0), \ldots, y_N^\varepsilon(0))$$

$$+ \int_0^t \frac{\partial \chi^\varepsilon}{\partial y_j} \left(\sum_{k=0}^{N+j} \frac{c_{jk}}{\varepsilon^k}\, ds + \frac{\sigma_j\, dw_j}{\varepsilon^j} \right)$$

$$+ \frac{1}{2} \int_0^t \operatorname{tr} \frac{\partial^2 \chi^\varepsilon}{\partial y_i \partial y_j} \frac{\sigma_i Q_{ij}\sigma_j^*}{\varepsilon^{i+j}}\, ds,$$

---

[35] The precise values of c_{jk} are easily defined.

[36] Except however the evolution problems with $k \neq 2$.

where $Q_{ij} = Ew_i(t)w_j^*(t)$, $(Q_{ij} = Q_{ji}^*)$, $(Q_{ii} = I)$. Therefore

$$\chi^\varepsilon(y_0^\varepsilon(t), y_1^\varepsilon(t), \dots, y_N^\varepsilon(t)) = \chi^\varepsilon(y_0^\varepsilon(0), \dots, y_N^\varepsilon(0))$$

$$quad + \int_0^t \left(\sum_{k=0}^N \left[\sum_{j=0}^N \frac{\partial \chi^\varepsilon}{\partial y_j} c_{jk} + \frac{1}{2} \sum_{i+j=k} \operatorname{tr} \frac{\partial^2 \chi^\varepsilon}{\partial y_i \partial y_j} \sigma_i Q_{ij} \sigma_j^* \right] \frac{1}{\varepsilon^k} \right.$$

$$\left. + \sum_{k=N+1}^{2N} \left[\sum_{j=k-N} \frac{\partial \chi^\varepsilon}{\partial y_j} c_{jk} + \frac{1}{2} \sum_{\substack{i+j=k \\ i,j \le N}} \operatorname{tr} \frac{\partial^2 \chi^\varepsilon}{\partial y_i \partial y_j} \sigma_i Q_{ij} \sigma_j^* \right] \frac{1}{\varepsilon^k} \right) dw$$

$$+ \int_0^t \frac{1}{\varepsilon^j} \sum_{j=0}^N \frac{\partial \chi^\varepsilon}{\partial y_j} \sigma_j \, dw_j,$$

and $\chi^\varepsilon = O(\varepsilon)$.

We look for χ^ε as follows:

(11.3.7) $\chi^\varepsilon = \varepsilon \chi_1(y_0, y_1) + \cdots + \varepsilon^{N-1} \chi_{N-1}(y_0, \dots, y_{N-1})$

$$+ \varepsilon^N \chi_N(y_0, \dots, y_N) + \cdots + \varepsilon^{2N-1} \chi_{2N-1}(y_0, \dots, y_N).$$

Let us set
(11.3.8)

$$-A_{2N+1-k} = \sum_{j=0}^N c_{jk} \frac{\partial}{\partial y_j} + \frac{1}{2} \sum_{\substack{i+j=k \\ i,j \le N}} \operatorname{tr} \frac{\partial^2}{\partial y_i \partial y_j} \sigma_i Q_{ij} \sigma_j^*, \qquad \text{for } 0 \le k \le N,$$

$$-A_{2N+1-k} = \sum_{j=k-N}^N c_{jk} \frac{\partial}{\partial y_j} + \frac{1}{2} \sum_{\substack{i+j=k \\ i,j \le N}} \operatorname{tr} \frac{\partial}{\partial y_i \partial y_j} \sigma_i Q_{ij} \sigma_j^*, \qquad \text{for } N+1 \le k \le 2N.$$

We try to realize the following conditions

$$A_1 \chi_N + A_2 \chi_{N-1} + \cdots + A_n \chi_1 = -c_{0N},$$

(11.3.9) $A_1 \chi_{N+1} + A_2 \chi_N + \cdots + A_{N+1} \chi_1 = -c_{0,N-1},$

$$A_1 \chi_{2N-1} + A_2 \chi_{2N-2} + \cdots + A_{2N-1} \chi_1 = -c_{0,1}.$$

If we define $\chi_1, \dots, \chi_{2N-1}$ in order that (11.3.9) be satisfied, then one can easily check that the family μ_x^ε remains in a weakly compact subset of $\mathcal{M}_+^1(\Psi)$. This will also be sufficient to obtain the limit. We omit the calculations which are quite tedious. It is easy to check that (11.3.9) corresponds to the conditions obtained homogenization, reiterated homogenization or averaging.

12. Comments and problems

As has been noted in the text of the Chapter, Freĭdlin was the first to consider the problem

(12.0.10) $$-a_{ij}\left(\frac{x}{\varepsilon}\right) \frac{\partial^2 u_\varepsilon}{\partial x_i \partial y_j} - b_i\left(\frac{x}{\varepsilon}\right) \frac{\partial u_\varepsilon}{\partial x_i} + a_0\left(\frac{x}{\varepsilon}\right) u_\varepsilon = f,$$

$$u_\varepsilon \mid_\Gamma = 0$$

and the limit of u_ε was obtained by probabilistic arguments. As stated in the general introduction, problem (12.0.10) is to some extent a natural extension of the averaging problem (where coefficients are rapidly varying in time in a periodic manner). In a completely independent way, analysts and numerical analysts have been working on "homogenization" with motivation coming from physics and mechanics as explained in the general introduction. In that context operators in divergence form appear more naturally than (12.0.10). The authors were the first to study the general framework (which includes (12.0.10) and divergence form operators)

(12.0.11)

$$- a_{ij} \left(\frac{x}{\varepsilon}, x \right) \frac{\partial^2 u_\varepsilon}{\partial x_i \partial x_j} - \frac{1}{\varepsilon} b_i \left(\frac{x}{\varepsilon} \right) \frac{\partial u_\varepsilon}{\partial x_i} - c_i \left(\frac{x}{\varepsilon}, x \right) \frac{\partial u_\varepsilon}{\partial x_i} + a_0 \left(\frac{x}{\varepsilon}, x \right) u_\varepsilon = f$$

$$u_\varepsilon \mid_\Gamma = 0.$$

To obtain a limit for u^ε, we need an assumption on the vector field $b(y)$, namely that $\int b(y) \, m(dy) = 0$, where $m(dy)$ is a suitable probability measure on the torus. The study of the problem when this assumption is not satisfied is largely open.

If we use the method of asymptotic expansions, it is easy to check that the ansatz should be of the form

(12.0.12)
$$u_\varepsilon \sim \varepsilon u_1(x) + \varepsilon^2 u_2 \left(\frac{x}{\varepsilon}, x \right) + \cdots$$

where u_1 is a solution of

(12.0.13)
$$-\bar{b}_i(x) \frac{\partial u_1}{\partial x_i} + \bar{a}_0(x) u_1 = f(x).$$

If $\bar{a}_0(x)$ is bounded below by a positive constant, the equation (12.0.13) makes sense, with some appropriate boundary conditions. The interesting problem is then the study of the convergence of $\frac{u_\varepsilon}{\varepsilon}$, especially the boundary layer analysis. Another quite interesting problem concerns the case when the support of f does not coincide with $\mathcal{O}$, and when the characteristics of (12.0.13) do not necessarily meet the support of f. These problems are related to the study of elliptic problems with small diffusion coefficients, as considered by Ventcel' and Freĭdlin [121]. Related considerations for the Cauchy problem corresponding to (12.0.11) will be given in Chapter 4.

The probabilistic methods deal with the study of the convergence of a family of measures μ^ε on $C([0, \infty); \mathbb{R}^n)$. We have obtained only weak convergence results. They are sufficient to study the convergence of the expected value of functionals which are continuous on $C[[0, \infty); \mathbb{R}^n)$. But nothing more can be said on the probabilistic behavior of the functional itself. It would be very interesting to approach the study of this behavior from the point of view of large deviations, as in Donsker-Varadhan [35] or as in Azencott-Ruget [7].

Along these lines, we can mention the problem of the behavior of the principal eigenvalue of the operator (12.0.11) (non self adjoint case). Does the principal eigenvalue converge toward the principal eigenvalue of the homogenized operator? This seems very likely, but it has not yet been proved.

The same problem can be posed for the transport operator. It is a very important problem for the applications, and one should look for methods which can be carried over from diffusion operators to transport operators.

Our methods (probabilistic or asymptotic expansions) can be used to recover classical results on the asymptotic behavior of random evolutions (see the survey by M. Pinsky [93]). A more general problem can be studied within the framework of Transport Theory (see A. Bensoussan, J.L. Lions, G.C. Papanicolaou [**13**] for details).

The study of the behavior of reflected diffusions with rapidly varying coefficients is also largely open. In the present chapter, we have considered only the averaging problem, i.e., we have not considered the case of reflected processes with unbounded drifts. We have also not considered the problem of reflection with rapidly varying directions. This prevents us from using probabilistic techniques to study the problem

$$(12.0.14) \qquad -\frac{\partial}{\partial x_i} a_{ij}\left(\frac{x}{\varepsilon}\right)\frac{\partial u_\varepsilon}{\partial x_j} + a_0\left(\frac{x}{\varepsilon}\right) u_\varepsilon = f$$

$$a_{ij}\left(\frac{x}{\varepsilon}\right)\frac{\partial u_\varepsilon}{\partial x_j} n_i \mid_\Gamma = 0.$$

To introduce a reflected process to study (12.0.14) is probably not a good approach since we know that the boundary conditions do not play a significant role in the convergence of operators given in divergence form. It would be interesting to see if this comment can be interpreted in probabilistic terms. In any case we know nothing about the behavior of $u_\varepsilon(x)$ near the boundary. It is likely that the behavior of the boundary process plays a fundamental role. This involves homogenization for Levy processes, which is an interesting problem by itself, probably not out of reach.

It is not easy to study nonlinear problems by probabilistic techniques because this involves the study of nonlinear functionals of probability measures μ^ε, which converge only in the weak sense.

For instance, one can consider stochastic control problems or stopping time problems for diffusions with rapidly varying coefficients. What happens to the optimal control or the optimal cost as ε tends to 0? It is not true in general that the optimal cost, for instance, will converge toward the optimal cost of the same stochastic control problem for the homogenized diffusion. This can be seen by studying the behavior of the Hamilton Jacobi equation, by analytic techniques, using the methods of Chapter 1, §16. The limit problem may not even be an Hamilton-Jacobi equation.

In some cases, we can assert by analytic techniques that the limit problem is indeed the optimal cost of the same stochastic control problem for the homogenized operator. This is the case for instance, for optimal stopping time problems, which are related to variational inequalities (see A. Bensoussan - J.L. Lions [**10**]). One expects that it should be possible to use probabilistic techniques to obtain the desired convergence results.

These questions are also related to the use of singular perturbation techniques in stochastic control problems. We refer to O'Malley and Kung [**89**], Delebecque-Quadrat [**32**] for a precise description of the problems and results.

Some nonlinear equations can be handled by probabilistic methods without using stochastic control techniques. These are the equations which can be interpreted in terms of branching processes.

As we have seen in this chapter, probabilistic techniques require a great deal of regularity on the coefficients of the diffusion. It would be interesting to relax these assumptions as much as possible. We have seen in §11 that this is possible for averaging. The technical proof of Lemma 11.2, is of course much more complicated than the proof which is available in the case of regular coefficients. It does not seem to extend easily to homogenization problems.

CHAPTER 4

High Frequency Wave Propagation in Periodic Structures

Orientation

In this chapter we study asymptotic problems in periodic structures involving wave propagation when the typical wavelength of the motion is comparable to the period of the structure and both are small. We also study several other related problems, including some long wavelength (or low frequency) cases that were analyzed in previous chapters. They are now seen in a somewhat different context which should provide more insight into their structure.

We begin in section one with a review of some notions and conventions regarding scaling. It is impossible to be absolutely precise about such things; we simply attempt to convey a point of view that we have found useful.

Section 2 is intended as a general introduction to the WKB or geometrical optics methods. It is hoped that by selecting for brief presentation only the basic ideas, or the ones that we have found particularly useful in periodic structures, it will be of assistance to the reader. It should be pointed out that there are ramifications, both regarding the mathematical techniques as well as the physical applications, that are not dealt with at all here (or dealt with in very simplified forms). For example we examine diffraction effects only in a very indirect way. The reason is simply that such problems are very difficult already in homogeneous media; in periodic media they seem to be impenetrable. In Sections 2.11–2.14 we present some results that deal with the asymptotic simplification (order reduction) of hyperbolic systems.

In Section 3 we analyze the spectral properties of differential operators with periodic coefficients and give the so-called Bloch expansion theorem.

The Bloch expansion is used directly in Section 4 to solve several problems and analyze their asymptotic behavior. We make frequent contact with results of previous chapters by specializing to low frequency. For the reader's convenience we provide the necessary information here in order to avoid tedious cross references.

Section 5 contains the general expansion process where the full force of the WKB method enters. It is hoped that the close connections between the results in this section and those of Section 2 have been made clear. There is a good deal of methodological and conceptual unity linking all these problems closely.

The phenomena, mathematical or physical, contemplated in this chapter come in much greater variety than they did in previous chapters. Consequently our results are much less complete or refined compared to the ones in Chapters 1, 2 or 3. There is an enormous number of unsolved problems, as far as we know, in connection with practically every topic discussed here.

1. Formulation of the problems

1.1. High frequency wave propagation. Consider the following form of the Klein-Gordon equation

$$(1.1.1) \qquad \frac{\partial^2 u(x,t)}{\partial t^2} = \nabla \cdot (c^2(x)\nabla u(x,t)) - W(x)u(x,t), \qquad t > 0,$$

$$u(x,0) = f(x),$$

$$\frac{\partial u}{\partial t}(x,0) = g(x), \qquad x \in \mathbb{R}^n,$$

with $f(x)$ and $g(x)$ smooth[1] functions of compact support and $c^2(x) > 0$, $W(x) \geq 0$ smooth. We know that (1.1.1) has a unique smooth solution in each finite time interval.[2]

We are interested in the behavior of the solution of (1.1.1) when the coefficient functions $c^2(x)$, $W(x)$ and the data have special forms. A convenient way to describe the forms of interest is by introducing a small parameter $\varepsilon > 0$. We now consider the following problems.

Suppose that c^2 and W are slowly varying functions of x so we replace c^2, W in (1.1.1) by $c^2(\varepsilon x)$, $W(\varepsilon x)$ with $\varepsilon > 0$ and ε small. Suppose also that the data depend on ε so we write f_ε and g_ε. We are interested in the behavior of u_ε, the solution of (1.1.1), under these changes when $\varepsilon \to 0$. However, this would be a trivial problem if we were interested in the behavior of u_ε as $\varepsilon \to 0$ with x and t **fixed**. The interesting and much more difficult problem is the analysis of u_ε as $\varepsilon \to 0$ when t and x also change and become large as $\varepsilon \to 0$. A **rescaling** of space and time variables is called for.

For the rescaling we introduce t' and x' by

$$(1.1.2) \qquad\qquad t' = \varepsilon t, \qquad x' = \varepsilon x,$$

and rewrite (1.1.1) in the new variables. Dropping the primes, this yields the problem

$$(1.1.3) \qquad \frac{\partial^2 u_\varepsilon(x,t)}{\partial t^2} = \nabla \cdot (c^2(x)\nabla u_\varepsilon(x,t)) - \frac{1}{\varepsilon^2}W(x)u_\varepsilon(x,t), \qquad t > 0,$$

$$u_\varepsilon(x,0) = f_\varepsilon(x),$$

$$\frac{\partial u_\varepsilon(x,t)}{\partial t} = g_\varepsilon(x), \qquad x \in \mathbb{R}^n.$$

We emphasize that, despite the appearance of a **large** term in (1.1.3), when ε is small, problem (1.1.3) is a rescaled version of one with slowly varying coefficients.

Of course the discussion is incomplete without specification of the dependence of f_ε and g_ε on ε. Said another way, the terms slowly varying, etc., are relative to scales of variation of the data since there are no other length scales in (1.1.1) (we have no boundaries). We now classify the data as follows.

---

[1]This means C^∞.

[2]A general result about regularity of solutions of hyperbolic systems, covering all the needs of this chapter, is given by Rauch and Massey [**97**]. Simpler theorems which suffice most of the time can be found in Courant-Hilbert [**29**].

CLASS A. $f_\varepsilon(x)$ and $g_\varepsilon(x)$ have asymptotic power series expansions in ε

$$(1.1.4) \qquad f_\varepsilon(x) \sim f_0(x) + \varepsilon f_1(x) + \cdots,$$
$$g_\varepsilon(x) \sim g_0(x) + \varepsilon g_1(x) + \cdots.$$

The terms in the expansions are C_0^∞ functions. Problem (1.1.3) with data (1.1.4), **class A** data by definition, is the static or low frequency problem. This problem is discussed further in Section 2.4 where it is shown that the terminology is **not** fully consistent (but merely convenient).

The expansions in (1.1.4) do not have to start with ε^0. Since the problem is linear this makes no difference; the solution is simply multiplied by a power of ε.

In this chapter we adopt frequently (but not always) the convention that the data are to be multiplied by a suitable power of ε (positive or negative) in order to make all terms in the appropriate energy identity of **equal strength in** ε, if possible, **and finite as** $\varepsilon \to 0$.

Let $(,)$ denote the inner product over $\mathbb{R}^n$ (complex-valued) and $|\ |$ the norm based on this inner product, i.e. L^2 norm. The energy identity for (1.1.3) is obtained by multiplying by $u_{\varepsilon t}$, integrating over $\mathbb{R}^n$, etc., which yields

$$(1.1.5) \quad |u_{\varepsilon t}|^2 + (c^2 \nabla u_\varepsilon, \nabla u_\varepsilon) + \frac{1}{\varepsilon^2}(W u_\varepsilon, u_\varepsilon)$$
$$= |g_\varepsilon|^2 + (c^2 \nabla f_\varepsilon, \nabla f_\varepsilon) + \frac{1}{\varepsilon^2}(W f_\varepsilon, f_\varepsilon).$$

We see that on the right side of (1.1.5) the term due to the potential W dominates (we assume $W > 0$). With data of the form (1.1.4), or (1.1.4) multiplied by a power of ε, one cannot change this lack of balance. To make the energy finite as $\varepsilon \to 0$ we must take $f_0(x) \equiv 0$ in (1.1.4).

CLASS B. $f_\varepsilon(x)$ and $g_\varepsilon(x)$ have the form $(i = \sqrt{-1})$

$$(1.1.6) \qquad f_\varepsilon(x) \sum e^{i\tilde{S}(x)/\varepsilon} \tilde{f}_\varepsilon(x),$$
$$g_\varepsilon(x) \sim \frac{1}{\varepsilon} e^{i\tilde{S}(x)/\varepsilon} \tilde{g}_\varepsilon(x),$$

where $\tilde{S}(x)$ is real-valued and C^∞, $\tilde{f}_\varepsilon$ and $\tilde{g}_\varepsilon$ are complex valued and C_0^∞ and have asymptotic power series expansions like (1.1.4). Note that the data (1.1.6) are complex-valued. Since (1.1.3) is linear, both the real and imaginary parts of u_ε are solutions but the complex notation is convenient and widely used. Data of the form (1.1.6) are called **class B**.

To understand the physical significance of (1.1.6) suppose

$$(1.1.7) \qquad \tilde{S}(x) = k \cdot x,$$

where k is a constant vector and dot stands for the inner product of vectors in $\mathbb{R}^n$. The function $\tilde{S}(x)$ in (1.1.6) is called the **phase** function. The particular phase function (1.1.7) gives data (1.1.6) that are **spatially modulated plane waves with rapidly varying phase**.

Note that all three terms on the right side of (1.1.5) are now of order ε^{-2}. To make the energy of order one as $\varepsilon \to 0$ we must either take $\tilde{f}_0 \equiv \tilde{g}_0 \equiv 0$ or replace u_ε by $\varepsilon u_\varepsilon$.

Initial data of the form (1.1.6) may seem at first too special. As we noted, **(1.1.3) with class B data** does correspond to **high frequency wave propagation in a slowly varying medium**. However, we shall see that by linear superposition of asymptotic solutions of such problems, solutions to more general problems can be obtained.

CLASS C. Consider the case where f_ε and g_ε have the form

$$(1.1.8) \qquad f_\varepsilon(x) = \varepsilon^{1-n/2} f\left(x, \frac{x}{\varepsilon}\right),$$
$$g_\varepsilon(x) = e^{-n/2} g\left(x, \frac{x}{\varepsilon}\right).$$

Here f and g are C^∞ functions of compact support in x and y ($\equiv x/\varepsilon$), $(x, y) \in \mathbb{R}^{2n}$.

We could have taken f and g depending on ε with an expansion like (1.1.4) but (1.1.8) will suffice. The scaling in (1.1.8) is chosen so that the terms on the right side of (1.1.5) are of order one as $\varepsilon \to 0$. This is easily seen by noting that for any function $h(x) \in C^\infty$ with compact support,

$$(1.1.9) \qquad \varepsilon^{-n} h(x/\varepsilon) \to \delta(x) \int_{\mathbb{R}^n} h(x)\, dx, \quad \text{as } \varepsilon \to 0,$$

where the convergence is in the distribution sense.

Problem (1.1.3) with data of the form (1.1.8), **class C data**, corresponds to **wave propagation in a slowly varying medium with spatially localized data**. Said another way, the scale of variation of the coefficients c^2 and W and the typical diameter of the support of the data are comparable and both are small.

Define[3]

$$(1.1.10) \qquad \hat{f}(x, \kappa) = \frac{1}{(2\pi)^{n/2}} \int e^{-i\kappa \cdot y} f(x, y)\, dy,$$
$$\hat{g}(x, \kappa) = \frac{1}{(2\pi)^{n/2}} \int e^{-i\kappa \cdot y} g(x, y)\, dy,$$

which are C^∞ functions of compact support in x. We may write (1.1.8) in the form

$$(1.1.11) \qquad f_\varepsilon(x) = \frac{\varepsilon^{1-n/2}}{(2\pi)^{n/2}} \int e^{i\kappa \cdot x/\varepsilon} \hat{f}(x, \kappa)\, d\kappa,$$
$$g_\varepsilon(x) = \frac{\varepsilon^{-n/2}}{(2\pi)^{n/2}} \int e^{i\kappa \cdot x/\varepsilon} \hat{g}(x, \kappa)\, d\kappa.$$

By linearity, the solution of (1.1.3) with data (1.1.11) is the integral over all κ of the solutions of (1.1.3) with data

$$(1.1.12) \qquad f_\varepsilon(x, \kappa) \equiv \frac{\varepsilon^{1-n/2}}{(2\pi)^{n/2}} e^{i\kappa \cdot x/\varepsilon} \hat{f}(x, \kappa),$$
$$g_\varepsilon(x, \kappa) \equiv \frac{\varepsilon^{-n/2}}{(2\pi)^{n/2}} e^{i\kappa \cdot x/\varepsilon} \hat{g}(x, \kappa), \qquad \kappa \in \mathbb{R}^n,\ x \in \mathbb{R}^n.$$

This is just a family of problems, parametrized by $\kappa \in \mathbb{R}^n$, with data of **class B** (i.e. (1.1.6)). We conclude that: the analysis of (1.1.3) with class C data reduces effectively to the analysis of (1.1.3) with class B data.

---

[3]When the domain of integration is not shown it is assumed to be $\mathbb{R}^n$.

So far we have restricted ourselves to the pure initial value problem (1.1.3). First, it is clear that the above scaling notions carry over to

$$(1.1.13) \qquad \frac{\partial^2 u_\varepsilon}{\partial t^2} = \nabla \cdot (c^2 \nabla u_\varepsilon) - \frac{1}{\varepsilon^2} W(x) u_\varepsilon + h_\varepsilon(x, t), \qquad t > 0,$$

$$u_\varepsilon(x, 0) = f_\varepsilon(x),$$

$$\frac{\partial u_\varepsilon}{\partial t}(x, t) = g_\varepsilon(x), \qquad x \in \mathbb{R}^n,$$

which is the **forced problem** corresponding to (1.1.3). The forcing function may be rapidly varying in both space and time. A typical example is

$$(1.1.14) \qquad h_\varepsilon(x, t) = \varepsilon^\alpha e^{i\omega t/\varepsilon} \tilde{h}\left(x, \frac{x}{\varepsilon}\right),$$

where $\omega > 0$ is fixed, $\tilde{h}(x, y)$, $y = x/\varepsilon$, is C^∞ and has compact support and α is suitably chosen to make u_ε behave in some prescribed way as $\varepsilon \to 0$ (for example, with energy $O(1)$).

Second, one can consider (1.1.3) in the presence of boundaries where homogeneous or inhomogeneous boundary conditions are satisfied. In the inhomogeneous case, the data may be rapidly varying in space or time. In the simplest cases the phenomena of interest are reflection and transmission. More complicated phenomena involve diffraction. We shall not discuss such problems in this chapter, as was mentioned in the introduction, except for some simple examples.

1.2. Propagation in periodic structures. We consider directly the scaled problem (1.1.3) and ask the following question. If c^2 and W are not only functions of x, i.e. slowly varying functions, but also functions of x/ε, $c^2 = c^2\left(x, \frac{x}{\varepsilon}\right)$, $W = W\left(x, \frac{x}{\varepsilon}\right)$, such that $c^2(x, y)$, $W(x, y)$ are periodic in y of period 2π in all directions,[4] how does the corresponding solution u_ε behave? Coefficients c^2 and W depending on ε (so that $W = W(x, y)$ evaluated at $y = x/\varepsilon$) as above are called **locally periodic**. If $c^2 = c^2(y)$, $W = W(y)$ with $y = x/\varepsilon$ then they are called **globally periodic**.

We point out that the new problem

$$(1.2.1) \qquad \frac{\partial^2 u_\varepsilon}{\partial t^2} = \nabla \cdot \left(c^2\left(x, \frac{x}{\varepsilon}\right) \nabla u_\varepsilon\right) - \frac{1}{\varepsilon^2} W\left(x, \frac{x}{\varepsilon}\right) u_\varepsilon, \qquad t > 0,$$

$$u_\varepsilon(x, 0) = f_\varepsilon(x),$$

$$\frac{\partial u_\varepsilon}{\partial t}(x, 0) = g_\varepsilon(x),$$

has **three** natural length scales associated with it: the **slow** scale of variation of the coefficients, the **fast** scale of variation of the coefficients which is proportional to $2\pi\varepsilon$ and the scale of variation of the data which has not yet been specified.

As in Section 1.1 we distinguish three classes of data:

CLASS A. Data of the form (1.1.1). Problem (1.2.1) with such, **class A data**, is the static or low frequency homogenization problem. The scale of variation of the data is comparable to the slow scale (cf. Section 2.4 for additional remarks regarding class A).

---

[4]It is convenient in this chapter to depart from the convention of previous chapters where the period was one. We denote by $Y = (]0, 1[)^n$, the n-dimensional torus (end points of intervals identified). The 2π-torus is denoted by $2\pi Y$.

CLASS B. Data of the form

(1.2.2)
$$f_\varepsilon(x) \sim e^{i\tilde{S}(x)/\varepsilon} f\left(x, \frac{x}{\varepsilon}\right),$$

$$g_\varepsilon(x) \sim \frac{1}{\varepsilon} e^{i\tilde{S}(x)/\varepsilon} g\left(x, \frac{x}{\varepsilon}\right),$$

where $\tilde{S}(x)$ is real-valued and C^∞ and $f(x,y)$, $g(x,y)$ are complex-valued, $C^\infty(\mathbb{R}^n \times 2\pi Y)$ (i.e. they are 2π-periodic in y) and of compact support in x.

As in Section 1.1 this class of data, class B, is called the **spatially modulated waves with rapidly varying phase**. Note that the modulation is not necessarily slowly varying, i.e. we have x/ε dependence. However, the rapid variation in the modulation is **periodic** and as such it does not induce propagation of energy (these matters are taken up in detail in the following sections). By convention, we do not call data of the form (1.2.2) with $\tilde{S} \equiv 0$ "high frequency" data: we call them **rapidly varying periodic data**.

CLASS C. Data of the form (1.1.8) with precisely the hypotheses introduced there. This class, **class C**, is called the class of **spatially localized data**.

Of course, the above distinctions are to some extent arbitrary. For example, why not data of the form, say,

$$f_\varepsilon(x) \sim e^{i\tilde{S}(x)/\varepsilon^k} f\left(x, \frac{x}{\varepsilon}\right), \qquad k \neq 1,$$

with the hypotheses of (1.2.2)? The reason is that if $k < 1$ then we are essentially dealing with rapidly varying **periodic** data. Similarly if $k > 1$ then the coefficients are relatively slowly varying relative to the phase so the rapidly varying periodic coefficients play a minor role, i.e. we are reduced to problems of the form discussed in Section 1.1. The "interesting" cases therefore is precisely the one we have singled out: namely, the one where the induced waves interact genuinely with the periodic structure.

A final remark about class A data (low frequency data). For many equations, including (1.1.3) and (1.2.1) when $W \not\equiv 0$, the fact that $\tilde{S}(x) \equiv 0$ does **not** imply that rapid phase variations do not appear at a later time. This again points out that **class B** data is a basic class among the three (cf. Section 2.4 for additional discussion).

2. The W. K. B. or geometrical optics method

2.1. Expansion for the Klein-Gordon equation. To illustrate the methods in the title we shall work consistently with one example (until Section 2.10) which is at first the following: To analyze as $\varepsilon \to 0$ the behavior of u_ε solution of the equation

(2.1.1)
$$\frac{\partial^2 u_\varepsilon(x,t)}{\partial t^2} = \nabla \cdot (c^2(x)\nabla u_\varepsilon(x,t)) - \frac{1}{\varepsilon^2} W(x) u_\varepsilon(x,t), \qquad t > 0,$$

$$u_\varepsilon(x,0) = f_\varepsilon(x),$$

$$\frac{\partial u_\varepsilon}{\partial t}(x,0) = g_\varepsilon(x), \qquad x \in \mathbb{R}^n,$$

where

(2.1.2)
$$f_\varepsilon(x) = \varepsilon e^{i\tilde{S}(x)/\varepsilon} f(x)$$

and

$$(2.1.3) \qquad g_\varepsilon(x) = e^{i\tilde{S}(x)/\varepsilon} g(x),$$

with $\tilde{S}(x)$, C^∞ and real-valued, f and g complex-valued C^∞ with compact support. Also $c^2(x) \geq c_0 > 0$ and $W(x) > 0$.

We begin by attempting to construct an expansion for u_ε of the form

$$(2.1.4) \qquad u_\varepsilon(x,t) = e^{iS(x,t)/\varepsilon} v_\varepsilon(x,t),$$

$$(2.1.5) \qquad v_\varepsilon(x,t) = \varepsilon v_0(x,t) + \varepsilon^2 v_1(x,t) + \varepsilon^3 v_2(x,t) + \cdots.$$

Inserting (2.1.4) into (2.1.1) yields the following equation for v_ε:

$$(2.1.6) \qquad (\varepsilon^{-2} A_1 + \varepsilon^{-1} A_2 + A_3) v_\varepsilon = 0,$$

where

$$(2.1.7) \qquad A_1 = -c^2 (\nabla S)^2 + (S_t)^2 - W,$$

$$(2.1.8) \qquad A_2 = -2iS_t \partial_t + i\nabla \cdot (c^2 \nabla S) + ic^2 \nabla S \cdot \nabla - iS_{tt},$$

$$(2.1.9) \qquad A_3 = \nabla \cdot (c^2 \nabla) - \partial_t^2.$$

Inserting (2.1.5) into (2.1.6) and equating coefficients of equal powers of ε yields the following sequence of problems

$$(2.1.10) \qquad A_1 v_0 = 0,$$

$$(2.1.11) \qquad A_1 v_1 + A_2 v_0 = 0,$$

$$(2.1.12) \qquad A_1 v_2 + A_2 v_1 + A_3 v_0 = 0,$$

etc. From (2.1.10) we conclude that if v_0 is not to be identically zero we must have

$$(2.1.13) \qquad S_t \pm [c^2(x)(\nabla S)^2 + W]^{1/2} = 0.$$

This is the eikonal equation for the phase function (actually two of them, one with $+$, one with $-$). Its solution is discussed further in the next section.

Note that this choice of S makes the operator A_1 identically zero. Hence (2.1.11) is directly an equation for $v_0(x,t)$, i.e.

$$(2.1.14) \qquad A_2 v_0 = 0.$$

This equation is called the **transport equation** and is discussed further in Section 2.3. Note the coefficients of A_2 depend on S. When the plus sign is used in (2.1.13) we write S^+ (and similarly S^-). The corresponding v_0 is denoted by $v_0^\pm$.

Equation (2.1.12) and the succeeding ones of higher order are called the **higher order transport equations**. Corresponding to S^+ or S^- we have v_1^+ or v_1^-, etc.

The solution now is represented in the form

$$(2.1.15) \qquad u_\varepsilon(x,t) \sim e^{iS^+(x,t)/\varepsilon} [\varepsilon v_0^+(x,t) + \varepsilon^2 v_1^+(x,t) + \cdots]$$

$$+ e^{iS^-(x,t)/\varepsilon} [\varepsilon v_0^i(x,t) + \varepsilon^2 v_1^-(x,t) + \cdots]$$

with $u_{\varepsilon t}(x,t)$ asymptotic to the formal derivative of the terms on the right side of (2.1.15). Comparison with the initial conditions in (2.1.1) yields initial conditions for $S^\pm$ and $v_k^\pm$, $k \geq 0$.

The validity of (2.1.15) as an asymptotic expansion, and the corresponding one for $u_{\varepsilon t}$, follows immediately after it has been shown that such an expansion, with smooth $S^{\pm}$ and $v_k^{\pm}$, $k \geq 0$, exists. The energy identity (1.1.5) shows in fact that

$$(2.1.16) \qquad \sup_{0 \leq t \leq T < \infty} \{|(u_{\varepsilon} - u_{\varepsilon}^N)_t| + |\nabla(u_{\varepsilon} - u_{\varepsilon}^N)|\} = O(\varepsilon^N),$$

where u_{ε}^N consists of the sum of terms including the one proportional to ε^N in (2.1.15).

2.2. Eikonal equation and rays. We now consider (2.1.13) in some detail. We pick the one with the $+$ sign and set

$$(2.2.1) \qquad S_t + \omega(\nabla S, x) = 0, \qquad S(0, x) = \tilde{S}(x), \qquad x \in \mathbb{R}^n,$$

where $\omega(k, x)$ is the **Hamiltonian** (or **radian frequency**)

$$(2.2.2) \qquad w(k, x) = [c^2(x)k^2 + W(x)]^{1/2}, \qquad k \in \mathbb{R}^n, \qquad x \in \mathbb{R}^n.$$

The **eikonal** (or Hamilton-Jacobi) equation (2.2.1) is a first order nonlinear PDE that controls the evolution of the phase function.[5]

To solve (2.2.1) we introduce Hamilton's system of ODEs for he **rays** $x(t)$ and **momental** (or **wavenumbers**) $k(t)$.

$$(2.2.3) \qquad \frac{dx_p(t)}{dt} = \frac{\partial \omega(k(t), x(t))}{\partial k_p}, \qquad\qquad x_p(0) = x_p,$$

$$\frac{dk_p(t)}{dt} = -\frac{\partial \omega(k(t), x(t))}{\partial x_p}, \qquad k_p(0) = \frac{\partial \tilde{S}(x)}{\partial x_p}, \qquad p = 1, 2, \ldots, n.$$

Recall that W is assumed positive and C^{∞} which makes $\omega(k, x)$ a C^{∞} function of k and x. System (2.2.3) has a unique solution for any finite time interval which depends smoothly on the initial value of x and k. Suppose (2.2.1) has a smooth solution. Then

$$(2.2.4) \qquad \frac{d}{dt} S(x(t), t) = \frac{\partial S}{\partial t}(x(t), t) + \sum_{p=1}^{n} \frac{\partial \omega(k(t), x(t))}{\partial k_p} \frac{\partial S(x(t), t)}{\partial x_p}$$

$$= -\omega(k(t), x(t)) + \sum_{p=1}^{n} \frac{\partial \omega(k(t), x(t))}{\partial k_p} k_p(t),$$

where we identify

$$(2.2.5) \qquad k(t) = \nabla S(x(t), t).$$

If we also introduce the Lagrangian

$$(2.2.6) \qquad L(x, \dot{x}) = -\omega(k, x) + \sum_{p=1}^{n} \frac{\partial \omega(k, x)}{\partial k_p} k_p,$$

where k is considered as a function of x and $\dot{x}$ obtained from[6] $\dot{x} = \nabla_k \omega(k, x)$, then we have the representation

$$(2.2.7) \qquad S(t, x(t)) = \tilde{S}(x) + \int_0^t L(x(s), \dot{x}(s)) \, ds.$$

---

[5] The S^+ is called the forward and S^- the backward moving phase by extending the terminology of the **plane wave case**: $\tilde{S}(x) = k \cdot x$, $W(x) = $ constant, $\omega = \omega(k)$ and $S^{\pm}(x, t) = k \cdot x \mp \omega(k)t$.
[6] This is possible because $\omega(k, x)$ is convex in k.

From (2.2.2) we find the following formulas:

$$(2.2.8) \qquad \frac{\partial \omega}{\partial k}(k, x) = \frac{c^2 k}{\sqrt{c^2 k^2 + W}} = \frac{c^2 k}{\omega} = \dot{x},$$

$$(2.2.9) \qquad c^2 - x^2 = c^2 W (c^2 k^2 + W)^{-1} > 0,$$

$$(2.2.10) \qquad k = \frac{\dot{x}\sqrt{W}}{c\sqrt{c^2 - x^2}},$$

$$(2.2.11) \qquad L(x, \dot{x}) = -\frac{\sqrt{W}}{c}\sqrt{c^2 - \dot{x}^2}, \qquad (|\dot{x}| \le c \text{ always}).$$

Consider the system (2.2.3) with the 2-point boundary conditions

$$(2.2.12) \qquad x_p(t) = x_p, \qquad k_p(0) = \frac{\partial \tilde{S}(x)}{\partial x_p}, \qquad p = 1, 2, \dots, n,$$

and suppose it has a unique solution for $0 \le t \le t_0$ for some $t_0 > 0$. The mapping $\tilde{x} \to x(t, \tilde{x})$, $x(0, \tilde{x}) = \tilde{x}$ is one to one for $0 \le t \le t_0$ and

$$(2.2.13) \qquad S(t, x) = \tilde{S}(x) + \int_0^t L(x(s), \dot{x}(s))\, ds, \qquad 0 \le t \le t_0,$$

is the unique smooth solution of (2.2.1). In (2.2.13) the path $x(s)$ is the one satisfying (2.2.3) with boundary conditions (2.2.12).

A point (x, t) at which the mapping $\tilde{x} \to x(t, \tilde{x})$, $x(0, \tilde{x}) = \tilde{x}$ becomes singular (its Jacobian vanishes) is a conjugate or focal point of the Hamiltonian system. At such points (2.2.1) fails to have solutions. In fact the whole expansion scheme (2.1.4), (2.1.5) is inappropriate there and different types of expansions are needed. We restrict discussion to the **local** expansion at present, i.e. up to a time $t_0 > 0$ without focal points.

2.3. Transport equations. We consider now equation (2.1.11) which in view of (2.1.10) and (2.1.8) becomes

$$(2.3.1) \qquad -2S_t v_{0t} + \nabla \cdot (c^2 \nabla S v_0) + c^2 \nabla S \cdot \nabla v_0 - S_{tt} v_0 = 0.$$

This equation is satisfied by both v_0^+ and v_0^-. Multiplying (2.3.1) by[7] $\bar{v}_0$ and adding to the conjugate of (2.3.1) multiplied by v_0, we obtain

$$(2.3.2) \qquad (\omega |v_0|^2)_t + \nabla \cdot (c^2 k |v_0|^2) = 0,$$

where we write

$$(2.3.3) \qquad k = k(x, t) = \nabla S(x, t), \qquad \omega = \omega(x, t) = \omega(k(x, t), x).$$

This abuses notation to some extent but the meaning will be clear from the context. Note in particular that $k(t) = k(x(t), t)$ with $k(t), x(t)$ as in (2.2.3).

From (2.2.8) it follows that (2.3.2) has the form

$$(2.3.4) \qquad (\omega |v_0|^2)_t + \nabla \cdot \left(\frac{\partial \omega}{\partial k} \omega |v_0|^2 \right) = 0.$$

---

[7]Bar denotes complex conjugate.

The quantity $\omega|v_0|^2$ will be identified in Section 2.5 with the first approximation to the **wave action**. Thus (2.3.4) is a conservation equation for the first approximation of the wave action. The velocity field is identified as the **group velocity**

$$(2.3.5) \qquad \bar{c}_p(x,t) = \frac{\partial \omega}{\partial k_p}(k(x,t), x), \qquad p = 1, 2, \ldots, n.$$

From (2.3.3) and (2.2.1) we obtain, in addition to (2.3.4), the following kinematic equations for the wavenumber and radian frequency fields $k(x,t), \omega(x,t)$:

$$(2.3.6) \qquad \frac{\partial k_p}{\partial x_q} = \frac{\partial k_q}{\partial x_p}, \qquad p, 1 = 1, 2, \ldots, n.$$

$$(2.3.7) \qquad \frac{\partial k_p}{\partial t} + \sum_{q=1}^{n} \frac{\partial \omega}{\partial k_q} \frac{\partial k_p}{\partial x_q} + \frac{\partial \omega}{\partial x_p} = 0, \qquad p = 1, 2, \ldots, n.$$

$$(2.3.8) \qquad \frac{\partial \omega}{\partial t} + \sum_{p,q=1}^{n} \frac{\partial \omega}{\partial k_p} \frac{\partial \omega}{\partial k_q} \frac{\partial k_p}{\partial x_q} + \sum_{p=1}^{n} \frac{\partial \omega}{\partial k_p} \frac{\partial \omega}{\partial k_p} = 0.$$

Using (2.3.8), we obtain from (2.3.4) an equation for $\omega^2|v_0|^2$ which will be identified later as the first approximation of the **wave energy**

$$(2.3.9) \qquad (\omega^2|v_0|^2)_t + \nabla \cdot \left(\frac{\partial \omega}{\partial k} \omega^2|v_0|^2 \right) = 0.$$

Consider an infinitesimal tube of rays between $t = t_1$ and $t = t_2$, $0 \le t_1 < t_2 \le t_0$ ($t_0 > 0$ as in Section 2.2). Integrating (2.3.9) over this tube and using the divergence theorem we obtain the usual **law of energy propagation of geometrical optics**:

The energy density at a point x is inversely proportional to the density of rays at the point.

If $\tilde{x} \to x(t, \tilde{x})$, $x(0, \tilde{x}) = \tilde{x}$ is the ray transformation, with $t \le t_0$, then the above statement translates to

$$(2.3.10) \qquad \omega^2|v_0|^2(x(t), t) = \omega^2|v_0|^2(x, 0)(J, t, \tilde{x}))^{-1},$$

where $J(t, \tilde{x})$ is the Jacobian determinant of the transformation. Note that the ω^2 cancels in (2.3.10) because $\omega = \omega(k, x)$ does not depend on t and is hence a constant of the motion (2.2.3).

From (2.3.10) it follows that $v_0(t, x)$, determined from (2.3.4), (2.3.9) or (2.3.10), becomes singular at focal points of (2.2.3). This is why such points are also called **caustic** points: the wave intensity at these points is higher on account of he focusing of the rays. At such points the expansion (2.1.4), (2.1.5) fails.

The higher order transport equation (2.1.12) may be treated in much the same way to obtain the second term $v_1(x, t)$. It satisfies the inhomogeneous equation

$$(2.3.11) \qquad 2\omega v_{1t} + \omega_t v_1 + \nabla \cdot \left(\frac{\partial \omega}{\partial k} \omega v_1 \right) + \omega \frac{\partial \omega}{\partial k} \cdot \nabla v_1 = i[\nabla \cdot (c^2 \nabla v_0) - v_{0tt}].$$

Multiplying by v_1, rearranging and integrating over an infinitesimal tube of rays we obtain[8]

$$(2.3.12) \qquad \omega v_1^2(x(t), t)J(t) = \omega v_1^2(x, 0) + i \int_0^t v_1(x(s), s)\Box v_0(x(s), s) \, ds,$$

---

[8] $J(t) = J(t, \tilde{x})$ is the Jacobian of $x \to x(t, x)$, $x(0, x) = x$ where $x(t)$ satisfies (2.2.3).

where

(2.3.13)
$$\Box \equiv \nabla \cdot (c^2 \nabla \cdot) - \partial_t^2.$$

Equation (2.3.12) is equivalent to

(2.3.14) $$v_1(x(t), t) = v_1(x, 0) J^{-1/2}(t) + \frac{i}{2\omega} \int_0^t \left(\frac{J(s)}{J(t)} \right)^{1/2} \Box v_0(x(s), s) \, ds,$$

as can be verified by direct calculation.

We see therefore that as long as $J(t)$ does not vanish (no caustics), the expansion (2.1.4) and (2.1.5) can be constructed by solving ordinary differential equations only.

We let us determine the initial conditions, i.e. of $v_0^\pm(x, 0), v_1^\pm(x, 0), \dots$ From (2.1.15) and (2.1.2), (2.1.3) we obtain, evaluating functions at $t = 0$,

$$\varepsilon(v_0^+ + v_0^-) + \varepsilon^2(v_1^+ + v_1^-) + \cdots = \varepsilon f,$$

and

$$-\frac{i\omega}{\varepsilon}(\varepsilon v_0^+ + \varepsilon v_1^+ + \cdots) + (\varepsilon v_{0t}^+ + \varepsilon^2 v_{1t}^+ + \cdots)$$

$$+\frac{i\omega}{\varepsilon}(\varepsilon v_0^- + \varepsilon^2 v_1^- + \cdots) + (\varepsilon v_{0t}^- + \varepsilon^2 v_{1t}^- + \cdots) = g.$$

Hence,

(2.3.15)
$$v_0^+(x, 0) = \frac{1}{2} \left(f(x) + \frac{ig(x)}{\omega(\nabla \tilde{S}(x), x)} \right),$$

$$v_0^-(x, 0) = \frac{1}{2} \left(f(x) - \frac{ig(x)}{\omega(\nabla \tilde{S}(x), x)} \right),$$

(2.3.16)
$$v_1^\pm = \mp \frac{i}{2\omega}(v_{0t}^+ + v_{0t}^-), \text{ etc.}$$

The specification of $v_0^\pm, v_1^\pm, \dots$, is now complete.

2.4. Connections with the static problem. Suppose first that the potential W is strictly positive as we have been assuming and that $\tilde{S}(x) \equiv 0$ in (2.1.2) and (2.1.3), i.e., the data has no rapid phase variation.

With ω as in (2.2.2) we see that for $t > 0$ $S(t, x) \not\equiv 0$ even when $\tilde{S}(x) \equiv 0$. That is: **the system itself creates rapid phase variations whether or not the data have them.**

This explains why we remarked that the classification of data in Section 1.1 is not entirely appropriate.

On the other hand:

(2.4.1)   If the Hamiltonian is a homogeneous function[9]

                 of k, then slowly varying data ($\tilde{S}(x) \equiv 0$) produce

                 slowly varying solutions ($S(t, x) \equiv 0$, $t > 0$).

This is immediate from (2.2.1) since by homogeneity

(2.4.2) $$\omega(0, x) \equiv 0.$$

---

[9]Of degree one, two, etc., this means $W(x) \equiv 0$ and $\omega(k, x) = c^2(x)k^2$ in (2.2.2).

Moreover

$$(2.4.3) \qquad \frac{\partial \omega}{\partial k}(0, x) \equiv 0,$$

and hence the transport equation (2.3.1) is trivially satisfied. Indeed the problem (2.1.1) is now trivial since $W \equiv 0$ and $\tilde{S} \equiv 0$ leave us with a problem without a small parameter, i.e. the wave equation which is to be solved exactly.

The example (2.1.1) at hand is too simple to illustrate some interested phenomena that take place when the Hamiltonians **are homogeneous** but at the same time the static problem is nontrivial. The problems with periodic structure have this feature. We return to this point in Section 2.13.

2.5. Propagation of energy. Consider the energy identity for (2.1.1) obtained by multiplying (2.1.1) by $\bar{u}_\varepsilon$ and integrating over a region $\mathcal{O}$, and by adding to it the conjugate of (2.1.1) multiplied by u_ε. This yields

$$(2.5.1) \quad \frac{\partial}{\partial t} \int_{\mathcal{O}} \left[|u_{\varepsilon t}|^2 + c^2 |\nabla u_\varepsilon|^2 + \frac{1}{\varepsilon^2} W |u_\varepsilon|^2 \right] \, dx$$
$$= \int_{\partial \mathcal{O}} \hat{n} \cdot \left[c^2 (u_{\varepsilon t} \nabla \bar{u}_\varepsilon + \bar{u}_{\varepsilon t} \nabla u_\varepsilon) \right] \, dS(x),$$

where $\hat{n}$ is the unit outward normal at the boundary $\partial \mathcal{O}$. Define the energy density $E_\varepsilon(x, t)$ and the flux density $F_\varepsilon(x, t)$ (a vector), respectively, as follows

$$(2.5.2) \qquad E_\varepsilon(x, t) = \frac{1}{2} \left(|u_{\varepsilon t}(x, t)|^2 + c^2(x) |\nabla u_\varepsilon(x, t)|^2 + \frac{1}{\varepsilon^2} W(x) |u_\varepsilon(x, t)|^2 \right)$$

$$(2.5.3) \qquad F_\varepsilon(x, t) = -\frac{c^2(x)}{2} \left[u_{\varepsilon t}(x, t) \nabla \bar{u}_\varepsilon(x, t) + \bar{u}_{\varepsilon t}(x, t) \nabla u_\varepsilon(x, t) \right].$$

The differential form of (2.5.1) is

$$(2.5.4) \qquad \frac{\partial E_\varepsilon}{\partial t} + \nabla \cdot F_\varepsilon = 0.$$

We now use the expansion (2.1.15) in (2.5.2) and (2.5.3). We find that[10]

$$(2.5.5) \qquad E_\varepsilon = \frac{1}{2} \left[\omega^2 |v_0^+|^2 + \omega |v_0^-|^2 + 2\Re\{\omega^2 v_0^+ v_0^- e^{i(S^+ + S^-)/\varepsilon}\} \right.$$
$$+ c^2 k^2 |v_0^+|^2 + c^2 k^2 |v_0^-|^2 + 2c^2 \Re\{-k^2 v_0^+ v_0^- e^{i(S^+ + S^-)/\varepsilon}\}$$
$$\left. + W |v_0^+|^2 + W |v_0^-|^2 + 2W \Re\{v_0^+ v_0^- e^{i(S^+ + S^-)\varepsilon}\} + O(\varepsilon) \right]$$
$$= \omega^2 |v_0^+|^2 + \omega^2 |v_0^-|^2 + O(\varepsilon) + \text{ rapidly oscillating terms},$$

$$(2.5.6) \qquad F_\varepsilon = \frac{\partial \omega^+}{\partial k} \omega^2 |v_0^+|^2 + \frac{\partial \omega}{\partial k} \omega^2 |v_0^-|^2 + O(\varepsilon) + \text{ rapidly oscillating terms}.$$

Comparing (2.5.4), (2.5.5), (2.5.6) and (2.3.9), we see that

(1) Each of the two notes ($\pm =$ forward, backward) satisfies its own energy equation to highest order so that (2.5.4) holds to highest order;

(2) The rapidly oscillating terms drop out because (2.5.4) is **always** interpreted in the integral form (2.5.1).

We also note that the terminology used in Section 2.3 is indeed appropriate if we recall that action is, by definition, energy divided by (radian) frequency.

---

[10]Here $\omega^\pm = \pm\omega = \pm(c^2 k^2 + W)^{1/2}$.

2.6. Spatially localized data. Consider now (2.1.1) with data f_ε and g_ε given by

$$(2.6.1) \qquad f_\varepsilon(x) = \varepsilon^{1-n/2} f\left(x, \frac{x}{\varepsilon}\right),$$

$$(2.6.2) \qquad g_\varepsilon(x) = \varepsilon^{-n/2} g\left(x, \frac{x}{\varepsilon}\right),$$

with $f(x,y)$ and $g(x,y)$, C^∞ functions on $\mathbb{R}^n \times \mathbb{R}^n$ and with compact support. This corresponds to class C data (cf. (1.1.8)).

Let $\hat{f}(x,\kappa)$ and $\hat{g}(x,\kappa)$ be defined by (1.1.10) and similarly let $f_\varepsilon(x,\kappa)$, $g_\varepsilon(x,\kappa)$ be defined by (1.1.12). With $\kappa \in \mathbb{R}^n$ **fixed** let $u_\varepsilon(x,t;\kappa)$ be the solution of (2.1.1) with initial data equal to $f_\varepsilon(x,\kappa)$, $g_\varepsilon(x,\kappa)$. Clearly in this case

$$(2.6.3) \qquad \tilde{S}(x) \equiv \kappa \cdot x,$$

and

$$(2.6.4) \qquad f(x) \to \varepsilon^{-n/2} \hat{f}(x,\kappa),$$
$$g(x) \to \varepsilon^{-n/2} \hat{g}(x,\kappa).$$

The asymptotic expansion of u_ε takes the form

$$(2.6.5) \qquad u_\varepsilon(x,t;\kappa) = \varepsilon^{-n/2} e^{iS^+(x,t;\kappa)/\varepsilon}$$
$$\cdot [\varepsilon v_0^+(x,t;\kappa) + \varepsilon^2 v_1^+(x,t;\kappa) + \cdots]$$
$$+ \varepsilon^{-n/2} e^{iS^-(x,t;\kappa)/\varepsilon}$$
$$\cdot [\varepsilon v_0^-(x,t;\kappa) + \varepsilon^2 v_1^-(x,t;\kappa) + \cdots],$$

with κ a parameter. Note that the factor $\varepsilon^{-n/2}$ has been pulled outside on the right in (2.6.5) so that v_0, v_1, etc. are determined by (2.3.15), (2.3.16), etc. with $f \to \hat{f}(x,\kappa)$, $g \to \hat{g}(x,\kappa)$.

The solution of (2.1.1) with data (2.6.1), (2.6.2) is then, by linearity,

$$(2.6.6) \qquad u_\varepsilon(x,t) = \frac{\varepsilon^{-n/2}}{(2\pi)^n} \int_{\mathbb{R}^n} d\kappa \left[e^{iS^+(x,t;\kappa)/\varepsilon}[\varepsilon v_0^+(x,t;\kappa) + \cdots] \right.$$
$$\left. + e^{iS^-(x,t;\kappa)/\varepsilon}[\varepsilon v_0^-(x,t;\kappa) + \cdots] \right].$$

It remains therefore to simplify the integral in (2.6.6), which is convergent since $\hat{f}$ and $\hat{g}$ decay rapidly as $|\kappa| \to \infty$ and this is inherited by the $v_0^\pm$, etc.

Consider one integral and let us drop the superscripts

$$(2.6.7) \qquad I_\varepsilon(x,t) = \frac{\varepsilon^{-n/2}}{(2\pi)^n} \int e^{iS(x,t;\kappa)/\varepsilon} v_0(x,t;\kappa)\, d\kappa.$$

We analyze this integral asymptotically as $\varepsilon \to 0$ by the **method of stationary phase**. The region of integration in (2.6.7) may be taken as any compact subset of $\mathbb{R}^n$ which is sufficiently large, in view of the rapid decay of v_0 as $|\kappa| \to \infty$.

We fix now[11] (x,t), $t > 0$, and consider the system of equations

$$(2.6.8) \qquad \frac{\partial S}{\partial \kappa_p}(x,t;\kappa) = 0 \qquad p = 1, 2, \ldots, n.$$

---

[11]Recall that $0 \le t \le t_0$ and $t_0 > 0$ is sufficiently small so that the phase function $S(x,t;\kappa)$ exists in $0 \le t \le t_0$ and is uniquely defined.

If there does not exist a κ satisfying (2.6.8), then I_ε at this point (x, t) decays to zero faster than any power of ε which means that at such points the integral I_ε is interpreted as zero.

If there is a unique point $\kappa^* = \kappa^*(x, t)$ such that (2.6.8) is satisfied and such that the matrix

$$(2.6.9) \qquad Q = \left(\frac{\partial^2 S(x, t; \kappa^*)}{\partial \kappa_p \partial \kappa_q} \right),$$

is nonsingular, then we have[12]

$$(2.6.10) \qquad I_\varepsilon(x, t) \sim \frac{(2\pi)^{-n/2}}{|\det Q(x, t)|^{1/2}} e^{iS(x,t;\kappa^*)/\varepsilon + i\sigma\pi/4} \cdot v_0(x, t; \kappa^*).$$

Here $\sigma = \sigma(x, t)$ is the signature of the matrix $Q(x, t)$, i.e. the number of positive minus the number of negative eigenvalues. If Q is singular at (x, t) then (2.6.10) is not valid.

Let us examine in detain the nature of equation (2.6.8) and its solutions κ^*. Define

$$(2.6.11) \qquad U_p(x, t; \kappa) = \frac{\partial S(x, t; \kappa)}{\partial \kappa_p}, \qquad p = 1, 2, \ldots, n.$$

From (2.2.1) it follows that[13]

$$(2.6.12) \qquad \frac{\partial U_p}{\partial t} + \sum_{q=1}^{n} \bar{c}_q \frac{\partial U_p}{\partial x_q} = 0, \qquad U_p(x, 0; \kappa) = x_p, \qquad p = 1, 2, \ldots, n.$$

To solve (2.6.12) we use $x(t; x, \kappa)$ which is the solution of the ray equations (2.2.3) with $x(0) = x$ and $k(0) = \kappa$. Then $U_p(x(t; x, \kappa), t; \kappa)$ is independent of t and hence

$$(2.6.13) \qquad U_p(x(t; x, \kappa), t; \kappa) = x_p, \qquad p = 1, 2, \ldots, n.$$

Thus

$$(2.6.14) \qquad U_p(x(t; 0, \kappa), t; \kappa) = 0.$$

Given (x, t) fixed with $t > 0$, if there exists $\kappa^* = \kappa^*(x, t)$ such that

$$x(t; 0, \kappa^*) = x,$$

then this κ^* also solves (2.6.11) and conversely. This follows immediately from (2.6.14). Note also that

$$(2.6.15) \qquad \dot{x}_p(0; 0, \kappa) = \frac{\partial \omega(\kappa, 0)}{\partial \kappa_p}, \qquad p = 1, 2, \ldots, n.$$

Consider now the pencil of rays $x(t; 0, \kappa)$ issuing out of the origin and such that $k(0; 0, \kappa) = \kappa$ (i.e. family of solutions to (2.2.3)). This family defines a conoid in (x, t) space $t > 0$ and since $W > 0$ and $0 < c(x) < \infty$,

$$\left| \frac{\partial \omega(k, x)}{\partial k} \right| \leq c(x);$$

the local speed along the rays **never exceeds** the local phase velocity $c(x)$. This conoid is called the **ray conoid** (or light cone or forward light cone). If (x, t) is not in the interior of the light cone there exists no κ^* and hence the solution u_ε is zero.

---

[12]R. M. Lewis [**67**], J. J. Duistermaat [**37**]
[13]The group velocity $\bar{c}$ is defined by (2.3.5)

If (x, t) is in the interior of the light cone and t is sufficiently small then there will be only one ray that connects it to the origin and so u_ε behaves asymptotically like (2.6.10) along that ray (with κ^* constant). Note that S^+ and S^- phases cannot be stationary at the same κ since in fact $S^-(x, t; \kappa) = -S^+(x, t; -\kappa)$. Said another way: If for a given κ^* corresponding to a fixed (x, t) and S^+ the ray with this κ^* is in the forward cone, the ray with this κ^* corresponding to S^- is in the backward cone and hence does not contribute.

In conclusion, **inside** the forward light cone the solution u_ε has the form

$$(2.6.16) \qquad u_\varepsilon(x, t) \sim \frac{(2\pi)^{-n/2}}{|\det Q(x, t)|^{1/2}} e^{iS^+(x, t; \kappa^*)/\varepsilon + i\sigma\pi/4} v_0^+(x, t; \kappa^*)$$

$$+ \frac{(2\pi)^{-n/2}}{|\det Q(x, t)|^{1/2}} e^{-S^+(x, t\kappa^*)/\varepsilon - i\sigma\pi/4}$$

$$\cdot v_0^-(x, t; -\kappa^*),$$

where $\kappa^* = \kappa^*(x, t)$ is the (assumed unique) solution of (2.6.11) for this (x, t) (and with $S = S^+$). The time ranges over some interval $0 < t \le t_0$ before caustics appear. Also, (2.6.16) is **not** valid at $t = 0$ since the right hand side is singular. Outside the light cone u_ε is zero. On the boundary of the light cone (2.6.16) is not valid because the integral in (2.6.7) is taken over a compact set of κ's.

The expansion (2.6.6) is valid up to some time $t_0 > 0$ and an estimate like (2.1.16) is easily obtained (with $T = t_0$). The subsequent approximation of u_ε by the stationary phase method is not uniformly valid as we have just explained. We note finally that the result (2.6.16) could have been obtained directly by constructing special solutions to (2.2.1) and then matching the singularity[14] in (2.6.16) with the singularity of the exact solution of (2.1.1) with constant coefficients $c^2 = c^2(0)$, $W = W(0)$.

2.7. Expansion for the fundamental solution. Consider the problem (2.1.1) with initial data

$$(2.7.1) \qquad\qquad\qquad u_\varepsilon(x, 0) = 0,$$

$$(2.7.2) \qquad\qquad\qquad \frac{\partial u_\varepsilon}{\partial t}(x, 0) = \varepsilon^{-1}\delta(x).$$

Here $\delta(x)$ is the delta function centered at the origin and the factor ε^{-1} is inserted to make u_ε of order one as $\varepsilon \to 0$. We note that this problem has distribution-valued initial data and consequently the solution itself will be a distribution. The question is how does this distribution solution behave as $\varepsilon \to 0$?

We clarify first what is meant by "behave." Suppose $f(x)$ and $g(x)$ are C_0^∞ functions. Let G_ε be the distribution solution of (2.1.1), (2.7.1) and (2.7.2). Then[15]

$$(2.7.3) \qquad\qquad\qquad u_\varepsilon = G_{\varepsilon t} * f + G_\varepsilon * g,$$

is the solution of (2.1.1) with slowly varying data $u_\varepsilon(x, 0) = f(x)$, $u_{\varepsilon t}(x, 0) = \varepsilon^{-1}g(x)$. For each f and $g \in C_0^\infty$ the expansion of $u_\varepsilon(x, t)$ (for t sufficiently small) was obtained in previous sections. However the estimate of the error depends each time on f and g and their derivatives. Here we want to construct a distribution

---

[14]This is frequently called the method of canonical problems, cf. R. M. Lewis [**67**], J. B. Keller [**55**].

[15]Here $*$ denotes convolution.

G_ε^N such that $G_\varepsilon - G_\varepsilon^N$ is smooth and becomes smoother as N increases. Moreover this is to hold uniformly as $\varepsilon \to 0$. As a consequence, G_ε^N contains the singular part of G_ε for each ε, i.e. we have precise information about the character of the singularity of G_ε.

There are a number of ways G_ε^N can be constructed.[16] Since we already know in sufficient detail the behavior of smooth solutions and since the nature of the singularities of G_ε in media with periodic structure appears to be very complicated, we shall not discuss the analysis of G_ε further.

2.8. Expansion near smooth caustics. The construction of asymptotic expansions of u_ε, solution of (2.1.1), (2.1.2) and (2.1.3), valid globally in t, and not merely until the first time t_0 that caustics appear, can be obtained by the general methods of Keller, Maslov and Hörmander.[17] This expansion reduces at each non-caustic point (x, t) to a sum of terms of the form (2.1.15) with their phases appropriately adjusted by factors of $\pi/4$. In the immediate vicinity of a smooth caustic the field does not behave like (2.1.15) but rather in a manner first described by Ludwig[18] and Kravtsov.

Instead of (2.1.4) and (2.1.5) one uses the ansatz

$$(2.8.1) \qquad u_\varepsilon = e^{iS(x,t)/\varepsilon}[V(\varepsilon^{-2/3}\rho(x,t))v_\varepsilon(x,t) \\ + i\varepsilon^{1/3}V'(\varepsilon^{-2/3}\rho(x,t))w_\varepsilon(x,t)],$$

where v_ε and w_ε have asymptotic power series expansions as in (2.1.5) and $\rho(x,t) = 0$ is assumed to be the equation defining the smooth caustic surface. The function $V(\zeta)$ is the Airy function of minus its argument and $V'' + \zeta V = 0$. The Airy function models locally the behavior of the field across the caustic from the illuminated side $\rho > 0$ to the shadow $\rho < 0$.

The expression (2.8.1) is inserted into (2.1.1) and coefficients of equal powers of ε of V and V' are set to zero separately. This leads to eikonal equations, rays, transport equations, etc. analogous to the ones of the usual WKB with some important differences. We refer to the paper of Ludwig for the complete analysis.

In media with periodic structure the ansatz of the form (2.8.1), suitably generalized, works just as well as is does for (2.1.1).

2.9. Impact problem. This is an initial-boundary value problem consisting of equation (2.1.1) for $t > 0$ and $x \in \mathbb{R}^n$ such that (with $x = (x_1, x_2, \ldots, x_n)$), $x_n > 0$, the initial conditions

$$(2.9.1) \qquad u_\varepsilon(x, 0) = 0, \qquad u_{\varepsilon t}(x, 0) = 0,$$

and the boundary condition

$$(2.9.2) \qquad u_\varepsilon(x', 0, t) = \phi(x', t), \qquad x = (x', x_n), \qquad t > 0,$$

with $\phi(x', t)$ a given smooth function such that $\phi(x', 0) \not\equiv 0$ or $\phi(x', 0) \equiv 0$ but some t derivative of ϕ at $t = 0$ is not zero. Notice that the impact function ϕ does not depend on ε.

For $n = 1$, one space dimension, this problem was analyzed in detail by Reiss.[19] These results have not been generalized to more dimensions.

---

[16]P. D. Lax [**65**], Zauderer [**129**], J. J. Duistermaat [**37**].

[17]cf. J. J. Duistermaat [**37**], J. Leray [**66**] and references herein. See also, J. B. Keller [**54**].

[18]D. Ludwig [**74**].

[19]E. L. Reiss [**98**], D. Ahluwalia, S. Stone and E. L. Reiss [**4**].

A simpler version of the impact problem consists of (2.1.1), (2.9.1) and

$$(2.9.3) \qquad u_\varepsilon(x',0,t) = e^{i\tilde{S}(x',t)/\varepsilon} \varepsilon f(x',t), \qquad t > 0,$$

where $\tilde{S}(x',t)$ is real and C^∞ and $f(x',t)$ is complex-valued C^∞ and of compact support in $\mathbb{R}^{n-1} \times [0,\infty)$. One can also consider **space-time localized data** on the hyperplane $x_n = 0$ by Fourier representation and stationary phase analysis as in Section 2.6. It follows that the impact problem with space-time data (2.9.3) is a basic problem and we now consider it briefly.

First we note that (2.1.1), (2.9.1) and (2.9.3) has a smooth solution for each $t < \infty$. This follows from the fact that the data (2.9.3) vanish identically in the neighborhood of the space-time corner $x_n = 0$, $t = 0$ (cf. Rauch and Massey [**97**]).

Next we note that if we obtain an expansion which satisfied (2.9.3) identically, then the energy identity (1.1.5) can be applied to the difference of the exact minus the approximate solution. Now the inner product is over $x = (x', x_n)$ with $x_n > 0$ and the boundary terms vanish. The approximate solution (i.e., the expansion) must be valid throughout $x_n > 0$ for this argument to work. This puts a serious restriction that one expects however should be removable.

Consider again the ansatz (2.1.4), (2.1.5). Everything goes as in Section 2.1 at first. It will be clear however that in (2.1.15) only the terms with superscript $+$ should appear. Consider (2.2.1). We prescribe initial conditions $\tilde{S}(x) \equiv 0$ and boundary condition

$$(2.9.4) \qquad S(x',t) = S(x',t) \text{ on } x_n = 0, \qquad t > 0.$$

We **assume** that the eikonal equation (2.2.1) with (2.9.4) has a smooth solution globally for $x_n > 0$, $0 \le t \le T < \infty$. This rules out the formation of caustics for rays issuing out of the hyperplane $x_n = 0$ ($t > 0$). With this assumption the construction of the expansion proceeds as before.

The transport equations for $v_0^+, v_0^-, \ldots$, etc., are as before. In particular, initial values for v_0^+ are given on $x_n = 0$ by the function f and for $v_1^+, \ldots$ zero. The quantities $v_0^-, v_1^-, \ldots$ are identically zero. This is the only way that (2.9.1) will be satisfied.

The above generalize to periodic media under the assumption that no caustics form (for each mode under consideration) and a few others as we discuss in detail later (such as modal nondegeneracy).

2.10. Symmetric hyperbolic systems. A number of equations of mathematical physics can be written as symmetric hyperbolic systems which admit an elegant theory.[20]   First we describe the general framework and then give some examples.

Let $A^1(x), A^2(x), \ldots, A^n(x)$ be hermitian $N \times N$ matrix functions of $x \in \mathbb{R}^n$ and let $A^0(x)$ be hermitian and positive definite and $B(x)$ an arbitrary $N \times N$ matrix. The entries are assumed to be smooth functions. Consider the system of partial differential equations

$$(2.10.1) \quad A^0 \frac{\partial u}{\partial t} = \sum_{p=1}^n A^p \frac{\partial u}{\partial x_p} + Bu, \qquad t > 0, \qquad x \in \mathbb{R}^n, \qquad u(x,0) = f(x),$$

where $u(x,t)$ and $f(x)$ are N-vector functions and matrix multiplication is implied in (2.10.1).

---

[20]Courant-Hilbert, Vol. II [**29**] and the many references cited therein.

Suppose (2.10.1) has a differentiable solution. We show that this solution is unique and define the **domain of dependence** and **range of influence** of a point.

Let $k \in \mathbb{R}^n$ be a fixed vector and consider the **characteristic equation** associated with (2.10.1)

$$(2.10.2) \qquad Q(\omega, k) = \det\left(A^0 \omega + \sum_{p=1}^{n} A^p k_p\right) = 0.$$

This is an eigenvalue equation for each k fixed with the eigenvalue parameter being ω. Note that ω depends on k and x since the matrices depend on x. From the assumption that the matrices A^p are hermitian and A^0 is positive definite it follows that (2.10.2) has N real solutions $\omega_1(k, x), \omega_2(k, x), \ldots, \omega_n(k, x)$, counting multiplicities. Moreover there exist for each x and k N linearly independent eigenvectors for the matrix on the right side of (2.10.2), unique up to a constant factor (which may depend on x and k).

Since (2.10.2) is an algebraic equation depending analytically on k, the roots $\omega_m(k, x)$ are differentiable functions of k and x at all points where they are **distinct**. Similarly the eigenvectors can be chosen to be differentiable functions of x and k **locally** in the neighborhood of a point where the corresponding ω is isolated. It is difficult to say, in general, much more about the eigenvalues ω_m and the corresponding eigenvectors as regards their global behavior with respect to k and x (except in the case $n = 1$). We shall assume that for all k and x the $\omega_m(k, x)|k|^{-1}$ are uniformly bounded in absolute value.

The mth **bicharacteristic** curves or **rays** are solutions of the Hamiltonian equations

$$(2.10.3) \qquad \frac{dx^{(m)}(t)}{dt} = \frac{\partial \omega_m}{\partial k}(k^{(m)}(t), x^{(m)}(t)), \qquad x^{(m)}(0) = x,$$

$$\frac{dk^{(m)}(t)}{dt} = -\frac{\partial \omega_m}{\partial x}(k^{(m)}(t), x^{(m)}(t)), \qquad k^{(m)}(0) = k.$$

The system of rays issuing out of a point x forms a multiply sheeting conoid. The convex hull of this conoid is called the (forward) **ray conoid** (or light cone). Similar definitions hold for the backward ray conoid.

Let G be a bounded region of space-time, $t \geq 0$. The energy identity associated with (2.10.1) is obtained by taking the inner product of (2.10.1) by $\bar{u}$, integrating over G and integrating by parts, etc. If we use for G the region between $t = 0$, $t = t_1 < \infty$ and the surface of the backward ray conoid issuing from a point (x, t_2), $t_2 > t$, we deduce[21] that the domain of dependence of the solution at a point is contained in the intersection of the backward ray conoid with $t = 0$. As a consequence data with compact support produce solutions with compact support and the way the support expands with time, i.e. the first arrival of a signal, is controlled by the ray conoid. The range of influence of a region $D \subset \mathbb{R}^n$ at time $t > 0$ is the union of the forward ray conoids issuing from each point of D and extended up to time t.

---

[21]Courant-Hilbert, Vol. II [**29**].

The energy identity for u of (2.10.1) over a fixed spatial region $\mathcal{O}$ takes the form

$$(2.10.4) \quad \frac{\partial}{\partial t} \int_{\mathcal{O}} (\bar{u}(x,t), A^0(x)u(x,t))\, dx$$

$$= \int_{\partial\mathcal{O}} \sum_{p=1}^{n} \hat{n}_p(\bar{u}(x,t), A^p(x)u(x,t))\, dS$$

$$+ \int_{\mathcal{O}} 2\Re\left(\bar{u}(x,t), \left(B(x) - \frac{1}{2}\sum_{p=1}^{n} \frac{\partial A^p(x)}{\partial x_p}\right)u(x,t)\right)\, dx$$

where $(,)$ denotes Euclidean inner product of vectors and $\hat{n}_p(x)$ are the components of the unit outward normal to $\partial\mathcal{O}$ at x. On the basis of this identity and the domain of dependence considerations just discussed we conclude that the norm

$$(2.10.5) \qquad \|u(t)\|_0^2 = \int_{\mathbb{R}^n} (\bar{u}(x,t), A^0(x)u(x,t))\, dx$$

of any solution u of (2.10.1) satisfies

$$(2.10.6) \qquad \|u(t)\|_0 \le c_1 e^{c_2 t}\|f\|_0,$$

where c_1 and c_2 are constants independent of f. Similarly L^2 norms (with respect to A^0) of derivatives of u satisfy inequalities such as (2.10.6) involving the corresponding derivatives of f. Thus, for symmetric hyperbolic systems we have simple a priori estimates which give uniqueness at once. Existence also follows by using these identities (cf. Courant-Hilbert, Vol. II). The situation for initial-boundary valued problems is more complicated and will not be discussed here.

We consider now two examples: the equations of acoustics and Maxwell's equation.

Let $u(x,t)$, with values in $\mathbb{R}^3$, denote the small variations in the velocity field and $p(x,t)$, scalar-valued, the excess negative pressure over a fixed reference state. The linear acoustic equations for u and p are

$$(2.10.7) \qquad \rho u_t = \nabla p,$$

$$p_t = \mu \nabla \cdot u, \qquad t > 0, \qquad x \in \mathbb{R}^3,$$

$$(2.10.8) \qquad u(x,0) = u^0(x), \qquad p(x,0) = p^0(x).$$

Here $\rho = \rho(x)$ and $\mu = \mu(x)$ are the material density and the density times the square of the sound speed, respectively. We shall assume that ρ and μ are **given smooth** functions and

$$(2.10.9) \qquad 0 < \rho_0 \le \rho(x) \le \rho_1 < \infty, \qquad 0 < \mu_0 \le \mu(x) \le \mu_1 < \infty.$$

A typical **initial-boundary** value problem for (2.10.7), (2.10.8) in the interior of a smooth, bounded domain $\mathcal{O} \subset \mathbb{R}^3$ is provided by the boundary conditions

$$(2.10.10) \quad \hat{n}(x)\cdot u(x,t) = 0, \qquad x \in \partial\mathcal{O}, \qquad t > 0, \qquad \hat{n} = \text{ unit outward normal.}$$

Equations (2.10.7) constitute a symmetric hyperbolic system with (u, p) the solution vector $(N = 4)$ and in the notation (2.10.1)

(2.10.11)
$$A^0(x) = \begin{pmatrix} \rho I & 0 \\ 0 & \mu^{-1} \end{pmatrix}, \qquad I = 3 \times 3 \text{ identity},$$

$$A^1(x) = \begin{pmatrix} 0 & 0 & 0 & 1 \\ 0 & 0 & 0 & 0 \\ 0 & 0 & 0 & 0 \\ 1 & 0 & 0 & 0 \end{pmatrix},$$

(2.10.12)
$$A^2(x) = \begin{pmatrix} 0 & 0 & 0 & 0 \\ 0 & 0 & 0 & 1 \\ 0 & 0 & 0 & 0 \\ 0 & 1 & 0 & 0 \end{pmatrix},$$

$$A^3(x) = \begin{pmatrix} 0 & 0 & 0 & 0 \\ 0 & 0 & 0 & 0 \\ 0 & 0 & 0 & 1 \\ 0 & 0 & 1 & 0 \end{pmatrix}.$$

We write in short

(2.10.13)
$$\begin{pmatrix} \rho & 0 \\ 0 & \mu^{-1} \end{pmatrix} \begin{pmatrix} u \\ p \end{pmatrix}_t = \begin{pmatrix} 0 & \nabla \\ \nabla \cdot & 0 \end{pmatrix} \begin{pmatrix} u \\ p \end{pmatrix}, \qquad t > 0, \qquad \begin{pmatrix} u \\ p \end{pmatrix}\Big|_{t=0} = \begin{pmatrix} u^0 \\ p^0 \end{pmatrix}.$$

The energy density is defined by

(2.10.14)
$$E(x, t) = \frac{1}{2} \rho(x) |u(x, t)|^2 + \frac{1}{2\mu(x)} |p(x, t)|^2,$$

while the energy flux density is defined by

(2.10.15)
$$F(x, t) = p(x, t) u(x, t).$$

The energy identity (2.10.4) becomes

(2.10.16)
$$\frac{\partial}{\partial t} \int_{\mathcal{O}} E(x, t) \, dx = \int_{\partial \mathcal{O}} \hat{n}(x) \cdot F(x, t) \, dS.$$

Maxwell's equations for the electric and magnetic fields $E(x, t)$ and $H(x, t)$, vector-valued, have the following form[22]

(2.10.17)
$$\begin{pmatrix} \varepsilon(x) & 0 \\ 0 & \mu(x) \end{pmatrix} \begin{pmatrix} E \\ H \end{pmatrix}_t = \begin{pmatrix} 0 & \nabla^\wedge \\ -\nabla^\wedge & 0 \end{pmatrix} \begin{pmatrix} E \\ H \end{pmatrix} + \begin{pmatrix} \sigma(x) & 0 \\ 0 & 0 \end{pmatrix} \begin{pmatrix} E \\ H \end{pmatrix},$$

(2.10.18)
$$E(x, 0) = E^0(x), \qquad H(x, 0) = H^0(x),$$

(2.10.19)
$$\nabla \cdot (\varepsilon E^0) = 0, \qquad \nabla \cdot (\mu H^0) = 0.$$

Here $\nabla^\wedge$ denotes the rot or curl operator, ε, μ and σ are, respectively, the dielectric tensor, magnetic permeability tensor and the conductivity tensor (3×3 matrices).

---

[22]Landau-Lifchitz [**64**]. We assume that there are no sources of charges or currents in (2.10.17) and that ε, μ and σ are **symmetric** and ε, μ **positive definite**.

Note that (2.10.19) imposes a restriction on the data which is preserved by the equation for all $t > 0$. Noting also that

$$V \wedge E = \begin{pmatrix} 0 & -\frac{\partial}{\partial x_3} & \frac{\partial}{\partial x_2} \\ \frac{\partial}{\partial x_3} & 0 & -\frac{\partial}{\partial x_1} \\ -\frac{\partial}{\partial x_2} & \frac{\partial}{\partial x_1} & 0 \end{pmatrix} \begin{pmatrix} E_1 \\ E_2 \\ E_3 \end{pmatrix}$$

it follows immediately that (2.10.17) is a symmetric hyperbolic system.

The usual boundary conditions across interfaces are:

(2.10.20)   tangential components of E are continuous;

tangential components of H differ by a surface current density;

normal components of εE differ by a surface charge density;

normal components of μH are continuous.

We shall deal mostly with unbounded media in the sequel (except in Section 4.11).

The energy identity takes the following form. Let Ω and P be the field energy density and energy flux respectively

(2.10.21) $$\Omega(x,t) = \frac{1}{2}\varepsilon(x)E(x,t) \cdot E(x,t) + \frac{1}{2}\mu(x)H(x,t) \cdot H(x,t),$$

(2.10.22) $$P(x,t) = E(x,t) \wedge H(x,t).$$

Then we have for any region $\mathcal{O} \subset \mathbb{R}^3$

(2.10.23) $$\frac{\partial}{\partial t} \int_{\mathcal{O}} \Omega(x,t) \, dx + \int_{\partial \mathcal{O}} \hat{n}(x) \cdot P(x,t) \, dS = - \int_{\mathcal{O}} \sigma(x)E(x,t) \cdot E(x,t) \, dx.$$

2.11. Expansions for symmetric hyperbolic systems (low frequency). As was discussed in Section 1.1, equations with slowly varying coefficients lead, upon rescaling, to equations with large terms. For symmetric hyperbolic systems with smooth coefficient matrices the scaled problem takes the form

(2.11.1) $$A^0(x)\frac{\partial u_\varepsilon(x,t)}{\partial t} = \sum_{p=1}^n A^p(x)\frac{\partial u_\varepsilon(x,t)}{\partial x_p} + \frac{1}{\varepsilon}B(x)u_\varepsilon(x,t), \qquad t > 0,$$

$$u_\varepsilon(x,0) = f_\varepsilon(x), \qquad x \in \mathbb{R}^n.$$

The data $f_\varepsilon(x)$ is N-vector valued and may have one of the three forms, according to the discussion in Section 1.1:

(2.11.2)

A. $f_\varepsilon(x) = f(x)$ with the components of $f(x)$, C_0^∞ functions;

(2.11.3)

B. $f_\varepsilon(x) = e^{i\tilde{S}(x)/\varepsilon}f(x)$ where $\tilde{S}(x)$ is real valued and C^∞

and the components of $f(x)$ are complex-valued and in C_0^∞;

(2.11.4)

C. $f_\varepsilon(x) = \varepsilon^{-n/2}f\left(x, \frac{x}{\varepsilon}\right)$ where the components of $f(x,y)$ are C_0^∞ functions.

Note that when $B(x) \equiv 0$ the equation does not depend on ε at all and in case A one must solve the problem exactly: no asymptotics are involved. In case B we recover the problem of oscillatory initial data[23] while in case C we are concerned

---

[23]P. D. Lax; cf. also Courant-Hilbert, [29], Vol. II.

with the approximation of the fundamental solution (scaled so that the energy is of order one as $\varepsilon \to 0$).

Consider now class A data with $B(x) \not\equiv 0$. We shall illustrate the asymptotic analysis with two theorems regarding model Boltzmann-like equations. Most of the content of theorems one and two below carries over to

(1) infinite dimensional hyperbolic systems ($N = \infty$) and
(2) mixed initial-boundary value problems.

The infinite dimensional case is not really difficult given enough information on B. The initial-boundary value problems are, however, harder.[24]

With no loss in generality we may take $A_0 = I$. We assume that $B(x)$ is **symmetric** and negative semidefinite:

$$(2.11.5) \qquad (v, Bv) \leq 0, \ \forall v.$$

Let $\phi_1(x), \phi_2(x), \ldots, \phi_{N_0}(x)$, $1 \leq N_0 < N$ be a set of vectors spanning the null space of $B(x)$ with N_0 independent of x and with $\phi_j(x)$ smooth in x. Moreover, we assume that

$$(2.11.6) \qquad (\phi_i(x), \phi_j(x)) = \delta_{ij}, \qquad i, j = 1, 2, \ldots, N_0.$$

Define the symmetric $N_0 \times N_0$ matrices $\bar{A}^p(x)$ by
$$(2.11.7)$$
$$\bar{A}^p_{jk}(x) = (\phi_j(x), A^p(x)\phi_k(x)), \qquad j, k = 1, 2, \ldots, N_0, \qquad p = 1, 2, \ldots, n.$$

Define also

$$(2.11.8) \qquad \bar{B}_{jk}(x) = \sum_{p=1}^{n} \left(\phi_j(x), A^p(x) \frac{\partial \phi_k(x)}{\partial x_p} \right), \qquad j, k = 1, 2, \ldots, N_0.$$

Let $(\bar{u}_j(x, t))$, $j = 1, 2, \ldots, N_0$ be the solution vector of the N_0 dimensional **symmetric hyperbolic system**

$$(2.11.9) \qquad \frac{\partial \bar{u}_j(x, t)}{\partial t} = \sum_{p=1}^{n} \sum_{k=1}^{N_0} \bar{A}^p_{jk}(x) \frac{\partial \bar{u}_k(x, t)}{\partial x_p} + \sum_{k=1}^{N_0} \bar{B}_{jk}(x) \bar{u}_k(x, t), \qquad t > 0,$$

$$\bar{u}_j(x, 0) = f_j(x),$$

with

$$(2.11.10) \qquad f_j(x) = (f(x), \phi_j(x)).$$

THEOREM 2.1. *Let* $\bar{u}(x, t)$ *be defined by*

$$(2.11.11) \qquad \bar{u}(x, t) = \sum_{j=1}^{N_0} \bar{u}_j(x, t) \phi_j(x).$$

Under the above conditions,[25] for any N-vector $f(x) = (f_\varepsilon(x))$ with C_0 components and any $t > 0$,

$$(2.11.12) \qquad \lim_{\varepsilon \downarrow 0} \int_{\mathbb{R}^n} |u_\varepsilon(x, t) - \bar{u}(x, t)|^2 \, dx = 0.$$

REMARK 2.2. Note that the result does not hold uniformly in $[0, T]$, $T < \infty$, because there is an initial layer involved.

---

[24] A. Bensoussan, J. L. Lions, G. C. Papanicolaou [13]. The term infinite dimensional hyperbolic systems really means transport equations here.

[25] Recall $A^p(x)$, $p = 1, \ldots, n$ and $B(x)$ are matrices with C^∞ and bounded entries.

REMARK 2.3. The point of Theorem 2.1 is the order-reduction which can be rather dramatic. For example, (2.11.1) could be ∞-dimensional ($N = \infty$) (which we have not set up here but is not difficult to do) whereas (2.11.9) could be finite dimensional. The most famous example of this is the hydrodynamical limit of the linearized Boltzmann equation[26] where (2.11.9) corresponds to the inviscid, linearized Euler equations and $N_0 = 5$.

PROOF OF THEOREM 2.1. Let $P(x)$ be the $N \times N$ matrix projecting into the null space of $B(x)$, i.e.

$$(2.11.13) \qquad P(x)v = \sum_{j=1}^{N_0} (v, \phi_j(x))\phi_j(x).$$

By a simple argument involving an initial layer[27] analysis we may assume that the data f satisfy

$$Pf = f$$

which means that

$$(2.11.14) \qquad f(x) = \sum_{j=1}^{N_0} f_j(x)\phi_j(x),$$

with $f_j(x)$ defined in (2.11.10).

Now we construct an expansion of $u_\varepsilon(x, t)$ of the form

$$(2.11.15) \qquad u_\varepsilon(x, t) = u_0(x, t) + \varepsilon u_1(x, t) + \varepsilon^2 u_2(x, t) + \cdots .$$

From (2.11.1) and (2.11.15) the following sequence of problems obtains for u_0, u_1, $u_2, \ldots,$

$$(2.11.16) \qquad Bu_0 = 0,$$

$$(2.11.17) \qquad Bu_1 + \sum_p A^p \frac{\partial u_0}{\partial x_p} - \frac{\partial u_0}{\partial t} = 0,$$

$$(2.11.18) \qquad label4.2.119 Bu_2 + \sum_p A^p \frac{\partial u_1}{\partial x_p} - \frac{\partial u_1}{\partial t} = 0, \qquad \text{etc.}$$

From (2.11.16) we conclude that $u_0(x, t)$ belongs to the null space of B, i.e., $Pu_0 = u_0$ and we write

$$(2.11.19) \qquad u_0(x, t) - \sum_{j=1}^{N_0} \bar{u}_j(x, t)\phi_j(x),$$

with the $\bar{u}_j(x, t)$ to be determined.

From (2.11.17) we see that for u_1 to exist the component of the inhomogeneous term in the null space of B must vanish. This provides restrictions on u_0 which are exactly equations (2.11.9) for $\bar{u}_j(x, t)$. Thus $u_1(x, t)$ is well determined up to a vector function in the null space of B, which we take identically zero, and satisfies (2.11.17).

Note that $\bar{u}(x, t) = u_0(x, t)$ now. Define

$$(2.11.20) \qquad w_\varepsilon(x, t) = u_\varepsilon(x, t) - \bar{u}(x, t) - \varepsilon u_1(x, t).$$

---

[26]H. Grad [**49**].

[27]This is discussed more fully in Section 2.13 (Theorem 2.8, Proof 1).

It is easily verified that

$$(2.11.21) \qquad \frac{\partial w_\varepsilon}{\partial t} = \sum_p A^p \frac{\partial w_\varepsilon}{\partial x_p} + \frac{1}{\varepsilon} B w_\varepsilon + \varepsilon g_1(x,t),$$

$$w_\varepsilon(x,0) = \varepsilon g_2(x),$$

where $g_1(x,t)$ and $g_2(x)$ are smooth and bounded functions of compact support. By the hyperbolicity of (2.11.1) and (2.11.9), w_ε has compact support. Hence the energy identity over $\mathbb{R}^n$ gives

$$\int_{\mathbb{R}^n} (w_\varepsilon(x,t), w_\varepsilon(x,t)) \, ds$$

$$- \frac{1}{\varepsilon} \int_0^t \int_{\mathbb{R}^n} \left(w_\varepsilon(x,s), \left(B(x) - \frac{\varepsilon}{2} \sum_p \frac{\partial A^p(x)}{\partial x_p} \right) w_\varepsilon(x,s) \right) dx \, ds$$

$$= \varepsilon \int_0^t \int_{\mathbb{R}^n} (w_\varepsilon(x,s), g_1(x,s)) \, dx \, ds + \varepsilon \int_{\mathbb{R}^n} |g_2(x)|^2 \, dx.$$

Since B is negative semidefinite it follows that

$$(2.11.22) \qquad \int_{\mathbb{R}^n} |w_\varepsilon(x,t)|^2 \, dx \le \varepsilon C_1 \int_{\mathbb{R}^n} |g_2(x)|^2 \, dx$$

$$+ \varepsilon^2 C_2 \int_0^t \int_{\mathbb{R}^n} |g_1(x,s)|^2 \, dx \, ds + C_3 \int_0^t \int_{\mathbb{R}^n} |w_\varepsilon(x,s)|^2 \, dx.$$

This implies that

$$\int_{\mathbb{R}^n} |w_\varepsilon(x,t)|^2 \, dx \le C_4 \varepsilon,$$

and hence (2.11.12) is proved.   ∎

REMARK 2.4. Note that our estimates are sharper than the regularity hypotheses on coefficients and data required. Note also that we have the error estimate $O(\sqrt{\varepsilon})$ for the norm $\|u_\varepsilon - \bar{u}\|$. If we had taken care of the initial layer terms, we could have easily obtained

$$\sup_{0 \le t \le T} \|u_\varepsilon - \bar{u} - \bar{u}_\varepsilon^{IL}\| = O(\varepsilon),$$

where $\bar{u}_\varepsilon^{IL}(x,t)$ denotes the initial layer corrections. We shall not carry out the details for this.

REMARK 2.5. Sometimes it is possible to approximate u_ε of (2.11.1) by the solution $\bar{u}_\varepsilon$ of a lower dimensional **parabolic system** where the coefficients of the second spatial derivatives are proportional to ε. Under favorable circumstances, far more special than in Theorem 2.1 above, the advantage gained by this is that the approximation $\bar{u}_\varepsilon$ is valid for a much longer time interval of order ε^{-1}. In the context of the linearized Boltzmann equation this is the Chapman-Enskog expansion[28] (up to second order) whereas Theorem 2.1 is the Hilbert expansion. We return to this in Section 4.6.

---

[28] In the optics and acoustic literature this corresponds to the Leontovich-Fock or parabolic approximation [**115**].

2.12. Expansions for symmetric hyperbolic systems (probabilistic).
Consider again $u_\varepsilon(x,t)$, solution of (2.11.1) with $f_\varepsilon(x) = f(x)$ an N-vector function
with C_0^∞ components. We shall prove a theorem similar to Theorem 2.1 of the
previous section under different conditions on B. Note that we are still in the
"low frequency" case, i.e. class A data. Most of what follows goes through for
∞-dimensional systems.

An important restriction in the probabilistic treatment is that the matrices
$A^p(x)$, $p = 0, 1, 2, \ldots, n$ are **diagonal**. We may take $A^0(x) \equiv I$, the identity, and

$$(2.12.1) \qquad (A_{jk}^p(x)) = (A_j^p(x)\delta_{jk}), \qquad p = 1, 2, \ldots, n, \qquad j, k = 1, 2, \ldots, N.$$

We assume that

$$(2.12.2) \qquad \text{the matrix } B \text{ has non-negative off-diagonal elements}$$

$$\text{with row sums normalized to equal one.}$$

The normalization of the row sums to one is not indispensable and is introduced
for convenience. Instead of taking B in (2.11.1) as independent of ε, in this section
we shall assume that $B(x)$ depends on ε also and that

$$(2.12.3) \qquad B_\varepsilon(x) = B^0(x) + \varepsilon B^1(x).$$

The matrices B^0 and B^1 are specified below where it is also explained why we
introduce (2.12.3).

A matrix B satisfying (2.12.3) is the infinitesimal generator of a continuous
time Markov chain with N states.[29] As a consequence, and because of (2.12.1),
$u_\varepsilon(x,t)$, solution of (2.11.1), has the following properties:

$$\text{if each component of } f(x) \text{ is nonnegative then each}$$

$$(2.12.4) \qquad \text{component of } u_\varepsilon(x,t), \, t > 0, \text{ is nonnegative,}$$

$$\sup_x |u_\varepsilon(x,t)| \leq \sup_x |f(x)|, \qquad t \geq 0,$$

where **in this section** $|\,|$ stands for the maximum absolute value of a row vector
(and **not** the Euclidean norm). We say now that the hyperbolic system satisfies the
maximum principle. Note that estimate (2.12.4) is independent of ε; it will allow
us to obtain an asymptotic expansion much as in Section 2.11.

A matrix B satisfying (2.12.2) has the vector 1 (all coefficients equal to one)
as a right null-vector. If the off-diagonal elements of B are strictly positive, B
has a one-dimensional null-space and the rest of the eigenvalues have negative real
parts.[30] The components of the, unique up to normalization, left null-vector of B
$\bar{p}_i$, $i = 1, 2, \ldots, N$, are strictly positive and we take them so that

$$(2.12.5) \qquad \sum_{i=1}^{N} \bar{p}_i = 1.$$

If some off-diagonal elements of B vanish then the set of states $\{1, 2, \ldots, N\}$
can be decomposed into classes of states such that by permutations of rows and
columns B has block diagonal form. Each block along the diagonal may or may not
have positive off-diagonal entries (within the block); however, the null-space of each

---

[29]The hyperbolic system (2.11.1) under (2.12.1), (2.12.2) is also called a random evolution (cf.
R. Hersh [**51**], M. Pinsky [**93**] and T. G. Kurtz [**63**]).

[30]This is a special case of the Perron-Frobenius theorem (cf. Gantmacher [**45**]).

block is one-dimensional without excluding pure imaginary eigenvalues (periodically moving states).[31] We shall assume that $B_\varepsilon(x)$ of (2.12.3) satisfies the following:

(2.12.6)   (i)   When ε is positive the set of states $\{1, 2, \ldots, N\}$ is irreducible with no periodically moving states.

(ii)   When $\varepsilon = 0$ (i.e., regarding B^0) the states $\{1, 2, \ldots, N\}$ decompose into r classes $S_1, S_2, \ldots, S_r$, $1 \le r < N$, of irreducible and aperiodic states with $N_1, N_2, \ldots, N_r$ components, respectively $(N_1 + N_2 + \cdots + N_r = N)$.

Note that as a result of (2.12.6) B_ε has no imaginary eigenvalues for $\varepsilon \ge 0$ and it has a null-space which is one-dimensional when $\varepsilon > 0$ and is r-dimensional when $\varepsilon = 0$.

Let $\bar{P}_i^\nu$, $i = 1, 2, \ldots, N$, $\nu = 1, 2, \ldots, r$ denote the components of vectors $\bar{p}^\nu$ spanning the null-space of B_0. These vectors can be chosen so that

$$(2.12.7) \qquad \bar{p}_j^\nu > 0, \qquad j \in S_\nu, \qquad (\bar{p}^\nu, 1^\nu) = \sum_{j \in S_\mu} \bar{p}_j^\nu = \delta_{\nu\mu},$$

where 1^ν has components $1_j^\nu = 1$, $j \in S_\nu$, $1_j^\nu = 0$, $j \notin S_\nu$. We say that the $\bar{p}^\nu$ have disjoint support with union $S = \{1, 2, \ldots, N\}$.

Define

$$(2.12.8) \qquad \bar{A}_\nu^p(x) = \sum_{j \in S_\nu} \bar{p}_j^\nu A_j^p(x), \qquad \nu = 1, 2, \ldots, r \qquad p = 1, 2, \ldots, n,$$

$$(2.12.9) \qquad \bar{f}_\nu(x) = \sum_j \bar{p}_j^\nu f_j(x), \qquad f_j = j^{\text{th}} \text{ component of } f,$$

$$(2.12.10) \qquad \bar{B}_{\nu\mu}(x) = \sum_{\substack{j \in S_\nu \\ k \in S_\mu}} \bar{p}_j^\nu B_{jk}^1(x), \qquad \nu, \mu = 1, 2, \ldots, r.$$

We have now the following theorem which is an asymptotic **order**-reduction result by **lumping of components**.

THEOREM 2.6. *Under the above hypotheses, each component $u_{j\varepsilon}(x, t)$, $j = 1, 2, \ldots, N$, of the solution of (2.11.1) with $f_\varepsilon(x) = f(x)$ converges for each $t > 0$ as $\varepsilon \to 0$ to $\bar{u}_{\nu(j)}(x, t)$, uniformly in x, where $\nu(j)$ is the index μ of the unique class S_μ to which j belongs. The functions $\bar{u}_\nu(x, t)$, $\nu = 1, 2, \ldots, r$ are solutions of the symmetric hyperbolic system*

$$(2.12.11) \qquad \frac{\partial \bar{u}_\nu(x, t)}{\partial t} = \sum_{p=1}^n \bar{A}_\nu^p(x) \frac{\partial \bar{u}_\nu(x, t)}{\partial x_p} + \sum_{\mu=1}^r \bar{B}_{\nu\mu}(x) \bar{u}_\mu(x, t), \qquad t > 0,$$

$$\bar{u}_\nu(x, 0) = \bar{f}_\nu(x), \qquad \nu = 1, 2, \ldots, r.$$

We shall not give the proof of Theorem 2.6 in detail since it does not differ in any essential way from the one of Theorem 2.1, Section 2.11. We point out a couple of things that are important in the proof. The first is smoothness of all coefficients and data which gives smoothness of solutions. The second is that B_ε

---

[31] W. Feller, Vol. I [**39**]. The states composing each block are said to form an irreducible set of states.

has at most an r-dimensional null-space with the rest of the eigenvalues having negative real parts independently of ε. This allows us to construct a simple initial layer expansion which gives the lumping of the data and the restriction $t > 0$ in Theorem 2.6 above.

In the event

$$(2.12.12) \qquad \bar{A}_\nu^p(x) \equiv 0, \qquad \nu = 1, 2, \ldots, r, \qquad p = 1, 2, \ldots, n,$$

Theorem 2.6 does not have much content, evidently. It turns out that $(2.12.12)$ occurs in some important applications in transport theory and, in any case, it is of interest to know if an improvement to Theorem 2.6 can be obtained.

It is convenient to rescale $(2.11.1)$ so that the final result appears in neat form. It is necessary to rescale time $(t \to t/\varepsilon)$ and B_ε $(B_\varepsilon \to B^0 + \varepsilon^2 B^1)$. Then $(2.11.1)$ becomes

$$(2.12.13) \qquad \frac{\partial u_{j\varepsilon}}{\partial t} = \frac{1}{\varepsilon} \sum_{p=1}^{n} A_j^p(x) \frac{\partial u_{j\varepsilon}}{\partial x_p} + \frac{1}{\varepsilon^2} \sum k = 1^N (B_{jk}^0(x) + \varepsilon^2 B_{jk}^1(x)) u_{k\varepsilon},$$

$$u_{j\varepsilon}(x, 0) = f_j(x), \qquad j = 1, 2, \ldots, N.$$

The matrix B^0 can be written as

$$(2.12.14) \qquad \qquad B^0 = \oplus_{\nu=1}^r B_\nu^0,$$

i.e. as the direct sum of matrices B_ν^0 for the transition within S_ν only. On vectors v such that $(p^\nu, v) = 0$ the matrix $(B_\nu^0)^{-1}$ is well defined up to a constant which we take to be zero (Fredholm alternative). We write $C_\nu = -(B_\nu^0)^{-1}$ for short.

For each $g(x)$, a C_0^∞ function, we define diffusion operators $\mathcal{A}_\nu$ as follows

$$(2.12.15) \quad \mathcal{A}_\nu g(x) = \sum_{p,q=1}^{n} \left(p^\nu, A^q \frac{\partial}{\partial x_q} \left(C_\nu A^p 1^\nu \frac{\partial g(x)}{\partial x_p} \right) \right), \qquad \nu = 1, 2, \ldots, r.$$

This is indeed a well defined diffusion operator[32] in view of $(2.12.12)$. Note that p_ν, A^p and C_ν all depend smoothly on x.

THEOREM 2.7. *Under the hypotheses of Theorem 2.6 and $(2.12.12)$ the solution $u_{j\varepsilon}(x, t)$ of $(2.12.13)$ for each $t > 0$ converges uniformly in x as $\varepsilon \to 0$ to the solution $\bar{u}_{\nu(j)}(x, t)$ (cf. Theorem 2.6) where $\bar{u}_\nu(x, t)$ is the solution of the parabolic system*

$$(2.12.16) \qquad \frac{\partial \bar{u}_\nu}{\partial t} = \mathcal{A}_\nu \bar{u}_\nu + \sum_{\mu=1}^r \bar{B}_{\nu\mu} \bar{u}_\nu, \qquad t > 0,$$

$$\bar{u}_\nu(x, 0) = \bar{f}_\nu(x), \qquad \nu = 1, 2, \ldots, r.$$

PROOF OF THEOREM 2.7. First we observe that since $\bar{f}_\nu(x) \in C_0^\infty$ and the coefficients in $(2.12.16)$ are C^∞, the solution $\bar{u}_\nu(x, t)$ is C^∞ for each $t > 0$ and tends to zero as $|x| \to \infty$ faster than any power of $|x|$. This is easily seen because the system $(2.12.16)$ is coupled only through the undifferentiated terms, i.e. it has particularly simple structure.

And in Theorems 2.1 and 2.6, we may assume that $f(x)$ has the form

$$(2.12.17) \qquad \qquad f(x) = \sum_{\nu=1}^r 1^\nu \bar{f}_\nu(x),$$

---

[32] We leave it to the reader to verify that $\mathcal{A}_\nu$ is elliptic, but not necessarily strictly elliptic.

since this can be achieved by a simple initial layer expansion (which explains why we must have $t > 0$).

The proof now goes much the same way as most proofs in the previous chapters as well as the one of Theorem 2.1. We construct an expansion

$$u_\varepsilon = u_0 + \varepsilon u_1 + \varepsilon^2 u_2 + \cdots,$$

by the usual process. Our hypotheses guarantee that

$$u_0(x, t) = \sum_{\nu=1}^{r} 1^\nu \bar{u}_\nu(x, t),$$

where $\bar{u}_\nu$ satisfy (2.12.16), and that $u_1(x, t)$ and $u_2(x, t)$ can be properly constructed and are smooth. We find finally that

$$u_\varepsilon - u_0 - \varepsilon u_1 - \varepsilon^2 u_2 \equiv w_\varepsilon$$

satisfies the inhomogeneous version of (2.12.13) with the inhomogeneous term of order ε and with initial data of order ε. The theorem then follows from the a priori estimate (2.12.4). ∎

2.13. Expansion for symmetric hyperbolic systems (high frequency). The systems of Sections 2.11 and 2.12 had one common feature regarding the matrix B: it had a null-space but we did not allow purely imaginary eigenvalues. It turns out that in the asymptotic analysis of systems with periodic structure, as we shall see in Sections 4 and 5, the B is involved has pure imaginary eigenvalues (both in static and high frequency cases). In general, systems that deal with waves also have such B's. It is appropriate therefore to call problems with B as in the previous sections **dissipative** or **Boltzmann-like** systems.

We now consider briefly (2.11.1) when $f_\varepsilon(x) = f(x)$ and $B(x)$ is **skew-symmetric** and has an N_0-dimensional null space as described below (2.11.5), including (2.11.6).

THEOREM 2.8. *Under the hypotheses of Theorem 2.1 (Section 2.11) but with $B(x)$ skew-symmetric, for any N-vector test function $h(x, t) \in (C^\infty(\mathbb{R}^n \times]0, \infty[))$ that vanishes identically for t and $|x|$ large and $Ph(x, t) = h(x, t)$ (cf. (2.11.13)), we have*

$$(2.13.1) \quad \lim_{\varepsilon \downarrow 0} \int_0^\infty \int_{\mathbb{R}^n} (h(x, t), u_\varepsilon(x, t)) \, dt \, dx = \int_0^\infty \int_{\mathbb{R}^n} (h(x, t), \bar{u}(x, t)) \, dt \, dx.$$

Here $\bar{u}(x, t)$ is the solution of the symmetric hyperbolic system (2.11.9) as in Theorem 2.1.

REMARK 2.9. The difference between Theorems 2.1 and 2.8 is the **mode of convergence**. In Theorem 2.1 we have strong convergence in $L^\infty(0, T; (L^2(\mathbb{R}^n))^N)$. In Theorem 2.8 we have weak star convergence in $L^\infty(0, T; (L^2(\mathbb{R}^n))_P^N)$, where $(L^2(\mathbb{R}^n))_P^N$ are square integrable N-vector functions $\bar{u}$ with $P\bar{u} = \bar{u}$.

We shall give **two** proofs of this theorem in order to illustrate some points that play an important role in periodic-structure problems later.

PROOF I. System (2.11.1) has a smooth solution for $\varepsilon > 0$; as before coefficients and data are C^∞ and the data have compact support.

We construct an asymptotic expansion for u_ε as follows. Let

$$(2.13.2) \qquad u_\varepsilon(x,t) = u_0\left(x,t,\frac{t}{\varepsilon}\right) + \varepsilon u_1\left(x,t,\frac{t}{\varepsilon}\right) + \varepsilon^2 u_2\left(x,t,\frac{t}{\varepsilon}\right) + \cdots,$$

and put

$$(2.13.3) \qquad\qquad\qquad \tau = t/\varepsilon.$$

We do **not** assume that $f(x)$ has the form (2.11.14) since there will be **no initial layer now**; in fact this is the basic difference here as we shall see. From (2.11.1) and (2.13.2) the following sequence of problems obtains for $u_0, u_1, u_2, \ldots$

$$(2.13.4) \qquad\qquad\qquad Bu_0 - \frac{\partial u_0}{\partial \tau} = 0,$$

$$(2.13.5) \qquad Bu_1 - \frac{\partial u_1}{\partial \tau} + \sum_{p=1}^{n} A^p \frac{\partial u_0}{\partial x_p} - \frac{\partial u_0}{\partial t} = 0, \qquad \text{etc.}$$

From (2.13.4) we get (cf. (2.11.11)),

$$(2.13.6) \qquad\qquad u_0(x,t,\tau) = \bar{u}(x,t) + e^{B\tau} v_0(x,t),$$

$$\bar{u}(x,t) = \sum_{j=1}^{N_0} \bar{u}_j(x,t)\phi_j(x).$$

Here $\bar{u}_j(x,t)$, $j = 1, 2, \ldots, N_0$ are to be determined so that

$$(2.13.7) \qquad\qquad \bar{u}_j(x,0) = f_j(x) = (f(x), \phi_j(x))$$

and $v_0(x,t)$ is a vector function satisfying

$$(2.13.8) \qquad\qquad (v_0(x,t), \phi_j(x)) = 0, \qquad t > 0,$$

$$(2.13.9) \qquad\qquad v_0(x,0) = f(x) - \sum_{j=1}^{N_0} f_j(x)\phi_j(x),$$

and is also to be determined.

We consider next (2.13.5). We write u_1 in the form

$$(2.13.10) \quad u_1(x,t,\tau) = e^{B\tau} v_1(x,t)$$
$$+ e^{B\tau} \int_0^\tau e^{-Bs} \left[\sum_{p=1}^{n} A^p \frac{\partial}{\partial x_p}(\bar{u} + e^{Bs} v_0) - \frac{\partial}{\partial t}(\bar{u} + e^{Bs} v) \right] ds.$$

We must impose conditions on the integral term in (2.13.10) so that it be **uniformly bounded** in τ. Since B is skew symmetric the components of $e^{B\tau}$ are quasi-periodic functions of τ. Hence the integrand on the right side of (2.13.10) is quasiperiodic in s and for u_1 to be uniformly bounded in τ (pointwise in x and t) it is necessary and sufficient that (for each x and t fixed)

$$(2.13.11) \qquad 0 = \lim_{\tau \uparrow \infty} \frac{1}{\tau} \int_0^\tau e^{-Bs} \left[\sum_{p=1}^{n} A^p \frac{\partial}{\partial x_p}(\bar{u} + e^{Bs} v_0) - \frac{\partial}{\partial t}(\bar{u} + e^{Bs} v_0) \right] ds.$$

We rewrite this as conditions on $\bar{u}$ and on v_0 separately as follows:

$$(2.13.12) \quad \lim_{\tau \uparrow \infty} \frac{1}{\tau} \int_0^\tau e^{-Bs} \left(\sum_{p=1}^{n} A^p \frac{\partial \bar{u}}{\partial x_p} - \frac{\partial \bar{u}}{\partial t} \right) ds = P\left(\sum_{p=1}^{n} A^p \frac{\partial u}{\partial x_p} - \frac{\partial \bar{u}}{\partial t} \right) = 0,$$

$$(2.13.13) \qquad \lim_{\tau \uparrow \infty} \frac{1}{\tau} \int_0^\tau e^{-Bs} \sum_{p=1}^n A^p \frac{\partial}{\partial x_p} (e^{Bs} v_0) \, ds = \frac{\partial v_0}{\partial t}.$$

But (2.13.12) is nothing more than system (2.11.9); hence (2.13.12) is satisfied if $\bar{u}$ is as in the statement of the theorem. The left hand side of (2.13.13) is a differential operator acting on $v_0(x, t)$ and the system (2.13.13) with initial condition (2.13.9) satisfies (2.13.8). Therefore $v_0(x, t)$ is well defined and thus $u_0(x, t, \tau)$ in (2.13.6) is completely determined.

One can similarly determine u_1 as in Theorem 2.1, Section 2.11. Exactly as it was proved in that theorem, we obtain that

$$(2.13.14) \qquad \sup_{0 \le t \le T} \int_{\mathbb{R}^n} \left| u_\varepsilon(x, t) - u_0 \left(x, t, \frac{t}{\varepsilon} \right) \right|^2 dx \to 0 \text{ as } \varepsilon \to 0$$

for any $T < \infty$. The proof of (2.13.1) from (2.13.14) is as follows. We have

$$\lim_{\varepsilon \downarrow 0} \int_0^T \int_{\mathbb{R}^n} (h(x, t), u_\varepsilon(x, t) - \bar{u}(x, t)) \, dx \, dt$$

$$= \lim_{\varepsilon \downarrow 0} \int_0^T \int_{\mathbb{R}^n} \left(h(x, t), u_\varepsilon(x, t) - u_0 \left(x, t, \frac{t}{\varepsilon} \right) \right) dx \, dt$$

$$+ \lim_{\varepsilon \downarrow 0} \int_0^T \int_{\mathbb{R}^n} (h(x, t), e^{Bt/\varepsilon} v_0(x, t)) \, dx \, dt.$$

The first term on the right goes to zero because of (2.13.14). The second goes to zero by the Riemann-Lebesgue lemma since, by (2.13.8), all terms in the integral have oscillatory factors. Proof I is complete. ∎

Although Proof I is direct, constructive and shows that we can compute even the rapidly oscillating parts of the solution and get strong convergence (2.13.14), it has the disadvantage that it uses too much information if all one wants is (2.13.1) **and nothing more**. We give next an indirect proof.

PROOF II. From the energy identity for (2.11.1) we conclude that

$$(2.13.15) \qquad \sup_{0 \le t \le T} \int_{\mathbb{R}^n} |u_\varepsilon(x, t)|^2 \, dx \le C, \qquad T < \infty$$

where C depends on f but not on ε. This means that the family u_ε, $0 < \varepsilon \le 1$ say, is weak star compact in $L^\infty(0, T; (L^2(\mathbb{R}^n))^N)$. Let u_ε denote a convergent subsequence. We shall identify the limit uniquely (so that then any sequence converges) as the solution $\bar{u}$ of (2.11.9).

For the identification we give a weak formulation of (2.11.1) and use adjoint expansion. Let $h(x, t)$ be as in the statement of the theorem.[33] Taking the inner product of (2.11.1), integrating over space and time and integrating by parts we obtain

$$(2.13.16) \quad \int_0^\infty \int_{\mathbb{R}^n} (h_t(x, t), u_\varepsilon(x, t)) \, dx \, dt + \int_{\mathbb{R}^n} (h(x, 0), f(x)) \, dx$$

$$= \int_0^\infty \int_{\mathbb{R}^n} \left(\sum_{p=1}^n \frac{\partial}{\partial x_p} (A^p h), u_\varepsilon \right) dx \, dt + \frac{1}{\varepsilon} \int_0^\infty \int_{\mathbb{R}^n} (Bh, u_\varepsilon) \, dx \, dt.$$

---

[33]But without $Ph = h$, yet.

Fix $h(t, x)$ as in the statement of the theorem and let $h(x, t0) = h_0(x)$. Now let $Ph(x, t) = h(x, t)$. Let

$$(2.13.17) \qquad A^\varepsilon = \frac{\partial}{\partial t} - \sum_{p=1}^{n} \frac{\partial}{\partial x_p} (A^p \cdot) - \frac{1}{\varepsilon} B,$$

which is the negative of the adjoint operator associated with (2.11.1). Let $\mathcal{A}$ be defined by

$$(2.13.18) \qquad \mathcal{A} = \frac{\partial}{\partial t} - \sum_{p=1}^{n} \frac{\partial}{\partial x_p} (\bar{A}^p \cdot) - \bar{B}.$$

This is meant to be a coordinate free representation of the adjoint of the operator (2.11.9) acting on $\bar{u}$. Acting on vector functions v with $Pv = v$, i.e. in the null space, it is exactly the formal adjoint of (2.11.9), by definition. Acting on v such that $Pv = 0$, i.e. off the null space, $\mathcal{A}v = 0$, by definition.

By perturbation expansion we construct a function $h_1(x, t)$ such that if we set

$$(2.13.19) \qquad h_\varepsilon(x, t) = h(x, t) + \varepsilon h_1(x, t),$$

then

$$(2.13.20) \qquad A^\varepsilon h_\varepsilon = \mathcal{A}h + O(\varepsilon), \qquad h_\varepsilon(x, 0) = h_0 + O(\varepsilon).$$

Note that $Ph_\varepsilon \neq h_\varepsilon$ here but $Ph = h$ by hypothesis. We leave it to the reader to verify that h_1 can be so constructed (a computation analogous to the one for u_1 in (2.11.15)). Now (2.13.16) (with $h \to h_\varepsilon$) and (2.13.19)–(2.13.20) give us

$$(2.13.21) \qquad 0 = \int_0^\infty \int_{\mathbb{R}^n} (A^\varepsilon h^\varepsilon, u^\varepsilon) \, dx \, dt + \int_{\mathbb{R}^n} (h_\varepsilon(x, 0), f(x)) \, dx$$
$$= \int_0^\infty \int_{\mathbb{R}^n} (\mathcal{A}h, u^\varepsilon) \, dx \, dt + \int_{\mathbb{R}^n} (h_0(x), f(x)) \, dx + O(\varepsilon),$$

where the constant in $O(\varepsilon)$ depends only on f and h and not on u^ε, in view of (2.13.15). Passing to the limit as $\varepsilon \to 0$ in (2.13.21) we obtain

$$(2.13.22) \qquad \int_0^\infty \int_{\mathbb{R}^n} (\mathcal{A}u, \bar{u}) \, dx \, dt + \int_{\mathbb{R}^n} (h_0(x), f(x)) \, dx = 0.$$

This is just the weak form of (2.11.9) and since $h(x, t)$ is in a determining class of test functions and (2.11.9) has a unique solution, the result (2.13.1) follows. ∎

REMARK 2.10. The second proof uses very few of our regularity hypotheses and only the weak existence and uniqueness for (2.11.1) and (2.11.9). This is a major advantage that is has over the first proof.

One objective of Theorem 2.8 is the clarification of what it really means to have static or high-frequency data, i.e. that in the wave propagation context nothing damps out and we must **either** construct the full oscillatory form of the expansion (as in (2.13.2)) **or** weaken the notion of convergence sufficiently to kill the oscillations. Notice that in Theorem 2.8, $f_\varepsilon(x) = f(x)$, i.e. we have still class A data.

We now pass to class B data (2.11.3) and consider (2.11.1) anew. We assume that $B(x)$ is skew symmetric but we do not specify its spectrum further for the moment. **We also assume for simplicity that $A^0 = A^0(x)$ but the A^p are constant matrices for $p = 1, 2, \ldots, N$.** Set

$$u_\varepsilon = e^{iS(x,t)/\varepsilon} v_\varepsilon(x, t).$$

Inserting this into (2.11.1) yields the following problem for v_ε:

$$(2.13.23) \quad \frac{1}{\varepsilon}\left[i\frac{\partial S}{\partial t} A^0 v_\varepsilon - i\sum_{p=1}^n \frac{\partial S}{\partial x_p} A^p v_\varepsilon - Bv_\varepsilon \right] + \left[A^0 \frac{\partial v_\varepsilon}{\partial t} - \sum_{p=1}^n A^p \frac{\partial v_\varepsilon}{\partial x_p} \right] = 0.$$

Let

$$(2.13.24) \qquad \frac{\partial S}{\partial t} = S_t = -\omega, \qquad \frac{\partial S}{\partial x_p} = S_p = k_p,$$

and consider the $N \times N$ matrix eigenvalue problem (ω is the eigenvalue parameter)

$$(2.13.25) \qquad \left(-\sum_{p=1}^n A^p k_p + iB \right) v = \omega A^0 v.$$

The variable $x \in \mathbb{R}^n$ is suppressed here; it is merely a parameter. Similarly, we regard $k = (k_p) \in \mathbb{R}^n$ as a parameter.

Recall that A^p, $p = 0, 1, 2, \ldots, N$ are real and symmetric, A^0 is positive definite and B is skew-symmetric so that iB is hermitian. Thus for each k and x there exist N real eigenvalues

$$\omega_1(k, x) \le \omega_2(k, x) \le \cdots \le \omega_N(k, x),$$

and N linearly independent eigenvectors

$$\phi_1(k, x), \phi_2(k, x), \ldots, \phi_N(k, x),$$

which can be chosen orthonormal relative to A^0

$$(2.13.26) \qquad (\phi_j(k, x), A^0(x)\phi_k(k, x)) = \delta_{kj}, \qquad k, j = 1, 2, \ldots, N.$$

The inner product here is the hermitian inner product

$$(2.13.27) \qquad (u, v) = \sum_{j=1}^N u_j \bar{v}_j.$$

As k and x vary so do the eigenvalues and the eigenfunctions. It is possible to show that the eigenvalues are continuous and the eigenfunctions can be constructed[34] so as to be measurable functions of x and k. In the neighborhood of a point (k, x) where an eigenvalue $\omega(k, x)$ is isolated, it and its corresponding eigenvector $\phi(k, x)$ (as constructed by Wilcox) are differentiable functions.

Now consider the eikonal equations
(2.13.28)
$$S_t^m + \omega_m(\nabla S^m, x) = 0, \qquad t > 0, \qquad S^m(x, 0) = \tilde{S}(x), \qquad m = 1, 2, \ldots, N.$$

---

[34]cf. C. Wilcox [**126**] and T. Kato [**53**].

We shall assume that the following holds regarding (2.13.28):

(2.13.29) The gradient of the initial data $\nabla \tilde{S}(x)$ varies smoothly within a neighborhood of a fixed vector k and in which the $\omega_m(k,x)$ are distinct. Moreover there is a $t_0 > 0$ such that (2.13.28) has a unique smooth solution in $0 \le t < t_0$ (no caustics in $0 \le t < t_0$) and $\nabla S^m(x,t)$ stays near k.

If we fix m and let[35] S satisfy (2.13.28) then (2.13.23) becomes

$$\frac{1}{\varepsilon}\left[-\sum_{p=1}^{n}\frac{\partial S}{\partial x_p}A^p + iB - \omega^0\right]v_\varepsilon - i\left[A^0\frac{\partial v_\varepsilon}{\partial t} - \sum_{p=1}^{n}A^p\frac{\partial v_\varepsilon}{\partial x_p}\right] = 0.$$

Letting $v_\varepsilon = v_0 + \varepsilon v_1 + \varepsilon^2 v_2 + \cdots$, we obtain, as usual,

(2.13.30) $$\left[-\sum_{p=1}^{n}\frac{\partial S}{\partial x_p}A^p + iB - \omega A^0\right]v_0 = 0,$$

(2.13.31) $$\left[-\sum_{p=1}^{n}\frac{\partial S}{\partial x_p}A^p + iB - \omega A^0\right]v_1 - i\left[A^0\frac{\partial v_0}{\partial t} - \sum_{p=1}^{n}A^p\frac{\partial v_0}{\partial x_p}\right] = 0, \qquad \text{etc.}$$

From (2.13.30) we conclude that

(2.13.32) $$v_0(x,t) = u_0(x,t)\phi(\nabla S(x,t),x),$$

where $u_0(x,t)$ is a scalar function to be determined and $\phi(x,k)$ is the eigenvector corresponding to the eigenvalue ω (subscript m omitted).

Next we use (2.13.32) in (2.13.31). Since ω is an isolated eigenvalue, the solvability condition yields one equation for $u_0(x,t)$, the usual transport equation:

(2.13.33) $$\left(\phi, A^0\frac{\partial}{\partial t}(u_0\phi)\right) - \sum_{p=1}^{n}\left(\phi, A^p\frac{\partial}{\partial x_p}(u_0\phi)\right) = 0.$$

Using the fact that

(2.13.34) $$\frac{\partial\omega(k,x)}{\partial k_p} = -(\phi(k,x), A^p\phi(k,x)), \qquad p = 1, 2, \ldots, N,$$

obtained by differentiating (2.13.25), it follows that $u_0(x,t)$ satisfies

(2.13.35) $$\frac{\partial u_0(x,t)}{\partial t} + \sum_{p=1}^{n}\frac{\partial\omega}{\partial k_p}(\nabla S(x,t),x)\frac{\partial u_0(x,t)}{\partial x_p}$$

$$+ \frac{1}{2}\sum_{p=1}^{n}\frac{\partial}{\partial x_p}\left(\frac{\partial\omega}{\partial k_p}(\nabla S(x,t),x)\right)u_0(x,t)$$

$$+ b(x,t)u_0(x,t) = 0, \qquad t > 0,$$

---

[35]We frequently drop the superscript m, to simplify notation, when discussing one mode at a time, i.e. when m is fixed.

where $b(x,t)$ is defined by

(2.13.36)
$$b(x,t) = \left(\phi(\nabla S(x,t), x), A^0(x) \frac{\partial}{\partial t} \phi(\nabla S(x,t), x) \right)$$
$$+ \frac{1}{2} \sum_{p=1}^{n} \left(\frac{\partial}{\partial x_p} \phi(\nabla S(x,t), x), A^p \phi(\nabla S(x,t), x) \right)$$
$$- \frac{1}{2} \sum_{p=1}^{n} \left(A^p \phi(\nabla S(x,t), x), \frac{\partial}{\partial x_p} \phi(\nabla S(x,t), x) \right).$$

It is easily verified that $b(x,t)$ is **purely imaginary**.

Let $C^m(k,x)$ be defined by

(2.13.37)
$$C^m(k,x) = \left[-\sum_{p=1}^{n} k_p A^p + iB(x) - \omega_m(k,x) A^0(x) \right]^{-1}$$

acting on vectors orthogonal to $\phi_m(k,x)$. We can write v_1, in view of (2.13.33), in the form

(2.13.38)
$$v_1(x,t) = iC(\nabla S(x,t), x) \left[A^0(x) \frac{\partial v_0(x,t)}{\partial t} - \sum_{p=1}^{n} A^p \frac{\partial v_0(x,t)}{\partial x_p} \right] + v_{10}(x,t)$$
$$\equiv v_{10}(x,t) + v_{11}(x,t),$$

where $v_{10}(x,t)$ is undetermined but we have

(2.13.39)
$$v_{10}(x,t) = u_1(x,t) \phi(\nabla S(x,t), x).$$

The function $u_1(x,t)$ is determined by the solvability condition for v_2 and it satisfies an **inhomogeneous transport equation** of the form (2.13.35), namely

(2.13.40)
$$\frac{\partial u_1(x,t)}{\partial t} + \sum_{p=1}^{n} \frac{\partial \omega}{\partial k_p}(\nabla S(x,t), x) \frac{\partial u_1(x,t)}{\partial x_p}$$
$$+ \sum_{p=1}^{n} \frac{\partial}{\partial x_p} \left(\frac{\partial \omega}{\partial k_p}(\nabla S(x,t), x) \right) u_1(x,t) + b(x,t) u_0(x,t)$$
$$+ i \left(\phi, \left(A^0 \frac{\partial}{\partial t} - \sum_{p=1}^{n} A^p \frac{\partial}{\partial x_p} \right) \cdot C \left(A^0 \frac{\partial v_0}{\partial t} - \sum_{q=1}^{n} A^q \frac{\partial v_0}{\partial x_q} \right) \right) = 0,$$

where v_0 is given by (2.13.32) with u_0 solution of (2.13.35).

Now we consider the assignment of initial data. The solution u_ε of (2.11.1) is asymptotically of the form

(2.13.41)
$$u_\varepsilon(x,t) = \sum_{m=1}^{N} e^{iS^m(x,t)/\varepsilon} [u_0^m(x,t) \phi_m(\nabla S^m(x,t), x)$$
$$+ \varepsilon(v_{11}^m(x,t) + u_1^m(x,t) \phi_m(\nabla S^m(x,t), x)) + O(\varepsilon^2)]$$

with entries as just described. To satisfy (2.11.3) we must have

$$(2.13.42) \qquad f(x) = \sum_{m=1}^{N} [u_0^m(x,0)\phi_m(\nabla \tilde{S}(x), x)$$
$$+ \varepsilon(v_{11}^m(x,0) + u_1^m(x,0)\phi_m(\nabla \tilde{S}(x), x)) + O(\varepsilon^2)].$$

Hence

$$(2.13.43) \qquad u_0^m(x,0) = (f(x), A^0(x)\phi_m(\nabla \tilde{S}(x), x)), \qquad m = 1,2,\ldots,N,$$

and

$$(2.13.44) \qquad u_1^m(x,0) = - \sum_{m'\neq m} (v_{11}^{m'}(x,0), A^0(x)\phi_{m'}(\nabla \tilde{S}(x), x)), \qquad \text{etc.}$$

We make two remarks regarding the asymptotic solution of (2.13.41) of (2.11.1), (2.11.3). First it is considerably more involved than the low frequency case as we must expect in view of the analysis of the Klein-Gordon equation in previous sections. **The proof of its validity is routine under (2.13.29)** which says that everything in sight is smooth so the energy estimate applies immediately.

Second, we would like to point out the significance of the inhomogeneous term in (2.13.40) and to carrying out the expansion correctly through order ε. Because of (2.13.29), each wave mode (i.e. term in (2.13.41) labeled by m) propagates independently of the other modes as far as the zero order term in ε goes. We have no mode coupling to zero order. The order ε term includes mode coupling effects and more specifically the way the energy is transferred from one mode to another; it is the lowest order term in which this occurs. Hence its significance. In Section 4.6 we shall see that sometimes one can elevate the mode coupling effect to order one in ε. The static case is, of course such an instance but there are also others.

2.14. WKB for dissipative symmetric hyperbolic systems. Consider the system of Section 2.12, i.e.

$$(2.14.1) \qquad \frac{\partial u_{j\varepsilon}}{\partial t} = \sum_{p=1}^{n} A_j^p(x)\frac{\partial u_{jt}}{\partial x_p} + \frac{1}{\varepsilon}\sum_{\ell=1}^{N} B_{j\ell}(x)u_{\ell\varepsilon}, \qquad t > 0,$$

with initial data

$$u_{j\varepsilon}(x,0) = f_j(x), \qquad j = 1,2,\ldots,N.$$

Here $f_j(x)$, $A_j^p(x)$ and $B_{j\ell}(x)$ are C^∞. Moreover the $f_j(x)$ have compact support and $0 \leq f_j(x) \leq 1$. We define

$$(2.14.2) \qquad 0 = \bigcup_{j=1}^{N}\{x \in \mathbb{R}^n \mid f_j(x) \neq 0\},$$

which is open with $\bar{G}$ compact. **The matrix $B_{j\ell}(x)$ is assumed to have**[36] **positive off diagonal elements and unit row sums.** Note also that the matrices A^p in (2.14.1) are assumed to be diagonal.

Theorem 2.6 of Section 2.12 implies that for each $t > 0$ and each $j = 1,2,\ldots,N$, $u_{j\varepsilon}(x,t) \to \bar{u}(x,t)$ uniformly in x where

$$(2.14.3) \qquad \frac{\partial \bar{u}(x,t)}{\partial t} = \sum_{p=1}^{n} \bar{A}^p(x)\frac{\partial \bar{u}(x,t)}{\partial x_p}, \qquad t > 0$$

---

[36]Hence a dissipative system (cf. beginning of Section 2.13)

and

(2.14.4)
$$\bar{u}(x,0) = \sum_{j=1}^{N} \bar{p}_j f_j(x).$$

Here $\bar{p}_j > 0$ ($\sum \bar{p}_j = 1$) are the components of the left null-vector of B (null-space has dimension one) and

(2.14.5)
$$\bar{A}^p(x) = \sum_{j=1}^{N} \bar{p}_j A_j^p(x).$$

Since $\bar{u}(x,0)$ has compact support G, the support G_t of $\bar{u}(x,t)$ will be precisely the set of points x that are translates of points of G along the orbits of the vector field $\bar{A}(x) = (\bar{A}^p(x))$. Note that G is the union of the supports of the $f_j(x)$, $j = 1, 2, \ldots, N$, i.e. as defined by (2.14.2).

On the other hand the union of the supports of $u_{j\varepsilon}(x,t)$ at $t > 0$, $\varepsilon > 0$ for $j = 1, 2, \ldots, N$, call it $\tilde{G}_t$, is in general much bigger than G_t (it is independent of ε). We have just seen that $u_{\varepsilon j}(x,t) \to 0$ as $\varepsilon \to 0$ for $x \in \tilde{G}_t - G_t$. How does $u_{j\varepsilon}(x,t)$ tend to zero in $\tilde{G}_t - G_t$? This is the question we examine in this section.[37]

The answer to this question is this. For each $x \in \tilde{G}_t$, $j = 1, 2, \ldots, N$ and $t > 0$

(2.14.6)
$$\lim_{\varepsilon \downarrow 0} \varepsilon \log u_{j\varepsilon}(x,t) = -\bar{S}(x,t),$$

where $\bar{S}(x,t)$ is the solution of a calculus of variations problem depending only on the support G of the initial data (and of course the coefficients in (2.14.1)). Moreover, $\bar{S}(x,t) > 0$ in $\bar{G}_t - G_t$ and $\bar{S} \equiv 0$ in G_t; hence $u_{j\varepsilon}$ decays to zero exponentially in $\tilde{G}_t - G_t$. **This result is valid for all $t > 0$ fixed, no matter what the support of the f_j's looks like.** Here we shall prove it only for ∂G smooth and t small. Our method fails in general on account of the formation of caustics as we have seen several times already. By employing certain probabilistic tools[38] a general proof can be given.

Recall that for problem (2.14.1) we have a maximum principle which will be used repeatedly below. The first step is to majorize and minorize the solution of (2.14.1) with the solution of two other problems; this may be called a penalty argument. The second step is to analyze these two problems by a WKB asymptotic expansion. In the end we remove the penalty and recover (2.14.6).

Let $h(x)$ be a C^∞ function such that

(2.14.7)
$$G = \{x \in \mathbb{R}^n \mid h(x) < 0\}, \qquad \partial G = \{x \in \mathbb{R}^n \mid h(x) = 0\},$$

and such that $|\nabla h(x)| \geq 1$ on ∂G.

Such a function exists because G is smooth by hypothesis. Fix a sufficiently large integer γ and for each $\alpha > 0$ put

$$g_\alpha(v) = \begin{cases} \alpha v^\gamma, & v \geq 0 \\ 0, & v < 0 \end{cases}$$

and

(2.14.8)
$$\tilde{S}_\alpha^{(1)}(x) = g_\alpha(h(x)).$$

---

[37]This problem is treated in great generality by M. I. Frĭdlin [**41**] by probabilistic methods.
[38]M. I. Freĭdlin [**41**], S. R. S. Varadhan [**118**], A. D. Ventcel [**119**] and [**120**], M. D. Donsker and S. R. S. Varadhan [**35**].

This is a function that is differentiable as often as desired (by taking γ large). Moreover

(2.14.9)
$$\tilde{S}_\alpha^{(1)}(x) = 0, \qquad \text{for } x \in G \cup \partial G,$$
$$\tilde{S}_\alpha^{(1)}(x) > 0, \qquad \text{for } x \notin G \cup \partial G.$$

Similarly let

(2.14.10)
$$\tilde{S}_{\alpha,\beta}^{(2)}(x) = g_\alpha(h(x) + \beta), \qquad \beta > 0,$$

which is also differentiable and has the property that if

$$G^{(\beta)} = \{x \in \mathbb{R}^n \mid h(x) + \beta < 0\} \subset G,$$

then

(2.14.11)
$$\tilde{S}_{\alpha,\beta}^{(2)}(x) = 0 \text{ for } x \in G^{(\beta)} \cup \partial G^{(\beta)},$$

and

(2.14.12)
$$\tilde{S}_{\alpha,\beta}^{(2)}(x) \to \infty \text{ as } \alpha \to \infty, \text{ with } x \in G - (G^{(\beta)} \cup \partial G^{(\beta)}) \text{ fixed and } \beta > 0 \text{ fixed.}$$

Consider next the eigenvalue problem[39]

(2.14.13)
$$\left(-\sum_{p=1}^n k_p A^p(x) + B(x) \right) \phi(k, x) = \omega(k, x)\phi(k, x),$$

for each $k \in \mathbb{R}^n$ and $x \in \mathbb{R}^n$ fixed. By the Perron-Frobenius theorem[40] we know that there exists an isolated maximal eigenvalue, which we call $\omega(k, x)$, which is smooth in k and x and

(2.14.14)
$$\omega(0, x) \equiv 0.$$

Moreover, $\omega(k, x)$ is a convex function of k for each x, being a maximal eigenvalue. There exists also a right eigenvector $\phi(k, x)$ whose components $\phi_j(k, x)$ are **positive** and smooth functions of k, x. A left eigenvector $\bar{p}(k, x)$ also exists with positive and smooth components. We have also that

(2.14.15)
$$\phi(0, x) \equiv 1, \qquad \bar{p}(0, x) = \bar{p}(x),$$

where 1 is the vector with all components equal to one and $\bar{p}(x) = (\bar{p}_j(x))$ is the left null vector of $B(x)$. We have the normalization

(2.14.16)
$$\sum_{j=1}^N p_j(k, x)\phi_j(k, x) = 1.$$

Let $S(x, t)$ denote the solution of the Hamilton-Jacobi equation

(2.14.17)
$$S_t + \omega(\nabla S, x) = 0, \qquad t > 0$$

with $S(x, 0)$ equal to $\tilde{S}^{(1)}(x)$ or $\tilde{S}_{\alpha,\beta}^{(2)}(x)$ and with superscripts and subscripts suppressed temporarily. This equation has a smooth solution up to some time $\tilde{t}_0$ when caustics form. We discuss $F(x, t)$ now in some detail.

Since $\omega(k, x)$ is convex we may define the Lagrangian function

(2.14.18)
$$L(x, \dot{x}) = \sup_k [(\dot{x}, x) - \omega(k, x)]$$

---

[39]Here $A^p(x) = (A_j^p(x)\delta_{ij})$ are diagonal matrices.
[40]Gantmacher [**45**], Vol. II.

which is the conjugate convex function corresponding to ω. Because of (2.14.14) and the convexity of ω it follows that

$$(2.14.19) \qquad L(x, \dot{x}) \geq 0 \text{ and } L(x, \dot{x}) = 0 \text{ only when } \dot{x} = \left.\frac{\partial \omega(k, x)}{\partial k}\right|_{k=0}.$$

From (2.14.13) we find by differentiation that (cf. (2.14.5))

$$(2.14.20) \qquad \left.\frac{\partial \omega(k, x)}{\partial k_p}\right|_{k=0} = -\bar{A}^p(x), \qquad p = 1, 2, \ldots, n.$$

Thus: the Lagrangian $L(x, \dot{x})$ is strictly positive along all trajectories $x(t)$ except along the characteristics of (2.14.3) (with reversed time in view of the minus sign in (2.14.20)) on which it vanishes. The Lagrangian $L(x, \dot{x}) = +\infty$ for directions $\dot{x}$ that are inaccessible and this follows automatically from (2.14.18).

Let $(x(t), k(t))$ be the solution of the Hamiltonian system

$$(2.14.21) \qquad \begin{aligned} \frac{dx(t)}{dt} &= \frac{\partial \omega(k(t), x(t))}{\partial k}, & x(0) &= x, \\ \frac{dk(t)}{dt} &= -\frac{\partial \omega(k(t), x(t))}{\partial x}, & k(0) &= \nabla \tilde{S}(x) \end{aligned}$$

where $\tilde{S}(x)$ is either $\tilde{S}_\alpha^{(1)}(x)$ or $\tilde{S}_{\alpha,\beta}^{(2)}(x)$. Then

$$(2.14.22) \qquad S(x(t), t) = \tilde{S}(x) + \int_0^t L(x(s), \dot{x}(s)) \, ds$$

satisfies the Hamilton-Jacobi equation (2.14.17). The function

$$(2.14.23) \qquad S(x, t) = \inf_{\xi(t)=x} \left\{ \tilde{S}(\xi(0)) + \int_0^t L(\xi(s), \dot{\xi}(s)) \, ds \right\},$$

with the infimum taken over all differentiable paths ξ that terminate at x at time t, is the unique smooth solution of (2.14.17) up to the time $\tilde{t}_0 > 0$ when caustics form.

Let

$$(2.14.24) \qquad \bar{S}(x, t) = \inf_{\substack{\xi(t)=x \\ \xi(0)\in G}} \left\{ \int_0^t L(\xi(s), \dot{\xi}(s)) \, ds \right\}, \qquad 0 \leq t < t_0$$

where $t_0 > 0$ is the first time uniqueness breaks term for this variational problem. If $\tilde{G}_t$ is the support set of (2.14.1) at time t, as defined before, then $\bar{S}(x, t) < \infty$ for $x \in \tilde{G}_t$, $\bar{S}(x, t) = +\infty$ for $x \notin \tilde{G}_t$ and $\bar{S}(x, t) > 0$ for $x \in \tilde{G}_t - G_t$, $\bar{S}(x, t) = 0$ for $x \in G_t$. We also note that, using superscript and subscript notation now,

$$(2.14.25) \qquad \lim_{\alpha \to \infty} S_\alpha^{(1)}(x, t) = \tilde{S}(x, t),$$

$$(2.14.26) \qquad \lim_{\beta \to 0} \lim_{\alpha \to \infty} S_{\alpha,\beta}^{(2)}(x, t) = \bar{S}(x, t),$$

for $t < t_0$.

We return now to the penalty argument. Let $t < t_0$ be fixed and let α sufficiently large be fixed also. We write the solution $u_{j\varepsilon}(x, t)$ of (2.14.1) in the form

$$(2.14.27) \qquad u_{j\varepsilon}(x, t) = e^{-S_\alpha^{(1)}(x,t)/\varepsilon} \phi_j(\nabla S_\alpha^{(1)}, x) v_{j\varepsilon}(x, t).$$

It follows that $v_{j\varepsilon}^\alpha(x,t)$ satisfies the problem

(2.14.28)

$$\frac{\partial v_{j\varepsilon}^\alpha}{\partial t} = \frac{1}{\phi_j} \sum_{p=1}^n A_j^p \frac{\partial(\phi_j v_{j\varepsilon}^\alpha)}{\partial x_p} + \frac{1}{\varepsilon} \frac{1}{\phi_j} \sum_{\ell=1}^N \left[B_j - \sum_{p=1}^n \frac{\partial S_\alpha^{(1)}}{\partial x_p} A_j^p \delta_{j\ell} - \omega \right] (\phi_\ell v_{\ell\varepsilon}^\alpha),$$

$$v_{j\varepsilon}^\alpha(x,0) = \frac{f_j(x)}{\phi_j(\nabla \tilde{S}_\alpha^{(1)}(x), x)}$$

From (2.14.27) we obtain

(2.14.29)
$$\varepsilon \log u_{j\varepsilon}(x,t) = -S_\alpha^{(1)}(x,t) + \varepsilon \log \phi_j(\nabla S_\alpha^{(1)}(x,t), x)$$
$$+ \varepsilon \log v_{j\varepsilon}^\alpha(x,t).$$

Because of (2.14.13) the system (2.14.28) obeys the maximum principle and therefore

(2.14.30)
$$\overline{\lim_{\varepsilon \downarrow 0}} \, \varepsilon \log u_{j\varepsilon}(x,t) \leq -S_\alpha^{(1)}(x,t).$$

Letting $\alpha \to \infty$ we obtain

(2.14.31)
$$\overline{\lim_{\varepsilon \downarrow 0}} \, \varepsilon \log u_{j\varepsilon}(x,t) \leq -\bar{S}(x,t), \qquad 0 < t < t_0$$

which is one half of the result (2.14.6).

To obtain the lower bound on $\varepsilon \log u_{j\varepsilon}$ we proceed as follows. Fix $0 < t < t_0$, $x \in \tilde{G}_t$ and $\beta > 0$. The parameter α is chosen large enough below and is then fixed also. If we replace $S_\alpha^{(1)}$ by $S_{\alpha,\beta}^{(2)}$ in (2.14.27) we obtain

(2.14.32)
$$\varepsilon \log u_{j\varepsilon}(x,t) \geq -S_{\alpha,\beta}^{(2)}(x,t) + \varepsilon \log \phi_j(\nabla S_{\alpha,\beta}^{(2)}(x,t), x)$$
$$+ \varepsilon \log v_{j\varepsilon}^{\alpha,\beta}(x,t).$$

Here $v_{j\varepsilon}^{\alpha,\beta}(x,t)$ is the solution of (2.14.28) with $S_\alpha^{(1)}$ replaced by $S_{\alpha,\beta}^{(2)}$.

Now choose α so large that if $x(s)$, $0 \leq s \leq t$, with $x(t) = x$, denotes the optimal path in (2.14.23) (with $\tilde{S}$ replaced by $\tilde{S}_{\alpha,\beta}^{(2)}$), then $x(0) \in G$. This is possible because β is positive and (2.14.25) holds. With α and β so fixed now we look at system (2.14.28). It is precisely in the form of Theorem 2.6, Section 2.12, so we can apply to it this theorem (an extra undifferentiated term of order one as $\varepsilon \to 0$ on the right side (2.14.28) is harmless). We find that $v_{j\varepsilon}^{\alpha,\beta}(x,t) \to \bar{v}^{\alpha,\beta}(x,t)$, $0 < t < t_0$ as $\varepsilon \to 0$ where $\bar{v}^{\alpha,\beta}(x,t)$ satisfies a system of the form (2.14.3). Now, however, the vector field $\bar{A}^p(x)$ is time dependent and instead of (2.14.20) we have (this is easily verified from (2.14.13))

(2.14.33)
$$\bar{A}^p(x,t) = -\left. \frac{\partial \omega(k,x)}{\partial k_p} \right|_{k=\nabla S_{\alpha,\beta}^{(2)}(x,t)}.$$

These observations and the selection of α large enough imply that $\bar{v}^{\alpha,\beta}(x,t) > 0$ for $x \in \tilde{G}_t$, $0 < t < t_0$, because the characteristics of the vector field $\bar{A}^p$ in (2.14.33) are precisely the optimal paths of (2.14.23).

Therefore from (2.14.32) we obtain

(2.14.34)
$$\varliminf_{\varepsilon \downarrow 0} \varepsilon \log u_{j\varepsilon}(x,t) \geq -S_{\alpha,\beta}^{(2)}(x,t).$$

In view of (2.14.26) this means that for $x \in \tilde{G}_t$ and $0 < t < t_0$

$$(2.14.35) \qquad \lim_{\varepsilon \downarrow 0} \varepsilon \log u_{j\varepsilon}(x, t) \geq -\bar{S}(x, t)$$

and hence, by (2.14.31)

$$(2.14.36) \qquad \lim_{\varepsilon \downarrow 0} \varepsilon \log u_{j\varepsilon}(x, t) = -\bar{S}(x, t), \qquad 0 < t < t_0.$$

We summarize the results as follows.

THEOREM 2.11. *The solution $u_{j\varepsilon}(x, t)$ of (2.14.1) under the hypothesis that G of (2.14.2) is smooth and $t < t_0$, satisfies (2.14.6) where the decay rate $\bar{S}(x, t)$ is the cost function of the variational problem (2.14.24). The Lagrangian $L(x, \dot{x})$ is defined by (2.14.18) and the Hamiltonian $\omega(k, x)$ is the maximal eigenvalue of (2.14.13)*

Recall that Theorem 2.6 of Section 2.12 dealt with the case where $B(x) = B_\varepsilon(x)$ and the null space became $r > 1$-dimensional when $\varepsilon = 0$. Theorem 2.11 can be extended to that more complicated situation.

2.15. Operator form of the WKB. Consider the symmetric hyperbolic system

$$(2.15.1) \qquad A^0(x) \frac{\partial u(x, t)}{\partial t} = \sum_{p=1}^{n} A^p \frac{\partial u(x, t)}{\partial x_p} + B(x)u, \qquad t > 0,$$

with some initial conditions. In Section 1.1 we saw how slowly varying coefficients lead to scaled systems like (2.15.1). The notion of slowly varying coefficients can take many other forms however. We shall now examine two of them.

The first corresponds to horizontally stratified media with slow variation of parameters in the horizontal direction. Let x_n be the coordinate perpendicular to the stratification and write $x = (x', x_n)$ with $x' \in \mathbb{R}^{n-1}$ the coordinates within a horizontal plane of stratification. Suppose that

$$(2.15.2) \qquad A^0 = A^0(\varepsilon x', x_n), \qquad B = B(\varepsilon x', x_n).$$

Define

$$(2.15.3) \qquad y = \varepsilon x', \qquad y \in \mathbb{R}^{n-1},$$

and rescale $t \to t/\varepsilon$. Then (2.15.1) becomes

$$(2.15.4)$$

$$A^0(y, x_n) \frac{\partial u_\varepsilon(y, x_n, t)}{\partial t} = \sum_{p=1}^{n-1} A^p \frac{\partial u_\varepsilon(y, x_n, t)}{\partial y_p}$$

$$+ \frac{1}{\varepsilon} \left[B(y, x_n) + A^n \frac{\partial}{\partial x_n} \right] u_\varepsilon(y, x_n, t), \qquad t > 0,$$

with some initial conditions and/or boundary conditions. Here as usual the matrices A^p are symmetric, A^0 is positive definite and B negative semi-definite.

Assume that

$$(2.15.5) \qquad A^n \text{ has } r \text{ positive eigenvalues and } N - r \text{ negative}$$
$$\text{eigenvalues (no zero eigenvalues).}$$

We may then, with no loss of generality, take A^n to be diagonal already with $+1$ in the first r diagonal positions and -1 in the last $N - r$ diagonal positions. A well posed mixed initial-boundary value problem for (2.15.4) over $\mathbb{R}^{n-1} \times [a, b]$, a, b finite, is the following:

(2.15.6) $\qquad u_\varepsilon(y, x_n, 0) = f_\varepsilon(y, x_n) \qquad \text{(given)}$

(2.15.7) $\qquad u_{j\varepsilon}(y, b, t) = 0, \qquad j = 1, 2, \ldots, r, \qquad y \in \mathbb{R}^{n-1},$

$\qquad\qquad u_{j\varepsilon}(y, a, t) = 0, \qquad j = r + 1, \ldots, N, \qquad y \in \mathbb{R}^{n-1}.$

One easily verifies that the following energy identity holds (when $f_\varepsilon(y, x_n)$ has compact support in y and satisfies (2.15.7)):

(2.15.8) $\quad \dfrac{1}{2} \displaystyle\int_a^b \int \sum_{j\ell=1}^N u_{j\varepsilon}(y, x_n, t) A_{j\ell}^0(y, x_n) u_{\ell\varepsilon}(y, x_n, t) \, dy \, dx_n$

$$- \frac{1}{\varepsilon} \int_0^t \int_a^b \int \sum_{j,\ell=1}^N u_{j\varepsilon}(y, x_n, s) B_{j\ell}(y, x_n) u_{\ell\varepsilon}(y, x_n, s) \, dy \, dx_n \, ds$$

$$+ \frac{1}{2\varepsilon} \int_0^t \int \left[\sum_{j=r+1}^N u_{j\varepsilon}^2(y, b, s) + \sum_{j=1}^r u_{j\varepsilon}^2(y, a, s) \right] dy \, ds$$

$$= \frac{1}{2} \int_a^b \int \sum_{j,\ell=1}^N f_{j\varepsilon}(y, x_n) A_j^0(y, x_n) f_{\ell\varepsilon}(y, x_n) \, dy \, dx_n.$$

Using this identity, uniqueness follows immediately from (2.15.4), (2.15.6) and (2.15.7) as well as an ε independent bound. Existence of a smooth solution also follows but is more complicated.[41]

A problem of the form (2.15.4), (2.15.6) and (2.15.7) arises in underwater acoustics and a detailed asymptotic analysis (including extensive numerical work) has been given by Weinberg-Burridge (J. Acoust. Soc. America, 55 (1974), 63-79). Other problems of the same form have been analyzed by Shen and J. B. Keller [105]. As we shall see the analysis of problems with periodic structure leads to an equation of the form (2.15.4). This explains why we consider this problem here.

Assume that B is skew-symmetric. Then the asymptotic analysis of (2.15.4) follows exactly the pattern of analysis in Section 2.13: both as regards the static problem described in Theorem 2.8 and the high frequency expansion (2.13.41). A major role is now played by the eigenvalue problem

(2.15.9) $\quad \left(\displaystyle\sum_{p=1}^{n-1} k_p A^p - iB(y, x_n) - iA^n \frac{\partial}{\partial x_n} \right) \phi(x_n; k, y)$

$$= \omega(k, y) A^0(y, x_n) \phi(x_n; k, y),$$

where $k \in \mathbb{R}^{n-1}$ and $y \in \mathbb{R}^{n-1}$ are parameters. The formal operator on the left is over $L^2(a, b; \mathbb{R}^N)$ with boundary conditions as in (2.15.7) and has a unique self-adjoint extension. For each k, y fixed, the spectrum is discrete and the eigenvalues

---

[41]K. O. Friedrichs [44], H. O. Kreiss [62].

depend smoothly on k, y near points where they are distinct. Eigenvectors can also be selected to be smoothly dependent on k, y locally.[42]

Before going to the second example we note that it is appropriate to refer to (2.15.4), (2.15.6), (2.15.7) as **horizontal propagation** in a **horizontally stratified medium**. It is also of interest to study **vertical propagation in a horizontally stratified medium**. This is a scattering problem whereby A^0 and B are assumed to be independent of x_n for $|x_n|$ large and at $t = 0$ a wave with support in the lower homogeneous region is prescribed traveling in the $+x_n$ direction thereby creating at time t later a reflected waveform (if t is large enough) in the upper homogeneous region.

The second problem corresponds to a **waveguide structure** with slow variation of parameters in the direction of its axis. Let x_n now be the coordinate along the waveguide structure and let $x' \in \mathbb{R}^{n-1}$ be the transverse coordinates. Suppose that

$$(2.15.10) \qquad A^0 = A^0(x', \varepsilon x_n), \qquad B = B(x', \varepsilon x_n).$$

Define

$$(2.15.11) \qquad y = \varepsilon x_n, \qquad y \in \mathbb{R}^1,$$

and rescale $t \to t/\varepsilon$. Then (2.15.1) becomes

(2.15.12)

$$A^0(x', y)\frac{\partial u_\varepsilon(x', y, t)}{\partial t} = A^n \frac{\partial u_\varepsilon(x', y, t)}{\partial y}$$
$$+ \frac{1}{\varepsilon}\left[B(x', y) + \sum_{p=1}^{n-1} A^p \frac{\partial}{\partial x'_p}\right] u_\varepsilon(x', y, t), \qquad t > 0,$$

with some initial and boundary conditions. Suppose that $x' \in \mathcal{O}$ where $\mathcal{O} \subset \mathbb{R}^{n-1}$ is a bounded open set. We introduce boundary conditions on $\partial\mathcal{O}$ so that the operator

$$(2.15.13) \qquad -i\sum_{p=1}^{n-1} A^p \frac{\partial}{\partial x'_p} - iB(x', y) + kA^n,$$

with $k \in \mathbb{R}^1$, $y \in \mathbb{R}^1$ parameters, has a unique selfadjoint extension on $L^2(\mathcal{O}; \mathbb{R}^N)$. Physically, there are many possibilities — metallic waveguides, dielectric waveguides, etc., and (2.15.13) may be considered on all of $\mathbb{R}^{n-1}$ with interface conditions on $\partial\mathcal{O}$. When the spectrum of (2.15.13) (relative to A^0) is **discrete** for each k, y, the asymptotic analysis goes exactly as in Section 2.13. Otherwise new phenomena arise which we shall not discuss here (since that has no immediate analog in periodic structures). An example of a waveguide problem with periodic structure is given inSection 4.12.

---

[42]The work of Wilcox [**127**] applies directly to (2.15.9) even though it is aimed at the analogous problem for Bloch waves.

3. Spectral theory for differential operators with periodic coefficients

3.1. The shifted cell problems for a second order elliptic operator.

Let A be the formal differential operator

$$(3.1.1) \qquad A = -\sum_{p,q=1}^{N} \frac{\partial}{\partial y_p} \left(a_{pq}(y) \frac{\partial}{\partial y_q} \cdot \right) + W(y),$$

defined on $C_0^\infty(\mathbb{R}^n)$ with $(a_{pq}(y))$ symmetric, smooth and uniformly positive definite matrix of periodic functions of period 2π in each variable (i.e. functions of $2\pi Y$ in the notation of Section 1.2). Let $W(y) \geq 0$ be a real-valued 2π-periodic smooth function. We shall find the spectral resolution of the closure of this operator in $L^2(\mathbb{R}^n)$ (complex-valued). The spectral resolution is given in terms of the Bloch waves associated with A as we now describe.

Let Y be the unit n-dimensional torus and let $2\pi\mathbb{Z}^n$ be the 2π-lattice in $\mathbb{R}^n$ centered at the origin. Let $H^1(2\pi Y)$ be the Hilbert space of functions on $2\pi Y$ that have square integrable derivatives. For each $k \in Y$ define

$$(3.1.2) \qquad A(k) = -\sum_{p,q=1}^{n} \left(\frac{\partial}{\partial y_p} + ik_p \right) \left[a_{pq}(y) \left(\frac{\partial}{\partial y_p} + ik_q \right) \cdot \right] + W(y).$$

Consider the eigenvalue problem, **the shifted cell problem**,

$$(3.1.3) \qquad A(k)\phi = \omega^2 \phi, \qquad \phi \in H^1(2\pi Y),$$

for each $k \in Y$. Since $W \geq 0$ the operator $A(k)$ is nonnegative and hence ω^2 in (3.1.3) is used consistently. For any $f \in H^1(2\pi Y)$ and with $(,)$ the usual inner product over $2\pi Y$ we have

$$(3.1.4) \quad (A(k)f, f) = \int_{2\pi Y} \sum_{p,q=1}^{n} a_{pq}(y) \left(\frac{\partial}{\partial y_p} + ik_p \right) f(y) \overline{\left(\frac{\partial}{\partial y_q} + ik_q \right) f(y)} \, dy$$

$$+ \int_{2\pi Y} W(y) f(y) \bar{f}(y) \, dy$$

$$\geq \alpha_1 \int_{2\pi Y} \sum_{p=1}^{n} \left| \frac{\partial f(y)}{\partial y_p} \right|^2 dy - \alpha_2 (f, f),$$

for all $k \in Y$ with α_1 and α_2 positive constants that do not depend on k.

From (3.1.4) it follows that the essentially selfadjoint operator $A(k)$ has for each $k \in Y$ a compact resolvent $(\lambda + A(k))^{-1}$, $\Re\lambda \neq 0$. As a consequence, for each k the eigenvalue problem (3.1.3) has a discrete sequence of eigenvalues

$$(3.1.5) \qquad \omega_0^2(k) \leq \omega_1^2(k) \leq \omega_3^2(k) \leq \cdots,$$

with corresponding eigenfunctions, the Bloch waves,

$$(3.1.6) \qquad \phi_1(y; k), \phi_2(y; k), \ldots,$$

which we take to be orthonormalized, i.e.

$$(3.1.7) \qquad (\phi_m, \phi_{m'}) = \delta_{mm'}.$$

The eigenfunctions are smooth functions y and they are complete in the space $L^2(2\pi Y)$ (complex-valued).

The dependence of $\omega_m^2(k)$ and $\phi_m(y; k)$, $m \geq 0$, on k requires special attention; it has been analyzed by Wilcox [127]. He finds that the ω_m's are real analytic functions of k everywhere except on Y subsets of measure zero where their multiplicity changes; they are continuous for all k. The eigenfunctions $\phi_m(y; k)$ can be constructed so as to be measurable functions of k, in general; in the component of null sets in Y they are analytic functions of k with values in $C(2\pi Y)$, the continuous functions on $2\pi Y$.

3.2. The Bloch expansion theorem. All considerations regarding asymptotics in this chapter are local in character. It is always assumed that the relevant k's are restricted so that the ω_m's and ϕ_m's are smooth. The following theorem, however, requires only the measurability of the ϕ_m's.

THEOREM 3.1 (Bloch expansion). *Let $f \in L^2(\mathbb{R}^n)$ (complex-valued). Then*

$$(3.2.1) \qquad f(y) = \lim_{N\uparrow\infty} \int_Y dy \sum_{m=0}^N \hat{f}_m(k) e^{ik\cdot y} \phi_m(y; k)$$

and

$$(3.2.2) \qquad \hat{f}_m(k) = \lim_{N\uparrow\infty} \int_{|y|\leq N} dy\, f(y) e^{-ik\cdot y} \bar{\phi}_m(y; k).$$

Moreover, Parseval's identity holds

$$(3.2.3) \qquad \int_{\mathbb{R}^n} |f(y)|^2\, dy = \int_Y dk \sum_{m=0}^\infty |\hat{f}_m(k)|^2.$$

PROOF (GELFAND [**46**], ODEH-KELLER [**88**]). In view of (3.2.3) it is enough to show that (3.2.1)–(3.2.3) hold with $f \in \mathscr{S}(\mathbb{R}^n)$, the space of C^∞ rapidly decreasing functions on $\mathbb{R}^n$.

For such an f, the function

$$(3.2.4) \qquad \tilde{f}(y; k) = \sum_{\gamma\in 2\pi\mathbb{Z}^n} f(y+\gamma) e^{-ik\cdot(y+\gamma)}, \qquad k \in Y$$

belongs to $2\pi Y$. We expand it in terms of $\{\phi_m(y; k)\}$

$$(3.2.5) \qquad \tilde{f}(y; k) = \sum_{m=0}^\infty f_m(k)\phi_m(y; k)$$

where

$$(3.2.6) \qquad \hat{f}_m(k) = \int_{2\pi Y} dy\, f(y; k)\bar{\phi}_m(y; k)$$

$$= \int_{2\pi Y} dy \sum_{\gamma\in 2\pi\mathbb{Z}^n} f(y+\gamma) e^{-i(y+\gamma)\cdot k} \bar{\phi}_m(y+\gamma; k)$$

$$= \int_{\mathbb{R}^n} dy\, f(y) e^{-ik\cdot y} \bar{\phi}_m(y; k).$$

From (3.2.4) and (3.2.5) we also have

$$(3.2.7) \qquad f(y) = \int_Y dk\, e^{ik\cdot y} \tilde{f}(y; k)$$

$$= \int_Y dk \sum_{m=0}^\infty \hat{f}_m(k) e^{ik\cdot y} \phi_m(y; k).$$

This proves (3.2.1) and (3.2.2) for $f \in \mathscr{S}(\mathbb{R}^n)$.

To prove (3.2.3) we note that from (3.2.5) we have

$$(3.2.8) \qquad \int_{2\pi Y} dy \, |\tilde{f}(y;k)|^2 = \sum_{m=0}^{\infty} |\hat{f}_m(k)|^2.$$

Using (3.2.4) on the left in (3.2.8), integrating over $k \in Y$ and rearranging yields (3.2.3). ∎

As a simple example of this theorem consider the case $a_{pq} = \delta_{pq}$ and $W = 0$ so that $A = -\Delta$. Evidently

$$(3.2.9) \qquad \phi_m(y;k) = (2\pi)^{-n/2} e^{im \cdot y}, \qquad m \in \mathbb{Z}^n,$$

and ϕ_m does not depend on k. The eigenvalues are

$$(3.2.10) \qquad \omega_m^2(k) = |k + m|^2, \qquad m \in \mathbb{Z}^n, \qquad k \in Y.$$

The theorem above is now the Plancherel-Parseval theorem for the Fourier transform in $\mathbb{R}^n$ and it is deduced from the corresponding result for Fourier series.

We note that (3.1.1) and (3.1.2) are connected by

$$Ae^{ik \cdot y} = e^{ik \cdot y} A(k),$$

which explains the terminology "shifted cell problem." Thus,

$$(3.2.11) \qquad Ae^{ik \cdot y} \phi_m(y;k) = e^{ik \cdot y} \omega_m^2(k) \phi_m(y;k).$$

The above theorem[43] then and (3.2.11) give the spectral resolutoin of A for which $f \in \mathscr{S}(\mathbb{R}^n)$ is

$$(3.2.12) \qquad Af(y) = \int_Y dk \sum_{m=0}^{\infty} \hat{f}_m(k) e^{ik \cdot y} \omega_m^2(k) \phi_m(y;k).$$

3.3. Bloch expansion for the acoustic equation. For the acoustic system (2.10.7) with ρ and μ 2π-periodic, the previous results apply with slight modifications. The operator at hand has the form

$$\begin{pmatrix} \rho^{-1} & 0 \\ 0 & \mu \end{pmatrix} \begin{pmatrix} 0 & \nabla \\ \nabla \cdot & 0 \end{pmatrix}$$

acting on 4-vector functions (u, p) on $\mathbb{R}^3$. If we replace (u, p) by $e^{ik \cdot y}(u, p)$, the shifted cell problem analogous to (3.1.3) becomes

$$(3.3.1) \qquad (\nabla + ik)\phi = -i\omega\rho\psi, \qquad\qquad \phi \in H^2(2\pi Y),$$
$$(\nabla + ik) \cdot \psi = -i\omega\mu^{-1}\phi, \qquad\qquad \psi \in (L^2(2\pi Y))^3.$$

In fact for ϕ we have the eigenvalue problem

$$(3.3.2) \qquad -(\nabla + ik) \cdot [\rho^{-1}(\nabla + ik)\phi] = \omega^2 \mu^{-1}\phi,$$

which is in a slightly more general form than (3.1.3) on account of the μ^{-1} on the right. however, upon defining a **weighted** inner product with weight μ^{-1}, all goes as before and we obtain the Bloch eigenfunctions and eigenvalues for (3.3.2)

$$\omega_0^2(k) \le \omega_1^2(k) \le \cdots,$$
$$\phi_0(y;k), \phi_1(y;k), \ldots$$

---

[43] See also section 3.6 for further remarks regarding this theorem.

From (3.3.1) we also find that

$$(3.3.3) \qquad \psi_m(y;k) = \frac{i}{\omega_m(k)\rho(y)}(\nabla + ik)\phi_m(y;k).$$

This definition of ψ_m holds for all $m = 0, 1, 2, \ldots$, and all $k \in Y$ **except** for $\psi_0(y;0)$ which is not defined since here $\omega_0^2(0) = 0$, and

$$(3.3.4) \qquad \phi_0(y;0) = \left(\int_{2\pi Y} \mu^{-1}(y)\,dy\right)^{-1/2}.$$

But from (3.3.1) we see that when $k = 0$, $\omega = 0$

$$(3.3.5) \qquad \phi_0(y;0) = \nabla \wedge \tilde{\psi}(y)$$

where $\tilde{\psi}(y)$ is an arbitrary differentiable vector function.

This abnormality at $k = 0$ for the $m = 0$ mode makes the static problem quite interesting as we shall see in Section 4.10.

3.4. Bloch expansion for Maxwell's equation. The shifted cell problem for Maxwell's system (2.10.17) (with $\sigma \equiv 0$) analogous to (3.1.3) has the following form (analogous to (3.3.1))

$$(3.4.1) \qquad (\nabla + ik) \wedge H = -i\omega\varepsilon E, \qquad\qquad (\nabla + ik) \cdot (\varepsilon E) = 0,$$
$$-(\nabla + ik) \wedge E = -i\omega\mu H, \qquad\qquad (\nabla + ik) \cdot (\mu H) = 0,$$

with $(E, H) \in (H^1(2\pi Y))^6$ and $\mu(y), \varepsilon(y)$ smooth symmetric positive definite 3×3 matrices of 2π-periodic functions.

Eliminating H in (3.4.1) we obtain the following eigenvalue problem for E ($k \in Y$ is fixed)

$$(3.4.2) \qquad (\nabla + ik) \wedge [\mu^{-1}(\nabla + ik) \wedge E] = \omega^2 \varepsilon E, \qquad E \in (H^1(2\pi Y))^3,$$
$$\text{and } (\nabla + ik) \cdot (cE) = 0.$$

Once E has been determined then

$$(3.4.3) \qquad H = \frac{\mu^{-1}}{i\omega}(\nabla + ik) \wedge E,$$

provided that $\omega \neq 0$. This can happen only when $k = 0$.

For $k \in Y$ fixed, the eigenvalue problem (3.4.2), relative to the inner product over $(L^2(2\pi Y))^3$ weighted by (the positive definite matrix) ε, has discrete eigenvalues and corresponding eigenfunctions as did (3.1.3). The static case $k = 0$, $\omega = 0$ is singular as it was for (3.3.1) and this leads to some interesting phenomena as we discuss in Section 4.11.

3.5. The dynamo problem. This problem is treated in detail by Childress [26] and G. O. Roberts [99] and is as follows (kinematic dynamo). Given a 2π-periodic vector function $u(y)$ in $(C^\infty(2\pi Y))^3$ the magnetic field $B(y, t)$ satisfies the initial value problem

$$(3.5.1) \qquad \frac{\partial B}{\partial t} = \nabla \wedge (u \wedge B) + \lambda \Delta B, \qquad t > 0, \qquad \lambda > 0,$$
$$B(y, 0) = B_0(y), \qquad \nabla \cdot B_0 = 0.$$

The shifted cell problem corresponding to (3.5.1) is now the eigenvalue problem

$$(3.5.2) \qquad \lambda(\nabla + ik)^2 U + (\nabla + ik) \wedge (u \wedge U) = \omega U,$$

$$\text{with } U \in (H^1(2\pi Y))^3, \qquad k \in Y \text{ fixed.}$$

The operator on the right of (3.5.2) has compact resolvent, hence discrete spectrum but is not selfadjoint. The dynamo problem consists of showing that (3.5.1) has solutions whose L^2 norm over $\mathbb{R}^n$ grows with t. While one can develop spectral results as is done by Roberts [99], the conclusion that one seeks is an essentially static phenomenon (in our terminology). It can be therefore treated directly by expansion processes which are much more flexible than spectral analysis and work, in particular, even in the presence of boundaries. We shall not, however, treat this problem in detail here.

3.6. Some nonselfadjoint problems. Consider the general elliptic operator

$$(3.6.1) \qquad A = \sum_{p,q=1}^{n} a_{pq}(y) \frac{\partial^2}{\partial y_p \partial y_q} + \sum_{p=1}^{n} b_p(y) \frac{\partial}{\partial y_p} + c(y),$$

and $(a_{pq}(y))$, $(b_p(y))$ and $c(y)$ smooth functions on $2\pi Y$ and (a_{pq}) positive definite and $c(y) \leq 0$. The shifted cell problem for this operator is not a selfadjoint problem and hence an expansion for it is not easily constructed. In Section 5.8 we shall be concerned however with the shifted cell problem for **imaginary** k, i.e.

$$(3.6.2) \qquad A(k) = \sum_{p,q=1}^{n} a_{pq} \left(\frac{\partial}{\partial y_p} - k_p \right) \left(\frac{\partial}{\partial y_q} - k_q \right)$$

$$+ \sum_{p=1}^{n} b_p(y) \left(\frac{\partial}{\partial y_p} - k_p \right) + c(y),$$

acting on $C^\infty(2\pi Y)$ with $k \in \mathbb{R}^n$ fixed. The operator (3.6.2) has a closure in $C(2\pi Y)$ which generates a continuous semigroup, but not of contractions, on $C(2\pi Y)$ which is however positivity preserving (strong maximum principle).

From the Perron-Frobenius theory[44] we conclude that $A(k)$ has an isolated maximal eigenvalue $\omega(k)$ (which is a convex function of k) and strictly positive right and left eigenfunctions corresponding to it. We shall use this result in Section 5.8.

In the remainder of this section we shall give a general representation theorem for the solution of equations with periodic coefficients that **does not depend on spectral theory**.

Let A be a differential operator of order m with 2π-periodic coefficients. Using multi-index notation we write

$$(3.6.3) \qquad A = \sum_{|\alpha| \leq m} a_\alpha(y) D^\alpha,$$

where

$$D^\alpha = \frac{\partial^{|\alpha|}}{\partial y_1^{\alpha_1} \cdots \partial y_n^{\alpha_n}}$$

and

$$(3.6.4) \qquad a_\alpha(y) \in C^\infty(2\pi Y).$$

---

[44]cf. for example Kreĭn-Rutman [61], T. Harris [50].

Let $f \in \mathscr{S}(\mathbb{R}^n)$, hence not a periodic function, and consider the problem

$$(3.6.5) \qquad\qquad Au = f.$$

Assume for the moment that (3.6.5) has a solution $u \in \mathscr{S}(\mathbb{R}^n)$ so that the calculations that follow make sense.

With $k \in Y$ fixed we define $A(k)$, as in Section 3.2, by

$$(3.6.6) \qquad\qquad e^{-ik \cdot y} A = A(k) e^{-ik \cdot y}$$

so that

$$(3.6.7) \qquad\qquad A(k) = \sum_{|\alpha| \le m} a_\alpha(y)(D + ik)^\alpha$$

with

$$(D + ik)^\alpha = \left(\frac{\partial}{\partial y_1} + ik_1 \right)^{\alpha_1} \cdots \left(\frac{\partial}{\partial y_m} + ik_n \right)^{\alpha_n}.$$

For each $z \in \mathbb{R}^n$ let T_z be the translation by z operator

$$(3.6.8) \qquad\qquad (T_z f)(y) = f(y + z).$$

Since $a_\alpha \in C^\infty(2\pi Y)$, i.e. are periodic and smooth, it follows that

$$(3.6.9) \qquad\qquad T_\gamma A = A T_\gamma \text{ and } T_\gamma A(k) = A(k) T_\gamma, \ \forall \gamma \in 2\pi \mathbb{Z}^n,$$

which means that A and $A(k)$ commute with translations by any lattice vector (vector γ in $2\pi \mathbb{Z}^n$).

With $k \in Y$ fixed we multiply (3.6.5) by $e^{-ik \cdot y}$ and use (3.6.6). This yields

$$(3.6.10) \qquad\qquad A(k) e^{-ik \cdot y} u(y) = e^{-ik \cdot y} f(y).$$

Now we apply T_γ to both sides of (3.6.10), we use (3.6.9) and then sum over all $\gamma \in 2\pi \mathbb{Z}^n$. We obtain the following equation

$$(3.6.11) \qquad\qquad A(k) \tilde{u}(y; k) = \tilde{f}(y; k).$$

Here

$$(3.6.12) \qquad\qquad \tilde{u}(y; k) = \sum_{\gamma \in 2\pi \mathbb{Z}^n} e^{-ik \cdot (y+\gamma)} u(y + \gamma)$$

and

$$(3.6.13) \qquad\qquad \tilde{f}(y; k) = \sum_{\gamma \in 2\pi \mathbb{Z}^n} e^{-ik \cdot (y+\gamma)} f(y + \gamma).$$

By hypothesis the sums are convergent and hence both $\tilde{u}$ and $\tilde{f}$ are 2π-periodic functions.

Let us assume that problem (3.6.11) as a problem on the torus $2\pi Y$ has a solution which we write as

$$(3.6.14) \qquad\qquad \tilde{u}(y; k) = (A^{-1}(k) \tilde{f}(\cdot; k))(y).$$

Multiplying (3.6.14) by $e^{ik \cdot y}$, integrating with respect to k over Y and using (3.6.12) we obtain

$$(3.6.15) \qquad u(y) = \int_Y e^{ik \cdot y} (A^{-1}(y) \tilde{f}(\cdot; k))(y) \, dk$$

$$= \int_Y e^{ik \cdot y} A^{-1}(k) \left[\sum_{\gamma \in 2\pi \mathbb{Z}^n} e^{-ik \cdot (y+\gamma)} f(y + \gamma) \right] dk.$$

We summarize and make the above precise in the following theorem which is typical of what one may expect in other similar situations.

THEOREM 3.2. *Let A be defined by (3.6.3) with $C^\infty(2\pi Y)$ coefficients, acting on $L^2(\mathbb{R}^n)$. For each $k \in Y$ let $A(k)$ be defined by (3.6.7) acting on $L^2(2\pi Y)$. Assume that there is a constant C independent of $k \in Y$ such that*

$$(3.6.16) \qquad |A^{-1}(k)|_{L^2(2\pi Y)} \le C.$$

Assume also that for some complex number λ, $(A-\lambda)^{-1}$ exists and is a bounded operator on $L^2(\mathbb{R}^n)$. Assume finally that $A^{-1}(k)$ is strongly measurable as a function of $k \in Y$.

Then equation (3.6.5) has a unique solution $u \in L^2(\mathbb{R}^n)$ for each $f \in L^2(\mathbb{R}^n)$. Moreover this solution is given by the representation formula (3.6.15).

PROOF. We note first that u of (3.6.15) is well defined as an element of $L^2(\mathbb{R}^n)$ because of our hypotheses. Moreover the operator A of (3.6.3) is closed since its resolvent set is non empty. Let $\mathcal{D}(A)$ denote the domain of A in $L^2(\mathbb{R}^n)$.

With $f \in L^2(\mathbb{R}^n)$ given, let $f_m \in \mathscr{S}(\mathbb{R}^n)$ be a sequence such that

$$(3.6.17) \qquad f_m \to f \text{ in } L^2(\mathbb{R}^n) \text{ as } m \to \infty.$$

Define $u_m(y)$ by (3.6.15) with f replaced by $f_m(y)$. It is easily seen, since A is closed so we can take it inside the integral, that

$$(3.6.18) \qquad Au_m = f_m.$$

Thus $u_m \in \mathcal{D}(A)$ and $Au_m \to f$ in $L^2(\mathbb{R}^n)$ as $m \to \infty$. Now

$$
|u - u_m|^2 = \int_{2\pi Y} \left| \int_Y dk\, e^{ik\cdot y} A^{-1}(k)(\tilde{f} - \tilde{f}_m) \right|^2 dy
$$

$$
\le C \int_{2\pi Y} \left(\int_Y dk\, |\tilde{f} - \tilde{f}_m|^2 \right) dy
$$

$$
= C \int_{\mathbb{R}^n} |f - f_m|^2\, dy
$$

and hence $|u - u_m|_{L^2(\mathbb{R}^n)} \to 0$ as $m \to \infty$. Since A is closed we conclude that $Au = f$ and u is indeed the solution of our problem (3.6.5). The proof is complete. ∎

The main point of this elementary result is this: problem (3.6.5) is effectively reduced to the solution of **a family of shifted cell problems** (3.6.11), parametrized by k which belongs to the compact set Y; the final result is obtained by integration over k, i.e. by (3.6.15).

4. Simple applications of the spectral expansion

4.1. Lattice waves. To motivate better the physical significance of the results in later sections we analyze here a very simple example for which all computatoins can be carried out explicitly by elementary means.[45]

---

[45]This is done in practically all solid state physics books, e.g. Ziman [130], as well as in Brillouin [22].

Let $u_p(t)$, $p \in \mathbb{Z}$, be the displacements from equilibrium of oscillators located at p, with mass M_p and coupled to each other by linear springs. The equations of motion are (in suitable units)

$$(4.1.1) \qquad M_p \frac{d^2 u_p(t)}{dt^2} = \frac{1}{2} u_{p+1}(t) - u_p(t) + \frac{1}{2} u_{p-1}(t), \qquad p \in \mathbb{Z},$$

$$u_p(0) = u_p, \qquad \frac{du_p(0)}{dt} = V_p, \qquad U_p, V_p \equiv 0 \text{ for } |p| \text{ large.}$$

We shall assume that there are two masses M_0 and M_1 periodically distributed: M_0 at the even locations, M_1 at the odd locations. The system has periodic structure with period 2, i.e. $M_{p+2} = M_p$.

First we construct the spectral resolution for the operator in (4.1.1). We put

$$u_p = e^{i(kp - \omega t)} \phi(p, k),$$

with $k \in [0,]1$ and $\phi(p + 2; k) = \phi(p; k)$ in (4.1.1). This leads to the "shifted cell" eigenvalue problem

$$(4.1.2) \qquad -M_0 \omega^2(k) \phi(0; k) = \phi(1; k) \cos \pi k - \phi(0; k),$$

$$-M_1 \omega^2(k) \phi(1; k) = \phi(0; k) \cos \pi k - \phi(1; k),$$

which is now a 2×2 matrix eigenvalue problem for each $k \in [0, 1]$. We rewrite it in the form

$$(4.1.3) \qquad \begin{pmatrix} 1 & -\cos \pi k \\ -\cos \pi k & 1 \end{pmatrix} \begin{pmatrix} \phi(0; k) \\ \phi(1; k) \end{pmatrix} = \omega^2(k) \begin{pmatrix} M_0 & 0 \\ 0 & M_1 \end{pmatrix} \begin{pmatrix} \phi(0; k) \\ \phi(1; k) \end{pmatrix}$$

The two eigenvalues $\omega_0^2(k) \le \omega_1^2(k)$ are usually called the **acoustical** and **optical** modes respectively given by ($-$ with 0, $+$ with 1)

$$(4.1.4) \qquad \omega_{0,1}^2(k) = \frac{M_0 + M_1 \mp \sqrt{(M_0 + M_1)^2 - 4 M_0 M_1 \sin^2 \pi k}}{2 M_0 M_1}.$$

The eigenfunctions $\phi_0(p; k)$, $\phi_1(p; k)$ are chosen orthonormal relative to M_0, M_1, i.e.

$$(4.1.5) \qquad \sum_{p=0,1} M_p \phi_m(p; k) \bar{\phi}_{m'}(p; k) = \delta_{mm'}, \qquad m, m' = 0, 1.$$

Given u_p with $u_p \equiv 0$ for $|p|$ large we can construct its expansion in discrete Bloch waves as in Section 4.2. We obtain

$$(4.1.6) \qquad u_p = \int_0^1 dk \sum_{m=0,1} \hat{u}_m(k) e^{i\pi kp} \phi_m(p; k),$$

$$\hat{u}_m(k) = \sum_{p \in \mathbb{Z}} u_p M_p e^{-i\pi kp} \phi_m(p; k).$$

Now we can use (4.1.6) to solve (4.1.1). This is elementary. We obtain
(4.1.7)

$$u_p(t) = \int_0^1 dk \sum_{m=0,1} \left[A_m(k) e^{i(\pi kp + \omega_m(k)t)} + B_m(k) e^{i(\pi kp - \omega_m(k)t)} \right] \phi_m(p; k),$$

with

(4.1.8)
$$A_m(k) = \frac{1}{2}\left[\hat{U}_m(k) + \frac{1}{i\omega_m(k)}\hat{V}_m(k)\right],$$
$$B_m(k) = \frac{1}{2}\left[\hat{U}_m(k) - \frac{1}{i\omega_m(k)}\hat{V}_m(k)\right].$$

Here $U_m(k)$ and $V_m(k)$ are given by (4.1.6) with u_p replaced by U_p and V_p respectively.

Formula (4.1.7) is of course a representation of the exact solution to (4.1.1) but the $\omega_m(k)$ (cf. (ref4.4.4)) are quite complicated. However the integral can be evaluated asymptotically by stationary phase and this is the primary utility of (4.1.7).

More general and multidimensional problems on a lattice can be treated the same way.

4.2. Schrödinger equation. We shall now introduce a problem which we shall investigate in some detail until Section 4.8. We have chosen this problem for computational convenience but the results, although simple, should be useful at least in understanding the various expansion processes that we have used so far.

Consider the Schrödinger equation with modulated plane wave initial data:

(4.2.1)
$$-\frac{i}{\varepsilon}\frac{\partial u_\varepsilon}{\partial t} - \sum_{p,q=1}^{n}\frac{\partial}{\partial x_p}\left(a_{pq}\left(\frac{x}{\varepsilon}\right)\frac{\partial u_\varepsilon}{\partial x_q}\right) + \frac{1}{\varepsilon^2}W\left(\frac{x}{\varepsilon}\right)u_\varepsilon = 0, \qquad t > 0, \qquad x \in \mathbb{R}^n,$$

$$u_\varepsilon(x,0) = e^{ik_0\cdot x/\varepsilon}\sum_{m'=0}^{N_0} f_{m'}(x)\phi_{m'}\left(\frac{x}{\varepsilon};k_0\right).$$

Here $(a_{pq}(y))$ and $W(y) \geq 0$ are smooth functions on $2\pi Y$, $f_m(x)$ are in $C_0(\mathbb{R}^n)$ and the (a_{pq}) matrix is positive definite. Equation (4.2.1) has a unique smooth solution with t and ε independent $L^2(\mathbb{R}^n)$ norm. The functions $\phi_m(y;k)$ are the Bloch eigenfunctions of Section 3.1. We assume that for some $N \geq N_0$

(4.2.2) k_0 is such that $\omega_0(k_0) \leq \cdots \leq \omega_N(k_0)$ are **distinct**.

Since the solution u_ε belongs to L^2 for each t it can be expanded in eigenfunctions using the theorem of Section 3.2. We have

(4.2.3) $$u_\varepsilon(x,t) = \int_Y dk \sum_{m=0}^{\infty} e^{i(k\cdot x - \omega_m^2(k)t)/\varepsilon}\hat{f}_m(k)\phi_m\left(\frac{x}{\varepsilon};k\right),$$

(4.2.4) $$\hat{f}_m(k) = \int_{\mathbb{R}^n} dy\left[\sum_{m'=0}^{N_0} f_{m'}(\varepsilon y)\phi_{m'}(y;k_0)e^{ik_0\cdot y}\right]\cdot e^{-ik\cdot y}\phi_m(y;k).$$

The sum in (4.2.3) is understood convergent in the L^2 sense.

We are interested in the behavior of the exact solution (4.2.3) as $\varepsilon \to 0$. Although writing the exact solution down and then expanding it is contrary to our general methodology,[46] we carry it out here to illustrate the more general results of Sections 5.1–5.8.

---

[46]We want methods that give the asymptotic behavior directly, especially when we do not have formulas for the exact solution.

We shall first carry out the expansion of u_ε in (4.2.3) formally. In Section 4.5 we show how one proves its validity.

By simple changes of variables we rewrite (4.2.3) and (4.2.4) as follows:

$$(4.2.5) \qquad u_\varepsilon(x,t) = \sum_{m=0}^{\infty} e^{i(k_0 \cdot x - \omega_m^2(k_0)t)/\varepsilon}$$

$$\cdot \int_{\frac{1}{\varepsilon}Y} \exp\left\{ ik \cdot x - \frac{it}{\varepsilon}(\omega_m^2(k_0 + \varepsilon k) - \omega_m^2(k_0)) \right\}$$

$$\cdot \varepsilon^n \hat{f}_m^\varepsilon(k_0 + \varepsilon k) \phi_m\left(\frac{x}{\varepsilon}; k_0 + \varepsilon k\right),$$

$$(4.2.6) \quad \varepsilon^n \hat{f}_m(k_0 + \varepsilon k) = \int_{\mathbb{R}^n} dy \sum_{m'=0}^{N_0} f_{m'}(y) e^{-ik \cdot y} \phi_{m'}\left(\frac{y}{\varepsilon}; k_0\right) \bar{\phi}_m\left(\frac{y}{\varepsilon}; k_0 + \varepsilon k\right).$$

We now expand formally in powers of ε. The exponential inside the integral in (4.2.5) becomes

$$\exp\left\{ ik \cdot x - i\sum_{p=1}^{n} c_p^{(m)}(k_0)k_p t - i\varepsilon \sum_{p,q=1}^{n} \tilde{a}_{pq}^{(m)}(k_0)k_p k_q t + O(\varepsilon^2 t) \right\},$$

where

$$(4.2.7) \qquad c_p^{(m)}(k_0) = \frac{\partial \omega_m^2(k_0)}{\partial k_p},$$

$$(4.2.8) \qquad \tilde{a}_{pq}^{(m)}(k_0) = \frac{1}{2}\frac{\partial^2 \omega_m^2(k_0)}{\partial k_p \partial k_q}, \qquad p,q = 1,2,\ldots,n.$$

It is assumed that these derivatives exist although hypothesis (4.2.2) covers only the case $m \leq N$. We shall clarify this in Section 4.5.

We expand next the terms outside the exponential in the integral in (4.2.5)

$$\varepsilon^n \hat{f}^\varepsilon(k_0 + \varepsilon k) \phi_m\left(\frac{x}{\varepsilon}; k_0 + \varepsilon k\right)$$

$$= \phi_m\left(\frac{x}{\varepsilon}; k_0 + \varepsilon k\right) \int_{\mathbb{R}^n} dy \sum_{m'=0}^{N_0} f_{m'}(y) e^{-ik \cdot y} \phi_{m'}\left(\frac{y}{\varepsilon}; k_0\right) \bar{\phi}_m\left(\frac{y}{\varepsilon}; k_0 + \varepsilon k\right)$$

$$= \sum_{m'=0}^{N_0} \hat{f}_{m'}(k) \left\{ \phi_m\left(\frac{x}{\varepsilon}; k_0\right) \delta_{mm'} + \varepsilon\left[\delta_{mm'} \sum_{p=1}^{n} \frac{1}{i}\frac{\partial \phi_m\left(\frac{x}{\varepsilon}; k_0\right)}{\partial k_p} ik_p \right.\right.$$

$$\left.\left. - \phi_m\left(\frac{x}{\varepsilon}; k_0\right) \sum_{p=1}^{n}\left(\phi_{m'}(\cdot; k_0), \frac{1}{i}\frac{\partial \phi_m(\cdot; k_0)}{\partial k_p} \right) ik_p \right] + O(\varepsilon^2) \right\}.$$

Here we use the notation $(\,,)$ for (complex) inner product over $2\pi Y$ and $\hat{f}_{m'}(k)$ is the Fourier transform over $\mathbb{R}^n$ of $f_{m'}(x)$, the coefficient of $\phi_{m'}$ in the data in (4.2.1). We also use the fact that if $g(x,y)$ is smooth on $\mathbb{R}^n \times 2\pi Y$ and has compact support in x then

$$(4.2.9) \qquad \int_{\mathbb{R}^n} g\left(x, \frac{x}{\varepsilon}\right) dx \to \int_{\mathbb{R}^n} dx \int_{2\pi Y} g(x,y)\, dy \text{ as } \varepsilon \to 0.$$

4.3. Nature of the expansion. We collect the results just obtained and note that u_ε of (4.2.5) has the form

$$(4.3.1) \qquad u_\varepsilon(x,t) = \sum_{m=0}^{\infty} e^{i(k_0 \cdot x - \omega_m^2(k_0)t)/\varepsilon}$$
$$\cdot \left[v_0^{(m,\varepsilon)}\left(x, \frac{x}{\varepsilon}, t\right) + \varepsilon v_1^{(m,\varepsilon)}\left(x, \frac{x}{\varepsilon}, t\right) + \cdots \right],$$

where

$$(4.3.2) \qquad v_0^{(m,\varepsilon)} = \phi_m\left(\frac{x}{\varepsilon}; k_0\right) v_{00}^{(m,\varepsilon)}(x,t),$$

and $v_{00}^{(m,\varepsilon)}$ satisfies the Schrödinger equation

$$(4.3.3) \qquad \frac{i}{\varepsilon} \left[\frac{\partial v_{00}^{(m,\varepsilon)}}{\partial t} + \sum_{p=1}^{n} c_p^{(m)}(k_0) \frac{\partial v_{00}^{(m,\varepsilon)}}{\partial x_p} \right]$$
$$= \sum_{p,q=1}^{n} \bar{a}_{pq}^{(m)}(k_0) \frac{\partial^2 v_{00}^{(m,\varepsilon)}}{\partial x_p \partial x_q}, \qquad t > 0, \qquad x \in \mathbb{R}^n,$$
$$v_{00}^{(m,\varepsilon)}(x,0) = f_m(x), \qquad 0 \le m \le N_0,$$
$$v_{00}^{(m,\varepsilon)}(x,0) \equiv 0, \qquad N_0 + 1 \le m.$$

Dependence of $v_0^{(m,\varepsilon)}$ on k_0 is not shown explicitly.

The terms $v_1^{(m,\varepsilon)}$ in (4.3.1) can be written, after some simple rearrangements, in the following form:

$$(4.3.4) \quad v_1^{(m,\varepsilon)} = \sum_{p=1}^{n} \chi_p^{(m)}\left(\frac{x}{\varepsilon}; k_0\right) \frac{\partial v_{00}^{(m,\varepsilon)}}{\partial x_p}$$
$$- \sum_{m'=0}^{N_0} \sum_{p=0}^{n} \left(\chi_p^{(m')}(\cdot; k_0), \phi_m(\cdot; k_0) \right) \phi_m\left(\frac{x}{\varepsilon}; k_0\right) \frac{\partial v_{00}^{(m',\varepsilon)}}{\partial x_p}.$$

Here $\chi_p^{(m)}(y; k_0)$ are functions on $2\pi Y$ given by

$$(4.3.5) \qquad \chi_p^{(m)}(y; k_0) = \frac{1}{i}\frac{\partial \phi_m(y; k_0)}{\partial k_p} - \left(\frac{1}{i}\frac{\partial \phi_m(\cdot; k_0)}{\partial k_p}, \phi_m(\cdot; k_0) \right) \phi_m(y; k_0).$$

The functions $\chi_p^{(m)}(y; k_0)$ are important so we shall now examine them in more detail. We begin with the shifted cell eigenvalue problem

$$(4.3.6) \qquad A(k)\phi_m(y; k) = \omega_m^2(k)\phi_m(y; k), \qquad 0 \le m \le N,$$

where $A(k)$ is given by (3.1.2). The vector $k \in Y$ is to be confined to a neighborhood of k_0 so that (4.2.2) holds and we can differentiate (4.3.6) with respect to k. Differentiating once we obtain

$$(4.3.7) \qquad \left[-\omega_m^2(k) + A(k) \right] \frac{\partial \phi_m(y; k)}{\partial k_p} + \left[A_{,p}(k) - \frac{\partial \omega_m^2(k)}{\partial k_p} \right] \phi_m(y; k) = 0,$$

where $A_{,p}(k)$ is an operator defined by

$$(4.3.8) \qquad \frac{1}{i} A_{,p}(k) = -\sum_{q=1}^{n} a_{pq}(y) \left(\frac{\partial y_q}{\partial +} ik_q \right)$$

$$- \sum_{q=1}^{n} \left(\frac{\partial}{\partial y_q} + ik_q \right) [a_{pq}(y)\cdot], \qquad p = 1, 2, \ldots, n.$$

Taking inner products with $\phi_{m'}$ in (4.3.7) we obtain

$$(4.3.9) \qquad \frac{\partial \omega_m^2(k)}{\partial k_p} = (A_{,p}(k)\phi_m(\cdot; k), \phi_m(\cdot; k))$$

and

$$(4.3.10) \qquad \left[\omega_m^2(k) - \omega_m^2(k') \right] \left(\frac{\partial \phi_m}{\partial k_p}, \phi_{m'} \right) = (A_{,p}(k)\phi_m, \phi_{m'}), \qquad m \neq m'.$$

Now from the definition (4.3.5) of $\chi_p^{(m)}$ and (4.3.7) we deduce that $\chi_p^{(m)}$ is the unique solution of
$$(4.3.11)$$
$$\left[-\omega_m^2(k_0) + A(k_0) \right] \chi_p^{(m)}(y; k_0) + \frac{1}{i} \left[A_{,p}(k_0) - \frac{\partial \omega_m^2(k_0)}{\partial k_0} \right] \phi_m(y; k_0) = 0 \text{ on } 2\pi Y,$$

subject to the normalization

$$(4.3.12) \qquad (\chi_p^{(m)}, \phi_m) = 0, \qquad p = 1, 2, \ldots, n.$$

The eigenvalue problem (4.3.6) is of course the **shifted cell eigenvalue problem**. We shall now show the following important facts:

(4.3.13)   In order to obtain the first two terms in (4.3.1)

one has to solve the eigenvalue problem (4.3.6)

only at k_0 and it is not necessary to differentiate

either ω_m^2 or ϕ_m at k_0. The derivatives can be

obtained by integration and by solving other cell

problems. Of course one must also solve (4.3.3)

which is not a cell problem but has **constant**

coefficients. Equation (4.3.3) is the **effective**

Schrödinger equation for the mth mode (or band)

at k_0.

Specifically:

(1) $\frac{\partial \omega_m^2(k_0)}{\partial k_p} = c_p^{(m)}(k_0)$ is given by (4.3.9) which does not involve k differentiation.

(2) $\chi_p^{(m)}$ is given by solving (4.3.11) subject to (4.3.12). No k differentiation is involved.

(3) $\tilde{a}_{pq}^{(m)}(k_0) = \frac{1}{2} \frac{\partial^2 \omega_m^2(k_0)}{\partial k_p \partial k_q}$ by (4.2.8). We now show how the matrix can be obtained without k differentiation (cf. (4.3.16)).

First we differentiate (4.3.7) with respect to k_q. This gives

$$(4.3.14) \quad [-\omega_m^2 + A(k)]\frac{1}{2}\frac{\partial^2\phi_m}{\partial k_p \partial k_q} + \frac{1}{2}\left[A_{,p} - \frac{\partial\omega_m^2}{\partial k_p}\right)\frac{\partial\phi_m}{\partial k_q}$$
$$+ \frac{1}{2}\left[A_{,q} - \frac{\partial\omega_m^2}{\partial k_q}\right]\frac{\partial\phi_m}{\partial k_p} + \left[a_{pq} - \frac{1}{2}\frac{\partial^2\omega_m^2}{\partial k_p \partial k_q}\right]\phi_m = 0.$$

Taking inner products with ϕ_m we obtain

$$(4.3.15) \qquad \frac{1}{2}\frac{\partial^2\omega_m^2}{\partial k_p \partial k_q} = (a_{pq}\phi_m, \phi_m) - \frac{1}{2}\left(\left(a_{,p} - \frac{\partial\omega}{\partial k_p}\right)\frac{\partial\phi_m}{\partial k_q}, \phi_m\right)$$
$$- \frac{1}{2}\left(\left(A_{,q} - \frac{\partial\omega_m^2}{\partial k_q}\right)\frac{\partial\phi_m}{\partial k_p}, \phi_m\right).$$

Using (4.3.5) and (4.3.3) we obtain

$$(4.3.16) \qquad \frac{1}{2}\frac{\partial^2\omega_m^2}{\partial k_q \partial k_q} = (a_{pq}\phi_m, \phi_m) - \frac{1}{2}\left(\left(A_{,p} - \frac{\partial\omega_m^2}{\partial k_p}\right)\chi_q^{(m)}, \phi_m\right)$$
$$- \frac{1}{2}\left(\left(A_{,q} - \frac{\partial\omega_m^2}{\partial k_q}\right)\chi_p^{(m)}, \phi_m\right).$$

This is the desired formula.

There are two important reasons why we do not want to differentiate with respect to k. First, the eigenfunction ϕ_m in (4.3.6) is determined only up to a phase factor that could depend on k. We have just shown that $v_0^{(m,\varepsilon)}$ in (4.3.1) are determined independently of how one chooses the ϕ_m's (still under hypothesis (4.2.2)). For numerical computations avoiding k-differentiations is a very important objective.

Second, **the above results can be easily modified to deal with the case that at k_0 the ω_m^2 are not distinct**, i.e. **when we have degeneracy at k_0.** The form (4.3.1) changes then and instead of decoupled effective Schrödinger equations for each mode (or band) m as in (4.3.3) we get **coupled** effective Schrödinger equations (their number being equal to the multiplicity of the particular degenerate ω_m^2 at k_0). Unfortunately we have not been able to generalize this observation to the inhomogeneous (locally periodic) case[47] of Sections 5.1–5.8. Therefore we shall not discuss the degenerate case further.

When it exists, the tensor

$$\left(\frac{\partial\omega_m^2}{\partial k_p \partial k_q}\right)^{-1} = (a_{pq}^{(m)})^{-1}$$

is called the **effective mass tensor**[48] of the mth band. The tensor along with the **group velocity** vector $c_p^{(m)}$ (cf. (4.2.7)) are basic quantities that one can use to study many complex phenomena associated with transport properties of solids.

---

[47] Where exact solutions are not available as they are here.
[48] cf. Ziman [**130**].

4.4. Connection with the static theory. Let us specialize the results of Section 4.3, i.e. the expansion (4.3.1) for problem (4.2.1), to the following case:

$$(4.4.1) \qquad -i\frac{\partial u_\varepsilon}{\partial t} - \sum_{p,q=1}^{n} \frac{\partial}{\partial x_p}\left(a_{pq}\left(\frac{x}{\varepsilon}\right)\frac{\partial u_\varepsilon}{\partial x_q}\right) = 0, \qquad t > 0, \qquad x \in \mathbb{R}^n,$$

$$u_\varepsilon(x,0) = (2\pi)^{-n/2}f_0(x).$$

This corresponds to setting $W \equiv 0$ in (4.2.1) and $f_{m'}(x) \equiv 0$ for $m' \geq 1$ with $k_0 = 0$. Note also that we have rescaled the time: $t \to t/\varepsilon$, because this is the scaling used throughout Chapters 2 and 3. Note also that in this case[49]

$$(4.4.2) \qquad\qquad \omega_0^2(0) = 0, \qquad \phi_0(y;0) \equiv (2\pi)^{-n/2},$$

which explains the factor $(2\pi)^{-n/2}$ in $u_\varepsilon(x,0)$ in (4.4.1).

The results of Section 4.3 specialize as follows. The expansion (4.3.1) has now the form

$$(4.4.3) \qquad u_\varepsilon(x,t) = v_{00}^{(0)}(x,t) + \varepsilon v_1^{(0)}\left(x,\frac{x}{\varepsilon},t\right)$$

$$- \varepsilon \sum_{m=1}^{\infty} e^{i\omega_m^2(0)t/\varepsilon^2} \sum_{p=0}^{n} (\chi_p^{(0)}(\cdot;0), \phi_m(\cdot;0))$$

$$\cdot \phi_m\left(\frac{x}{\varepsilon};0\right)\frac{\partial v_{00}^{(0)}(x,t)}{\partial x_p} + \cdots.$$

Note the effect of rescaling of $t \to t/\varepsilon$ in (4.4.3). Note also that $\chi_p^{(0)}(y;0) = \chi_p(y)$ satisfies, from (4.3.11) and (4.4.2),

$$(4.4.4) \qquad\qquad A\chi_p(y) - \sum_{q=1}^{n} \frac{\partial a_{pq}(y)}{\partial y_q} = 0, \qquad p = 1, 2, \dots, n,$$

$\chi_p(y)$ 2π-periodic and

$$(\chi_p, \phi_0) = \frac{1}{(2\pi)^{n/2}} \fint_{2\pi Y} \chi_p(y)\, dy = 0.$$

Here $A = A(0)$, clearly (cf. (3.1.1), (3.1.2) with $W \equiv 0$).

In (4.4.3) $v_{00}^{(0)}$ satisfies the Schrödinger equation (cf. (4.3.3))

$$(4.4.5) \qquad u\frac{\partial v_{00}^{(0)}}{\partial t} = -\sum_{p,q=1}^{n} \bar{a}_{pq}^{(0)}(0)\frac{\partial^2 v_{00}^{(0)}}{\partial x_p \partial x_q}, \qquad t > 0, \qquad x \in \mathbb{R}^n,$$

$$v_{00}^{(0)}(x,0) = f_0(x).$$

The coefficients $(\bar{a}_{pq}^{(0)}(0))$ are given by (4.3.16) as follows:

$$(4.4.6) \qquad \bar{a}_{pq}^{(0)}(0) = \frac{1}{(2\pi)^n} \fint_{2\pi Y} \sum_{\ell=1}^{n} a_{p\ell}(y)\left[\delta_{\ell q} + \frac{\partial \chi_q(y)}{\partial y_\ell}\right] dy.$$

The function $v_1^{(0)}(x,y,t)$ in (4.4.3) is given by

$$(4.4.7) \qquad\qquad v_1^{(0)} = \sum_{p=1}^{n} \chi_p(y)\frac{\partial v_{00}^{(0)}}{\partial x_p}.$$

---

[49]Recall that we have $(\phi_m, \phi_{m'}) = \delta_{mm'}$. This explains normalization in (4.4.2).

The reader can verify easily that the above results agree precisely with the ones obtained by direct expansion in previous chapters. Of course we do not have boundary conditions; the spectral expansions do not work in that case, at least not in the form given here. The expansion processes of previous chapters work just as well, however, even in the presence of boundaries.

REMARK 4.1. In previous chapters we constructed expansions of the form

$$(4.4.8) \qquad u_\varepsilon(x,t) = v_{00}^{(0)}(x,t) + \varepsilon v_1^{(0)}\left(x, \frac{x}{\varepsilon}, t\right) + \varepsilon^2 v_2^{(0)}\left(x, \frac{x}{\varepsilon}, t\right) + \cdots,$$

but we did **not** have infinite sums with oscillatory time factors like in (4.4.3) (note that we have no $v_2^{(0)}$ term in (4.4.3) since we stopped with $O(\varepsilon)$ terms). The reason is that expansions like (4.4.8) cannot satisfy initial conditions beyond the ε^0 level. Since the problem does not have initial layers the oscillatory terms like in (4.4.3) are necessary to satisfy initial conditions. As we saw in Section 2.13, if we do not look for expansions or results valid pointwise in t the oscillatory terms drop out[50] and we recover the ansatz (4.4.8).

REMARK 4.2. The time rescaling in the analysis of the static problem (i.e., $k_0 = 0$, which is all that counts) is a consequence of the fact that

$$c_p^{(m)} = \left.\frac{\partial \omega_m^2(k)}{\partial k_p}\right|_{k=0} = 0$$

in the case of distinct ω_m^2 since they are even about $k = 0$ and differentiable. When the group velocity $c_p^{(m)}$ vanishes we can have oscillatory solutions but they are stationary; they do not propagate. The dispersive effects that come from $\tilde{a}_{pq}^{(m)}$ (the reciprocal of the effective mass tensor) are one order of magnitude smaller than the propagation effects except when $k = 0$ (or near zero).

4.5. Validity of the expansion. Consider problem (4.2.1) with condition (4.2.2). The solutions $\chi_p^{(m)}(y; k_0)$, $p = 0, 1, \ldots, n$, $m = 0, 1, \ldots, N$ of (4.3.11) are smooth bounded functions. This smoothness and the smoothness of the data $f_m(x)$ ($\in C_0^\infty(\mathbb{R}^n)$) suffice for the following theorems which summarize results of previous sections.

THEOREM 4.3. *Let* $v_{00}^{(m)}(x,t)$ *be the solution of*

$$(4.5.1) \qquad \frac{\partial v_{00}^{(m)}}{\partial t} + \sum_{p=1}^n c_p^{(m)}(k_0)\frac{\partial v_{00}^{(m)}}{\partial x_p} = 0, \qquad t > 0, \qquad x \in \mathbb{R}^n,$$

$$v_{00}^{(m)}(x,0) = f_m(x), \qquad 0 \le m \le N_0.$$

Let $\chi_p^{(m)}(y; k_0)$ *be defined by* (4.3.11), (4.3.12) *and let* $v_{10}^{(m)}(x,t)$ *be the solution of*

$$(4.5.2) \qquad \frac{\partial v_{10}^{(m)}}{\partial t} + \sum_{p=1}^n c_p^{(m)}(k_0)\frac{\partial v_{10}^{(m)}}{\partial x_p} = i \sum_{p,q=1}^n \tilde{a}_{pq}^{(m)}(k_0)\frac{\partial^2 v_{00}^{(m)}}{\partial x_p \partial x_q}, \qquad t > 0,$$

$$v_{10}^{(m)}(x,0) = -\sum_{m'=0}^{N_0}\sum_{p=0}^n (\chi_p^{(m')}, \phi_m)\frac{\partial v_{00}^{(m')}}{\partial x_p}, \qquad 0 \le m \le N.$$

---

[50]In particular if we consider weak convergence results only.

Define

(4.5.3) $v_0^{(m)}(x, y, t) = \phi_m(y; k_0) v_{00}^{(m)}(x, t),$

(4.5.4) $v_1^{(m)}(x, y, t) = \sum_{p=0}^{n} \chi_p^{(m)}(y; k_0) \dfrac{\partial v_{00}^{(m)}(x, t)}{\partial x_p} + \phi_m(y; k_0) v_{10}^{(m)}(x, t),$

and

(4.5.5) $z^{(\varepsilon, N)}(x, t) = \sum_{m=0}^{N} e^{i(k_0 \cdot x - \omega_m^2(k_0)t)/\varepsilon} \cdot \left[v_0^{(m)}\left(x, \dfrac{x}{\varepsilon}, t\right) + \varepsilon v_1^{(m)}\left(x, \dfrac{x}{\varepsilon}, t\right) \right].$

Then if $u_\varepsilon(x, t)$ is the solution of (4.2.1), under the hypothesis stated there including (4.2.2), we have

(4.5.6) $|u_\varepsilon(\cdot, t) - z^{\varepsilon, N}(\cdot, t)|_{L^2(\mathbb{R}^n)} \leq \varepsilon[C_N(f) + \varepsilon \tilde{C}(f)], \qquad 0 \leq t \leq T < \infty.$

Here $C_N(f)$, $\tilde{C}(f)$ are constants that depend on the data $f_m(x)$, $0 \leq m \leq N_0$ but do not depend on ε. Moreover, the constant $\tilde{C}_N(f)$ can be made as small as desired, for given data f_m, $0 \leq m \leq N_0$, by taking N sufficiently large.

PROOF. The proof is routine under our assumptions and the elementary L^2 estimate (energy estimate) for problem (4.2.1). It is possible to construct a smooth function $v_2^{(m)}(x, y, t)$ such that

$$\int_{\mathbb{R}^n} \left[\sup_{0 \leq t \leq T} \sup_{y \in 2\pi Y} |v_2^{(m)}(x, y, t)| \right]^2 dx < \infty$$

and if we define

$$\tilde{z}^{\varepsilon, N}(x, t) = \sum_{m=0}^{N} e^{i(k_0 \cdot x - \omega_m^2(k_0)t)/\varepsilon} \cdot \left[v_0^{(m)} + \varepsilon v_1^{(m)} + \varepsilon^2 v_2^{(m)} \right]_{y=x/\varepsilon}$$

then

(4.5.7) $\left[-\dfrac{i}{\varepsilon} \dfrac{\partial}{\partial t} - \sum_{p,q=1}^{n} \dfrac{\partial}{\partial x_p} \left(a_{pq}\left(\dfrac{x}{\varepsilon}\right) \dfrac{\partial}{\partial x_q} \cdot \right) + \dfrac{1}{\varepsilon^2} W\left(\dfrac{X}{\varepsilon}\right) \right] \cdot (u_\varepsilon - \tilde{z}^{\varepsilon, N}) = g_1^\varepsilon,$

$(u_\varepsilon - \tilde{z}^{\varepsilon, N})\big|_{t=0} = g_2^\varepsilon.$

The functions g_1^ε and g_2^ε are such that

$$\left(\int_{\mathbb{R}^n} \left[\sup_{0 \leq t \leq T} \sup_{y \in 2\pi Y} |g_1^\varepsilon(x, y, t)| \right]^2 dx \right)^{1/2} = O(\varepsilon),$$

$$\left(\int_{\mathbb{R}^n} \left[\sup_{y \in 2\pi Y} |g_2^\varepsilon(x, y)| \right]^2 dx \right)^{1/2} \leq C_N(f)\varepsilon.$$

From these facts (4.5.6) follows immediately and the proof is complete. ∎

The explicit construction of $v_2^{(m)}$ (as well as $v_0^{(m)}$ and $v_1^{(m)}$) will be given in a more general context in Section 5.1. It is, in any case, simply the "next term" in the expansion (4.3.1) (slightly modified).

We should explain why we have chosen to state our results in the form of Theorem 4.3; in particular we must explain the role of N. This is because hypothesis (4.2.2) is a restrictive one and it is unlikely that it would be valid for all N. This necessitates working with only finitely many modes. Theorem 4.3 can be generalized to deal with **degenerate modes**, i.e. without (4.2.2), as we mentioned below (4.3.16).

The second feature of Theorem 4.3 that departs from (4.3.1) is the fact that we do not use $v_{00}^{(m,\varepsilon)}(x,t)$ of (4.3.3) but rather expand it further. Thus, in Theorem 4.3, (4.5.1) and (4.5.2) are simple first order linear (homogeneous and inhomogeneous, respectively) partial differential equations which can be solved explicitly trivially. For example

$$v_{00}^{(m)}(x,t) = f_m(x - c^{(m)}t),$$

where $c^{(m)} = (c_p^m(k_0))$ is the group velocity vector. The reason for doing this is that in the more general context of Section 5.1 an analog for Theorem 4.4 below does not exist in general. Therefore 4.3 is the prototype of what we can get in general.

THEOREM 4.4. *Let* $v_{00}^{(m,\varepsilon)}(x,t)$ *be the solution of (4.3.3) and define*

$$(4.5.8) \qquad z^\varepsilon(x,t) = \sum_{m=0}^{N_0} e^{i(k_0 \cdot x - \omega_m^2(k_0)t)/\varepsilon} \phi_m\left(\frac{x}{\varepsilon}; k_0\right) v_{00}^{(m,\varepsilon)}(x,t).$$

Then

$$(4.5.9) \qquad |u_\varepsilon(\cdot,t) - z(\cdot,t)|_{L^2(\mathbb{R}^n)} \le C\varepsilon$$

for $0 \le t \le T/\varepsilon$, $T < \infty$, *with* C *a constant depending on* $f_m(x)$, $0 \le m \le N_0$, *but not on* ε. *Hypothesis (4.2.2) is, of course, assumed here.*

REMARK 4.5. Note that (4.5.9) is valid on a much larger time interval than (4.5.6). On the other hand the error in (4.5.6) is actually better than $O(\varepsilon)$ since $C_N(f)$ can be controlled with N. Anyway, Theorem 4.4 is a better theorem. **It also generalizes to degenerate modes (i.e., without (4.2.2)).**

PROOF. The proof is again routine as in Theorem 4.3; here also because $v_{00}^{(m,\varepsilon)}(x,t)$ of (4.3.3) has a smooth solution with L^2 norm and L^2 norm of derivatives independent of ε ((4.3.3) has **constant** coefficients).

As is shown in Section 5.1 we can construct functions $v_1^{(m,\varepsilon)}(x,y,t)$ and $v_2^{(m,\varepsilon)}(x,y,t)$ such that

$$\int_{\mathbb{R}^n} \left[\sup_{0 \le t \le T/\varepsilon} \sup_{y \in 2\pi Y} |v_j^{(m,\varepsilon)}(x,y,t)|\right]^2 dx < \infty, \qquad j =, 1, 2,$$

and such that if

$$\tilde{z}^\varepsilon(x,t) = \sum_{m=0}^{N_0} e^{i(k_0 \cdot x - \omega_m^2(k_0)t)/\varepsilon} \cdot \left[v_0^{(m,\varepsilon)} + \varepsilon v_1^{(m,\varepsilon)} + \varepsilon^2 v_2^{(m,\varepsilon)}\right]_{y=x/\varepsilon},$$

then (4.5.7) holds again. The new functions g_1^ε and g_2^ε satisfy

$$\int_{\mathbb{R}^n} \left[\sup_{0 \le t \le T/\varepsilon} \sup_{y \in 2\pi Y} |g_1^\varepsilon(x,y,t)|\right]^2 dx = O(\varepsilon)$$

and

$$\int_{\mathbb{R}^n} \left[\sup_{y \in 2\pi Y} |g_2^\varepsilon(x,y)|\right]^2 dx = O(\varepsilon).$$

From these facts (4.5.9) follows immediately and the proof is complete.   ■

4.6. Relation between the Hilbert and Chapman-Enskog expansion.
Theorem 4.3 of the previous section is an example of a Hilbert expansion while
Theorem 4.4 is an example of a Chapman-Enskog expansion. This terminology is
not standard and readers are cautioned not to look for very clear analogies with
problems in other areas.[51] In acoustics and optics Theorem 4.3 is the usual geometrical optics (WKB) expansion while Theorem 4.4 is the Leontovich-Fock[52] or
parabolic approximation.

In the globally periodic case of problem (4.2.1) one can have either theorem;
in fact one can even write down the exact solution (4.2.3). In the locally periodic
case when

$$a_{pq} = a_{pq}\left(x, \frac{x}{\varepsilon}\right), \qquad W = W\left(x, \frac{x}{\varepsilon}\right),$$

as described in Section 5.1, Theorem 4.3 is valid again — up to the first time
caustics form — but Theorem 4.4 is more delicate and in general is false.

It is of interest to know when Theorem 4.4 can be obtained more generally. One
reason for this is that Theorem 4.4 is valid **uniformly** through the static regime
(cf. Section 4.4) and includes diffraction and mode coupling effects.

A look at the proof of Theorem 4.4 shows that in the locally periodic case two
things break down:

(1) caustics may form and
(2) $v_{00}^{(\varepsilon,m)}$ of (4.3.3) will not have derivatives with ε-independent norms over
    $0 \le t \le T/\varepsilon$.

These two things are not unrelated.

The rays, solutions of

$$\frac{dx^{(m)}}{dt} = c^{(m)}, \qquad \text{(cf. (4.2.7))},$$

are straight lines in the globally periodic case. In general they are curved. **If they
curve very slowly and if they have no envelopes in the relevant time
intervals then Theorem 4.4 will be valid in general.** Because of the way we
have set up our problems here this statement cannot be made completely precise
since either the coefficients in (4.3.3) will be constant or variable; no in between
situation can arise. To have (4.3.3) with slowly varying coefficients requires a more
elaborate scaling of the problems. We shall not do this here.

The general idea, however, that Theorem 4.4 should hold when the rays are
nearly straight lines is very useful.

4.7. Spatially localized data and stationary phase.
We shall now con-
sider the following problem:

(4.7.1)

$$-\frac{i}{\varepsilon}\frac{\partial u_\varepsilon}{\partial t} - \sum_{p,q=1}^{n} \frac{\partial}{\partial x_p}\left(a_{pq}\left(\frac{x}{\varepsilon}\right)\frac{\partial u_\varepsilon}{\partial x_q}\right) + \frac{1}{\varepsilon^2}W\left(\frac{x}{\varepsilon}\right)u_\varepsilon = 0, \qquad t > 0, \qquad x \in \mathbb{R}^n,$$

$$u_\varepsilon(x,0) = \varepsilon^{-n/2}f\left(\frac{x}{\varepsilon}\right),$$

---

[51]For the Boltzmann equation see Chapman and Cowling [**25**], (Chapter 7). What is called
Chapman-Enskog here is the second order case there. (p. 116).

[52]F. Tappert [**115**].

where $f(y) \in C_0^\infty$ (not periodic and complex valued). We shall assume specifically, in order to employ (4.2.2), that $f(y)$ has a finite-mode Bloch expansion

$$(4.7.2) \qquad f(y) = \int_Y d\kappa \sum_{m=0}^{N_0} \hat{f}_m(\kappa) e^{i\kappa \cdot y} \phi_m(y; \kappa), \qquad N_0 \leq N.$$

The scaling in (4.7.1) is chosen so that the initial probability amplitude $\varepsilon^{-n} |f(x/\varepsilon)|^2$ is approximately a δ-function for ε small and N_0 large (cf. Section 1.2).

Using the Bloch expansion theorem we can write the exact solution of (4.7.1) as follows:

$$(4.7.3) \qquad u_\varepsilon(x,t) = \varepsilon^{-n/2} \int_Y \sum_{m=0}^{N_0} e^{i(\kappa \cdot x - \omega_m^2(\kappa)t)/\varepsilon} \hat{f}_m(\kappa) \phi_m\left(\frac{x}{\varepsilon}; \kappa\right) d\kappa,$$

$$(4.7.4) \qquad \hat{f}_m(\kappa) = \int_{\mathbb{R}^n} f(y) e^{i\kappa \cdot y} \bar{\phi}_m(y; \kappa) \, dy.$$

The analysis we shall give for (4.7.3) is entirely analogous to the one given in Section 2.6.

The solution (4.7.3) is a finite sum of integrals of the form

$$(4.7.5) \qquad I_\varepsilon^{(m)}(x,t) = \varepsilon^{-n/2} \int_Y e^{iS^{(m)}(x,t;\kappa)} a_\varepsilon^{(m)}(x,t;\kappa) \, d\kappa,$$

where

$$(4.7.6) \qquad S^{(m)}(x,t;\kappa) = \kappa \cdot x - \omega_m^2(\kappa)t,$$
$$a_\varepsilon^{(m)}(x,t;\kappa) = \hat{f}_m(\kappa) \phi_m\left(\frac{x}{\varepsilon}; \kappa\right).$$

For each $t > 0$ and x fixed we consider the system of equations

$$(4.7.7) \qquad \frac{\partial S^{(m)}(x,t;\kappa)}{\partial \kappa_r} = 0, \qquad r = 1, 2, \ldots, n, \qquad 0 \leq m \leq N_0.$$

If there is no $\kappa \in Y$ satisfying (4.7.7) then the corresponding $I_\varepsilon^{(m)}$ goes to zero faster than any power of ε as $\varepsilon \to 0$. If there is a unique point $\kappa_m^* = \kappa_m^*(x,t)$ in Y satisfying (4.7.7) and such that

$$(4.7.8) \qquad Q^{(m)}(x,t) = \left(\frac{\partial^2 S^{(m)}(x,t;\kappa_m^*)}{\partial \kappa_r \partial \kappa_s}\right)$$

is nonsingular, then we have

$$(4.7.9) \quad I_\varepsilon^{(m)}(x,t) \sim \frac{(2\pi)^{n/2}}{|\det Q^{(m)}(x,t)|^{1/2}} e^{iS^{(m)}(x,t;\kappa_m^*)/\varepsilon + i\sigma_m \pi/4}$$

$$\cdot \hat{f}_m(\kappa_m^*) \phi_m\left(\frac{x}{\varepsilon}; \kappa_m^*\right) \equiv \tilde{I}_\varepsilon^{(m)}(x,t).$$

Here $\sigma_m = \sigma_m(x,t)$ is the signature of the matrix $Q^{(m)}$ i.e., the number of positive minus the number of negative eigenvalues of $Q^{(m)}$. If there are a finite number of isolated solutions to (4.7.7) then (4.7.9) is replaced by a finite sum of terms of the same form. If however $Q^{(m)}$ is singular then (4.7.9) is not valid and other types of expansions are necessary (using the Airy function in many cases).

In the particular case where $S^{(m)}$ is given by (4.7.6), equations (4.7.7) take the form

$$(4.7.10) \qquad \frac{\partial \omega_m^2(\kappa_m^*)}{\partial \kappa} = \frac{x}{t}.$$

Since the $\omega_m^2(\kappa)$, $0 \le m \le N$, are bounded and smooth, when (x, t) is inside the forward line cone, (4.7.10) will have a solution.

We conclude that: the asymptotically localized disturbance at the origin creates N_0 modes (because of the special form (4.7.2)) and each mode has waves (possibly several depending on the shape of $\omega_m^2(k)$) which propagate out of the origin with their group velocities as described by (4.7.9). The solution (4.7.3) has then the same asymptotic form as (4.3.1). This is the main point of this section (as it was in Section 2.6).

4.8. Behavior of probability amplitudes. Consider the problem

$$(4.8.1)$$

$$-\frac{i}{\varepsilon} \frac{\partial u_\varepsilon}{\partial t} - \sum_{p,q=1}^{n} \frac{\partial}{\partial x_p} \left(a_{pq} \left(\frac{x}{\varepsilon} \right) \frac{\partial u_\varepsilon}{\partial x_q} \right) + \frac{1}{\varepsilon^2} W \left(\frac{x}{\varepsilon} \right) u_\varepsilon = 0, \qquad t > 0, \qquad x \in \mathbb{R}^n,$$

$$u_\varepsilon(x, 0) = e^{ik_0 \cdot x / \varepsilon} \sum_{m'=0}^{N_0} f_{m'}(x) \phi_{m'} \left(\frac{x}{\varepsilon}; k_0 \right),$$

as in Section 4.2, including (4.2.2). Assume that

$$(4.8.2) \qquad \sum_{m=0}^{N_0} \int_{\mathbb{R}^n} |f_m(x)|^2 \, dx = 1.$$

Then,

$$(4.8.3) \qquad \int_{\mathbb{R}^n} |u_\varepsilon(x, t)|^2 \, dx$$

tends asymptotically to one as $\varepsilon \to 0$ at $t = 0$. Since (4.8.3) is independent of t, the same holds for any t. Note that $|u_\varepsilon(x, t)|^2$ is the probability density that the quantum particle will be at x at time t. We shall use Theorem 4.3 of Section 4.5 and (4.2.2) to obtain the following result.

THEOREM 4.6. *Let $\psi(x, t)$ be in $C_0^\infty((0, T) \times \mathbb{R}^n)$, $T < \infty$. Then*

$$(4.8.4) \qquad \int_0^T \int_{\mathbb{R}^n} \psi |u_\varepsilon|^2 \, dx \, dt$$

tends asymptotically as $\varepsilon \to 0$ to

$$(4.8.5) \qquad \int_0^T \int_{\mathbb{R}^n} dx \, dt \, \psi \left[\sum_{m=0}^{N_0} v_{00}^{(m)} \bar{v}_{00}^{(m)} + \varepsilon \sum_{m=0}^{N} (v_{00}^{(m)} \bar{v}_{10}^{(m)} + \bar{v}_{00}^{(m)} v_{10}^{(m)}) \right].$$

Here $v_{00}^{(m)}$ and $v_{10}^{(m)}$ are as in Theorem 4.3, Section 4.5.

PROOF. Let $z^{\varepsilon, N}$ be as in (4.5.5). Using (4.5.6) it follows easily that

$$(4.8.6) \qquad \int_0^T \int_{\mathbb{R}^n} dx \, dt \, \psi \left[|u_\varepsilon|^2 - |z^{\varepsilon, N}|^2 \right] \to 0, \text{ as } \varepsilon \to 0.$$

It suffices therefore to show that

$$\int_0^T \int_{\mathbb{R}^n} dx\, dt\, \psi |z^{\varepsilon,N}|^2$$

behaves like (4.8.5). This follows easily by using (4.2.9), the orthogonality of the ϕ_m's and the normalization of $\chi_p^{(m)}$ (cf. (4.3.11), (4.3.12)). ∎

4.9. The acoustic equations. Let $\rho(y)$ and $\mu(y)$ be smooth 2π-periodic functions on $\mathbb{R}^3$ (i.e., on $2\pi Y$) such that

(4.9.1) $$0 < \rho_0 \le \rho(y) \le \rho_1 < \infty, \qquad 0 < \mu_0 \le \mu(y) \le \mu_1 < \infty.$$

We shall analyze the acoustic equations (cf. Section 2.10)

(4.9.2) $$\rho\left(\frac{x}{\varepsilon}\right) u_{\varepsilon t} = \nabla p_\varepsilon,$$

$$p_{\varepsilon t} = \mu\left(\frac{x}{\varepsilon}\right) \nabla \cdot u_\varepsilon, \qquad t > 0, \qquad x \in \mathbb{R}^3.$$

We shall assume that

(4.9.3) $$p_\varepsilon(x,0) = e^{ik \cdot x/\varepsilon} \sum_{m'=0}^{N_0} f_{m'}(x)\phi_{m'}\left(\frac{x}{\varepsilon}; k\right).$$

It will turn out that **when** $k \ne 0$ asymptotic solutions of (4.9.2) satisfying (4.9.3) must necessarily have $u_\varepsilon(x,0)$ given in terms of p_ε; one **cannot** give $u_\varepsilon(x,0)$ arbitrarily.

In (4.9.3), $\phi_m(y;k)$ are the eigenfunctions corresponding to ω_m^2 in the shifted cell problem

(4.9.4) $$(\nabla_y + ik) \cdot \left[\frac{1}{\rho(y)}(\nabla_y + ik)\phi_m(y;k)\right] = -\omega^2(k)\frac{1}{\mu(y)}\phi_m(y;k)$$

in $H^1(2\pi Y)$, subject to (4.2.2) and $k \ne 0$ so that $\omega_m^2(k) \ne 0$. The normalization here is

(4.9.5) $$(\mu^{-1}\phi_m, \phi_{m'}) = \int_{2\pi Y} \mu^{-1}(y)\phi_m(y;k)\bar{\phi}_{m'}(y;k)\, dy = \delta_{mm'}.$$

We also have
(4.9.6)
$$\left(\frac{1}{\omega^2 \rho}(\nabla_y + ik)\phi_{m'} (\nabla_y + ik)\phi_m\right) = \frac{1}{\omega^2}\int_{2\pi Y} \rho^{-1}(y)|(\nabla_y + ik)\phi_m(y;k)|^2\, dy = 1.$$

We look for solutions of (4.9.2) of the following form where each term in the sum is a solution ($N \ge N_0$):

(4.9.7) $$u_\varepsilon = \sum_{m=0}^N e^{i(k \cdot x - \omega_m(k)t)/\varepsilon} \left[u_0^{(m)}\left(x, \frac{x}{\varepsilon}, t\right) + \varepsilon u_1^{(m)}\left(x, \frac{x}{\varepsilon}, t\right) + \cdots\right],$$

$$p_\varepsilon = \sum_{m=0}^N e^{i(k \cdot x - \omega_m(k)t)/\varepsilon} \left[p_0^{(m)}\left(x, \frac{x}{\varepsilon}, t\right) + \varepsilon p_1^{(m)}\left(x, \frac{x}{\varepsilon}, t\right) + \cdots\right].$$

Define operators $A_1(k)$ and A_2 by

(4.9.8)
$$A_1(k) = \begin{pmatrix} 0 & (\nabla_y + ik) \\ (\nabla_y + ik)\cdot & 0 \end{pmatrix} + i\omega \begin{pmatrix} \rho(y) & 0 \\ 0 & \mu^{-1}(y) \end{pmatrix},$$

(4.9.9) $$A_2 = \begin{pmatrix} 0 & \nabla_x \\ \nabla_{x'} & 0 \end{pmatrix} - \begin{pmatrix} \rho & 0 \\ 0 & \mu^{-1} \end{pmatrix} \frac{\partial}{\partial t}, \qquad \rho = \rho I, \qquad I = 3 \times 3 \text{ identity.}$$

Then (4.9.2) and (4.9.7) yield the following sequence of problems (we fix attention on one term in the sum)

(4.9.10)
$$A_1 \begin{pmatrix} u_0^{(m)} \\ p_0^{(m)} \end{pmatrix} = \begin{pmatrix} 0 \\ 0 \end{pmatrix},$$

(4.9.11)
$$A_1 \begin{pmatrix} u_1^{(m)} \\ p_1^{(m)} \end{pmatrix} + A_2 \begin{pmatrix} u_0^{(m)} \\ p_0^{(m)} \end{pmatrix} = \begin{pmatrix} 0 \\ 0 \end{pmatrix}, \qquad \text{etc.}$$

These equations (4.9.10) and (4.9.11) are more explicitly as follows:

(4.9.12)
$$(\nabla_y + ik)p_0^{(m)} + i\omega_m \rho u_0^{(m)} = 0,$$
$$(\nabla_y + ik)\cdot u_0^{(m)} + i\omega_m \mu^{-1} p_0^{(m)} = 0,$$

(4.9.13)
$$(\nabla_y + ik)p_1^{(m)} + i\omega_m \rho u_1^{(m)} + \nabla_x p_0^{(m)} - \rho u_{0t}^{(m)} = 0,$$
$$(\nabla_y + ik)\cdot u_1^{(m)} + i\omega_m \mu^{-1} p_1^{(m)} + \nabla_x \cdot u_0^{(m)} - \mu^{-1} p_{0t}^{(m)} = 0.$$

From (4.9.12) it follows that

(4.9.14)
$$p_0^{(m)} = \phi_m(y;k)p_{00}^{(m)}(x,t),$$
$$u_0^{(m)} = \frac{1}{i\omega_m(k)\rho(y)}(\nabla_y + ik)\phi_m(y;k)p_{00}^{(m)}(x,t).$$

We see therefore that when $k \neq 0$ and $p_\varepsilon(x,0)$ has the form (4.9.3), expansions of the form (4.9.7) leave no freedom for $u_0^{(m)}$; it is completely determined by $p_{00}^{(m)}$. The solvability condition for (4.9.13) yields a transport equation for $p_{00}^{(m)}(x,t)$.

Assume that (4.2.2) holds, i.e. that the $\omega_m^2(k)$ are distinct for $0 \leq m \leq N$. Multiplying the first equation in (4.9.13) by $\overline{(i\omega\rho)^{-1}(\nabla_y + ik)\phi_m}$, the second by $\bar{\phi}_m$, integrating over $y \in 2\pi Y$ and subtracting we obtain

(4.9.15) $$2p_{00t}^{(m)} + \frac{1}{\omega_m(k)}\left[\left(\phi_{m'}\frac{1}{i\rho}(\nabla_y + ik)\phi_m\right)\right.$$
$$\left. + \left(\phi_{m'}(\nabla_y + ik)\left(\frac{1}{i\rho}\phi_m\right)\right)\right] \cdot \nabla_x p_{00}^{(m)} = 0,$$

which is the same thing as (cf. (4.3.9))

$$p_{00t}^{(m)} + \frac{\partial \omega_m(k)}{\partial k} \cdot \nabla_x p_{00}^{(m)} = 0, \qquad t > 0$$

(4.9.16)
$$p_{00}^{(m)}(x,0) = f_m(x), \qquad 0 \leq m \leq N_0,$$

$$p_{00}^{(m)}(x,t) \equiv 0, \qquad N_0 + 1 \leq m \leq N.$$

To get the next terms we define $\chi_p^{(m)}(y;k)$, $p = 1, 2, 3$ by

$$(4.9.17) \quad (\nabla_y + ik) \cdot \left[\frac{1}{\rho}(\nabla_y + ik)\chi_p^{(m)} \right] + \omega_m^2 \mu^{-1}\chi_p^{(m)}$$

$$+ \left[\frac{1}{\rho}(\nabla_y + ik)\phi_m + (\nabla_y + ik)\left(\frac{1}{\rho}\phi_m \right) - i\frac{\partial \omega_m^2}{\partial k} \right] \cdot e_p = 0,$$

$$(4.9.18) \qquad\qquad (\phi^m, \chi_p^{(m)}) = 0,$$

where e_p, $p = 1, 2, 3$ are the column vectors $(1, 0, 0)$, $(0, 1, 0)$ and $(0, 0, 1)$, respectively. This is the same as (4.3.11), (4.3.12). In terms of $\chi_p^{(m)}$ we can write the solutions $p_1^{(m)}$ and $u_1^{(m)}$ of (4.9.13) as follows:

$$(4.9.19) \qquad p_1^{(m)} = \sum_{p=1}^{3} \chi_p^{(m)}(y;k)\frac{\partial p_{00}^{(m)}(x,t)}{\partial x_p} + \phi^{(m)}(y;k)p_{10}^{(m)}(x,t),$$

$$(4.9.20) \qquad u_1^{(m)} = \frac{1}{i\omega_m \rho}\left[\rho u_{0t}^{(m)} - \nabla_x p_0^{(m)} - (\nabla_y + ik)p_1^{(m)} \right].$$

Note again that $u_1^{(m)}$ is completely determined by $p_1^{(m)}$ and $p_0^{(m)}$. The functions $p_{10}^{(m)}(x,t)$ satisfy inhomogeneous transport equations obtained by the solvability condition for $p_2^{(m)}$, $u_2^{(m)}$, etc.

In conclusion: expansions for (4.9.2) of the form (4.9.7) can be constructed as usual when $k \neq 0$. The proof of their asymptotic validity in the form of Theorem 4.3 is essentially the same as the proof of that theorem. The system form of the problem plays no role at all. In particular the velocities u_ε are completely determined by the pressure p_ε and initial values cannot be prescribed for u_ε. This is a special feature of asymptotic solutions of the form (4.9.7).

4.10. Dual homogenization formulas. The case $k = 0$ requires special treatment. It is the static case studied in detail in Chapters 1 and 2. For comparison purposes with the previous section we give the results here in somewhat different form.

We consider the problem

$$(4.10.1) \qquad \rho\left(\frac{x}{\varepsilon} \right)u_{\varepsilon t} = \nabla p_\varepsilon, \qquad p_{\varepsilon t} = \mu\left(\frac{x}{\varepsilon} \right)\nabla \cdot u_\varepsilon, \qquad t > 0, \qquad x \in \mathbb{R}^3,$$

$$(4.10.2) \qquad p_\varepsilon(x,0) = f(x), \qquad u_\varepsilon(x,0) = g(x).$$

Here ρ and μ satisfy (4.9.1) and are smooth and $f \in C_0^\infty(\mathbb{R}^3)$, $g \in (C_0^\infty(\mathbb{R}^3))^3$.

If we look for an expansion of the form

$$(4.10.3) \qquad p_\varepsilon(x,t) = p^{(0)}\left(x, \frac{x}{\varepsilon}, t \right) + \varepsilon p^{(1)}\left(x, \frac{x}{\varepsilon}, t \right) + \varepsilon^2 p^{(2)}\left(x, \frac{x}{\varepsilon}, t \right) + \cdots,$$

$$u_\varepsilon(x,t) = u^{(0)}\left(x, \frac{x}{\varepsilon}, t \right) + \varepsilon u^{(1)}\left(x, \frac{x}{\varepsilon}, t \right) + \varepsilon^2 u^{(2)}\left(x, \frac{x}{\varepsilon}, t \right) + \cdots,$$

we find that we cannot satisfy the initial conditions $u_\varepsilon(x,0) = g(x)$ even to principal order in ε. Another ansatz will not work here; it is necessary to **either** consider expansions for special classes of data (not (4.10.2)) as in the previous section or to seek weak convergence results with general data like (4.10.2). We shall obtain here weak convergence results.

We begin with some constructions and then state and prove the main theorem. The results go through even for (4.10.1) **with boundary conditions** as follows.

Let $\mathcal{O} \subset \mathbb{R}^3$ be a bounded open set with $\partial\mathcal{O}$ its **smooth** boundary. We suppose that

$$(4.10.4) \qquad \text{equations (4.10.1) hold in } \mathcal{O} \text{ and } p_\varepsilon(x, t) = 0,$$
$$x \in \partial\mathcal{O} \text{ with } f \in C_0^\infty(\mathcal{O}).$$

The weak form of (4.10.1), (4.10.2), (4.10.4) is as follows. Let $v(x, t)$, $\pi(x, t)$ be in $(C^\infty(\mathbb{R}^3 \times [0, \infty)))^3$ and $C^\infty(\mathbb{R}^3 \times [0, \infty))$, respectively, such that they vanish identically for t large and $\pi(x, t)$ vanishes outside a compact subset of $\mathcal{O}$. **Denote by V the class of all such functions.** Multiplying by v and π, integrating and integrating by parts we obtain

$$(4.10.5) \quad \int_0^\infty \int_\mathcal{O} \left[\rho\left(\frac{x}{\varepsilon}\right) v_t \cdot u_\varepsilon + \mu^{-1}\left(\frac{x}{\varepsilon}\right) \pi_t p_\varepsilon - \nabla \cdot v p_\varepsilon - \nabla\pi \cdot u_\varepsilon \right] \, dx \, dt$$
$$+ \int_\mathcal{O} \left[\rho\left(\frac{x}{\varepsilon}\right) v(x, 0)g(x) + \mu^{-1}\left(\frac{x}{\varepsilon}\right) \pi(x, 0)f(x) \right] \, dx = 0, \qquad (v, \pi) \in V.$$

If we let

$$(4.10.6) \qquad A^\varepsilon = \begin{pmatrix} \rho\left(\frac{x}{\varepsilon}\right) & 0 \\ 0 & \mu^{-1}\left(\frac{x}{\varepsilon}\right) \end{pmatrix} \frac{\partial}{\partial t} - \begin{pmatrix} 0 & \nabla \\ \nabla\cdot & 0 \end{pmatrix},$$

then we can write (4.10.5) in the form $(\rho_\varepsilon = \rho\left(\frac{x}{\varepsilon}\right), \mu = \mu\left(\frac{x}{\varepsilon}\right))$

$$(4.10.7) \qquad \int_0^\infty \int_\mathcal{O} A^\varepsilon \begin{pmatrix} v \\ \pi \end{pmatrix} \cdot \begin{pmatrix} u_\varepsilon \\ p_\varepsilon \end{pmatrix} \, dx \, dt + \int_\mathcal{O} \begin{pmatrix} \rho_\varepsilon v(0) \\ \mu_\varepsilon^{-1}\pi(0) \end{pmatrix} \cdot \begin{pmatrix} g \\ f \end{pmatrix} \, dx = 0.$$

We introduce two sets of cell problems for vector functions $\bar{\chi}_j(y)$ and scalar functions $\chi_j(y)$, $j = 1, 2, 3$ defined on $2\pi Y$ as follows[53]

$$(4.10.8) \qquad \nabla \wedge [\rho(e_j + \nabla\tilde{\chi}_j)] = 0, \qquad \nabla \cdot \tilde{\chi}_j = 0, \qquad \mathcal{M}[\tilde{\chi}_j] = 0,$$
$$(4.10.9) \qquad \nabla \cdot [\rho^{-1}(e_j + \nabla\chi_j)] = 0, \qquad \mathcal{M}[\chi_j] = 0, \qquad j = 1, 2, 3.$$

These functions are uniquely defined by (4.10.8) and (4.10.9).

Define further[54] the constant matrix q^{-1} with entries

$$(4.10.10) \qquad q_{ij}^{-1} = \mathcal{M}\left[\rho(\delta_{ij} + \nabla \wedge \tilde{\chi}_j)_i \right].$$

By direct computation we find that the following identities hold (pointwise on $2\pi Y$)

$$(4.10.11) \qquad \rho\left(\delta_{ij} + (\nabla \wedge \tilde{\chi}_j)_i \right) = \sum_{k=1}^3 \left(\delta_{ik} + (\nabla\chi_k)_i \right) q_{kj}^{-1},$$

$$(4.10.12) \qquad \sum_{k=1}^3 \left(\delta_{ik} + (\nabla \wedge \tilde{\chi}_k)_i \right) q_{kj} = \rho^{-1} \left(\delta_{ij} + (\nabla\chi_j)_i \right).$$

From these identities it follows that q is also given by **the dual formula to (4.10.10):**

$$(4.10.13) \qquad q_{ij} = \mathcal{M}\left[\rho^{-1}(\delta_{ij} + (\nabla\chi_j)_i) \right].$$

It is easily verified that q^{-1} (or q) is positive definite (we use (4.9.1)).

---

[53]Here e_j are column vectors $e_1 = (1, 0, 0)$, $e_2 = (0, 1, 0)$, $e_3 = (0, 0, 1)$ and, as in previous chapters, $\mathcal{M}(f) = \frac{1}{(2\pi)^3} \int_{2\pi Y} f(y) \, dy$.

[54]$(\nabla \wedge \tilde{\chi}_j)_i$ is the ith component of $\nabla \wedge \tilde{\chi}_j$.

Define the operator

$$(4.10.14) \qquad \mathcal{A} = \begin{pmatrix} q^{-1} & 0 \\ 0 & \mathcal{M}[\mu^{-1}] \end{pmatrix} \frac{\partial}{\partial t} - \begin{pmatrix} 0 & 0 \\ \nabla \cdot & 0 \end{pmatrix}.$$

This is the homogenized operator corresponding to (4.10.6). Let (u, p) be the solution[55] of the problem

$$(4.10.15) \qquad \mathcal{A} \begin{pmatrix} u \\ p \end{pmatrix} = 0, \qquad t > 0, \qquad x \in \mathcal{O},$$

$$p(x, t) = 0, \qquad x \in \partial\mathcal{O}, \qquad u(x, 0) = g(x), \qquad p(x, 0) = f(x).$$

THEOREM 4.7. *For each $(v, \pi) \in V$ fixed the solution $(u_\varepsilon, p_\varepsilon)$ of (4.10.1), (4.10.2) and (4.10.4) satisfies*

$$(4.10.16) \qquad \lim_{\varepsilon \downarrow 0} \int_0^\infty \int_\mathcal{O} (v \cdot u_\varepsilon + \pi p_\varepsilon) \, dx \, dt = \int_0^\infty \int_\mathcal{O} (v \cdot u + \pi p) \, dx \, dt,$$

where (u, p) is the solution of (4.10.15).

REMARK 4.8. The convergence is weak star convergence. Note also that very little regularity is needed here but we shall not attempt to give the optimal conditions.

PROOF. From the energy identity for (4.10.1) and (4.9.1) we conclude there is a subsequence of $(u_\varepsilon, p_\varepsilon)$, also denoted by $(u_\varepsilon, p_\varepsilon)$, which converges weak star in $L^\infty([0, T]; (L^2(\mathcal{O}))^4)$ for each $T < \infty$. It remains, as usual, to identify the limit as the unique solution (u, p) of (4.10.15) which completes the proof. This is done by adjoint expansion as follows.

Fix (v, π) in V. We define $v^{(1)}(x, y, t)$ and $\pi^{(1)}(x, y, t)$ as follows:

$$(4.10.17) \qquad \pi^{(1)} = \sum_{k,j=1}^3 \chi_k(y) q_{kj}^{-1} \frac{\partial v_j(x, t)}{\partial t},$$

$$(4.10.18) \qquad v_i^{(1)} = \sum_{k,j=1}^3 (\nabla_x \wedge (\tilde{\chi}_k(y) v_k(x, t)))_i + M_i(y) \pi_t(x, t), \qquad i = 1, 2, 3,$$

where $M_i(y)$ are periodic functions such that

$$(4.10.19) \qquad \sum_{i=1}^3 \frac{\partial M_i(y)}{\partial y_i} = \mu^{-1}(y) - \mathcal{M}(\mu^{-1}).$$

Define further

$$(4.10.20) \qquad v_{i\varepsilon}(x, t) = v_i(x, t) + \sum_{j=1}^3 (\nabla_y \wedge \tilde{\chi}_j(y))_i v_j(x, t) \Big|_{y=x/\varepsilon} + v_i^{(1)} \left(x, \frac{x}{\varepsilon}, t \right),$$

$$(4.10.21) \qquad \tilde{\pi}_\varepsilon(x, t) = \pi(x, t) + \varepsilon \pi^{(1)} \left(x, \frac{x}{\varepsilon}, t \right).$$

Even though π vanishes near $\partial\mathcal{O}$ this is not true for $\tilde{\pi}_\varepsilon$ so we must cut it off as follows. Since $\partial\mathcal{O}$ is smooth there exists a function $m_\varepsilon(x)$ such that

$$(4.10.22) \qquad m_\varepsilon(x) = 0, \text{ when } \mathrm{dist}(x, \partial\mathcal{O}) \le \sqrt{\varepsilon},$$

$$m_\varepsilon(x) = 1, \text{ when } \mathrm{dist}(x, \partial\mathcal{O}) \ge 2\sqrt{\varepsilon},$$

---

[55]This problem has a unique smooth solution ($\partial\mathcal{O}$ is smooth).

$$|\nabla m_\varepsilon(x)| \leq C\varepsilon^{-1/2}, \qquad x \in \overline{\mathcal{O}}, \qquad C \text{ a constant.}$$

Now put

(4.10.23) $$\pi_\varepsilon(x,t) = \pi(x,t) + \varepsilon m_\varepsilon(x)\pi^{(1)}\left(x,\frac{x}{\varepsilon},t\right).$$

It can be verified by direct computation that the above constructions produce the following result (which was the objective of the constructions all along):

(4.10.24) $$A^\varepsilon\begin{pmatrix}v_\varepsilon \\ \pi_\varepsilon\end{pmatrix} = \mathcal{A}\begin{pmatrix}v \\ \pi\end{pmatrix} + \begin{pmatrix}h_1^\varepsilon \\ h_2^\varepsilon\end{pmatrix},$$

where A^ε and $\mathcal{A}$ are defined by (4.10.6) and (4.10.14) and

(4.10.25) $$h_1^\varepsilon = \varepsilon(\rho_\varepsilon v_t^{(1)} - m_\varepsilon \nabla_x \pi^{(1)}) - (m_\varepsilon - 1)\nabla_y \pi^{(1)} - \varepsilon\nabla m_\varepsilon \pi^{(1)},$$

(4.10.26) $$h_2^\varepsilon = \varepsilon(\mu^{-1}m_\varepsilon \pi_t^{(1)} - \nabla_x \cdot v^{(1)}).$$

Note that

(4.10.27) $$\sup_{0 \leq t} \int |h_1^\varepsilon(x,t)|^2 \, dx \to 0 \text{ as } \varepsilon \to 0,$$

and

(4.10.28) $$\sup_{0 \leq t} \int |h_2^\varepsilon(x,t)|^2 \, dx \to 0 \text{ as } \varepsilon \to 0.$$

Now we go to (4.10.7) and use it with (v,π) replaced by $(v_\varepsilon,\pi_\varepsilon)$. This yields,[56] along with (4.10.24), (4.10.20) and (4.10.23),

(4.10.29) $$\int_0^\infty \int_{\mathcal{O}} \mathcal{A}\begin{pmatrix}v \\ \pi\end{pmatrix} \cdot \begin{pmatrix}u_\varepsilon \\ p_\varepsilon\end{pmatrix} \, dx \, dt + \int_0^\infty \int_{\mathcal{O}} \begin{pmatrix}h_1^\varepsilon \\ h_2^\varepsilon\end{pmatrix} \cdot \begin{pmatrix}u_\varepsilon \\ p_\varepsilon\end{pmatrix} \, dx \, dt$$
$$+ \int_{\mathcal{O}} \begin{pmatrix}\rho_\varepsilon(\delta_{ij} + (\nabla \wedge \tilde{\chi}_j)_i^\varepsilon)v_j(0) \\ \mu_\varepsilon^{-1}\pi(0)\end{pmatrix} \cdot \begin{pmatrix}g \\ f\end{pmatrix} \, dx$$
$$+ \int_{\mathcal{O}} \begin{pmatrix}\varepsilon f_\varepsilon v^{(1)\varepsilon}(0) \\ \varepsilon\mu_\varepsilon^{-1}m_\varepsilon \pi^{(1)\varepsilon}(0)\end{pmatrix} \cdot \begin{pmatrix}g \\ f\end{pmatrix} \, dx = 0.$$

Passing to the limit $\varepsilon \to 0$ and using (4.10.27), (4.10.28) **and** (4.10.10), we obtain

(4.10.30) $$\int_0^\infty \int_{\mathcal{O}} \mathcal{A}\begin{pmatrix}v \\ \pi\end{pmatrix} \cdot \begin{pmatrix}u \\ p\end{pmatrix} \, dx \, dt + \int_{\mathcal{O}} \begin{pmatrix}q^{-1}v(0) \\ \mathcal{M}[\mu^{-1}]\pi(0)\end{pmatrix} \cdot \begin{pmatrix}g \\ f\end{pmatrix} \, dx = 0.$$

This is precisely the weak form of (4.10.15) so the proof of Theorem 4.7 is complete. ∎

4.11. Maxwell's equations. We shall now consider Maxwell's equations in a periodic medium, i.e., when the dielectric "constant" and the magnetic permeability are rapidly varying periodic functions of space.

To avoid confusion with the small parameter ε we depart the traditional notation and

(4.11.1)   denote by $\rho(y)$ the dielectric "constant" and by $\mu(y)$ the
            magnetic permeability; we assume that (4.9.1) holds.

---

[56]We use summation convention in (4.10.29) and the notation
$$(g)^\varepsilon = g(y)|_{y=x/\varepsilon} = g(x/\varepsilon).$$

Maxwell's equations are then (cf. (2.10.17)–(2.10.19))

$$(4.11.2) \qquad \begin{pmatrix} \rho_\varepsilon & 0 \\ 0 & \mu_\varepsilon \end{pmatrix} \frac{\partial}{\partial t} \begin{pmatrix} E_\varepsilon \\ H_\varepsilon \end{pmatrix} = \begin{pmatrix} 0 & \nabla\wedge \\ -\nabla\wedge & 0 \end{pmatrix} \begin{pmatrix} E_\varepsilon \\ H_\varepsilon \end{pmatrix}, \qquad t > 0,$$

where $\rho_\varepsilon(x) = \rho(x/\varepsilon)$ times the 3×3 identity and similarly for μ_ε.

We shall deal with the static case only here; the analogs of the results of Section 4.9 carry over without much change. In the static case we can also handle boundary conditions. We now describe the initial and boundary conditions for (4.11.2) that we shall adopt.

Let $\mathcal{O} \subset \mathbb{R}^3$ be a bounded open set with **smooth** boundary $\partial\mathcal{O}$. Let $\hat{n}$ denote the unit outward normal to $\partial\mathcal{O}$. We take (4.11.2) to hold in $\mathcal{O}$ subject to the boundary condition (conducting walls)

$$(4.11.3) \qquad \hat{n} \wedge E_\varepsilon = 0 \text{ for } x \in \partial\mathcal{O}.$$

For initial conditions we take

$$(4.11.4) \qquad E_\varepsilon(x,0) = \rho_\varepsilon^{-1}(x)f(x), \qquad H_\varepsilon(x,0) = \mu_\varepsilon^{-1}(x)g(x),$$

with $f \in (C_0^\infty(\mathcal{O}))^3$ and $g \in (C^\infty(\mathcal{O}))^3$ and satisfying

$$(4.11.5) \qquad \nabla \cdot f(x) = \nabla \cdot g(x) = 0.$$

Conditions (4.11.5) are imposed so that

$$(4.11.6) \qquad \nabla \cdot (\rho_\varepsilon(x)E_\varepsilon(x,0)) = 0, \qquad \nabla \cdot (\mu_\varepsilon(x)H_\varepsilon(x,0)) = 0,$$

which are the usual complementing conditions on data and are preserved by (4.11.2) at later times.

Let V denote the class of all functions $(u(x,t), v(x,t))$ where u and v are vector-valued and smooth, i.e. in $(C^\infty([0,\infty) \times \mathcal{O}))^3$, such that u and v are identically zero for t large and $\hat{n} \wedge u$ vanishes identically near $\partial\mathcal{O}$.

With $(u,v) \in V$ we introduce the weak form of (4.11.2), (4.11.3), (4.11.4) (this problem has a unique smooth solution as it stands when $\varepsilon > 0$ is fixed). Multiplying by (u,v), integrating and integrating by parts as usual we obtain

$$(4.11.7) \quad \int_0^\infty \int_\mathcal{O} [\rho_\varepsilon u_t \cdot E_\varepsilon + \mu_\varepsilon v_t \cdot H_\varepsilon - \nabla \wedge v \cdot E_\varepsilon + \nabla \wedge u \cdot H_\varepsilon] \, dx \, dt$$

$$+ \int_\mathcal{O} [u(0,x) \cdot f(x) + v(0,x) \cdot g(x)] \, dx = 0, \qquad (u,v) \in V.$$

This is the weak form of (4.11.2), (4.11.3), (4.11.4). If we define

$$(4.11.8) \qquad A^\varepsilon = \begin{pmatrix} \rho_\varepsilon & 0 \\ 0 & \mu_\varepsilon \end{pmatrix} \frac{\partial}{\partial t} - \begin{pmatrix} 0 & \nabla\wedge \\ -\nabla\wedge & 0 \end{pmatrix},$$

we can write it in operator form as follows:

$$(4.11.9) \qquad \int_0^\infty \int_\mathcal{O} A^\varepsilon \begin{pmatrix} u \\ v \end{pmatrix} \cdot \begin{pmatrix} E_\varepsilon \\ H_\varepsilon \end{pmatrix} \, dx \, dt + \int_\mathcal{O} \begin{pmatrix} u(0,x) \\ v(0,x) \end{pmatrix} \cdot \begin{pmatrix} f(x) \\ g(x) \end{pmatrix} \, dx = 0.$$

Now we continue with the necessary construction of cell functions which will enter in the proof of the theorem that follows, i.e. in the adjoint expansion (the constructions are, of course, dictated by the usual perturbation expansions).

We define two sets of cell functions; one associated with ρ and one with μ:

$$(4.11.10) \qquad \chi_j(y;\rho), \qquad \chi_j(y;\mu), \qquad \tilde{\chi}_j(y;\rho), \qquad \tilde{\chi}_j(y;\mu), \qquad j = 1,2,3.$$

The χ's are scalar and the $\tilde{\chi}$ are vector-valued functions on $2\pi Y$. They are defined uniquely by (cf. (4.10.8), (4.10.9))

$$\nabla \cdot [\rho(e_j + \nabla \chi_j(y;\rho))] = 0, \qquad \mathcal{M}[\chi_j] = 0,$$

(4.11.11)

$$\nabla \cdot [\mu(e_j + \nabla \chi_j(y;\mu))] = 0, \qquad \mathcal{M}[\chi_j] = 0,$$

$$\nabla \wedge \left[\rho^{-1}(e_j + \nabla \wedge \tilde{\chi}_j(y;\rho))\right] = 0, \qquad \nabla \cdot \tilde{\chi}_j = 0, \qquad \mathcal{M}[\tilde{\chi}_j] = 0,$$

(4.11.12)

$$\nabla \wedge \left[\mu^{-1}(e_j + \nabla \wedge \tilde{\chi}_j(y;\mu))\right] = 0, \qquad \nabla \cdot \tilde{\chi}_j = 0, \qquad \mathcal{M}[\chi_j] = 0, \qquad j = 1, 2, 3.$$

We define matrices $q(\rho) = (q_{ij}(\rho))$ and $q(\mu) = (q_{ij}(\mu))$ by

$$(4.11.13) \qquad q_{ij}(\rho) = \mathcal{M}\left[\rho\left(\delta_{ij} + \frac{\partial \chi_j(y;\rho)}{\partial y_i}\right)\right],$$

$$(4.11.14) \qquad q_{ij}(\mu) = \mathcal{M}\left[\mu\left(\delta_{ij} + \frac{\partial \chi_j(y;\mu)}{\partial y_i}\right)\right], \qquad i, j = 1, 2, 3.$$

It can be verified easily that $q(\rho)$ and $q(\mu)$ are positive definite matrices.

We also have flux identities (cf. (4.10.11), (4.10.12)),

$$(4.11.15) \qquad \rho^{-1}(y)\left(\delta_{ij} + (\nabla \wedge \tilde{\chi}_j(y;\rho))_i\right) = \sum_{k=1}^{3}\left(\delta_{ik} + (\nabla \chi_k(y;\rho))_i\right) q_{kj}^{-1}(\rho),$$

$$(4.11.16) \qquad \mu^{-1}(y)\left(\delta_{ij} + (\nabla \wedge \tilde{\chi}_j(y;\mu))_i\right) = \sum_{k=1}^{3}\left(\delta_{ik} + (\nabla \chi_k(y;\mu))_i\right) q_{kj}^{-1}(\mu),$$

and the **dual homogenization formulas**

$$(4.11.17) \qquad q_{ij}^{-1}(\rho) = \mathcal{M}\left[\rho^{-1}(\delta_{ij} + (\nabla \wedge \tilde{\chi}_j(y;\rho))_i)\right],$$

$$q_{ij}^{-1}(\mu) = \mathcal{M}\left[\mu^{-1}(\delta_{ij} + (\nabla \wedge \tilde{\chi}_j(y;\mu))_i)\right], \qquad i, j = 1, 2, 3.$$

Given $(u, v) \in V$, define $u^{(1)}(x, y, t)$ and $v^{(1)}(x, y, t)$ as follows:

$$(4.11.18) \qquad u_i^{(1)}(x, y, t) = \sum_{k=1}^{3} \chi_k(y;\rho)\frac{\partial u_k(x,t)}{\partial x_i}$$

$$- \sum_{k,j=1}^{3} \tilde{\chi}_k(y;\mu) q_{kj}(\mu)\frac{\partial v_j(x,t)}{\partial t},$$

$$(4.11.19) \qquad v_i^{(1)}(x, y, t) = \sum_{k=1}^{3} \chi_k(y;\mu)\frac{\partial v_k(x,t)}{\partial x_i}$$

$$+ \sum_{k,j=1}^{3} \tilde{\chi}_k(y;\rho) q_{kj}(\rho)\frac{\partial u_j(x,t)}{\partial t}, \qquad i = 1, 2, 3.$$

Let $m_\varepsilon(x)$ be the cutoff function defined by (4.10.22) and define

$$(4.11.20) \qquad u_{i\varepsilon}(x,t) = u_i(x,t) + \left[\sum_{k=1}^{3} \frac{\partial \chi_k(y;\rho)}{\partial y_i} u_k(x,t)\right]_{y=x/\varepsilon}$$

$$+ m_\varepsilon(x) u_i^{(1)}\left(x, \frac{x}{\varepsilon}, t\right),$$

$$(4.11.21) \qquad v_{i\varepsilon}(x,t) = v_i(x,t) + \left[\sum_{k=1}^{3} \frac{\partial \chi_k(y;\mu)}{\partial y_i} v_k(x,t)\right]_{y=x/\varepsilon}$$

$$+ v^{(1)}\left(x, \frac{x}{\varepsilon}, t\right), \qquad i = 1,2,3.$$

Define also the **homogenized operator** $\mathcal{A}$ by

$$(4.11.22) \qquad \mathcal{A} = \begin{pmatrix} q(\rho) & 0 \\ 0 & q(\mu) \end{pmatrix} \frac{\partial}{\partial t} - \begin{pmatrix} 0 & \nabla\wedge \\ -\nabla\wedge & 0 \end{pmatrix}.$$

Note that $q(\rho)$ and $q(\mu)$ are 3×3 matrices now given by (4.11.13), (4.11.14).

The above constructions lead to the following facts which can be verified by direct computation. For $(u,v) \in V$ given,

(1) $(u_\varepsilon, v_\varepsilon)$ belongs also to V, and
(2)

$$(4.11.23) \qquad A^\varepsilon \begin{pmatrix} u_\varepsilon \\ v_\varepsilon \end{pmatrix} = \begin{pmatrix} u \\ v \end{pmatrix} + \begin{pmatrix} h_1^\varepsilon \\ h_2^\varepsilon \end{pmatrix},$$

where h_i^ε, $i = 1, 2$ are such that

$$(4.11.24) \qquad \limsup_{\varepsilon \downarrow 0} \int |h_i^\varepsilon(x,t)|^2 \, dx = 0, \qquad i = 1,2.$$

Let (E, H) be the solution of the **homogenized** problem corresponding to (4.11.2), (4.11.3), (4.11.4):

$$\mathcal{A}\begin{pmatrix} E \\ H \end{pmatrix} = \begin{pmatrix} 0 \\ 0 \end{pmatrix}, \qquad t > 0, \qquad x \in \mathcal{O},$$

$$(4.11.25) \qquad \hat{n} \wedge E = 0, \qquad x \in \partial\mathcal{O},$$

$$E(x,0) = q^{-1}(\rho)f(x), \qquad H(x,0) = q^{-1}(\mu)g(x), \qquad x \in \mathcal{O}.$$

Note that problem (4.11.25) satisfies the constraints

$$(4.11.26) \qquad \nabla \cdot (q(\rho)E(x,0)) = 0, \qquad \nabla \cdot (q(\mu)H(x,0)) = 0$$

as shouild be in view of (4.11.6) and (4.11.5).

THEOREM 4.9. *The solution $(E_\varepsilon(x,t), H_\varepsilon(x,t))$ of (4.11.2), (4.11.3), (4.11.4) converges weak star in $L^\infty([0,T]; L^6(\mathcal{O}))$, $T < \infty$ and arbitrary, to $(E(x,t), H(x,t))$ solution of (4.11.25) i.e., for each $(u,v) \in V$*

$$(4.11.27) \qquad \int_0^\infty \int_\mathcal{O} (u \cdot E_\varepsilon + v \cdot H_\varepsilon) \, dx \, dt \to \int_0^\infty \int_\mathcal{O} (u \cdot E + v \cdot H) \, dx \, dt,$$

as $\varepsilon \to 0$.

PROOF. The proof is a verbatim repetition of the one of Theorem 4.7 of the previous section. The construction of the cell functions χ and $\tilde{\chi}$ that lead to (4.11.23) is the basic element of the proof and, as was pointed out already, this is straight perturbation theory. ∎

4.12. A one dimensional example. It is too difficult to deal in the high frequency context with boundary value problems even in one dimension. This is due to the complexity of the phenomena that take place near boundaries. **If the variations of the coefficients from a uniform background** (i.e., constant values) **are small** then a good deal can be said; even for higher dimensions and random media.[57] It is very likely that good progress will be made in such cases but we confine the discussion here to the following simple example.

In a one dimensional medium let the inhomogeneities occupy the interval $[0, L]$. With the time factor $e^{-i\omega t/\varepsilon}$ omitted the wave field $u_\varepsilon(x) = u_\varepsilon(x; k)$ satisfies

$$(4.12.1) \qquad \frac{d^2 u_\varepsilon}{dx^2} + \left(\frac{k}{\varepsilon}\right)^2 n^2\left(\frac{x}{\varepsilon}\right) u_\varepsilon = 0, \qquad 0 < x < L,$$

$$u_\varepsilon = T_\varepsilon e^{-ikx/\varepsilon}, \qquad x < 0,$$

$$u_\varepsilon = R_\varepsilon e^{ikx/\varepsilon} + e^{-ikx/\varepsilon}, \qquad x > L.$$

$$u_\varepsilon \text{ and } \frac{du_\varepsilon}{dx} \text{ continuous at } x = 0, \ x = L.$$

Here $n(y)$ is the index of refraction. We assume it is a real-valued **almost periodic** function such that

$$(4.12.2) \qquad\qquad n^2(y) = 1 + \varepsilon\mu(y),$$

with $|\mu(y)| \leq 1$, $\mu(y)$ real-valued and almost periodic.

To solve (4.12.1) we write u_ε and du_ε/dx in terms of new functions $A_\varepsilon, B_\varepsilon$ as follows:

$$(4.12.3) \qquad\qquad u_\varepsilon = A_\varepsilon(x) e^{i kx/\varepsilon} + B_\varepsilon(x) e^{-ikx/\varepsilon},$$

$$\frac{du_\varepsilon}{dx} = \frac{ik}{\varepsilon}\left(A_\varepsilon(x) e^{ikx/\varepsilon} - B_\varepsilon(x) e^{-ikx/\varepsilon}\right).$$

It follows then that

$$(4.12.4) \qquad \frac{dA_\varepsilon}{dx} = \frac{ik}{2}\mu\left(\frac{x}{\varepsilon}\right)\left(A_\varepsilon(x) + B_\varepsilon(x) e^{-2ikx/\varepsilon}\right),$$

$$\frac{dB_\varepsilon}{dx} = -\frac{ik}{2}\mu\left(\frac{x}{\varepsilon}\right)\left(A_\varepsilon(x) e^{2ikx/\varepsilon} + B_\varepsilon(x)\right), \qquad 0 < x < L,$$

with boundary conditions

$$(4.12.5) \qquad\qquad A_\varepsilon(0) = 0, \qquad B_\varepsilon(L) = 1.$$

We also have that R_ε and T_ε, the **reflection** and **transmission** coefficients respectively ($R_\varepsilon = R_\varepsilon(L; k)$ and $T_\varepsilon = T_\varepsilon(L; k)$ are complex valued), are given by[58]

$$(4.12.6) \qquad\qquad A_\varepsilon(L) = R_\varepsilon, \qquad B_\varepsilon(0) = T_\varepsilon.$$

One can easily analyze directly the asymptotic behavior of the solution of the two-point boundary value problem (4.12.4) as $\varepsilon \to 0$. We are primarily interested in $R_\varepsilon(L; k)$, the reflection coefficient. This can be handled directly as follows.

---

[57]See Kohler-Papanicolaou [**58**] and [**59**] and references therein for random wave problems. Formally one can treat even multidimensional random wave problems as in Burridge-Papanicolaou [**90**] but techniques for proofs have not been developed yet.

   [58]Note that because μ is real valued and $|\mu| \leq 1$, for $\varepsilon < 1$ we have $|R_\varepsilon|^2 + |T_\varepsilon|^2 = 1$; there is no loss or gain in energy.

From (4.12.4) we find that $U_\varepsilon(x) \equiv A_\varepsilon(x)/B_\varepsilon(x)$ satisfies that following equation
(4.12.7)
$$\frac{d}{dx}U_\varepsilon(x) = \frac{ik}{2}\mu\left(\frac{x}{\varepsilon}\right)\left[U_\varepsilon(x)e^{ikx/\varepsilon} + e^{-ikx/\varepsilon}\right]^2, \qquad 0 < x \le L, \qquad U_\varepsilon(0) = 0,$$
and

(4.12.8)
$$\frac{A_\varepsilon(L)}{B_\varepsilon(L)} = U_\varepsilon(L) = R_\varepsilon(L).$$

Therefore to study $R_\varepsilon(L; k)$ as $\varepsilon \to 0$ it is enough to analyze the asymptotic behavior of the nonlinear initial value problem (4.12.7).

The asymptotic analysis of (4.12.7) amounts to a direct and simple application of the method of averaging.[59] It is quite reasonable to assume that

(4.12.9)
$$\lim_{T\uparrow\infty} \frac{1}{T}\int_0^T \mu(y)\,dy = 0,$$

while we define

(4.12.10)
$$\lim_{T\uparrow\infty} \frac{1}{T}\int_0^T \mu(y)e^{2iky}\,dy \equiv S(2k),$$

which is the Fourier coefficient of μ at frequency $2k$. Since μ is real

(4.12.11)
$$S(-2k) = \bar{S}(2k).$$

We also have that

(4.12.12) $S(2k) = 0$ except when $2k$ equals one of the discrete values that form the spectrum of the almost periodic function μ.

Applying the averaging method to (4.12.7) we find that for any $L < \infty$

(4.12.13)
$$\lim_{\varepsilon\downarrow 0}\sup_{0\le x\le L} |U_\varepsilon(x) - U(x)| = 0,$$

where $U(x)$ satisfies

(4.12.14)
$$\frac{dU(x)}{dx} = \frac{ik}{2}\left[S(2k)U^2(x) + \bar{S}(2k)\right], \qquad 0 < x \le L, \qquad U(0) = 0,$$

and hence

(4.12.15) $R_\varepsilon(L; k) \to U(L) = U(L; k)$, as $\varepsilon \to 0$.

From (4.12.14) and (4.12.12) we conclude immediately that

(4.12.16) $R_\varepsilon(L; k)$ **is asymptotically zero whenever $2k$ is not a point in the spectrum of μ.** Physically, this says that the medium appears asymptotically to be transparent **except** at resonant frequencies (frequencies that match, i.e. resonate with, the inhomogeneities in the above way).

---

[59]N. N. Bogoliubov and Ju. A. Mitropolsky [**20**]. For (4.12.7), the limit equation (4.12.14) is obtained by fixing U_ε and averaging the right side of (4.12.7) with respect to $y = x/\varepsilon$.

At the resonant frequencies when $S(2k) \neq 0$ we do get reflection. In fact by solving (4.12.14) we find that

$$R_\varepsilon(L; k) \to R(L; k) \quad \text{as } \varepsilon \to 0,$$

where

(4.12.17) $$R(L; k) = -\frac{\tilde{a}(k)}{|a(k)|} \tanh(|a(k)|L), \qquad a(k) = \frac{ik}{2} S(2k).$$

This is the main result of this section.

The power reflection coefficient in the limit $\varepsilon \to 0$ has the form

(4.12.18) $$|R(L; k)|^2 = \tanh^2(|a(k)|L),$$

which shows that for L large compared to $|a(k)|^{-1}$ the amount of power reflected is quite substantial.

5. The general geometrical optics expansion

5.1. Expansion for Schrödinger's equation. We shall construct an asymptotic expansion for the solution $u_\varepsilon(x, t)$ of the problem

(5.1.1) $$\tilde{A}^\varepsilon u_\varepsilon = -\frac{i}{\varepsilon} \frac{\partial u_\varepsilon}{\partial t} - \sum_{p,q=1}^{n} \frac{\partial}{\partial x_p} \left(a_{pq} \left(x, \frac{x}{\varepsilon} \right) \frac{\partial u_\varepsilon}{\partial x_q} \right)$$

$$+ \frac{1}{\varepsilon^2} W \left(x, \frac{x}{\varepsilon} \right) u_\varepsilon = 0, \qquad t > 0, \qquad x \in \mathbb{R}^n,$$

(5.1.2) $$u_\varepsilon(x, 0) = e^{i\tilde{S}(x)/\varepsilon} f \left(x, \frac{x}{\varepsilon} \right), \qquad \tilde{S}(x) \text{ real-valued},$$

(5.1.3) $$f \left(x, \frac{x}{\varepsilon} \right) = \sum_{m=0}^{N_0} f_m(x) \phi_m \left(\frac{x}{\varepsilon}; \nabla \tilde{S}(x), x \right).$$

Here $W(x, y) \geq 0$ and $a_{pq}(x, y)$ are smooth functions[60] on $\mathbb{R}^n \times 2\pi Y$, a_{pq} is positive definite

(5.1.4) $$\sum_{p,q=1}^{n} a_{pq} \xi_p \xi_q \geq \alpha |\xi|, \qquad \alpha \text{ a positive constant},$$

and the functions $f_m(x)$, $0 \leq m \leq N_0$ are in $C_0^\infty(\mathbb{R}^n)$, complex-valued, and $\tilde{S}(x) \in C^\infty(\mathbb{R}^n)$ real-valued.

The functions $\phi_m(y; k, x)$ are the eigenfunctions of the shifted cell problem (cf. Section 3.1)

(5.1.5) $$A(k, x) \phi_m = \omega_m^2(k, x) \phi_m,$$

$$(\phi_m, \phi_{m'}) = \int_{2\pi Y} \phi_m(y; k, x) \bar{\phi}_m(y; k, x) \, dx = \delta_{mm'},$$

where

(5.1.6) $$A(k, x) = -\sum_{p,q=1}^{n} \left(\frac{\partial}{\partial y_p} + ik_p \right) \left[a_{pq}(x, y) \left(\frac{\partial}{\partial y_q} + ik_q \right) \cdot \right] + W(x, y),$$

---

[60] $2\pi Y$ the n-dimensional torus with basic set the parallelepiped all of whose sides are 2π.

where $k \in Y$ and $x \in \mathbb{R}^n$ are parameters. **It will be assumed throughout what
follows that:**

(5.1.7)
    for $0 \leq m \leq N$ with some $N \geq N_0$, the eigenvalues $\omega_m^2(k,x)$ are distinct
        for all $k \in K \subset Y$ and all $x \in \mathbb{R}^n$.

    Moreover, the eigenfunctions $\phi_m(y; k, x)$ are differentiable functions of y, k, x,
        $y \in 2\pi Y$, $k \in K$ and $x \in \mathbb{R}^n$.

This assumption is quite restrictive[61] and makes the expansion process rather simple. It would be interesting to find appropriate expansion forms in general when the multiplicity is variable.

    We shall construct an asymptotic expansion for u_ε in the form

$$(5.1.8) \qquad u_\varepsilon(x,t) \sim \sum_{m=0}^{N} e^{iS^{(m)}(x,t)/\varepsilon} v^{(m)}\left(x, \frac{x}{\varepsilon}, t; \varepsilon\right)$$

where each term in the sum is to satisfy (5.1.1). First we construct the phases $S^{(m)}(x,t)$, $0 \leq m \leq N$. They are to satisfy the Hamilton-Jacobi equations

$$(5.1.9) \qquad \frac{\partial S^{(m)}(x,t)}{\partial t} + \omega_m^2(\nabla S^{(m)}(x,t), x) = 0, \qquad t > 0, \qquad x \in \mathbb{R}^n,$$

$$S^{(m)}(x,0) = \tilde{S}(x), \qquad 0 \leq m \leq N.$$

We shall assume that $\tilde{S}(x)$ is such that

(5.1.10)   there exists a time $t_0 > 0$ for which (5.1.9) have a unique
                 smooth solution and $\nabla S^{(m)}(x,t) \in K$ for all
                 $x \in \mathbb{R}^n$, $0 \leq t < t_0$, $0 \leq m \leq N$.

Thus all our results are local in time and in wave-number space. The closer $\nabla \tilde{S}(x)$ is to being a constant (i.e., the more "plane" the initial wavefunction (5.1.2) is) the longer the interval of existence $[0, t_0)$ for (5.1.9) will be if $\nabla \tilde{S}(x) \in K$, $x \in \mathbb{R}^n$. In Section 5.2 we discuss (5.1.9) a bit further.

    With the $S^{(m)}$ so determined we find that

$$0 = \tilde{A}^\varepsilon u_\varepsilon = \sum_{m=0}^{N} e^{iS^{(m)}(x,t)/\varepsilon} A^{(m,\varepsilon)} v^{(m)}\left(x, \frac{x}{\varepsilon}, t; \varepsilon\right)$$

and hence for $0 \leq m \leq N$,

$$(5.1.11) \quad 0 = A^{(m,\varepsilon)} v^{(m)}(x,y,t;\varepsilon) = \left(\varepsilon^{-2} A_1^{(m)} + \varepsilon^{-1} A_2^{(m)} + A_3\right) \cdot v^{(m)}(x,y,t;\varepsilon),$$

---

[61]Recall that it is not necessary in the globally periodic case, Section 4.

and

$$(5.1.12) \qquad A_1^{(m)} = A(\nabla S^{(m)}(x,t), x) - \omega_m^2(\nabla S^{(m)}(x,t), x),$$

$$(5.1.13) \qquad A_2^{(m)} = i\frac{\partial}{\partial t} - \sum_{p,q=1}^{n} \left(\frac{\partial}{\partial y_p} + i\frac{\partial S^{(m)}}{\partial x_p} \right) \left[a_{pq}(x,t) \frac{\partial}{\partial x_q} \right]$$

$$- \sum_{p,q=1}^{n} \frac{\partial}{\partial x_p} \left[a_{pq}(x,t) \left(\frac{\partial}{\partial y_q} + i\frac{\partial S^{(m)}}{\partial x_q} \right) \cdot \right],$$

$$(5.1.14) \qquad A_e = - \sum_{p,q=1}^{n} \frac{\partial}{\partial x_p} \left(a_{pq}(x,t) \frac{\partial}{\partial x_q} \cdot \right).$$

We assume that the $v^{(m)}$ are 2π-periodic as functions of y and that they have asymptotic power series expansions

$$(5.1.15) \qquad v^{(m)}(x,y,t;\varepsilon) = v_0^{(m)}(x,y,t) + \varepsilon v_1^{(m)}(x,y,t)$$
$$+ \varepsilon^2 v_2^{(m)}(x,y,t) + \cdots.$$

Inserting (5.1.15) into (5.1.11) and equating coefficients of equal powers of ε yields the following sequence of problems

$$(5.1.16) \qquad A_1^{(m)} v_0^{(m)} = 0,$$

$$(5.1.17) \qquad A_1^{(m)} v_1^{(m)} + A_2^{(m)} v_0^{(m)} = 0,$$

$$(5.1.18) \qquad A_1^{(m)} v_2^{(m)} + A_2^{(m)} v_1^{(m)} + A_3 v_0^{(m)} = 0, \qquad \cdots.$$

Equation (5.1.16) implies that

$$(5.1.19) \qquad v_0^{(m)} = \phi_m(y; \nabla S^{(m)}(x,t), x) v_{00}^{(m)}(x,t), \qquad 0 \leq m \leq N,$$

where the function $v_{00}^{(m)}(x,t)$ is to be determined. From (5.1.3) we see, however, that

$$(5.1.20) \qquad \begin{aligned} v_{00}^{(m)}(x,0) &= f_m(x), & 0 \leq m \leq N_0, \\ v_{00}^{(m)}(x,0) &= 0, & N_0 + 1 \leq m \leq N. \end{aligned}$$

We consider next equation (5.1.17). It is an inhomogeneous equation for $v_1^{(m)}$. Since $A_1^{(m)}$ has a one-dimensional null-space the inhomogeneous term must satisfy a solvability condition. From (5.1.5) and (5.1.7) we conclude that this condition is

$$(5.1.21) \qquad \mathcal{A}_0^{(m)} v_{00}^{(m)} \equiv (A_2^{(m)} \phi_m v_{00}^{(m)}, \phi_m) = 0, \qquad 0 \leq m \leq N,$$

where the inner product is over $2\pi Y$ with (x,t) parameters. Equation (5.1.21) for $v_{00}^{(m)}(x,t)$ along with (5.1.20) constitutes the **homogenized transport equation** for the slowly varying amplitudes $v_{00}^{(m)}$ in the expansion (5.1.8), (5.1.15), (5.1.19), i.e.,

$$(5.1.22) \quad u_\varepsilon(x,t) \sim \sum_{m=0}^{N} e^{iS^{(m)}(x,t)/\varepsilon} \left[\phi_m \left(\frac{x}{\varepsilon}; \nabla S^{(m)}(x,t), x \right) v_{00}^{(m)}(x,t) + O(\varepsilon) \right].$$

Note that

$$(5.1.23) \qquad v_{00}^{(m)}(x,t) \equiv 0 \text{ for } N_0 + 1 \leq m \leq N,$$

in view of (5.1.21) and (5.1.20). A more explicit form for the operators $\mathcal{A}_0^{(m)}$ is given in Section 5.3. They are homogeneous, first order linear partial differential operators.

We return to (5.1.17) where now the solvability condition is satisfied. We have that

$$(5.1.24) \qquad v_1^{(m)} = -(A_1^{(m)})^{-1} A_2^{(m)} \phi_m v_{00}^{(m)} + \phi_m v_{10}^{(m)}$$

where the functions $v_{10}^{(m)}(x,t)$, $0 \leq m \leq N$, are to be determined again. A more explicit form for the first term on the right side of (5.1.24), in terms of cell problems, is given in Section 5.3.

We continue with (5.1.18) and apply the solvability condition to its inhomogeneous term. This yields an inhomogeneous transport equation for $v_{10}^{(m)}(x,t)$ of the form

$$(5.1.25) \qquad \mathcal{A}_0^{(m)} v_{10}^{(m)} + \mathcal{A}_1^{(m)} v_{00}^{(m)} = 0, \qquad 0 \leq m \leq N.$$

Here $\mathcal{A}_0^{(m)}$ is defined as in (5.1.21) and

$$(5.1.26) \qquad \mathcal{A}_1^{(m)} v_{00}^{(m)} = -(A_2^{(m)}(A_1^{(m)})^{-1} A_2^{(m)} \phi_m v_{00}^{(m)}, \phi_m) + (A_3 \phi_m v_{00}^{(m)}, \phi_m).$$

Initial conditions for $v_{10}^{(m)}(x,t)$, solution of (5.1.25), are obtained as follows. From (5.1.8), (5.1.15) and (5.1.2), (5.1.3) it follows that

$$\sum_{m=0}^{N} v_1^{(m)} \bigg|_{t=0} = 0.$$

Using (5.1.24) this condition takes the form
(5.1.27)

$$0 = \sum_{m=0}^{N} \sum_{m'=0}^{\infty} \left[v_{10}^{(m)} \delta_{mm'} - (1 - \delta_{mm'})((A_1^{(m)})^{-1} A_2^{(m)} \phi_m v_{00}^{(m)}, \phi_{m'}) \right] \cdot \phi_{m'} \bigg|_{t=0}.$$

We must therefore set

$$(5.1.28) \qquad v_{10}^{(m)} \bigg|_{t=0} = \sum_{m'=0}^{N} ((A_1^{(m')})^{-1} A_1^{(m')} \phi_{m'} v_{00}^{(m)}, \phi_m) \Bigg]_{t=0}, \qquad 0 \leq m \leq N.$$

This provides initial conditions for $v_{10}^{(m)}$ which are now determined completely from (5.1.25). However, these conditions do not suffice for (5.1.27). This is a natural limitation since we are working with only N modes while infinitely many modes are required to satisfy (5.1.27). We are precisely in the situation of Theorem 4.3, Section 4.5.

Let

$$(5.1.29) \quad z^{(\varepsilon,N)}(x,t) = \sum_{m=0}^{N} e^{iS^{(m)}(x,t)/\varepsilon}$$

$$\left\{ \phi_m v_{00}^{(m)} + \varepsilon \left[(A_1^{(m)})^{-1} A_2^{(m)} \phi_m v_{00}^{(m)} + \phi_m v_{10}^{(m)} \right] \right\}_{y=x/\varepsilon}.$$

THEOREM 5.1. *The statement of Theorem 4.3, Section 4.5, is valid here again provided $t \in [0, t_0)$ and (5.1.7), (5.1.10) hold. Explicitly,*

$$(5.1.30) \qquad |u_\varepsilon(\cdot, t) - z^{(\varepsilon,N)}(\cdot, t)|_{L^2(\mathbb{R}^n)} \leq \varepsilon[C_N(f) + \varepsilon \tilde{C}(f)], \qquad 0 \leq t < t_0,$$

where $C_N(f)$ and $C(f)$ are constants that depend on the data $f_m(x)$ ($\in C_0(\mathbb{R}^n)$), $0 \le m \le N_0$ but do not depend on ε. Moreover the constant $C_N(f)$ can be made as small as desired by taking N sufficiently large.

The function $v_2^{(m)}$ required in the proof of Theorem 4.3 is, of course, the next term $v_2^{(m)}$ in (5.1.15) and is obtained by carrying the above expansion process one step further.

5.2. Eikonal equations and rays. The analysis of the eikonal or Hamilton-Jacobi equations (5.1.9) parallels almost word for word the one given in Section 2.2, except that the Hamiltonians are now $\omega_m^2(k, x)$, $0 \le m \le N$, the eigenvalues of the shifted cell problems (5.1.5).

Let $(x^{(m)}(t), k^{(m)}(t))$ be solutions of the Hamiltonian system

(5.2.1)
$$\frac{dx_p^{(m)}(t)}{dt} = \frac{\partial \omega_m^2(k^{(m)}(t), x^{(m)}(t))}{\partial k_p}, \qquad x_p^{(m)}(0) = x_p,$$
$$\frac{dk_p^{(m)}(t)}{dt} = -\frac{\partial \omega_m^2(k^{(m)}(t), x^{(m)}(t))}{\partial x_p}, \qquad k_p^{(m)}(0) = \frac{\partial \tilde{S}(x)}{\partial x_p}, \qquad p = 1, 2, \ldots, n.$$

Locally, the solution of (5.1.9) can be written by using the rays $x^{(m)}(t; x)$

(5.2.2)
$$S^{(m)}(x^{(m)}(t), t) = \tilde{S}(x) + \int_0^t L^{(m)}(x^{(m)}(s), \dot{x}^{(m)}(s)) \, ds.$$

Here $L^{(m)}(x, x)$ is the Lagrangian associated with $\omega_m^2(k, x)$ as in (2.2.6).

Absence of caustics means that for $0 \le m \le N$ the mappings $x \to x^{(m)}(t; x)$ (where $x^{(m)}(0; x) = x$) are one to one on $\mathbb{R}^n$ and hence $S^{(m)}$ can be found at any point if it is known along rays by (5.2.2).

5.3. Transport equations. We now give more explicit forms for the operators $\mathcal{A}_0^{(m)}$ and $\mathcal{A}_1^{(m)}$ defined by (5.1.21) and (5.1.26), respectively.

We observe that $A_1^{(m)}$ has the form

(5.3.1)
$$A_1^{(m)} = \left[-\omega_m^2(k, x) + A(k, x) \right]_{k = \nabla S^{(m)}}$$

where $\nabla S^{(m)} = \nabla S^{(m)}(x, t)$ and $A(k, x)$ is given by (5.1.6). Define the **local group velocity of the mth mode** (or band) by

(5.3.2)
$$c_p^{(m)}(x, t) = \left. \frac{\partial \omega_m^2(k, x)}{\partial k_p} \right|_{k = \nabla S^{(m)}}, \qquad p = 1, 2, \ldots, n, \qquad 0 \le m \le N.$$

Direct computation yields now

(5.3.3)
$$\mathcal{A}_0^{(m)} v_{00}^{(m)}(x, t) = -i \left[\frac{\partial v_{00}^{(m)}(x, t)}{\partial t} + \sum_{p=1}^n c_p^{(m)}(x, t) \frac{\partial v_{00}^{(m)}(x, t)}{\partial x_p} \right.$$
$$\left. + \frac{1}{2} \sum_{p=1}^n \frac{\partial c_p^{(m)}(x, t)}{\partial x_p} v_{00}^{(m)}(x, t) \right]$$
$$+ b^{(m)}(x, t) v_{00}^{(m)}(x, t) = 0, \qquad 0 \le m \le N.$$

In (5.3.3) the functions $b^{(m)}(x,t)$ are real-valued and are given by

$$(5.3.4) \qquad b^{(m)} = -i(\phi_{mt}, \phi_m) - \frac{1}{2} \sum_{p=1}^{n} \frac{\partial}{\partial x_p} \left(\frac{\partial a_{pq}}{\partial y_q} \phi_m, \phi_m \right)$$

$$+ \sum_{p,q=1}^{n} \left(\left(a_{pq} \left(\frac{\partial}{\partial y_q} + i \frac{\partial S^{(m)}}{\partial x_q} \right) \phi_m, \phi_{mp} \right) \right.$$

$$\left. - \left(\left(\frac{\partial}{\partial y_q} + i \frac{\partial S^{(m)}}{\partial x_p} \right) (a_{pq} \phi_{mq}), \phi_m \right) \right)$$

where

$$(5.3.5) \qquad \phi_{mt} = \sum_{p=1}^{n} \frac{\partial \phi_m(y; \nabla S^{(m)}, x)}{\partial k_p} \frac{\partial^2 S^{(m)}(x,t)}{\partial x_p \partial t}$$

and

$$(5.3.6) \qquad \phi_{mp} = \frac{\partial \phi_m(y; \nabla S^{(m)}, x)}{\partial x_p} + \sum_{q=1}^{n} \frac{\partial \phi_m(y; \nabla S^{(m)}, x)}{\partial k_q} \frac{\partial^2 S^{(m)}(x,t)}{\partial x_q \partial x_q}.$$

Equation (5.3.3) has the usual form of the transport equation of geometrical optics. It can be written as a conservation law as follows:

$$(5.3.7) \qquad -\frac{i}{2} \left[\frac{\partial (v_{00}^{(m)})^2}{\partial t} + \sum_{p=1}^{n} \frac{\partial}{\partial x_p} \left(c_p (v_{00}^{(m)})^2 \right) \right] + b(v_{00}^{(m)})^2 = 0.$$

We continue with the operator $\mathcal{A}_1^{(m)}$ defined by (5.1.26). Let $\chi_p^{(m)}(y; k, x)$ be the solutions of the (parametrically x-dependent) cell problems (cf. (4.3.11))

(5.3.8)

$$\left[-\omega_m^2(k,x) + A(k,x) \right] \chi_p^{(m)}(y; k, x) + \frac{1}{i} \left[A_{,p}(k,x) - \frac{\partial \omega_m^2(k,x)}{\partial k_p} \right] \phi_m(y; k, x) = 0,$$

$$(\chi_p^{(m)}, \phi_m) = 0, \qquad p = 1, 2, \ldots, n, \qquad 0 \le m \le N.$$

Here $A_{,p}(k,x)$ is defined by

$$(5.3.9) \qquad \frac{1}{i} A_{,p}(k,x) = -\sum_{q=1}^{n} a_{pq}(x,y) \left(\frac{\partial y_q}{\partial +} i k_q \right)$$

$$- \sum_{q=1}^{n} \left(\frac{\partial}{\partial y_q} + i k_q \right) [a_{pq}(x,y) \cdot]$$

and, as in (4.3.9),

$$(5.3.10) \qquad \frac{\partial \omega_m^2(k,x)}{\partial k_p} = \left(A_{,p}(k,x) \phi_m(\cdot; k, x), \phi_m(\cdot; k, x) \right).$$

The cell functions $\chi_m^p(y; k, x)$ are well defined by (5.3.8).

Define also the cell functions $\chi^m(y; x, t)$ by

$$(5.3.11) \qquad A_1^{(m)} \chi^{(m)} + B^{(m)} = 0, \qquad 0 \le m \le N,$$

$$(\chi^m, \phi_m) = 0,$$

where $B^{(m)}(x,t)$ are given by

$$(5.3.12) \qquad B^{(m)} = i\sum_{p=1}^{n} \frac{\partial}{\partial x_p}\left(\frac{\partial \omega_m^2}{\partial k_p}\right) - b^{(m)} - i\phi_{mt}$$

$$- \sum_{p,q=1}^{n}\left[\left(\frac{\partial}{\partial y_p} + i\frac{\partial S^{(m)}}{\partial x_p}\right)(a_{pq}\phi_{mq})\right.$$

$$\left. - \frac{\partial}{\partial x_p}\left(a_{pq}\left(\frac{\partial}{\partial y_q} + i\frac{\partial S^{(m)}}{\partial x_q}\right)\phi_m\right)\right].$$

Note that $\chi^{(m)}$ of (5.3.11) is well defined because $(B^{(m)}, \phi_m) = 0$.

Now we can write $v_1^{(m)}$ of (5.1.24) using the cell functions $\chi_p^{(m)}$ and $\chi^{(m)}$ as follows:

$$(5.3.13) \qquad v_1^{(m)} = \sum_{p=1}^{n}\chi_p^{(m)}(y; \nabla S^{(m)}, x)\frac{\partial v_{00}^{(m)}(x,t)}{\partial x_p}$$

$$+ \chi^{(m)}(y; x, t)v_{00}^{(m)}(x,t)$$

$$+ \phi_m(y; \nabla S^{(m)}, x)v_{10}^{(m)}(x,t).$$

From this it is easily seen that we can express $\mathcal{A}_1^{(m)}$ of (5.1.26) as follows:

$$(5.3.14) \qquad \mathcal{A}_1^{(m)}v_{00}^{(m)} = \sum_{p=1}^{n}\left(A_2^{(m)}\chi_p^{(m)}\frac{\partial v_{00}^{(m)}}{\partial x_p}, \phi_m\right)$$

$$+ (A_2^{(m)}\chi^{(m)}v_{00}^{(m)}, \phi_m) + (A_3\phi_m v_{00}^{(m)}, \phi_m).$$

Using the identity (4.3.16), with additional parametric x-dependence which changes nothing, we can rewrite (5.3.14) in the following more revealing form:

$$(5.3.15) \qquad \mathcal{A}_1^{(m)}v_{00}^{(m)} = -\sum_{p,q=1}^{n}\frac{\partial}{\partial x_p}\left(\tilde{a}_{pq}^{(m)}(x,t)\frac{\partial v_{00}^{(m)}(x,t)}{\partial x_q}\right)$$

$$+ \sum_{p=1}^{n}\tilde{b}_p^{(m)}(x,t)\frac{\partial v_{00}^{(m)}(x,t)}{\partial x_p}$$

$$+ \tilde{c}^{(m)}(x,t)v_{00}^{(m)}(x,t),$$

where (cf. (4.2.8), (4.3.16))

$$(5.3.16) \qquad \tilde{a}_{pq}^{(m)}(x,t) = \frac{1}{2}\left.\frac{\partial^2 \omega_m^2(k,x)}{\partial k_p \partial x_q}\right|_{k=\nabla S^{(m)}(x,t)}.$$

The coefficients $\tilde{b}_p^{(m)}$ and $\tilde{c}^{(m)}$ are given by certain easily obtained but long expressions which we do not write since they do not seem to be particularly interesting.

We close this section by observing that all present results specialize, as they should, to those of Sections 4.1–4.5 in the globally periodic case ($a_{pq} = a_{pq}(y)$, $W = W(y)$ in 5.1).

5.4. Connections with the static theory. We now specialize problem (5.1.1)–(5.1.3) to the following

$$(5.4.1) \qquad -i\frac{\partial u_\varepsilon}{\partial t} - \sum_{p,q=1}^{n} \frac{\partial}{\partial x_p}\left(a_{pq}\left(x,\frac{x}{\varepsilon}\right)\frac{\partial u_\varepsilon}{\partial x_q}\right) = 0, \qquad t > 0, \qquad x \in \mathbb{R}^n,$$

$$(5.4.2) \qquad u_\varepsilon(x,0) = (2\pi)^{-n/2} f_0(x) \in C_0^\infty(\mathbb{R}^n).$$

We assume again that $(a_{pq}(x,y))$ is smooth on $\mathbb{R}^n \times 2\pi Y$ and uniformly elliptic (cf. (5.1.4)).

Note that (5.4.1), (5.4.2) is obtained from (5.1.1)–(5.1.3) by setting

$$(5.4.3) \qquad W(x,y) \equiv 0, \qquad \tilde{S}(x) \equiv 0, \qquad N_0 = 0,$$

and by **rescaling time** $t \to t/\varepsilon$. The rescaling of the time was explained in Section 4.4. The new time scaling is the one used in Chapter 2. Since $W \equiv 0$, it follows that

$$(5.4.4) \qquad \omega_0^2(0,x) \equiv 0, \qquad \phi_0(y;k,x) \equiv (2\pi)^{-n/2}.$$

This explains the factor $(2\pi)^{-n/2}$ in (5.4.2).

By going through the steps in the expansion in Section 5.1 we see that:

$$(5.4.5) \qquad \text{the operator } -i\frac{\partial}{\partial t} \text{ is now part of } A_3 \text{ (5.1.14)}$$

and not of (5.1.13), because of the change $t \to t/\varepsilon$,

$$(5.4.6) \qquad \text{the phase function } S^{(0)}(x,t) \equiv 0,$$

$$(5.4.7) \qquad \text{the transport equation (5.1.21) is } \textbf{automatically}$$

$$\text{satisfied for } m = 0,$$

$$(5.4.8) \qquad \text{equation (5.1.25) with } m = 0 \text{ is the determining}$$

$$\text{equation for } v_{00}^{(0)}(x,t).$$

The determining equation[62] has explicitly, in view of (5.3.15), the following form

$$(5.4.9) \qquad -i\frac{\partial v_{00}^{(0)}(x,t)}{\partial t} - \sum_{p,q=1}^{n} \frac{\partial}{\partial x_p}\left(\tilde{a}_{pq}^{(0)}(x)\frac{\partial v_{00}^{(0)}(x,t)}{\partial x_q}\right), \qquad t > 0,$$

$$v_{00}^{(0)}(x,0) = f_0(x).$$

This is the **homogenized** Schrödinger equation and the coefficients $\tilde{a}_{pq}^{(0)}(x)$ are given by (5.3.16). Going back to the form (4.3.16) of the second derivatives of $\omega_m^2(k,x)$ and specializing to ω_0^2 at $k = 0$ (and using (5.4.4)) we find that

$$\tilde{a}_{pq}^{(0)}(x) = \frac{1}{(2\pi)^n} \sum_{\ell=1}^{n} \int_{2\pi Y} a_{p\ell}(x,y)\left(\frac{\partial \chi_p^{(0)}(y;0,x)}{\partial y_\ell} + \delta_{\ell q}\right) dy, \qquad p,q = 1,2,\ldots,n.$$

This is precisely the result obtained in Chapters 1 and 2. The cell functions $\chi_p^{(0)}(y;0,x)$ satisfy (5.3.8) with $m = 0$, $k = 0$ (and we make use of (5.4.4)).

---

[62]It is easily seen that $\tilde{b}^{(0)}$ and $\tilde{c}^{(0)}$ in (5.3.15) are now zero.

5.5. Spatially localized data. Consider (5.1.1) but now assume that

$$(5.5.1) \qquad u_\varepsilon(x,0) = \varepsilon^{-n/2} \int_Y d\kappa \sum_{m=0}^{N_0} f_m(\kappa,x) e^{i\kappa\cdot x/\varepsilon} \phi_m\left(\frac{x}{\varepsilon}; \kappa, x\right)$$

where $f_m(\kappa,x)$ are smooth functions on $Y \times \mathbb{R}^n$ with compact support in x. In view of the Bloch equation theorem in Section 3.2 the initial wavefunction (5.5.1) corresponds to data of class C (Section 1.2) but we have specialized them further to simplify the analysis. Problem (5.1.1), (5.5.1) is the locally periodic analog of (4.7.1), (4.7.2) in Section 4.7.

Essentially all the analysis of Section 4.7 carries over with minor modifications. These minor modifications introduced by the locally periodic nature of the problem are actually spelled out in Section 2.6. They now have to be applied to each term in the sum

$$(5.5.2) \qquad u_\varepsilon(x,t) \sim \varepsilon^{-n/2} \sum_{m=0}^{N} \int_Y d\kappa \, e^{iS^{(m)}(x,t;\kappa)/\varepsilon} v^{(m)}\left(x, \frac{x}{\varepsilon}, \kappa; \varepsilon\right)$$

which forms an asymptotic solution for (5.1.1), (5.5.1) when the integrands in (5.5.2) are solutions for each κ and m.

5.6. Behavior of probability amplitudes. Given the result (5.1.30), Theorem 4.6, Section 4.8, extends without any change, including the proof, to the locally periodic problem (5.1.1)–(5.1.3).

5.7. Expansion for the wave equation. We consider now the initial value problem

$$(5.7.1) \qquad \frac{\partial^2 u_\varepsilon}{\partial t^2} - \sum_{p,q=1}^{n} \frac{\partial}{\partial x_p}\left(a_{pq}\left(x, \frac{x}{\varepsilon}\right) \frac{\partial u_\varepsilon}{\partial x_q}\right) + \frac{1}{\varepsilon^2} W\left(x, \frac{x}{\varepsilon}\right) u_\varepsilon = 0, \qquad t > 0,$$

$$u_\varepsilon(0,x) = f_\varepsilon(x), \qquad \frac{\partial u_\varepsilon(x,0)}{\partial t} = g_\varepsilon(x), \qquad x \in \mathbb{R}^n,$$

under the same hypotheses as in Section 5.1. For initial data we take

$$(5.7.2) \qquad f_\varepsilon(x) = \varepsilon e^{i\tilde{S}(x)/\varepsilon} \sum_{m=0}^{N_0} f_m(x) \phi_m\left(\frac{x}{\varepsilon}; \nabla\tilde{S}(x), x\right),$$

$$g_\varepsilon(x) = e^{i\tilde{S}(x)/\varepsilon} \sum_{m=0}^{N_0} g_m(x) \phi_m\left(\frac{x}{\varepsilon}; \nabla\tilde{S}(x), x\right),$$

with f_m and g_m in $C_0^\infty(\mathbb{R}^n)$. We note that f_ε is proportional to ε so that the initial energy

$$(5.7.3) \qquad \int_{\mathbb{R}^n} \left[|g_\varepsilon(x)|^2 + \sum_{p,q=1}^{n} a_{pq}\left(x, \frac{x}{\varepsilon}\right) \frac{\partial f_\varepsilon(x)}{\partial x_p} \overline{\frac{\partial f_\varepsilon(x)}{\partial x_q}}\right] dx$$

is of order one as $\varepsilon \to 0$. The energy at time t

$$(5.7.4) \qquad \int_{\mathbb{R}^n} \left[\left|\frac{\partial u_\varepsilon(x,t)}{\partial t}\right|^2 + \sum_{p,q=1}^{n} a_{pq}\left(x, \frac{x}{\varepsilon}\right) \frac{\partial u_\varepsilon(x,t)}{\partial x_p} \overline{\frac{\partial u_\varepsilon(x,t)}{\partial x_q}}\right] dx$$

is bounded independently of $t \geq 0$ and ε (recall that $W \geq 0$). This gives the basic a priori estimate for the asymptotic analysis.

The expansion corresponding to (5.1.8) takes now the form

$$(5.7.5) \qquad u_\varepsilon(x,t) \sum \varepsilon \sum_{m=0}^{N} \left[e^{iS^{+(m)}(x,t)/\varepsilon} v^{+(m)}\left(x, \frac{x}{\varepsilon}, t; \varepsilon\right) \right.$$

$$\left. + e^{iS^{-(m)}(x,t)/\varepsilon} v^{-(m)}\left(x, \frac{x}{\varepsilon}, t; \varepsilon\right) \right].$$

The phases $S^{\pm(m)}$ satisfy the Hamilton-Jacobi equations

$$(5.7.6) \qquad \frac{\partial S^{\pm(m)}}{\partial t} \pm \omega_m(\nabla S^{\pm(m)}, x) = 0, \qquad t > 0,$$

$$S^{\pm(m)}(x,0) = \tilde{S}(x), \qquad x \in \mathbb{R}^n.$$

The functions $v^{\pm(m)}$ have expansions like (5.1.15) and each term in the sum (5.7.5) is to satisfy (5.7.1) (the $+$ and $-$ terms are to be separate solutions). Nothing special, not encountered in Section 5.1, arises for the present problem.

5.8. Expansion for the heat equation. Consider the heat equation

$$(5.8.1) \qquad \frac{1}{\varepsilon} \frac{\partial u_\varepsilon}{\partial t} = \sum_{p,q=1}^{n} a_{pq}\left(x, \frac{x}{\varepsilon}\right) \frac{\partial^2 u_\varepsilon}{\partial x_p \partial x_q}$$

$$+ \sum_{p=1}^{n} \left(\frac{1}{\varepsilon} b_p\left(x, \frac{x}{\varepsilon}\right) + c_p\left(x, \frac{x}{\varepsilon}\right) \right) \frac{\partial u_\varepsilon}{\partial x_p},$$

$$u_\varepsilon(x,0) = f(x) \geq 0$$

where the coefficients $a_{pq}(x,y)$, $b_p(x,y)$, $c_p(x,y)$ are smooth functions on $\mathbb{R}^n \times 2\pi Y$, (a_{pq}) is uniformly positive definite and $f \in C_0^\infty(\mathbb{R}^n)$. We point out the factor ε^{-1} in front of $\partial u_\varepsilon/\partial t$ in (5.8.1); this makes it a problem different from the ones studied in Chapter 3. The actual differences which necessitate time rescaling are explained in detail below.

Consider the operator

$$(5.8.2) \qquad A_1 = \sum_{p,q=1}^{n} a_{pq}(x,y) \frac{\partial^2}{\partial y_p \partial y_q} + \sum_{p=1}^{n} b_p(x,y) \frac{\partial}{\partial y_p},$$

acting on smooth functions on $2\pi Y$ (i.e., 2π-periodic in each direction) with $x \in \mathbb{R}^n$ a parameter. This operator is the generator of a contraction semigroup on $C(2\pi Y)$, the continuous functions on $2\pi Y$. By the strong maximum principle (implied by the uniform ellipticity) this semigroup maps nonnegative functions that do not vanish identically to strictly positive functions. We saw in Chapter 3 that the strong maximum principle and the smoothness of solutions of $\partial v/\partial t = A_1 v$, $v_1(y,0) = g(y)$, (x a parameter) for $t > 0$, independently of the smoothness of g, imply that the only nonzero solutions of

$$(5.8.3) \qquad A_1 \phi = 0$$

are the y-independent functions, say $\phi(y;x) \equiv 1$, and the only solution of the adjoint equation

$$(5.8.4) \qquad A_1^* \tilde{\phi} = 0$$

is a strictly positive function $\tilde{\phi}(y; x)$, parametrically dependent on x, and rendered unique by the normalization

$$(5.8.5) \qquad \int_{2\pi Y} \tilde{\phi}(y; x) \, dy = 1.$$

It was shown in Chapter 3 in several different ways and in such greater generality that under the present conditions

$$(5.8.6) \qquad u_\varepsilon(x, t) \to u(x, t) \text{ as } \varepsilon \to 0,$$

uniformly in $x \in \mathbb{R}^n$, $0 \le t \le T$, $T < \infty$ but arbitrary, where $u(t, x)$ is the solution of

$$(5.8.7) \qquad \frac{\partial u(x, t)}{\partial t} = \sum_{p=1}^{n} \tilde{b}_p(x) \frac{\partial u(x, t)}{\partial x_p},$$

$$u(x, 0) = f(x),$$

and

$$(5.8.8) \qquad \tilde{b}_p(x) = \int_{2\pi Y} b_p(x, y) \tilde{\phi}(y; x) \, dy, \qquad p = 1, 2, \dots, n.$$

Most of the analysis of Chapter 3 dealt with the case

$$(5.8.9) \qquad \tilde{b}_p(x) \equiv 0.$$

Indeed, if the right side of (5.8.1) is a formally self-adjoint operator then $\tilde{\phi}(y; x) \equiv 1$ and (5.8.9) is automatically satisfied since

$$(5.8.10) \qquad b_p(x, y) = \sum_{q=1}^{n} \frac{\partial a_{pq}(x, y)}{\partial y_q}.$$

In this section we shall study (5.8.1) when $\tilde{b}_p(x) \not\equiv 0$ so that (5.8.6)–(5.8.8) holds; more specifically we shall concentrate on the following problem which is analogous to the one in Section 2.14.

Let

$$(5.8.11) \qquad G = \{x \in \mathbb{R}^n \mid f(x) > 0\}.$$

We know that G is an open set with compact closure and smooth boundary. **We shall restrict $t < t_0$ to avoid caustics.**[63] This condition is not necessary for the validity of the theorem that follows; it results from a weakness in the method of analysis that we use. Let G_t denote the support of $u(x, t)$. It is precisely the set of points x that are translates of points of G along the orbits of the vector field $\tilde{b}(x) = (\tilde{b}_p(x))$ for a time of length t. G_t is also an open set with compact closure.

Now when ε is positive $u_\varepsilon(x, t)$ is strictly positive for any x and any $t > 0$ by the strong maximum principle. The question we shall answer here is this: how does $u_\varepsilon(x, t)$ go to zero as $\varepsilon \to 0$ when (x, t) is fixed, $t > 0$ and $x \in \mathbb{R}^n - G_t$? **References and remarks made in Section 2.14 are relevant here also.**

---

[63] For the "smooth" Hamilton-Jacobi equation as in Section 2.14.

Define the operator $A(k, x)$ by

(5.8.12)

$$A(k, x) = \sum_{p,q=1}^{n} a_{pq}(x, y) \left(\frac{\partial}{\partial y_p} - k_p \right) \left(\frac{\partial}{\partial y_q} - k_q \right) + \sum_{p=1}^{n} b_p(x, y) \left(\frac{\partial}{\partial y_p} - k_p \right)$$

$$= \sum_{p,q=1}^{n} a_{pq}(x, y) \frac{\partial^2}{\partial y_p \partial y_q}$$

$$+ \sum_{p=1}^{n} \left(b_p(x, y) - 2 \sum_{q=1}^{n} a_{pq}(x, y) k_q \right) \frac{\partial}{\partial y_p}$$

$$+ \left(\sum_{p,q=1}^{n} a_{pq}(x, y) k_p k_q - \sum_{p=1}^{n} b_p(x, y) k_p \right),$$

acting on smooth functions on $2\pi Y$ with $k \in \mathbb{R}^n$ and $x \in \mathbb{R}^n$ parameters. This operator, like A_1 of (5.8.2), generates a semigroup on $C(2\pi Y)$, which is not a contraction in general but is positivity preserving. From this fact one can deduce rather easily that

$$A(k, x) \text{ has an isolated maximal eigenvalue,}$$

$$\omega(k, x) \text{ with strictly positive eigenfunction } \phi(y; k, x),$$

(5.8.13) $\qquad A(k, x)\phi(y; k, x) = \omega(k, x)\phi(y; k, x), \qquad \omega \text{ real-valued.}$

The eigenfunction is unique up to a multiplicative factor and can be chosen to be smooth in $(k, x) \in \mathbb{R}^{2n}$ because the maximal eigenvalue is isolated. Similarly there is a strictly positive function $\tilde{\phi}(y; k, x)$ such that

(5.8.14) $\qquad\qquad A^*(k, x)\tilde{\phi}(y; k, x) = \omega(k, x)\tilde{\phi}(y; k, x),$

where $A^*(k, x)$ is the formal adjoint of $A(k, x)$. The function $\tilde{\phi}(y; k, x)$ (defined on $2\pi Y \times \mathbb{R}^n \times \mathbb{R}^n$) is unique up to a multiplicative factor.

Because of the uniform ellipticity assumption we have that

(5.8.15) $\qquad\qquad \omega(k, x) \text{ is strictly convex in } k.$

Let $L(x, \dot{x})$ be its conjugate convex function

(5.8.16) $\qquad\qquad L(x, \dot{x}) = \sup_{k}(k \cdot \dot{x} - \omega(k, x)).$

Since $\omega(0, x) \equiv 0$ we have that $L(x, \dot{x}) \geq 0$ and $L(x, \dot{x}) = 0$ only when $\dot{x} = \partial\omega(k, x)/\partial k$ at $k = 0$. But from (5.8.12) and (5.8.13) we find that

(5.8.17) $\qquad\qquad \left. \frac{\partial\omega(k, x)}{\partial k_p} \right|_{k=0} = -\tilde{b}_p(x), \qquad p = 1, 2, \ldots, n.$

For $t < t_0$ the following variational problem has a unique solution

(5.8.18) $\qquad\qquad S(x, t) = \sup_{\substack{x(t)=x \\ x(0) \in G}} \left\{ \int_0^t L(x(s), \dot{x}(s)) \, ds \right\}.$

From the observations just made we conclude that

(5.8.19) $\quad S(x, t) \geq 0$ and $S(x, t) = 0$ only when $x \in G_t$ (defined below (5.8.11)).

It is also clear that we may replace G by ∂G in (5.8.18) when $x \notin G_t$.

THEOREM 5.2. *The solution $u_\varepsilon(x,t)$ tends to zero as $\varepsilon \to 0$ with $t > 0$, $t < t_0$, $x \in \mathbb{R}^n - G_t$ fixed, and we have*

$$(5.8.20) \qquad \lim_{\varepsilon \downarrow 0} \varepsilon \log u_\varepsilon(x,t) = -S(x,t), \qquad \text{for each } x, t, \ t < t_0,$$

where $S(x,t)$ is defined by (5.8.18).

PROOF. The proof of this theorem is entirely analogous to the one of Theorem 2.11, Section 2.14, so it will be omitted. ∎

REMARK 5.3. The physical significance of this theorem is the following. We may take (5.8.1) in the form

$$(5.8.21)$$

$$\frac{\partial u_\varepsilon}{\partial t} - \sum_{p=1}^{n} \left(b_p\left(x, \frac{x}{\varepsilon}\right) + \varepsilon c_p\left(x, \frac{x}{\varepsilon}\right) \right) \frac{\partial u_\varepsilon}{\partial x_p} = \varepsilon \sum_{p,q=1}^{n} \frac{\partial}{\partial x_p}\left(a_{pq}\left(x, \frac{x}{\varepsilon}\right) \frac{\partial u_\varepsilon}{\partial x_q} \right),$$

$$u_\varepsilon(x,0) = f(x) \geq 0,$$

and suppose $u_\varepsilon(x,t)$ represents the temperature of a composite medium[64] which can sustain heat transfer by conduction and convection (like a porous medium). Suppose that ε measures the relative size of the period of the structure and is small and **also the conductivity is small, i.e., of order** ε. To principal order in ε, heat is transferred by convection along the average flow field $\tilde{b}(x)$ (cf. (5.8.6)–(5.8.8)). The theorem above gives the precise rate at which heat is transferred by conduction into regions that are inaccessible to transfer of heat by convection. The amount of heat so transferred is, of course, exponentially small in ε.

REMARK 5.4. The formal result in Theorem 5.2 is easily obtained by the WKB as in Section 2.14. Specifically, the solution u_ε of (5.8.1) is expanded (formally) in the form

$$u_\varepsilon(x,t) = e^{-S(x,t)/\varepsilon} \left[v_0\left(x, \frac{x}{\varepsilon}, t\right) + \varepsilon v_1\left(x, \frac{x}{\varepsilon}, t\right) + \cdots \right].$$

To obtain a proof one must use this expansion **after** one has carried out a "smoothing" of the region G as is done in the proof of Theorem 2.11. The restriction $t < t_0$ is a very serious disadvantage of this, rather simple, proof. The result is valid for any finite t.

REMARK 5.5. In the **globally periodic** case, where a_{pq} and b_p are functions of $y \in 2\pi Y$ only, the Hamiltonian $\omega = \omega(k)$ does not depend on x. If the support of $f = G$ is **convex** then the above theorem is valid globally in t since there will be no caustic or conjugate points for (5.8.18).

6. Comments and problems

In reviewing the material of this chapter we would like to point out again the principal defects in the methods presented and discuss them briefly.

The first defect is the appearance of caustics and the necessity to restrict discussion up to the time that they first appear. One knows how to overcome this difficulty at present (cf. references in Section 2.8) in many cases of interest. Periodic media problems are not different from other propagation problems as far as this point is concerned. Therefore at least in principle difficulties with caustics can be overcome. For problems admitting a probabilistic formulation, as in Sections 2.14

---

[64]With periodic structure.

and 5.1.8, the probabilistic methods (cf. references in Section 2.14) are, in our view, truly superior to anything else.

The second defect, the most serious one in our view, concerns the necessity to impose mode nondegeneracy in the locally periodic case (Section 5.1, assumption (5.1.7)). Recall that in the globally periodic case, Sections 4.2–4.8, this is frequently **not** necessary. This is not surprising since in the globally periodic case one can in fact write down the exact solution in terms of Bloch waves. The problem of mode degeneracy and mode coupling in the geometrical optics limit is essentially an open problem in any context even in relatively simple examples (compared to media with periodic structure). We refer to the paper of Ludwig and Granoff [**75**] for some first steps in the solution of this problem. There is currently a great deal of activity in this area.

The third defect concerns the absence of results in bounded regions (except in one dimension) with general boundaries. We pointed out repeatedly in previous Chapters that behavior near boundaries is very much a mystery even in the low frequency or static case. The only general problem that one expects may be solvable is the periodic half-space problem i.e. the determination of reflection and transmission operators for periodic half-spaces. Unfortunately even if this was solved it would not be of very much help in general. It is important here to introduce new simplifying assumptions, like small periodic perturbations from uniform media, as we pointed out in Section 4.12.

The fourth and last major defect concerns the constant use of smoothness assumptions on coefficients and to a lesser degree on data. In previous Chapters a good deal of effort was spent in removing regularity hypotheses and as a result it was possible to obtain a number of theorems with optimal conditions in this respect. Experience with geometrical optics methods in the usual setting (cf. for example J.B. Keller and R. Lewis [**56**]) shows that essentially new phenomena arise in the presence of discontinuities (diffraction effects) that require a great deal of insight into special, canonical, problems for their solution (cf. J.B. Keller [**55**]). It appears therefore at present that geometrical optics results in periodic media with discontinuous (non smooth) structure are difficult to obtain.

If one restricts attention to the propagation of energy only, which is different from a geometrical optics analysis, one expects that discontinuity effects would then be blurred out leading to tractable analysis. We have not discussed this viewpoint here and as far as we know it constitutes an open problem in media with periodic structure.

Bibliography

[1] Robert A. Adams. *Sobolev spaces.* Academic Press [A subsidiary of Harcourt Brace Jovanovich, Publishers], New York-London, 1975. Pure and Applied Mathematics, Vol. 65.

[2] S. Agmon, A. Douglis, and L. Nirenberg. Estimates near the boundary for solutions of elliptic partial differential equations satisfying general boundary conditions. I. *Comm. Pure Appl. Math.*, 12:623–727, 1959.

[3] S. Agmon, A. Douglis, and L. Nirenberg. Estimates near the boundary for solutions of elliptic partial differential equations satisfying general boundary conditions. II. *Comm. Pure Appl. Math.*, 17:35–92, 1964.

[4] Daljit S. Ahluwalia, Edward L. Reiss, and Stephen E. Stone. A uniform asymptotic analysis of dispersive wave motion. *Arch. Rational Mech. Anal.*, 54:340–355, 1974.

[5] Michel Artola and Georges Duvaut. Homogénéisation d'une plaque renforcée. *C. R. Acad. Sci. Paris Sér. A-B*, 284(12):A707–A710, 1977.

[6] Hédy Attouch and Yoshio Konishi. Convergence d'opérateurs maximaux monotones et inéquations variationnelles. *C. R. Acad. Sci. Paris Sér. A-B*, 282(9):Ai, A467–A469, 1976.

[7] R. Azencott and G. Ruget. Mélanges d'équations différentielles et grands écarts à la loi des grands nombres. *Z. Wahrscheinlichkeitstheorie und Verw. Gebiete*, 38(1):1–54, 1977.

[8] I. Babuska. Solution of interface problems by homogenization. *SIAM Journal on Math. Analysis*, 8:923–937, 1977.

[9] A. Bensoussan and J.-L. Lions. Inéquations quasi variationnelles dépendant d'un paramètre. *Ann. Scuola Norm. Sup. Pisa Cl. Sci. (4)*, 4(2):231–255, 1977.

[10] A. Bensoussan and J.-L. Lions. *Contrôle impulsionnel et inéquations quasi variationnelles*, volume 11 of *Méthodes Mathématiques de l'Informatique [Mathematical Methods of Information Science]*. Gauthier-Villars, Paris, 1982.

[11] A. Bensoussan, J.-L. Lions, and G. Papanicolaou. Perturbations et "augmentation" des conditions initiales. In *Singular perturbations and boundary layer theory (Proc. Conf., École Centrale, Lyon, 1976)*, pages 10–29. Lecture Notes in Math., Vol. 594. Springer, Berlin, 1977.

[12] A. Bensoussan, J.-L. Lions, and G. Papanicolaou. Homogenization and ergodic theory. In *Probability theory (Papers, VIIth Semester, Stefan Banach Internat. Math. Center, Warsaw, 1976)*, volume 5 of *Banach Center Publ.*, pages 15–25. PWN, Warsaw, 1979.

[13] Alain Bensoussan, Jacques-L. Lions, and George C. Papanicolaou. Boundary layers and homogenization of transport processes. *Publ. Res. Inst. Math. Sci.*, 15(1):53–157, 1979.

[14] Alain Bensoussan, Jacques-Louis Lions, and Georges Papanicolaou. Sur quelques phénomènes asymptotiques d'évolution. *C. R. Acad. Sci. Paris Sér. A-B*, 281(10):Ai, A317–A322, 1975.

[15] Alain Bensoussan, Jacques-Louis Lions, and Georges Papanicolaou. Homogénéisation, correcteurs et problèmes non-linéaires. *C. R. Acad. Sci. Paris Sér. A-B*, 282(22):Aii, A1277–A1282, 1976.

[16] Alain Bensoussan, Jacques-Louis Lions, and Georges Papanicolaou. Sur la convergence d'opérateurs différentiels avec potentiel fortement oscillant. *C. R. Acad. Sci. Paris Sér. A-B*, 284(11):A587–A592, 1977.

[17] Marco Biroli. Sur la G-convergence pour des inéquations quasi-variationnelles. *C. R. Acad. Sci. Paris Sér. A-B*, 284(16):A947–A950, 1977.

[18] L. Boccardo and I. Capuzzo Dolcetta. G-convergenza e problema di Dirichlet unilaterale. *Boll. Un. Mat. Ital. (4)*, 12(1-2):115–123, 1975.

[19] Lucio Boccardo and Paolo Marcellini. Sulla convergenza delle soluzioni di disequazioni variazionali. *Ann. Mat. Pura Appl. (4)*, 110:137–159, 1976.

[20] N. N. Bogoliubov and Y. A. Mitropolsky. *Asymptotic methods in the theory of non-linear oscillations.* Translated from the second revised Russian edition. International Monographs on Advanced Mathematics and Physics. Hindustan Publishing Corp., Delhi, Gordon and Breach Science Publishers, New York, 1961.

[21] J. F. Bourgat. Numerical experiments of the homogenization method for operators with periodic coefficients. In *Computing methods in applied sciences and engineering (Proc. Third Internat. Sympos., Versailles, 1977), I*, volume 704 of *Lecture Notes in Math.*, pages 330–356. Springer, Berlin, 1979.

[22] L. Brillouin. *Wave propagation in periodic structures. Electric filters and crystal lattices.* Dover Publications Inc., New York, N. Y., 1953. 2d ed.

[23] Luciano Carbone. Γ⁻-convergence d'intégrales sur des fonctions avec des contraintes sur le gradient. *Comm. Partial Differential Equations*, 2(6):627–651, 1977.

[24] Robert Wayne Carroll and Ralph E. Showalter. *Singular and degenerate Cauchy problems.* Academic Press [Harcourt Brace Jovanovich Publishers], New York, 1976. Mathematics in Science and Engineering, Vol. 127.

[25] Sydney Chapman and T. G. Cowling. *The mathematical theory of nonuniform gases.* Cambridge Mathematical Library. Cambridge University Press, Cambridge, third edition, 1990. An account of the kinetic theory of viscosity, thermal conduction and diffusion in gases, In co-operation with D. Burnett, With a foreword by Carlo Cercignani.

[26] Stephen Childress. New solutions of the kinematic dynamo problem. *J. Mathematical Phys.*, 11:3063–3076, 1970.

[27] Ferruccio Colombini. On the regularity of solutions of hyperbolic equations with discontinuous coefficients variable in time. *Comm. Partial Differential Equations*, 2(6):653–677, 1977.

[28] Ferruccio Colombini and Sergio Spagnolo. Sur la convergence de solutions d'équations paraboliques. *J. Math. Pures Appl. (9)*, 56(3):263–305, 1977.

[29] R. Courant and D. Hilbert. *Methods of mathematical physics. Vol. II.* Wiley Classics Library. John Wiley & Sons Inc., New York, 1989. Partial differential equations, Reprint of the 1962 original, A Wiley-Interscience Publication.

[30] E. De Giorgi and S. Spagnolo. Sulla convergenza degli integrali dell'energia per operatori ellittici del secondo ordine. *Boll. Un. Mat. Ital. (4)*, 8:391–411, 1973.

[31] Luciano de Simon. Sull'equazione delle onde con termine noto periodico. *Rend. Ist. Mat. Univ. Trieste*, 1:150–162, 1969.

[32] François Delebecque and Jean-Pierre Quadrat. Contribution of stochastic control singular perturbation averaging and team theories to an example of large-scale systems: management of hydropower production. *IEEE Trans. Automatic Control*, AC-23(2):209–222, 1978.

[33] Jacques Deny and Jacques Louis Lions. Espaces de Beppo Levi et applications. *C. R. Acad. Sci. Paris*, 239:1174–1177, 1954.

[34] C. Doléans-Dade and P.-A. Meyer. Intégrales stochastiques par rapport aux martingales locales. In *Séminaire de Probabilités, IV (Univ. Strasbourg, 1968/69)*, pages 77–107. Lecture Notes in Mathematics, Vol. 124. Springer, Berlin, 1970.

[35] M. D. Donsker and S. R. S. Varadhan. Asymptotic evaluation of certain Markov process expectations for large time. I. II. *Comm. Pure Appl. Math.*, 28:1–47; ibid. 28 (1975), 279–301, 1975.

[36] J. L. Doob. *Stochastic processes.* Wiley Classics Library. John Wiley & Sons Inc., New York, 1990. Reprint of the 1953 original, A Wiley-Interscience Publication.

[37] J. J. Duistermaat. *Fourier integral operators*, volume 130 of *Progress in Mathematics.* Birkhäuser Boston Inc., Boston, MA, 1996.

[38] G. Duvaut and J.-L. Lions. *Les inéquations en mécanique et en physique.* Dunod, Paris, 1972. Travaux et Recherches Mathématiques, No. 21.

[39] William Feller. *An introduction to probability theory and its applications. Vol. II.* John Wiley & Sons Inc., New York, 1966.

[40] M. I. Freĭdlin. The Dirichlet problem for an equation with periodic coefficients depending on a small parameter. *Teor. Verojatnost. i Primenen.*, 9:133–139, 1964.

[41] M. I. Freĭdlin. Fluctuations in dynamical systems with averaging. *Dokl. Akad. Nauk SSSR*, 226(2):273–276, 1976.

[42] Avner Friedman. *Partial differential equations of parabolic type.* Prentice-Hall Inc., Englewood Cliffs, N.J., 1964.

[43] Avner Friedman. *Stochastic differential equations and applications*. Dover Publications Inc., Mineola, NY, 2006. Two volumes bound as one, Reprint of the 1975 and 1976 original published in two volumes.

[44] K. O. Friedrichs. Symmetric positive linear differential equations. *Comm. Pure Appl. Math.*, 11:333–418, 1958.

[45] F. R. Gantmacher. *The theory of matrices. Vols. 1, 2*. Translated by K. A. Hirsch. Chelsea Publishing Co., New York, 1959.

[46] I. M. Gel′fand. Expansion in characteristic functions of an equation with periodic coefficients. *Doklady Akad. Nauk SSSR (N.S.)*, 73:1117–1120, 1950.

[47] I. I . Gikhman and A. V. Skorokhod. *Stokhasticheskie differentsialnye uravneniya i ikh prilozheniya*. "Naukova Dumka", Kiev, 1982.

[48] I. V. Girsanov. On Ito's stochastic integral equation. *Soviet Math. Dokl.*, 2:506–509, 1961.

[49] Harold Grad. Singular and nonuniform limits of solutions of the Boltzmann equation. In *Transport Theory (Proc. Sympos. Appl. Math., New York, 1967), SIAM-AMS Proc., Vol. I*, pages 269–308. Amer. Math. Soc., Providence, R.I., 1969.

[50] Theodore E. Harris. *The theory of branching processes*. Dover Phoenix Editions. Dover Publications Inc., Mineola, NY, 2002. Corrected reprint of the 1963 original [Springer, Berlin; MR0163361 (29 #664)].

[51] Reuben Hersh. Random evolutions: a survey of results and problems. *Rocky Mountain J. Math.*, 4:443–477, 1974. Based on lectures given by Richard Griego, Reuben Hersh, Tom Kurtz and George Papanicolaou, Papers arising from a Conference on Stochastic Differential Equations (Univ. Alberta, Edmonton, Alta., 1972).

[52] Kiyosi Itô and Henry P. McKean, Jr. *Diffusion processes and their sample paths*. Springer-Verlag, Berlin, 1974. Second printing, corrected, Die Grundlehren der mathematischen Wissenschaften, Band 125.

[53] Tosio Kato. *Perturbation theory for linear operators*. Classics in Mathematics. Springer-Verlag, Berlin, 1995. Reprint of the 1980 edition.

[54] Joseph B. Keller. Corrected Bohr-Sommerfeld quantum conditions for nonseparable systems. *Ann. Physics*, 4:180–188, 1958.

[55] Joseph B. Keller. Geometrical theory of diffraction. *J. Opt. Soc. Amer.*, 52:116–130, 1962.

[56] Joseph B. Keller and Robert M. Lewis. Asymptotic methods for partial differential equations: the reduced wave equation and Maxwell's equations. In *Surveys in applied mathematics, Vol. 1*, volume 1 of *Surveys Appl. Math.*, pages 1–82. Plenum, New York, 1995.

[57] Srinivasan Kesavan. Homogénéisation et valeurs propres. *C. R. Acad. Sci. Paris Sér. A-B*, 285(4):A229–A232, 1977.

[58] W. Kohler and G. C. Papanicolaou. Power statistics for wave propagation in one dimension and comparison with radiative transport theory. *J. Mathematical Phys.*, 14:1733–1745, 1973.

[59] W. Kohler and G. C. Papanicolaou. Power statistics for wave propagation in one dimension and comparison with radiative transport theory. II. *J. Mathematical Phys.*, 15:2186–2197, 1974.

[60] S. M. Kozlov. Averaging of differential operators with almost periodic, fast oscillating coefficients. *Dokl. Akad. Nauk SSSR*, 236(5):1068–1071, 1977.

[61] M. G. Kreĭn and M. A. Rutman. Linear operators leaving invariant a cone in a Banach space. *Amer. Math. Soc. Translation*, 1950(26):128, 1950.

[62] Heinz-Otto Kreiss. Initial boundary value problems for hyperbolic systems. *Comm. Pure Appl. Math.*, 23:277–298, 1970.

[63] Thomas G. Kurtz. A limit theorem for perturbed operator semigroups with applications to random evolutions. *J. Functional Analysis*, 12:55–67, 1973.

[64] L. Landau and E. Lifchitz. *Physique théorique ("Landau-Lifshits"). Tome 8*. Traduit du Russe. [Translations of Russian Works]. "Mir", Moscow, 1990. Électrodynamique des milieux continus. [Electrodynamics of continuous media], Second Russian edition revised by Lifchitz [Lifshits] and L. Pitayevski [L. P. Pitaevskiĭ], Translated from the second Russian edition by Anne Sokova.

[65] Peter D. Lax. Asymptotic solutions of oscillatory initial value problems. *Duke Math. J.*, 24:627–646, 1957.

[66] Jean Leray. Solutions asymptotiques et groupe symplectique. In *Fourier integral operators and partial differential equations (Colloq. Internat., Univ. Nice, Nice, 1974)*, pages 73–97. Lecture Notes in Math., Vol. 459. Springer, Berlin, 1975.

[67] Robert M. Lewis. Asymptotic methods for the solution of dispersive hyperbolic equations. In *Asymptotic Solutions of Differential Equations and Their Applications (Proc. Sympos., Math. Res. Center, U.S. Army, Univ. Wisconsin, Madison, Wis., 1964)*, pages 53–107. Wiley, New York, 1964.

[68] J.-L. Lions. *Équations différentielles opérationnelles et problèmes aux limites*. Die Grundlehren der mathematischen Wissenschaften, Bd. 111. Springer-Verlag, Berlin, 1961.

[69] J.-L. Lions. Lectures on elliptic partial differential equations. In *Tata Institute of Fundamental Research Lectures on Mathematics, No. 10*, pages iii+130+vi. Tata Institute of Fundamental Research, Bombay, 1967.

[70] J.-L. Lions and E. Magenes. Problemi ai limiti non omogenei. III. *Ann. Scuola Norm. Sup. Pisa (3)*, 15:41–103, 1961.

[71] J.-L. Lions and E. Magenes. *Problèmes aux limites non homogènes et applications. Vol. 1.* Travaux et Recherches Mathématiques, No. 17. Dunod, Paris, 1968.

[72] J.-L. Lions and E. Magenes. *Problèmes aux limites non homogènes et applications. Vol. 2.* Travaux et Recherches Mathématiques, No. 18. Dunod, Paris, 1968.

[73] J.-L. Lions and G. Stampacchia. Variational inequalities. *Comm. Pure Appl. Math.*, 20:493–519, 1967.

[74] Donald Ludwig. Uniform asymptotic expansions at a caustic. *Comm. Pure Appl. Math.*, 19:215–250, 1966.

[75] Donald Ludwig and Barry Granoff. Propagation of singularities along characteristics with nonuniform multiplicity. *J. Math. Anal. Appl.*, 21:556–574, 1968.

[76] Paolo Marcellini and Carlo Sbordone. An approach to the asymptotic behaviour of elliptic-parabolic operators. *J. Math. Pures Appl. (9)*, 56(2):157–182, 1977.

[77] Paolo Marcellini and Carlo Sbordone. Sur quelques questions de G-convergence et d'homogénéisation non linéaire. *C. R. Acad. Sci. Paris Sér. A-B*, 284(10):A535–A537, 1977.

[78] Paolo Marcellini and Carlo Sbordone. Homogenization of nonuniformly elliptic operators. *Applicable Anal.*, 8(2):101–113, 1978/79.

[79] V. G. Markov and O. A. Oleĭnik. On propagation of heat in one-dimensional disperse media. *Prikl. Mat. Meh.*, 39(6):1073–1081, 1975.

[80] Paul-André Meyer. *Probabilités et potentiel*. Publications de l'Institut de Mathématique de l'Université de Strasbourg, No. XIV. Actualités Scientifiques et Industrielles, No. 1318. Hermann, Paris, 1966.

[81] Norman G. Meyers. An L^pe-estimate for the gradient of solutions of second order elliptic divergence equations. *Ann. Scuola Norm. Sup. Pisa (3)*, 17:189–206, 1963.

[82] È. Muhamadiev. The invertibility of partial differential operators of elliptic type. *Dokl. Akad. Nauk SSSR*, 205:1292–1295, 1972.

[83] François Murat. Sur l'homogénéisation d'inéquations elliptiques du 2ème ordre, relatives au convexe $k(\psi_1, \psi_2) = \{v \in H_0^1(\Omega) \mid \psi_1 \leq v \leq \psi_2 \text{ p.p. dans } \Omega\}$, 1976.

[84] François Murat. Compacité par compensation. *Ann. Scuola Norm. Sup. Pisa Cl. Sci. (4)*, 5(3):489–507, 1978.

[85] Jindřich Nečas. *Les méthodes directes en théorie des équations elliptiques*. Masson et Cie, Éditeurs, Paris, 1967.

[86] Jacques Neveu. *Bases mathématiques du calcul des probabilités*. Préface de R. Fortet. Deuxième édition, revue et corrigée. Masson et Cie, Éditeurs, Paris, 1970.

[87] Louis Nirenberg. Remarks on strongly elliptic partial differential equations. *Comm. Pure Appl. Math.*, 8:649–675, 1955.

[88] Farouk Odeh and Joseph B. Keller. Partial differential equations with periodic coefficients and Bloch waves in crystals. *J. Mathematical Phys.*, 5:1499–1504, 1964.

[89] R. E. O'Malley, Jr. and C. F. Kung. The singularly perturbed linear state regulator problem. II. *SIAM J. Control*, 13:327–337, 1975.

[90] G. C. Papanicolaou and R. Burridge. Transport equations for the Stokes parameters from Maxwell's equations in a random medium. *J. Mathematical Phys.*, 16(10):2074–2085, 1975.

[91] K. R. Parthasarathy. *Probability measures on metric spaces*. AMS Chelsea Publishing, Providence, RI, 2005. Reprint of the 1967 original.

[92] Jaak Peetre. Another approach to elliptic boundary problems. *Comm. Pure Appl. Math.*, 14:711–731, 1961.

[93] Mark A. Pinsky. Multiplicative operator functionals and their asymptotic properties. In *Advances in probability and related topics, Vol. 3*, pages 1–100. Dekker, New York, 1974.

[94] Yu. V. Prohorov. Convergence of random processes and limit theorems in probability theory. *Teor. Veroyatnost. i Primenen.*, 1:177–238, 1956.

[95] Giuseppe Pulvirenti. Sulla sommabilità L^p delle derivate prime delle soluzioni deboli del problema di Cauchy-Dirichlet per le equazioni lineari del secondo ordine di tipo parabolico. *Matematiche (Catania)*, 22:250–265, 1967.

[96] Giuseppe Pulvirenti. Ancora sulla sommabilità L^p delle derivate prime delle soluzioni deboli del problema di Cauchy-Dirichlet. *Matematiche (Catania)*, 23:160–165, 1968.

[97] Jeffrey B. Rauch and Frank J. Massey, III. Differentiability of solutions to hyperbolic initial-boundary value problems. *Trans. Amer. Math. Soc.*, 189:303–318, 1974.

[98] Edward L. Reiss. The impact problem for the Klein-Gordon equation. *SIAM J. Appl. Math.*, 17:526–542, 1969.

[99] G. O. Roberts. Spatially periodic dynamos. *Philos. Trans. Roy. Soc. London Ser. A*, 266:535–558, 1970.

[100] Enrique Sánchez-Palencia. Comportements local et macroscopique d'un type de milieux physiques hétérogenes. *Internat. J. Engrg. Sci.*, 12:331–351, 1974.

[101] Carlo Sbordone. Su alcune applicazioni di un tipo di convergenza variazionale. *Ann. Scuola Norm. Sup. Pisa Cl. Sci. (4)*, 2(4):617–638, 1975.

[102] Carlo Sbordone. Sulla G-convergenza di equazioni ellittiche e paraboliche. *Ricerche Mat.*, 24(1):76–136, 1975.

[103] L. Schwartz. *Théorie des distributions. Tome I.* Actualités Sci. Ind., no. 1091 = Publ. Inst. Math. Univ. Strasbourg 9. Hermann & Cie., Paris, 1950.

[104] Laurent Schwartz. *Théorie des distributions. Tome II.* Actualités Sci. Ind., no. 1122 = Publ. Inst. Math. Univ. Strasbourg **10**. Hermann & Cie., Paris, 1951.

[105] M. C. Shen and J. B. Keller. Uniform ray theory of surface, internal and acoustic wave propagation in a rotating ocean or atmosphere. *SIAM J. Appl. Math.*, 28:857–875, 1975.

[106] R. E. Showalter and T. W. Ting. Asymptotic behavior of solutions of pseudo-parabolic partial differential equations. *Ann. Mat. Pura Appl. (4)*, 90:241–258, 1971.

[107] I. B. Simonenko. A justification of the method of averaging for abstract parabolic equations. *Dokl. Akad. Nauk SSSR*, 191:33–34, 1970.

[108] I. B. Simonenko. Justification of the averaging method for the convection problem in a field of rapidly oscillating forces and for other parabolic equations. *Mat. Sb. (N.S.)*, 87(129):236–253, 1972.

[109] S. L. Sobolev. *Nekotorye primeneniya funkcional'nogo analiza v matematičeskoĭ fizike.* Izdat. Leningrad. Gos. Univ., Leningrad, 1950.

[110] S. Spagnolo. Sulla convergenza di soluzioni di equazioni paraboliche ed ellittiche. *Ann. Scuola Norm. Sup. Pisa (3) 22 (1968), 571-597; errata, ibid. (3)*, 22:673, 1968.

[111] Sergio Spagnolo. Sul limite delle soluzioni di problemi di Cauchy relativi all'equazione del calore. *Ann. Scuola Norm. Sup. Pisa (3)*, 21:657–699, 1967.

[112] Daniel W. Stroock and S. R. S. Varadhan. Diffusion processes with continuous coefficients. I. *Comm. Pure Appl. Math.*, 22:345–400, 1969.

[113] Daniel W. Stroock and S. R. S. Varadhan. Diffusion processes with continuous coefficients. II. *Comm. Pure Appl. Math.*, 22:479–530, 1969.

[114] Daniel W. Stroock and S. R. S. Varadhan. Diffusion processes with boundary conditions. *Comm. Pure Appl. Math.*, 24:147–225, 1971.

[115] Fred D. Tappert. The parabolic approximation method. In *Wave propagation and underwater acoustics (Workshop, Mystic, Conn., 1974)*, pages 224–287. Lecture Notes in Phys., Vol. 70. Springer, Berlin, 1977.

[116] Luc Tartar. Cours Peccot, 1977.

[117] Luc Tartar. Compensated compactness and applications to partial differential equations. In *Non Linear Mechanics and Analysis (Heriot-Watt Symposium, Volume IV R. J. Knops Editor*, pages 136–212. Pitman Research Notes in Mathematics, Vol. 39. Pitman, Boston, 1979.

[118] S. R. S. Varadhan. Asymptotic probabilities and differential equations. *Comm. Pure Appl. Math.*, 19:261–286, 1966.

[119] A. D. Ventcel'. Rough limit theorems on large deviations for Markov random processes. I. *Teor. Verojatnost. i Primenen.*, 21(2):235–252, 1976.

[120] A. D. Ventcel'. Rough limit theorems on large deviations for Markov random processes. II. *Teor. Verojatnost. i Primenen.*, 21(3):512–526, 1976.

[121] A. D. Ventcel′ and M. I. Freĭdlin. Small random perturbations of dynamical systems. *Uspehi Mat. Nauk*, 25(1 (151)):3–55, 1970.

[122] V. N. Vragov. The mixed problem for a certain class of hyperbolic-parabolic equations. *Dokl. Akad. Nauk SSSR*, 224(2):273–276, 1975.

[123] Shinzo Watanabe. On stochastic differential equations for multi-dimensional diffusion processes with boundary conditions. *J. Math. Kyoto Univ.*, 11:169–180, 1971.

[124] Shinzo Watanabe. On stochastic differential equations for multi-dimensional diffusion processes with boundary conditions. II. *J. Math. Kyoto Univ.*, 11:545–551, 1971.

[125] Shinzo Watanabe and Toshio Yamada. On the uniqueness of solutions of stochastic differential equations. II. *J. Math. Kyoto Univ.*, 11:553–563, 1971.

[126] Calvin H. Wilcox. Measurable eigenvectors for Hermitian matrix-valued polynomials. *J. Math. Anal. Appl.*, 40:12–19, 1972.

[127] Calvin H. Wilcox. Theory of Bloch waves. *J. Analyse Math.*, 33:146–167, 1978.

[128] Toshio Yamada and Shinzo Watanabe. On the uniqueness of solutions of stochastic differential equations. *J. Math. Kyoto Univ.*, 11:155–167, 1971.

[129] Erich Zauderer. On a modification of Hadamard's method for obtaining fundamental solutions for hyperbolic and parabolic equations. *J. Inst. Math. Appl.*, 8:8–15, 1971.

[130] J. M. Ziman. *Principles of the theory of solids*. Cambridge University Press, London, second edition, 1972.

ISBN 978-0-8218-5324-5

CHEL/374.H

About this book

This is a reprinting of a book originally published in 1978. At that time it was the first book on the subject of homogenization, which is the asymptotic analysis of partial differential equations with rapidly oscillating coefficients, and as such it sets the stage for what problems to consider and what methods to use, including probabilistic methods. At the time the book was written the use of asymptotic expansions with multiple scales was new, especially their use as a theoretical tool, combined with energy methods and the construction of test functions for analysis with weak convergence methods. Before this book, multiple scale methods were primarily used for non-linear oscillation problems in the applied mathematics community, not for analyzing spatial oscillations as in homogenization.

In the current printing a number of minor corrections have been made, and the bibliography was significantly expanded to include some of the most important recent references. This book gives systematic introduction of multiple scale methods for partial differential equations, including their original use for rigorous mathematical analysis in elliptic, parabolic, and hyperbolic problems, and with the use of probabilistic methods when appropriate. The book continues to be interesting and useful to readers of different backgrounds, both from pure and applied mathematics, because of its informal style of introducing the multiple scale methodology and the detailed proofs.